# ENVIRONMENTAL AUDITS

## 7TH EDITION

**Lawrence B. Cahill**
Raymond W. Kane
Jennifer L. Karas
James C. Mauch
Courtney M. Price
Brian P. Riedel
Dawne P. Schomer
Thomas R. Vetrano

**Government Institutes**
**Rockville, Maryland**

Government Institutes, Inc., 4 Research Place, Suite 200, Rockville,
Maryland 20850

Copyright ©1996 by Government Institutes. All rights reserved.

00  99  98  97  96          5    4    3    2    1

ISBN: 0-86587-525-1

Printed in the United States of America

# SUMMARY OF CONTENTS

About the Editor and Principal Author . . . . . . . . . . . . . . . . . . xxi
Contributing Authors . . . . . . . . . . . . . . . . . . . . . . . . . . . xxiii
Preface . . . . . . . . . . . . . . . . . . . . . . . . . . . . . . . . . . xxvii


## PART I. MANAGING A PROGRAM

Chapter 1: Perspectives in Environmental Auditing . . . . . . . . . . . 5
Chapter 2: Government Perspective . . . . . . . . . . . . . . . . . . . 31
Chapter 3: Legal Issues . . . . . . . . . . . . . . . . . . . . . . . . . 51
Chapter 4: Elements of a Successful Program . . . . . . . . . . . . . . 65
Chapter 5: Impact of International Standards on Environmental
    Audit Programs . . . . . . . . . . . . . . . . . . . . . . . . . . 115
Chapter 6: Potential Effect of ISO 14000 Standards on
    Environmental Audit Training in the United States . . . . . . . 143
Chapter 7: A Review of Some Typical Programs . . . . . . . . . . . 157
Chapter 8: Benchmarking Environmental Audit Programs:
    Best Practices and Biggest Challenges . . . . . . . . . . . . . . 169
Chapter 9: Environmental Auditor Qualifications:
    Great Expectations . . . . . . . . . . . . . . . . . . . . . . . . 187
Chapter 10: Training Auditors . . . . . . . . . . . . . . . . . . . . . 195
Chapter 11: Environmental Audit Training in the Far East . . . . . 203
Chapter 12: Information Management . . . . . . . . . . . . . . . . . 213
Chapter 13: Using Groupware to Manage the EH&S Audit
    Program Documentation Process . . . . . . . . . . . . . . . . . 223
Chapter 14: Managing and Critiquing an Audit Program . . . . . . 235

## PART II: CONDUCTING THE AUDIT

Chapter 15: Conducting the Environmental Audit . . . . . . . . . . . 253
Chapter 16: Typical Compliance Problems Found During Audits 287
Chapter 17: Conducting Effective Opening and Closing
Conferences . . . . . . . . . . . . . . . . . . . . . . . . . . 333
Chapter 18: Preparing Quality Audit Reports . . . . . . . . . . . . 349
Chapter 19: Environmental Auditing: The Good and the Bad—
A Modern Fable . . . . . . . . . . . . . . . . . . . . . . . 381

## PART III: SPECIAL AUDITING TOPICS

Chapter 20: Property Transfer Assessments . . . . . . . . . . . . . 401
Chapter 21: Top Ten Reasons Why Phase I Environmental
Assessment Reports Miss the Mark . . . . . . . . . . . . . . . 433
Chapter 22: Waste Contractor Audits . . . . . . . . . . . . . . . . 447
Chapter 23: Waste Minimization or Pollution Prevention Audits . 473
Chapter 24: Evaluating Management Systems on
Environmental Audits . . . . . . . . . . . . . . . . . . . . . 483
Chapter 25: International Environmental Audits . . . . . . . . . . . 491

## APPENDICES

Appendix A: References . . . . . . . . . . . . . . . . . . . . . . . 509
Appendix B: U.S. EPA's Audit Policies: July 9, 1986
and December 22, 1995 . . . . . . . . . . . . . . . . . . . . . 512
Appendix C: Sample Hazardous Waste Audit Protocol . . . . . . . 529
Appendix D: Sample Training Tools . . . . . . . . . . . . . . . . . 628
Appendix E: Sample Regulatory Inspection Protocol . . . . . . . . . 689
Appendix F: Major Regulations to be Reviewed Prior to an Audit 694
Appendix G: Summary of Audit Programs . . . . . . . . . . . . . . 702

# TABLE OF CONTENTS

List of Figures and Exhibits . . . . . . . . . . . . . . . . . . . . . . . . xviii
About the Editor and Principal Author . . . . . . . . . . . . . . . . . xxi
Contributing Authors . . . . . . . . . . . . . . . . . . . . . . . . . . . xxiii
Preface . . . . . . . . . . . . . . . . . . . . . . . . . . . . . . . . . . . . xxvii
Acknowledgments . . . . . . . . . . . . . . . . . . . . . . . . . . . . xxix

## PART I: MANAGING A PROGRAM

### Chapter 1: Perspectives in Environmental Auditing . . . . . . . . . 5

Why Audit? . . . . . . . . . . . . . . . . . . . . . . . . . . . . . . . . . . 5
Impact on the Regulated Community . . . . . . . . . . . . . . . . . . . . 13
Evolution of Environmental Auditing . . . . . . . . . . . . . . . . . . . 18
Environmental Auditing Roundtable . . . . . . . . . . . . . . . . . . . . 19
The Institute for Environmental Auditing . . . . . . . . . . . . . . . . . 20
Defining Environmental Audits . . . . . . . . . . . . . . . . . . . . . . 22
Advantages and Disadvantages . . . . . . . . . . . . . . . . . . . . . . 23
Trends in Environmental Auditing . . . . . . . . . . . . . . . . . . . . 25

### Chapter 2: Government Perspective . . . . . . . . . . . . . . . . . . . 31

Introduction . . . . . . . . . . . . . . . . . . . . . . . . . . . . . . . . . 31
EPA Policy Encourages Use of Environmental Auditing . . . . . . . . 34
EPA's 1986 Auditing Policy . . . . . . . . . . . . . . . . . . . . . . . . 35
Elements of Effective Audit Programs . . . . . . . . . . . . . . . . . . 35
EPA's 1995 Final Self-Policing Policy . . . . . . . . . . . . . . . . . . 37
   Policy Incentives: Full and 75 Percent Gravity Mitigation of
   Civil Penalties; No Criminal Referral to DOJ . . . . . . . . . . . . 38
   Safeguards . . . . . . . . . . . . . . . . . . . . . . . . . . . . . . . . 38
   Incentives and Behavior . . . . . . . . . . . . . . . . . . . . . . . . 39
   Open and Inclusive Process Utilized to Develop the Policy . . . . 40

Policy Conditions  . . . . . . . . . . . . . . . . . . . . . . 41
   Relationship to State Laws and Regulations . . . . . . . . . . . . . 46
   Applicability and Effective Date  . . . . . . . . . . . . . . . . . . 46
   Tracking of Cases under the Policy  . . . . . . . . . . . . . . 47
   Policy Redresses Industry Concerns . . . . . . . . . . . . . . . . 47
   Privileges and Immunities . . . . . . . . . . . . . . . . . . . 49
Conclusion  . . . . . . . . . . . . . . . . . . . . . . . . 50

**Chapter 3: Legal Issues** . . . . . . . . . . . . . . . . . . **51**

Confidentiality . . . . . . . . . . . . . . . . . . . . . . . . 51
   Attorney-Client Privilege  . . . . . . . . . . . . . . . . . . . 52
   Work-Product Rule  . . . . . . . . . . . . . . . . . . . . . . 55
   Self Evaluative Privilege . . . . . . . . . . . . . . . . . . . . 57
   State Action on Environmental Audit Privilege  . . . . . . . . . . 58
Potential Liability Arising Out of Environmental Audits . . . . . . . . 60
   Liability for Violations Discovered . . . . . . . . . . . . . . . . 60
   Who is Liable for Violations Discovered . . . . . . . . . . . . . . 62
   Liability for Improperly Performed Audit . . . . . . . . . . . . . 63
Conclusion  . . . . . . . . . . . . . . . . . . . . . . . . 64

**Chapter 4: Elements of a Successful Program**  . . . . . . . . . . . **65**

Principles of an Audit Program . . . . . . . . . . . . . . . . . . 65
Planning the Program . . . . . . . . . . . . . . . . . . . . . . 67
   Program Objectives  . . . . . . . . . . . . . . . . . . . . . 71
   Roles and Responsibilities . . . . . . . . . . . . . . . . . . . 72
   Legal Protections . . . . . . . . . . . . . . . . . . . . . . 76
   Audit Confidentiality? Industry's Response . . . . . . . . . . . . 76
   Scope and Coverage . . . . . . . . . . . . . . . . . . . . . 77
   Facility Audit Schedules . . . . . . . . . . . . . . . . . . . . 79
   Auditor Selection and Training  . . . . . . . . . . . . . . . . . 85
   Audit Procedures . . . . . . . . . . . . . . . . . . . . . . 87
   Audit Reports and Documentation  . . . . . . . . . . . . . . . 87
   Audit Program Management & Evaluation  . . . . . . . . . . . . 90
   Audit Support Tools . . . . . . . . . . . . . . . . . . . . . 91

A Key Program Issue: Corporate Standards and Guidelines ..... 100
    Development of Waste Discharge Inventories ........... 100
    Secondary Containment for Hazardous Materials and Wastes . 101
    Internal Reporting of Environmentally Related Incidents ..... 101
    Environmental Recordkeeping/Records Retention Procedures . 105
    External Regulatory Inspection Procedures ............. 105
Additional Sampling ..................................... 106
Continuous Review .................................... 107
Evaluating the Results and Implementing the Solutions ......... 107

**Chapter 5: Impact of International Standards on**
**Environmental Audit Programs** ................... **115**

Seven Initiatives ..................................... 116
    The CERES Principles ............................. 117
    International Standards Organization (ISO) Environmental
        Standards ..................................... 121
    The European Community Eco-Audit Program .......... 123
    ICC Business Charter's Environmental Management Principles 125
    Responsible Care® ................................ 127
    British Standards Institution Standard BS 7750 .......... 128
    U.S. EPA's Environmental Leadership Program .......... 130
Corporate Responses .................................. 130
    Appointment of Environmental Experts to Corporate Boards .. 131
    Voluntary Public Reporting ......................... 133
    Third-Party Program Evaluations and Certifications ........ 136
    Benchmarking Studies ............................. 137

**Chapter 6: Potential Effect of ISO 14000 Standards on**
**Environmental Audit Training in the United States** ..... **143**

Introduction ......................................... 143
Environmental Audit Training in the U.S. ................. 144
ISO 14012 Guidelines ................................. 146
U.S. Implementation of the ISO 9000 Standards: A Standing
    Precedent ........................................ 147

Potential Effects . . . . . . . . . . . . . . . . . . . . . . . . . . 150
   Accreditation, Oversight and Certification: The Lack of
      Autonomy . . . . . . . . . . . . . . . . . . . . . . . . . . 150
   Length and Type of Courses: A Variety of Demands . . . . . . 150
   Class Size: Too Small to be Economical? . . . . . . . . . . . . 151
   Course Content: The Shift to Management Systems . . . . . . . 152
   Course Content: Teaching Relevant Regulatory Requirements . 154
   Course Content: Field Exercises . . . . . . . . . . . . . . . . . 154
   Course Examinations: Evaluating Personal Attributes . . . . . . 155
   Documentation: New Expectations . . . . . . . . . . . . . . . . 156
Conclusions . . . . . . . . . . . . . . . . . . . . . . . . . . . . . 156

**Chapter 7: A Review of Some Typical Programs** . . . . . . . . **157**

Program Overview and Scope . . . . . . . . . . . . . . . . . . . . 158
   Program Name . . . . . . . . . . . . . . . . . . . . . . . . . . 158
   Start Date . . . . . . . . . . . . . . . . . . . . . . . . . . . . 158
   Purpose . . . . . . . . . . . . . . . . . . . . . . . . . . . . . 161
   Program Organization . . . . . . . . . . . . . . . . . . . . . . 161
   Program Scope . . . . . . . . . . . . . . . . . . . . . . . . . 162
Program Methodology . . . . . . . . . . . . . . . . . . . . . . . 163
   Audit Methodology . . . . . . . . . . . . . . . . . . . . . . . 163
   Reporting Findings . . . . . . . . . . . . . . . . . . . . . . . 166
   Follow-Up Mechanisms . . . . . . . . . . . . . . . . . . . . . 167
Program Operations . . . . . . . . . . . . . . . . . . . . . . . . 167
   Audit Staffing . . . . . . . . . . . . . . . . . . . . . . . . . . 167
   Audit Duration . . . . . . . . . . . . . . . . . . . . . . . . . 168
   Number of Plants Per Year Audited . . . . . . . . . . . . . . . 168
   Frequency of Audits . . . . . . . . . . . . . . . . . . . . . . . 168
Conclusion . . . . . . . . . . . . . . . . . . . . . . . . . . . . . 168

**Chapter 8: Benchmarking Environmental Audit Programs:**
   **Best Practices and Biggest Challenges** . . . . . . . . . . . . . **169**

Overview of Benchmarking . . . . . . . . . . . . . . . . . . . . . 170
   Precisely Define the Scope . . . . . . . . . . . . . . . . . . . . 170

Select Target Companies Using a Variety of Techniques . . . . 172
Create Participation Incentives . . . . . . . . . . . . . . . . . . . . . 172
Develop Measurable Criteria . . . . . . . . . . . . . . . . . . . . . 172
Utilize Focus-Group Sessions . . . . . . . . . . . . . . . . . . . . 173
Best Practices . . . . . . . . . . . . . . . . . . . . . . . . . . . . . . . . 173
Reports to Management . . . . . . . . . . . . . . . . . . . . . . . . 173
Relationship to Compensation . . . . . . . . . . . . . . . . . . . . 174
Use of Spill Drills . . . . . . . . . . . . . . . . . . . . . . . . . . . 175
Red Tag Shutdowns . . . . . . . . . . . . . . . . . . . . . . . . . . 176
Community Participation . . . . . . . . . . . . . . . . . . . . . . . . 176
Next Site Participation . . . . . . . . . . . . . . . . . . . . . . . . 177
Use of Portable Computers ' . . . . . . . . . . . . . . . . . . . . . 177
Assessment of Ancillary Operations . . . . . . . . . . . . . . . . . 178
Use of Verification Audits . . . . . . . . . . . . . . . . . . . . . . . 179
Site-Satisfaction Questionnaire . . . . . . . . . . . . . . . . . . . 180
Periodic Third-Party Evaluations . . . . . . . . . . . . . . . . . . . 180
Development of a Program Newsletter . . . . . . . . . . . . . . . 181
Biggest Challenges . . . . . . . . . . . . . . . . . . . . . . . . . . . . 182
The Program Manual . . . . . . . . . . . . . . . . . . . . . . . . . 182
Protocol Updating . . . . . . . . . . . . . . . . . . . . . . . . . . . 182
State Regulatory Review . . . . . . . . . . . . . . . . . . . . . . . 183
Misleading Closing Conferences . . . . . . . . . . . . . . . . . . . 183
Timeliness and Quality of the Report . . . . . . . . . . . . . . . . 184
Insufficient Follow-up . . . . . . . . . . . . . . . . . . . . . . . . . 184

**Chapter 9: Environmental Auditor Qualifications: Great
Expectations** . . . . . . . . . . . . . . . . . . . . . . . . . . . . . **187**

Core Skills . . . . . . . . . . . . . . . . . . . . . . . . . . . . . . . . . 188
Learned Skills . . . . . . . . . . . . . . . . . . . . . . . . . . . . . . 188
Inherent Skills . . . . . . . . . . . . . . . . . . . . . . . . . . . . . 189
Observations in the Field . . . . . . . . . . . . . . . . . . . . . . . . 190
The Worst Attributes . . . . . . . . . . . . . . . . . . . . . . . . . 190
The Best Attributes . . . . . . . . . . . . . . . . . . . . . . . . . . 192
Making Good Things Happen . . . . . . . . . . . . . . . . . . . . . 193
Summary . . . . . . . . . . . . . . . . . . . . . . . . . . . . . . . . . . 193

**Chapter 10: Training Auditors** . . . . . . . . . . . . . . . . . . . . . . **195**

The Need . . . . . . . . . . . . . . . . . . . . . . . . . . . . . . . . . . . . . 195
Selecting the Trainers . . . . . . . . . . . . . . . . . . . . . . . . . . . . . 197
Selecting the Trainees . . . . . . . . . . . . . . . . . . . . . . . . . . . . 198
Selecting the Setting . . . . . . . . . . . . . . . . . . . . . . . . . . . . . 199
Selecting the Techniques . . . . . . . . . . . . . . . . . . . . . . . . . . . 199
    Lecture . . . . . . . . . . . . . . . . . . . . . . . . . . . . . . . . . . . 199
    Problem Solving . . . . . . . . . . . . . . . . . . . . . . . . . . . . . 200
    Role Playing . . . . . . . . . . . . . . . . . . . . . . . . . . . . . . . 200
    Keys to Making It Work . . . . . . . . . . . . . . . . . . . . . . . . 201

**Chapter 11: Environmental Audit Training in the Far East** . . **203**

First Stop: Singapore . . . . . . . . . . . . . . . . . . . . . . . . . . . . 204
The Transition: Singapore to Japan . . . . . . . . . . . . . . . . . . 208
Second Stop: Japan . . . . . . . . . . . . . . . . . . . . . . . . . . . . . 208

**Chapter 12: Information Management** . . . . . . . . . . . . . . . . **213**

Regulatory Databases . . . . . . . . . . . . . . . . . . . . . . . . . . . . 214
Automated Checklists . . . . . . . . . . . . . . . . . . . . . . . . . . . . 217
    Advantages . . . . . . . . . . . . . . . . . . . . . . . . . . . . . . . . 217
    Disadvantages . . . . . . . . . . . . . . . . . . . . . . . . . . . . . . 218
Evaluating Auditing Software Systems . . . . . . . . . . . . . . . . . 220

**Chapter 13: Using Groupware to Manage the EH&S Audit
Program Documentation Process** . . . . . . . . . . . . . . . . . **223**

Importance of Sound Document Management . . . . . . . . . . . . . 223
Traditional Responses to Audit Document Management . . . . . . . 225
The Groupware Solution . . . . . . . . . . . . . . . . . . . . . . . . . . 229
Conclusion . . . . . . . . . . . . . . . . . . . . . . . . . . . . . . . . . . . 233

**Chapter 14: Managing and Critiquing an Audit Program** . . . **235**

Beginning a Program . . . . . . . . . . . . . . . . . . . . . . . . 236
    Develop an Environmental Audit Policy and Procedure . . . . . 236
    Set Up the Organization . . . . . . . . . . . . . . . . . . . . 239
    Develop Tools . . . . . . . . . . . . . . . . . . . . . . . . 240
    Train Staff . . . . . . . . . . . . . . . . . . . . . . . . . . 240
    Test the Program . . . . . . . . . . . . . . . . . . . . . . . 241
    Set Review Schedule . . . . . . . . . . . . . . . . . . . . . 241
    Conduct Audits . . . . . . . . . . . . . . . . . . . . . . . . 242
    Implement Full Program . . . . . . . . . . . . . . . . . . . . 242
Managing a Program . . . . . . . . . . . . . . . . . . . . . . . . 242
    Assessment Tools . . . . . . . . . . . . . . . . . . . . . . . 243
    Management Reports . . . . . . . . . . . . . . . . . . . . . . 245
    Staff Management . . . . . . . . . . . . . . . . . . . . . . . 246
    Quality Assurance . . . . . . . . . . . . . . . . . . . . . . . 247
    Maintaining Awareness . . . . . . . . . . . . . . . . . . . . . 248
Twenty Tips for Achieving a Successful Audit . . . . . . . . . . . . 249

## PART II: CONDUCTING THE AUDIT

**Chapter 15: Conducting the Environmental Audit** . . . . . . . . **253**

Introduction . . . . . . . . . . . . . . . . . . . . . . . . . . . . 253
Pre-Audit Activities . . . . . . . . . . . . . . . . . . . . . . . . 254
    Team Selection and Formation . . . . . . . . . . . . . . . . . 254
    Completion and Review of Pre-Audit Questionnaire . . . . . . . 255
    Review of Relevant Regulations . . . . . . . . . . . . . . . . 256
    Definition of Audit Scope and Establishment of Team
        Responsibilities . . . . . . . . . . . . . . . . . . . . . . . 257
    Development of Detailed Audit Agenda . . . . . . . . . . . . . 261
    Review of Audit Protocols . . . . . . . . . . . . . . . . . . . 261
On-Site Activities . . . . . . . . . . . . . . . . . . . . . . . . . 265
    Opening Conference . . . . . . . . . . . . . . . . . . . . . . 265
    Orientation Tour . . . . . . . . . . . . . . . . . . . . . . . 266
    Records/Documentation Review . . . . . . . . . . . . . . . . . 267

Interviews With Facility Staff . . . . . . . . . . . . . . . . . . . . . . . . 270
Physical Inspection of Facilities . . . . . . . . . . . . . . . . . . . . . . 274
Daily Reviews . . . . . . . . . . . . . . . . . . . . . . . . . . . . . . . . . 284
Conduct Audit Debriefing . . . . . . . . . . . . . . . . . . . . . . . . . . 284
The Audit Report . . . . . . . . . . . . . . . . . . . . . . . . . . . . . . . 286

**Chapter 16: Typical Compliance Problems Found
During Audits** . . . . . . . . . . . . . . . . . . . . . . . . . . . . . . **287**

Common Compliance Problems . . . . . . . . . . . . . . . . . . . . . . . 288
PCB Management . . . . . . . . . . . . . . . . . . . . . . . . . . . . . 288
Wastewater Discharge . . . . . . . . . . . . . . . . . . . . . . . . . . . 289
Air Emissions . . . . . . . . . . . . . . . . . . . . . . . . . . . . . . . . 290
Oil Spill Control . . . . . . . . . . . . . . . . . . . . . . . . . . . . . . 291
Hazardous Waste Generation . . . . . . . . . . . . . . . . . . . . . . . 292
Community Right-To-Know . . . . . . . . . . . . . . . . . . . . . . . . 293
Worker Health and Hazard Communication . . . . . . . . . . . . . 293
Employee Safety . . . . . . . . . . . . . . . . . . . . . . . . . . . . . . 294
Pesticide Management . . . . . . . . . . . . . . . . . . . . . . . . . . . 295
Underground Storage Tanks . . . . . . . . . . . . . . . . . . . . . . . . 296
Hazardous Materials Storage . . . . . . . . . . . . . . . . . . . . . . . 296
Drinking Water . . . . . . . . . . . . . . . . . . . . . . . . . . . . . . . 297
Solid Waste Disposal . . . . . . . . . . . . . . . . . . . . . . . . . . . . 298
Typical Inspection Areas . . . . . . . . . . . . . . . . . . . . . . . . . . . 299
Auditor's Checklist . . . . . . . . . . . . . . . . . . . . . . . . . . . . . . 328
Preparing for the Audit . . . . . . . . . . . . . . . . . . . . . . . . . . 328
Conducting the Audit . . . . . . . . . . . . . . . . . . . . . . . . . . . 330

**Chapter 17: Conducting Effective Opening and Closing
Conferences** . . . . . . . . . . . . . . . . . . . . . . . . . . . . . **333**

Opening Conferences . . . . . . . . . . . . . . . . . . . . . . . . . . . . . 334
Take Charge of the Meeting . . . . . . . . . . . . . . . . . . . . . . . 334
Be Organized . . . . . . . . . . . . . . . . . . . . . . . . . . . . . . . . 335
Discuss Logistics . . . . . . . . . . . . . . . . . . . . . . . . . . . . . . 336
Find Out What's Happening at the Site . . . . . . . . . . . . . . . . 337

Schedule Additional Conferences . . . . . . . . . . . . . . . . . . . . 338
Address Important Topics . . . . . . . . . . . . . . . . . . . . . . . 338
Closing Conferences . . . . . . . . . . . . . . . . . . . . . . . . . . . 340
Take Charge of the Meeting . . . . . . . . . . . . . . . . . . . . . 340
Be Organized . . . . . . . . . . . . . . . . . . . . . . . . . . . . . . 341
Be Appreciative but Don't Bury the Message . . . . . . . . . . . 342
Set Priorities . . . . . . . . . . . . . . . . . . . . . . . . . . . . . . 343
Minimize Praise for "Acceptable" Performance . . . . . . . . . 343
Respond Professionally to Challenges . . . . . . . . . . . . . . . . 343
Focus on Root Causes but Avoid Evaluations of
    Staff Performance . . . . . . . . . . . . . . . . . . . . . . . . . 344
Understand How to Handle Repeat Findings . . . . . . . . . . . 345
Avoid Comparisons . . . . . . . . . . . . . . . . . . . . . . . . . . 345
Avoid Guarantees . . . . . . . . . . . . . . . . . . . . . . . . . . . 346
Leave Written Findings . . . . . . . . . . . . . . . . . . . . . . . . 347
Discuss the Next Steps . . . . . . . . . . . . . . . . . . . . . . . . 348

**Chapter 18: Preparing Quality Audit Reports** . . . . . . . . . . **349**

Field Preparation . . . . . . . . . . . . . . . . . . . . . . . . . . . . . 350
Keep the Customer in Mind . . . . . . . . . . . . . . . . . . . . . 350
Look for Underlying Causes . . . . . . . . . . . . . . . . . . . . . 351
Organize Daily . . . . . . . . . . . . . . . . . . . . . . . . . . . . . 351
Bottom-Line Interviews . . . . . . . . . . . . . . . . . . . . . . . . 351
Develop an Annotated Outline . . . . . . . . . . . . . . . . . . . . 352
Challenge Each Other . . . . . . . . . . . . . . . . . . . . . . . . . 353
Develop a Consistent Debriefing Approach . . . . . . . . . . . . 353
Report Preparation . . . . . . . . . . . . . . . . . . . . . . . . . . . . 353
Organize for Monitoring . . . . . . . . . . . . . . . . . . . . . . . 354
Start Early . . . . . . . . . . . . . . . . . . . . . . . . . . . . . . . . 355
Establish a Report Format . . . . . . . . . . . . . . . . . . . . . . 355
Pay Attention to Repeat Findings . . . . . . . . . . . . . . . . . . 358
Be Careful of "Good Practices" . . . . . . . . . . . . . . . . . . . 358
Set Priorities . . . . . . . . . . . . . . . . . . . . . . . . . . . . . . 359
Be Clear and Concise . . . . . . . . . . . . . . . . . . . . . . . . . 360
De-emphasize Numbers . . . . . . . . . . . . . . . . . . . . . . . . 362

Use Evidence in the Discussion of Findings . . . . . . . . . . . . 363
Avoid Common Pitfalls . . . . . . . . . . . . . . . . . . . . . . . . . . 364
Report Follow-Up . . . . . . . . . . . . . . . . . . . . . . . . . . . . . . . 367
Assure Legal Review of Reports . . . . . . . . . . . . . . . . . . 367
Limit Distribution of the Report . . . . . . . . . . . . . . . . . . . 367
Accept No Mistakes . . . . . . . . . . . . . . . . . . . . . . . . . . . . 368
Remove Barriers to Efficiency . . . . . . . . . . . . . . . . . . . . 368
Develop Action Plans . . . . . . . . . . . . . . . . . . . . . . . . . . . 369
Train the Auditors . . . . . . . . . . . . . . . . . . . . . . . . . . . . . 370

**Chapter 19: Environmental Auditing: The Good and the Bad—**
**A Modern Fable** . . . . . . . . . . . . . . . . . . . . . . . . . . . **381**

The Company . . . . . . . . . . . . . . . . . . . . . . . . . . . . . . . . . . 381
The Audit . . . . . . . . . . . . . . . . . . . . . . . . . . . . . . . . . . . . 382
The Beginning . . . . . . . . . . . . . . . . . . . . . . . . . . . . . . . . . 384
The First Meeting . . . . . . . . . . . . . . . . . . . . . . . . . . . . . . 384
The Field Work . . . . . . . . . . . . . . . . . . . . . . . . . . . . . . . 386
That Night . . . . . . . . . . . . . . . . . . . . . . . . . . . . . . . . . . . 387
The Next Day . . . . . . . . . . . . . . . . . . . . . . . . . . . . . . . . . 387
Finishing . . . . . . . . . . . . . . . . . . . . . . . . . . . . . . . . . . . . 389
The Briefing . . . . . . . . . . . . . . . . . . . . . . . . . . . . . . . . . . 390
The Report . . . . . . . . . . . . . . . . . . . . . . . . . . . . . . . . . . . 394
Epilogue . . . . . . . . . . . . . . . . . . . . . . . . . . . . . . . . . . . . 397

## PART III: SPECIAL AUDITING TOPICS

**Chapter 20: Property Transfer Assessments** . . . . . . . . . . . . . **401**

Introduction . . . . . . . . . . . . . . . . . . . . . . . . . . . . . . . . . . 401
Approach Overview . . . . . . . . . . . . . . . . . . . . . . . . . . . . . . 404
Scope and the ASTM Standard . . . . . . . . . . . . . . . . . . . . . 404
Assessment Team . . . . . . . . . . . . . . . . . . . . . . . . . . . . . . 405
External Contacts . . . . . . . . . . . . . . . . . . . . . . . . . . . . . . 406
The Three Phases . . . . . . . . . . . . . . . . . . . . . . . . . . . . . . 407
Phase I . . . . . . . . . . . . . . . . . . . . . . . . . . . . . . . . . . . 410

Phase II . . . . . . . . . . . . . . . . . . . . . . . . . . . 417
Phase III . . . . . . . . . . . . . . . . . . . . . . . . . . 418
Assessment Issues . . . . . . . . . . . . . . . . . . . . . . . 419
How Do I Know What Rules and Regulations Apply? . . . . . . 419
How Can I Be Sure That the Consultant I Retain is Qualified? 419
Do I Need a Big Firm or Can an Individual Consultant Suffice? 420
Can I Get a "Clean Bill of Health" Certification? . . . . . . . . 420
If There is Some Contamination, How Do I Know if There is a
Problem? In Other Words, "How Clean is Clean?" . . . . . 420
Some Deals Develop Quickly. How Much Advance Warning
Do I Have to Give the Consultant? . . . . . . . . . . . . . . . 421
What if the Assessment Identifies that Corrective Actions are
Necessary? . . . . . . . . . . . . . . . . . . . . . . . . . 421
What Situations Might Constitute Deal Killers? . . . . . . . . . . 422
What Do I Do With the Report if the Deal is Killed? . . . . . . . 424
Who Should Hire the Assessor? . . . . . . . . . . . . . . . . . . 424
What if an Imminent Hazard is Identified During the
Assessment? . . . . . . . . . . . . . . . . . . . . . . . 425
What are the General Approaches for Assessing Asbestos? . . . 425
How is Radon Assessed? . . . . . . . . . . . . . . . . . . . . . 427
Are Wetlands Important? . . . . . . . . . . . . . . . . . . . . . 428
Can Farmland Pose Liabilities? . . . . . . . . . . . . . . . . . 428
What About Urea Formaldehyde? . . . . . . . . . . . . . . . . . 429
What About Lead-Based Paint? . . . . . . . . . . . . . . . . . . 430
What About Drinking Water? . . . . . . . . . . . . . . . . . . . 430
Conclusions . . . . . . . . . . . . . . . . . . . . . . . . . . . 431

**Chapter 21: Top Ten Reasons Why Phase I Environmental
Assessment Reports Miss the Mark** . . . . . . . . . . . . . . **433**

Failure to Maintain Independence and Objectivity . . . . . . . . . . . 435
Failure to Define the Exact Scope of Work . . . . . . . . . . . . . . . 436
Use of Conjecture in Report Findings . . . . . . . . . . . . . . . . . 437
Use of Imprecise Language . . . . . . . . . . . . . . . . . . . . . . 438
Failure to Distinguish "Compliance" Findings from "Liability"
Findings . . . . . . . . . . . . . . . . . . . . . . . . . . . . . 439

Documented Sources . . . . . . . . . . . . . . . . . . . . . . . . . . . . . 440
Failure to State Assumptions Regarding Cost Estimates in
    Assessment Reports . . . . . . . . . . . . . . . . . . . . . . . . . . . . . 441
Lack of Editing and a Quality Assurance Review . . . . . . . . . . . 443
Disputes Over Disclaimers . . . . . . . . . . . . . . . . . . . . . . . . . . 444
Failure to Write the Report as a Business-Decision Tool . . . . . . . 445

**Chapter 22: Waste Contractor Audits** . . . . . . . . . . . . . . . . . **447**

Objective and Scope of a TSD Facility Audit Program . . . . . . . 450
Internal Programs *vs.* External Programs . . . . . . . . . . . . . . . . 451
Conducting the TSD Facility Audit . . . . . . . . . . . . . . . . . . . . 453
Pre-Audit Preparation . . . . . . . . . . . . . . . . . . . . . . . . . . . . . 454
Special Pre-Audit Considerations . . . . . . . . . . . . . . . . . . . . . 457
Selecting the Audit Team . . . . . . . . . . . . . . . . . . . . . . . . . . 462
The On-Site Audit . . . . . . . . . . . . . . . . . . . . . . . . . . . . . . . 464
The TSD Facility Audit Report . . . . . . . . . . . . . . . . . . . . . . . 470
Summary . . . . . . . . . . . . . . . . . . . . . . . . . . . . . . . . . . . . . 472

**Chapter 23: Waste Minimization or Pollution
    Prevention Audits** . . . . . . . . . . . . . . . . . . . . . . . . . . . . . **473**

**Chapter 24: Evaluating Management Systems on
    Environmental Audits** . . . . . . . . . . . . . . . . . . . . . . . . . . **483**

Why Evaluate Management Systems? . . . . . . . . . . . . . . . . . . . 483
What is a Management System? . . . . . . . . . . . . . . . . . . . . . . 485
How Do You Do It? . . . . . . . . . . . . . . . . . . . . . . . . . . . . . . 486
Why is It So Hard? . . . . . . . . . . . . . . . . . . . . . . . . . . . . . . 488
Closure . . . . . . . . . . . . . . . . . . . . . . . . . . . . . . . . . . . . . . 490

**Chapter 25: International Environmental Audits** . . . . . . . . . **491**

Role of Corporate Management . . . . . . . . . . . . . . . . . . . . . . 494
Design of International Audit Programs . . . . . . . . . . . . . . . . . 495
    Two-Phased Approach . . . . . . . . . . . . . . . . . . . . . . . . . . . 495

Audit Standards and Criteria . . . . . . . . . . . . . . . . . . . . . . 496
Conducting International Audits . . . . . . . . . . . . . . . . . . . . . 500
    Pre-Audit Activities . . . . . . . . . . . . . . . . . . . . . . . . . . 500
    On-Site Activities . . . . . . . . . . . . . . . . . . . . . . . . . . . 502
    Post-Audit Activities . . . . . . . . . . . . . . . . . . . . . . . . . 503
Summary of Key Elements in International Audits . . . . . . . . . . 504

## APPENDICES

Appendix A: References . . . . . . . . . . . . . . . . . . . . . . . . . 509
Appendix B: U.S. EPA's Audit Policies: July 9, 1986
    and December 22, 1995 . . . . . . . . . . . . . . . . . . . . . . . 512
Appendix C: Sample Hazardous Waste Audit Protocol . . . . . . . 529
Appendix D: Sample Training Tools . . . . . . . . . . . . . . . . . . 628
Appendix E: Sample Regulatory Inspection Protocol . . . . . . . . 689
Appendix F: Major Regulations to be Reviewed Prior
    to an Audit . . . . . . . . . . . . . . . . . . . . . . . . . . . . . . 694
Appendix G: Summary of Audit Programs . . . . . . . . . . . . . . 702

# LIST OF FIGURES AND EXHIBITS

Figure 1.1. Growth of Environmental Regulations . . . . . . . . . . . 8
Figure 1.2. EPA's Short-Term Agenda . . . . . . . . . . . . . . . . . 8
Figure 1.3. State Regulatory Stringency . . . . . . . . . . . . . . . . 10
Figure 1.4. Concentration of Lawyers in Developed Countries . . . 10
Figure 1.5. Warning of Illegal Disposal . . . . . . . . . . . . . . . 14
Figure 1.6. U.S. EPA Environmental Enforcement . . . . . . . . . . 15
Figure 1.7. Incidents at Walt Disney Co. . . . . . . . . . . . . . . . 15
Figure 1.8. Financial Impacts of Industry Incidents . . . . . . . . . 16
Figure 1.9. Guidelines for Audit Programs . . . . . . . . . . . . . . 26
Figure 1.10. EPA *vs.* OSHA: Two Comparisons . . . . . . . . . . . 29
Figure 4.1. Elements of an Audit Program . . . . . . . . . . . . . . 66
Figure 4.2. Why Audit? Industry's Response . . . . . . . . . . . . . 70
Figure 4.3. Sample Audit Program Organization . . . . . . . . . . . 74
Figure 4.4. Frequency of Audits? Industry's Response . . . . . . . 80
Figure 4.5. Site-Audit Frequency . . . . . . . . . . . . . . . . . . . 81
Figure 4.6. Assessing Risk in a Multi-Plant or Multi-Unit Setting . 83
Figure 4.7. Sample Site Visit Schedule . . . . . . . . . . . . . . . . 84
Figure 4.8. Example Site Environmental Audit Process Timeline . 88
Figure 4.9. Schedule and Disposition for Audit Documentation . . . 89
Figure 4.10. Sample Audit Checklist . . . . . . . . . . . . . . . . . 94
Figure 4.11. Tradeoffs in Questionnaire Development . . . . . . . . 97
Figure 4.12. Sample Waste Disposition . . . . . . . . . . . . . . . . 99
Figure 4.13. Model Regulatory Compliance System . . . . . . . . . 103
Figure 4.14. Environmental Audit Program-
  Draft Guidance Manual . . . . . . . . . . . . . . . . . . . . . . 108
Figure 5.1. Consequences of the Exxon Valdez . . . . . . . . . . . 118
Figure 5.2. Environmentalists on Boards of Directors . . . . . . . . 132
Figure 5.3. Toxics Release Rankings . . . . . . . . . . . . . . . . . 134
Figure 5.4. Third-Party Assessment of the Hoechst Celanese
  Corporation . . . . . . . . . . . . . . . . . . . . . . . . . . . . 138
Figure 5.5. Environmental Audit Benchmarking Study . . . . . . . 139

Figure 6.1. Accreditation and Registration Process . . . . . . . . . . 149
Figure 6.2. Evaluating Management Shifts . . . . . . . . . . . . . . 153
Figure 7.1. Summary of the Review of Twenty Environmental
    Audit Programs . . . . . . . . . . . . . . . . . . . . . . . . . 159
Figure 7.2. Standard Environmental Audit Methodology . . . . . . 165
Figure 8.1. Total Quality Management—The Benchmarking
    Process . . . . . . . . . . . . . . . . . . . . . . . . . . . . . 171
Figure 13.1. Audit Reporting Process Organizational
    Responsibilities . . . . . . . . . . . . . . . . . . . . . . . . . 225
Figure 13.2. Audit Reporting Process Critical Data Elements for
    Tracking . . . . . . . . . . . . . . . . . . . . . . . . . . . . . 228
Figure 13.3. Important Program Characteristics . . . . . . . . . . . 231
Figure 14.1. Compliance Monitoring . . . . . . . . . . . . . . . . . 244
Figure 15.1. Key Compliance Areas and Regulatory Authority
    Jurisdiction . . . . . . . . . . . . . . . . . . . . . . . . . . . 258
Figure 15.2. Typical Compliance Areas on an Environmental,
    Health and Safety Audit . . . . . . . . . . . . . . . . . . . . 259
Figure 15.3. Itinerary for Typical Audit . . . . . . . . . . . . . . . 262
Figure 15.4. Environmental Compliance Audit Program Documents
    to Review . . . . . . . . . . . . . . . . . . . . . . . . . . . . 269
Exhibit 15.1. Inspecting Remote Areas . . . . . . . . . . . . . . . 276
Exhibit 15.2. Inspecting Buildings . . . . . . . . . . . . . . . . . . 277
Exhibit 15.3. Timing Inspections . . . . . . . . . . . . . . . . . . 279
Exhibit 15.4. Inspecting the Fence Line . . . . . . . . . . . . . . . 280
Exhibit 15.5. Emergency Response Drills . . . . . . . . . . . . . . 282
Exhibit 15.6. Auditing Sampling Procedures . . . . . . . . . . . . 283
Exhibit 16.1. In-Service PCB Transformer . . . . . . . . . . . . . 300
Exhibit 16.2. PCB Transformers Out of Service and Awaiting
    Disposal . . . . . . . . . . . . . . . . . . . . . . . . . . . . . 302
Exhibit 16.3. Wastewater Treatment Facilities . . . . . . . . . . . 304
Exhibit 16.4. Stormwater Discharges . . . . . . . . . . . . . . . . 306
Exhibit 16.5. Hazardous Waste Accumulation in Drums . . . . . . 308
Exhibit 16.6. Hazardous Waste Storage Facility . . . . . . . . . . . 310
Exhibit 16.7. Above-Ground Oil Storage Tank . . . . . . . . . . . 312
Exhibit 16.8. Solid Waste Incinerator . . . . . . . . . . . . . . . . 314
Exhibit 16.9. Solvent Metal Cleaner . . . . . . . . . . . . . . . . . 316

Exhibit 16.10. Pesticide Storage Facility . . . . . . . . . . . . . . . 318
Exhibit 16.11. Hazardous Materials Dispensing Area
    for Shop Use . . . . . . . . . . . . . . . . . . . . . . . . . . . . . . . 320
Exhibit 16.12. Flammable/Combustible Materials Storage Facility  322
Exhibit 16.13. Solid Waste Disposal Facility . . . . . . . . . . . . . 324
Exhibit 16.14. Drinking Water Sampling Point . . . . . . . . . . . 326
Figure 18.1. Facility Environmental Compliance Status. . . . . . . 371
Figure 20.1. Approach to Conducting Site Environmental
    Assessments . . . . . . . . . . . . . . . . . . . . . . . . . . . . . . . 409
Figure 20.2. Potential Risks Posed by Real Estate Transactions . .  413
Figure 22.1. Sources for Obtaining the Pre-Audit Information  . .  458
Figure 22.2. Areas Most Often Addressed in Commercial
    TSD Facility Audit Programs . . . . . . . . . . . . . . . . . . . . . 460
Figure 22.3. Proposed TSD Facility Audit Agenda . . . . . . . . . . 465
Figure 22.4. Typical Problems Encountered During
    TSD Facility Audits . . . . . . . . . . . . . . . . . . . . . . . . . . . 469

# ABOUT THE EDITOR AND
# PRINCIPAL AUTHOR

Mr. Lawrence B. Cahill is a Senior Program Director in ERM's Management Consulting Group located in Exton, PA. Mr. Cahill has over twenty years of professional environmental experience with industry and consulting. Principally for Fortune 500 Companies such as DuPont, Exxon, Hoechst Celanese, BFGoodrich, Colgate-Palmolive, Hercules, Bristol-Myers Squibb, and Hughes Aircraft, he has developed, evaluated, and certified audit programs, conducted hundreds of audits, and has trained literally thousands of people in auditing skills all over the world, including in Asia-Pacific, South America, Africa, and throughout Europe, and North America. He has provided an "auditor's opinion letter" formally certifying the efficacy of EH&S audit programs for DuPont, Hoechst Celanese, Eastman Kodak, and Bristol-Myers Squibb, among others. These letters have appeared in the companies' annual environmental reports.

Mr. Cahill has been awarded *Distinguished Instructor* status by Government Institutes. He has taught GI's Environmental Audits course since 1983. He is co-chairperson of the Auditor Qualifications Workgroup of the Environmental Auditing Roundtable and was a member of the Standards Conformance Registration Advisory Group (SCRAG), commissioned by the U.S. Technical Advisory Group to develop a recommendation on the organization or organizations that would meet the objectives for company registrations and auditor certification under ISO 14000 in the U.S.

Mr. Cahill holds a BS in Mechanical Engineering from Northeastern University, an MS in Environmental Health Engineering from Northwestern University, and an MBA in Public Management from the Wharton School of the University of Pennsylvania. He previously was a Project Engineer with Exxon Research and Engineering where he audited plants in the U.S., Canada, and Europe.

# CONTRIBUTING AUTHORS

**RAYMOND W. KANE**
**Independent Consultant**
**Environmental Management Consulting**
Mr. Kane wrote all or part of Chapters 5, 15, 16, and 21.
Mr. Kane is currently a consultant with Environmental Management Consulting in Wayne, PA. He deals with EH&S compliance audits, audit program development, third party review of EH&S audit and management programs, EH&S training, ISO 14000 environmental management consulting, and environmental due diligence. Previously, Mr. Kane was vice president at McLaren/Hart Environmental Engineering Corporation. There he developed and conducted major environmental audit and compliance programs for private industry and federal organizations such as the National Institutes of Health and the U.S. Air Force. Mr. Kane received his Bachelor's and Master's degrees in Civil/Environmental Engineering from Villanova University and served for four years in the U.S. Navy as a submarine officer.

**THOMAS R. VETRANO**
**Principal**
**ENVIRON Corporation**
Mr. Vetrano wrote all or part of Chapters 22 and 25.
Mr. Vetrano is a principal at ENVIRON Corporation in Princeton, NJ, which provides international environmental and health risk management consulting services. He has over twelve years experience in strategic international environmental, health, safety, and risk management consulting. He has managed and conducted numerous environmental due diligence, site assessment, and compliance audit programs for Fortune 500 companies in over twenty-five countries on six continents. Mr. Vetrano has extensive experience in designing environmental compliance, auditing, and risk management programs for a wide range of companies, including aerospace, electronics/computer, chemical, pharmaceutical, and consumer

products companies. Mr. Vetrano has directed environmental due diligence programs for hundreds of merger/acquisitions, diventures, and corporate transactions totaling over $30 billion on behalf of corporations, law firms, and financial institutions. He has also designed and implemented commercial waste management vendor assessment programs for multinational clients. Prior to joining ENVIRON, Mr. Vetrano served as vice president and managing director of Kroll International, an international environment consulting and risk management firm. He also served as western regional manager for HART Environmental Management Corporation, directing HART's nationwide environmental assessment and auditing practice. Mr. Vetrano received a B.S. in Environmental Studies from Rutgers University and an M.S. in Environmental Engineering and Toxicology from the New Jersey Institute of Technology.

**COURTNEY M. PRICE**
**Partner**
**Reid & Priest**
Mrs. Price wrote Chapter 3.
Courtney M. Price is a partner in the Washington, D.C. office of Reid & Priest and the head of their Environmental Practice Group. She served for three years as the Assistant Administrator for Enforcement and Compliance Monitoring at the U.S. EPA. In this position, she was responsible for all of the EPA's judicial and administrative enforcement activities and played a significant role in the development of environmental statutes and regulations. Previously, Mrs. Price was the associate administrator for rulemaking and the deputy chief counsel of the National Highway Traffic Safety Administration at the U.S. Department of Transportation and staff attorney for regulatory litigation at the U.S. Department of Energy. Mrs. Price also has been in private practice in Los Angeles. She is a graduate of the University of Alabama (A.B.), the University of Southern California Law School (J.D.), and is a member of the Washington, D.C. and California Bars.

## JAMES C. MAUCH
**Vice President**
**Vista Environmental Information, Inc.**
Mr. Mauch co-authored Chapter 20 with Mr. Cahill.
Mr. Mauch is vice president for Vista Information Solutions, Inc., in Louisville, KY. He is also an attorney and was in private practice for five years with a Midwest general practice and litigation law firm. His practice involved trial work and commercial litigation, and focused in part on environmental and mineral rights issues affecting land use transactions. Since joining Vista, Mr. Mauch has been involved in federal legislative initiations and has published articles on the appropriate standards for Phase I environmental real estate assessments. Mr. Mauch is an active member of ASTM's E-50.02 subcommittee which developed the Site Assessment Standards, and he is currently chairman of the Public Records Task Group of that subcommittee.

## BRIAN P. RIEDEL
**U.S. Environmental Protection Agency**
Mr. Riedel wrote Chapter 2.
Brian P. Riedel is counsel for EPA's Office of Planning and Policy Analysis (OPPA) which serves the Assistant Administrator for Enforcement and Compliance Assurance (OECA). Mr. Riedel is co-author of EPA's interim and final environmental self-policing policies. He is co-chair of the Quick Response Team responsible for making recommendations regarding interpretation and application of the policies. In addition, Mr. Riedel is the OECA lead on enforcement and compliance matters relating to ISO 14001 standards for environmental management systems. He is a member of the U.S. Technical Advisory Group (TAG) to ISO Technical Committee 207 on environmental management. Before moving to EPA, Mr. Riedel practiced environmental law with the Washington D.C. law firm of Newman & Holtzinger. He received his law degree from the University of Wisconsin and A.B. from the University of Michigan.

**DAWNE P. SCHOMER**
**Corporate ESH Audit Program Manager**
**Texas Instruments**
Ms. Schomer co-authored Chapter 6 with Mr. Cahill.

Ms. Schomer is a member of the Corporate Environmental, Safety and Health Team of Texas Instruments, Inc., of Dallas, Texas. She currently manages the EHS Audit Program and the EHS Systems Program. She is also serving as a co-leader of the workgroup on Auditor Qualifications for the U.S. delegation of International Standards Organization (ISO)/ Technical Committee 207. Previously, she vice-chaired the Standards Conformance and Registration Advisory Group, which provided research and advice to the U.S. ISO/TC207 Technical Advisory Group on the issues of ISO 14000 certification and registration. Ms. Schomer participates in the Environmental Auditing Roundtable (EAR), and is co-chair of the EAR Auditor Qualifications/Training Workgroup.

**JENNIFER L. KARAS**
**Associate**
**Reid & Priest**
Ms. Reid assisted Courtney Price in the writing of Chapter 3.

Jennifer L. Karas is an associate in the Washington, DC office of Reid & Priest. She works with the Environmental Practice Group on a variety of environmental matters including counseling clients such as electric utilities, independent power producers, manufacturers, and lenders with regard to compliance with state and federal environmental statutes and regulations; defending client interests in private cost recovery actions under the Comprehensive Environmental Response, Compensation and Liability Act and in enforcement actions under the Clean Water Act; and advising the Real Estate and Project Development and Finance Practice Groups on environmental issues such as environmental due diligence, state property transfer laws, and permitting requirements for new facilities. Ms. Karas earned her Bachelor of Arts degree from the University of Michigan at Ann Arbor and is currently working on completed her Masters of Law in environmental law at George Washington University's National Law Center.

# PREFACE

It has been about six years since the publication of the Sixth Edition of *Environmental Audits*. In the preface to that edition I stated that it was quite possible that the sixth would be the last edition of the book. I followed that remark with the presumptuous statement that "there simply isn't that much more to say." Why am I so wrong so often?

In the past few years there have been tremendous and significant developments in the environmental auditing field, including the very recent development of ISO 14000 auditing guidelines. As a consequence, the Sixth Edition was becoming very outdated. This was problematic for me as the book is the principal tool used in the Government Institutes Environmental Audits training course, which I have been leading since 1983.

I have always said to the course attendees that once the supplementary materials become thicker than the course text, then I know it's time for a new edition. Well, for the past two years that has been the case as we've attempted to keep the course current. However, it has taken me these same two years to store up enough energy to put out another edition. Those of you who have attempted to write and edit a technical book while maintaining a full-time job and family responsibilities will know why it took a while to mobilize. The work on this book was done late at night and on weekends and not always in the U.S.

The Seventh Edition maintains the same basic three-part structure of the Sixth: Managing a Program, Conducting the Audit, and Special Auditing Topics. As in the past, the new edition stands alone as a complete document; where information from the previous edition was determined to be no longer relevant it was dropped. There are now twenty-five chapters in the book as opposed to seventeen, indicating the insertion of a considerable amount of new material. Moreover, no chapter in the Sixth Edition went untouched as we attempted to improve both the quality and the currency of the information.

Through the efforts of my co-authors, we believe we have addressed most of the issues currently impacting environmental audits and audit programs. It is a very dynamic setting in late 1995 and further changes likely are ahead of us. Sadly, for me this means that an Eighth Edition is probable, unless I buy that farm in Vermont...........

*Lawrence B. Cahill*
*November 1995*

# ACKNOWLEDGMENTS

I wish to thank the contributing authors for their efforts in delivering a quality product on time. Well, at least in delivering a quality product. Working with the staff at Government Institutes in preparing the manuscript was a pleasure. All the authors had to do was submit rough electronic versions of the material and, believe me, that was challenge enough. Thank you Martin and Jeff at GI for the encouragement (or was it harassment?) it took to get to the Seventh Edition. And finally, I wish to thank Dotty at ERM who has guided me through the wonders of the Macintosh world and file conversions and made sure that all diskettes got to where they were supposed to go. Nothing gets done well without help.

*To Claire, Brendon and Bryan who help me see what is important and never let me get too full of myself........*

# ENVIRONMENTAL

# AUDITS

## 7TH EDITION

# PART I


# MANAGING A PROGRAM

# 1

# PERSPECTIVES IN ENVIRONMENTAL AUDITING

---

*"I am an Environmentalist. "*[1]

---

## WHY AUDIT?

Managing compliance in today's regulatory setting has become an almost overwhelming exercise, involving more and more regulations, and affecting more and more organizations. In the United States, we have gone well beyond the straightforward regulation of air emissions and wastewater effluents, which were the prevalent initiatives in the legislation of the 1970s. Hazardous wastes and hazardous materials are now tightly controlled, and their discharges are freely and openly reported through community and worker right-to-know regulations. In the near future, more expansive controls probably will be based on meeting global, not national, objectives.

The traditional approach of regulating only major industry groups has also long passed us by. In this day and age, environmental rules affect neighborhood dry cleaners, liquor stores, auto parts stores and body shops, and, in some states, your local supermarket.

---

[1] Bush, George, Statement made when a candidate for President of the United States, Summer, 1988.

And even more disturbing, we find no safe harbor from these complicated and pervasive issues while sitting at home. Consider the following:

■ **Could your home contain friable asbestos?** Although asbestos insulation was outlawed around 1974, the U.S. EPA estimates that over 750,000 public buildings and innumerable homes still contain asbestos.

■ **Could it have been built over an abandoned landfill?** Love Canal is the classic case, and there are now 30,000 documented abandoned waste disposal sites across the country.

■ **Could there be radon in the basement?** U.S. EPA has reported that eight million U.S. homes are potentially tainted with unsafe radon levels. Scientists estimate that between 5,000 and 20,000 lung cancer deaths a year may be caused by radon.

■ **Could your drinking water supply be contaminated?** A study by the Center for Responsive Law found that nearly one in five public drinking water systems is contaminated with chemicals, some of them toxic. Many in-home supplies are contaminated with excessive levels of lead, abraded from the lead solder used to connect copper piping.

■ **Could fluorescent lights be a source of PCBs and mercury?** Light ballasts manufactured before 1976 will likely contain PCBs. Tubes can contain trace amounts of mercury.

This continuous emergence of new environmental risks has resulted in the regular enactment of laws designed to protect both public health and the environment. As Figure 1.1 demonstrates, we have seen a dramatic growth in federal environmental regulations alone, even during periods of "regulatory reform." Codified federal environmental regulations now total over 13,000 pages and have grown by thirty percent in the 1990s alone. And remarkably, solid waste regulations represent only twelve percent of the total. Included under this generic heading are hazardous waste regulations promulgated under the Resource Conservation and Recovery Act (RCRA) and the Comprehensive Environmental Response, Compensation and Liability Act (CERCLA, or Superfund)—two landmark pieces of legislation.

The stifling federal regulatory oversight is not likely to abate for some time, even with the "new" approach to regulation evolving in the mid-1990s. Current rhetoric is remarkably similar to that of the early 1980s and the first Reagan administration. As of May 1995, the U.S. EPA had some 429 regulations under development, revision, or review (*See* Figure 1.2). Over one-third of these are Clean Air Act regulations.

Sometimes lost in the glare of the national regulatory setting is the increasing impact that states have on compliance. Although states generally are required to adopt federal standards as the bare minimum of regulation, many states go much beyond what is required by federal regulations. For example, Proposition 65 in California and the Industrial Site Restoration Act (ISRA) in New Jersey not only carve out new regulatory frontiers but are often viewed as model legislation that could be adopted by other states. For organizations with facilities in more than one state, compliance assurance can be difficult. As can be seen in Figure 1.3, states will vary in their regulatory stringency; thus, it is typically quite a challenge to stay current on regulations promulgated in multiple states. And in the last few years, with the proliferation of community right-to-know legislation, local fire departments and county agencies have become players in the game.

**Figure 1.1**
**Growth of Environmental Regulations**

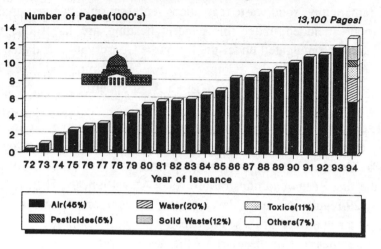

**Figure 1.2**
**EPA's Short-Term Agenda**

# 429(!) Regulations Under Development

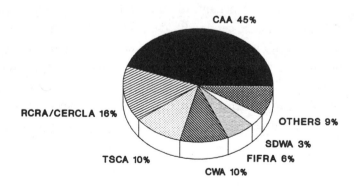

Source: May 1995 EPA Agenda

Moreover, the regulatory environment varies significantly from country to country. Figure 1.4 begs the question: Are there fewer lawyers per capita in other countries because the countries are less litigious by nature, or are other countries less litigious because they have fewer lawyers? The United States is known to be a very litigious society with an overwhelming regulatory framework. Other countries have a more cooperative relationship between the regulators and the regulated, and are much less litigious. For multi-national companies, the need for different rulebooks can be very challenging.

In this regulatory setting, one can only wonder if "fail-safe" management of compliance is, indeed, an achievable objective. Yet, the penalties for not complying are far too intimidating to risk applying anything but extreme diligence towards meeting standards and applying good management practices. Civil penalties of up to $25,000 per day are common in most statutes.

In addition, the U.S. EPA has reemphasized its enforcement role and has moved toward a policy of taking enforcement action against corporate officials as well as their companies. The following cases are examples of that policy:[2]

- *U.S. v. Wietzenhoff.* Michael Wietzenhoff and Thomas Mariani appealed their felony convictions for conspiracy and knowing violations of the Clean Water Act (CWA). A jury convicted the two plant managers of six felony counts. The judge sentenced Wietzenhoff and Mariani to twenty-one months and thirty-three months in prison, respectively. The Appeals Court agreed with the District Court that the felony provisions of the CWA do *not* require proof that defendants know that their conduct violates their NPDES Permit.

---

[2] Taken from U.S. EPA's *Enforcement and Compliance Assurance Accomplishments Report—FY 1994*, EPA 300-R-95-004, May 1995.

## Figure 1.3
## State Regulatory Stringency

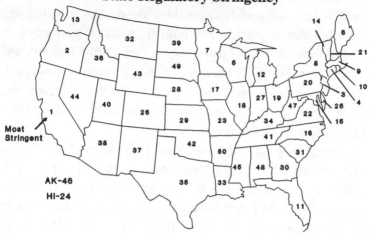

Institute for Southern Studies, 1991

## Figure 1.4
## Concentration of Lawyers in Developed Countries

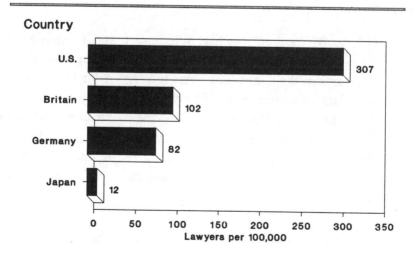

■ *U.S. v. Advance Plating Works, Inc.* Advance Plating, an electroplating and metal finishing shop located in Indianapolis, Indiana, was fined and its owner and president was sentenced to jail and also fined on October 8, 1993. The defendants engaged in the tampering of samples and the illegal discharging of company wastes into the Indianapolis sewer system.

■ *U.S. v. Giacomo Catucci.* Catucci, former president of Post-Tron, Inc., a computer software company, was sentenced on February 15, 1994 to twenty-seven months in prison for unlawful disposal of PCBs. The violations occurred after Catucci gave employees permission to scrap two PCB transformers, knowing the transformers contained PCBs.

■ *U.S. v. Craven Laboratories, Inc.* Don Craven and his company pleaded guilty on December 1, 1993 to various charges including FIFRA misdemeanors and criminal conspiracy. Craven directed his employees to use testing short cuts that produced false data which were used for pesticide registration. Numerous employees knowingly followed Craven's instructions (and were often paid bonuses for doing so), and understood the data were false and misleading. Craven was sentenced to a maximum of sixty months imprisonment and, along with the company, paid $30 million in fines and restitution.

■ *U.S. v. Charles A. Eidson and Sandra A. Eidson.* Sandra Eidson, former owner and officer of Cherokee Oil Company, Ltd., was sentenced on April 27, 1994, to serve thirty-seven months in prison, and her husband, Charles, was sentenced on March 11, 1994, to serve seventy months in prison for federal crimes committed while operating an oil recycling business. An investigation of the Eidson's oil recycling business in Tampa, Florida, revealed that the company represented to clients that it would dispose of the wastes in a lawful manner. However, they illegally disposed of the wastes into storm sewers instead.

■ *U.S. v. Robert Pardi.* On May 25, 1994, Pardi, an architect and the former director of the Asbestos Task Force of the New York City Board of Education, was sentenced to thirty months of imprisonment for falsely reporting that school buildings were free of asbestos contamination. Pardi was responsible for reporting to the EPA concerning the inspecting and testing of New York City public schools for the presence of asbestos.

■ *U.S. v. John Pizzuto.* In his second environmental prosecution, Pizzuto pleaded guilty, on December 16, 1993, in Huntington, W est Virginia, to a three-count indictment for violating TSCA by illegally storing PCBs in Nitro, WV. As a result of the West Virginia crimes, which occurred during Pizzuto's probation in Ohio, the Ohio federal judge revoked Pizzuto's probation on July 18, 1994, and ordered him jailed for an additional eighteen months, resulting in a total of thirty-six months of incarceration.

■ *U.S. v. William C. Whitman and Duane C. Whitman.* On July 28, 1994, William Whitman, a plant manager, and Duane Whitman, a shop foreman, of Durex Industries were found guilty of treating and storing hazardous waste without a permit. The prosecution of the defendants was initiated following the deaths of two nine-year-old boys from toluene fume asphyxiation on June 13, 1992. The two children had been playing in a dumpster in which toluene had been discarded.

Placement of notices in local newspapers is another approach that regulatory agencies are using with increasing frequency. As an example, Figure 1.5 is a copy of a full-page notice placed in the Sunday *Los Angeles Times*, as mandated by the Los Angeles Toxic Waste Strike Force.

The sum total of U.S. EPA's enforcement efforts is shown in Figure 1.6. Note that the amount of fines and the number of defendants charged has grown substantially over the past eighteen years. And no company, no matter how apparently innocuous its operations, is immune: note Walt Disney's record in the late 1980s and early 1990s (Figure 1.7) .

As if all this weren't enough to heighten the anxiety of corporate managers, they may also face the ire of stockholders if the company's stock price is adversely affected by compliance problems. This is a real risk, as indicated by Figure 1.8, which shows the short-term price drops of four public companies suffering through environmental incidents. The classic case in this regard is Union Carbide's Bhopal tragedy in late 1984.

## IMPACT ON THE REGULATED COMMUNITY

What are the implications of these policies and trends for management? With the high probability of continued controversy and complexity in the environmental arena, it is very important to avoid complacency and to pay attention to compliance matters. Management's ability to anticipate and respond to government and private actions aimed at redressing real or perceived environmental problems must remain undiminished. Yet, corporate outlays for environmental management, including those for staffing, generally have been reduced as have expenditures for many other business activities in the competitive economic climate of the 1990s. Thus, many companies may be less than ideally prepared to face a future regulatory environment that will continue to be complex and onerous.

The task of business, then, in this climate of uncertainty, is to assure minimal vulnerability in compliance-related matters and to do so in the most efficient and cost-effective way. One solution to this problem is the development and/or vigorous maintenance of a formal environmental audit program that evaluates compliance with regulatory requirements, good management practices, and corporate policies and procedures at all facilities.

**Figure 1.5**
**Warning of Illegal Disposal**

# WARNING

## THE ILLEGAL DISPOSAL OF TOXIC WASTES WILL RESULT IN JAIL.

### WE SHOULD KNOW WE GOT CAUGHT!

February 12, 1985

American          Corporation

Dear Businesses & Residents of the City & County of Los Angeles

Pollution of our environment has become a crisis.

Intentional clandestine acts of illegal disposal of hazardous waste, or "midnight dumping" are violent crimes against the community.

Over the past 2 years almost a dozen Chief Executive Officers of both large and small corporations have been sent to jail by the L.A. Toxic Waste Strike Force.

They have also been required to pay huge fines; pay for cleanups; speak in public about their misdeeds; and in some cases place ads publicizing their crime and punishment.

THE RISKS OF BEING CAUGHT ARE TOO HIGH —
AND THE CONSEQUENCES IF CAUGHT ARE NOT WORTH IT!

We are paying the price. *TODAY*, while you read this ad our President and Vice President are serving time in *JAIL* and we were forced to place this ad.

PLEASE TAKE THE LEGAL ALTERNATIVE AND PROTECT OUR ENVIRONMENT.

Very Truly Yours,

American          Corporation

141     AVENUE
LOS ANGELES, CA 90031

### Figure 1.6
### U.S. EPA Environmental Enforcement

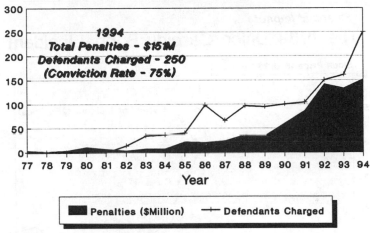

*1994*
*Total Penalties - $151M*
*Defendants Charged - 250*
*(Conviction Rate - 75%)*

Year

■ Penalties ($Million)    —＋— Defendants Charged

*Source: EPA Annual Enforcement Reports*

### Figure 1.7
### Incidents at Walt Disney Co.

# WALT DISNEY CO.
## Incidents Affect Mickey

| DATE | INCIDENT | FINE |
|------|----------|------|
| 7/90 | Waste Disposal | $550,000 |
| 4/90 | Sewage Discharge | $300,000 |
| 1/90 | Vulture Kill | $10,000 |
| 1989 | PCB Transport | $2,500 |
| 1988 | Waste Storage | $150,000 |
|      | TOTAL | $1,012,500 |

**Figure 1.8**
**Financial Impacts of Industry Incidents**

*Financial Impacts ...*
# The 1984 Union Carbide Bhopal Incident

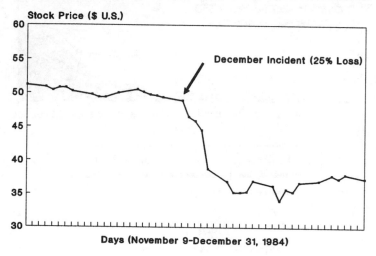

*Financial Impacts ...*
# The 1989 Phillips Houston Incident

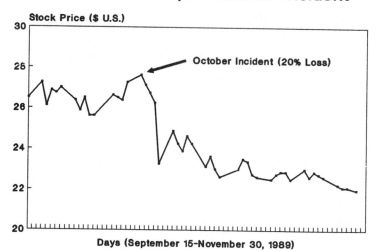

**Figure 1.8** *(cont'd)*

*Financial Impacts ...*
# The 1986 Sandoz Rhine River Incident

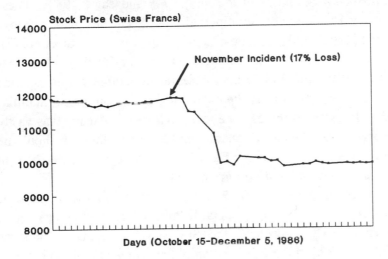

*Financial Impacts ...*
# The 1990 Arco Channelview Incident

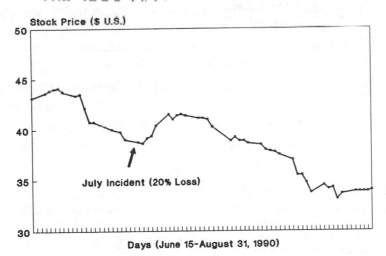

# EVOLUTION OF ENVIRONMENTAL AUDITING

As a separate and distinct compliance management tool, environmental auditing had its beginnings in the late 1970s and early 1980s. These beginnings were stimulated principally by the Securities and Exchange Commission (SEC) actions against three companies: U.S. Steel (1977), Allied Chemical (1979), and Occidental Petroleum (1980).[3] The SEC required each of these public companies to undertake corporate-wide audits to determine accurately the extent of the environmental liabilities they faced. In essence, the SEC believed that each company was vastly understating its liabilities in its annual report to stockholders. It should be noted that, since that original SEC audit, each company has had an effective environmental audit program in place.

Interestingly, in the 1990s, the SEC has once again raised this issue of inaccurate reporting of environmental liabilities by public companies in annual reports. Most recently, the SEC has stated that it believes companies are not portraying their potential Superfund liabilities properly. It is possible that another round of SEC actions could occur in the next couple of years, unless the perceived systemic discrepancies are remedied.

On the heels of the SEC audits came the implementation of major hazardous rules which were promulgated in the late 1970s and put into effect beginning in 1980. These rules were comprehensive, administratively complex, and potentially costly if not adhered to. Because of the nature of the industry, the large chemically-intensive corporations were the first to develop audit programs to better respond to these rules. In recent years, many Fortune 200 diversified manufacturing companies have begun to adopt the concept. And even government agencies have seen the light and are developing programs that focus both on compliance

---

[3] *SEC v. Allied Chemical Corp.*, No. 77-0373 (D.D.C., March 4, 1977); *SEC v. United States Steel Corp.*, (1979-1980 Transfer Binder) Fed. Sec. L. Rep. (CCH) 82,319 (1979); *SEC v. Occidental Petroleum Corp.*, (1980 Transfer Binder) Fed. Sec. L. Rep. (CCH) 82,622 (1980).

for operating facilities and defining liabilities where property is bought or sold.

Today, in the mid 1990s, environmental auditing has reached a certain level of maturity. Applicability has spread beyond the basic chemical industry to all types of industries and even government agencies.[4] Even simple properties are undergoing audits or assessments prior to sale. Additionally, most generators of hazardous waste are auditing the sites to which their wastes are being transported for handling and disposal by third parties.

Although now a widespread practice, auditing is still an evolving discipline. As a result, several associations have been organized to further the profession of environmental auditing. Two of the more well known are discussed below. The discussions are based on descriptions provided by the associations.

## ENVIRONMENTAL AUDITING ROUNDTABLE

The Environmental Auditing Roundtable (EAR) is a professional organization dedicated to furthering the development and professional practice of environmental auditing. The EAR dates back to January 1982, when managers of several environmental audit programs met informally to discuss their auditing programs and practices, as well as policy and regulatory actions related to auditing.

The original group soon increased to ten managers who held quarterly meetings during 1982 and 1983. Meetings were opened to all interested persons in September 1984, with the original ten members serving as the organization's first steering committee. At that time, the organization adopted the name Environmental Auditing Roundtable and developed a

---

[4] Government agencies certainly have reason to pause over environmental contamination. It has been reported that it may cost up to $110 billion and take nearly 60 years to clean up the Department of Energy's nuclear weapons manufacturing facilities.

statement of purpose and organizational principles.

In June 1987, EAR participants adopted a code of ethics and bylaws which, among other things, established general membership criteria and a five-member board of directors elected by the membership at large. Following peer review and vote by the membership, EAR adopted Standards for the Performance of Environmental Audits in 1993.

Most EAR members are practicing environmental, health, and safety auditors with extensive field experience. While EAR focuses primarily on meeting the needs of industry, membership is open to anyone with a professional interest in the practice of environmental auditing. Since June 1984, EAR has continued to meet quarterly and participation has risen steadily, with up to 200 participants attending each meeting. In 1995, the EAR has more than 500 members.

For more information contact:

> Environmental Auditing Roundtable
> Kathy Reith, Administrator
> 35888 Mildred Avenue
> North Ridgeville, OH 44039
> (216)327-6605
> Fax: (216)327-6609

# THE INSTITUTE FOR ENVIRONMENTAL AUDITING

The Institute for Environmental Auditing (IEA) is a formal association of environmental auditors and environmental managers founded in 1986 and incorporated as a nonprofit organization in Washington, DC. The Institute's objectives are to:

- Participate in and encourage the development of guidelines for environmental audits and environmental auditing programs.

- Encourage the use of environmental auditing by business and industry as a tool for business planning, for management, and for business transactions.

- Provide information to assist governmental agencies in understanding the proper role of environmental audits in regulatory affairs.

- Seek to promote recognition of the qualifications of those engaged in the field of environmental auditing.

- Participate in training and other activities that enable individuals engaged in environmental auditing to enhance their qualifications.

- Provide forums for its membership for the exchange of information, knowledge and ideas toward the end of advancing the overall quality of the practice of environmental auditing.

- Seek to establish ethical standards by which those engaged in the field should be guided, and seek to gain acceptance and recognition of those engaged in environmental auditing.

- Represent its membership before public and private institutions for the purpose of advancing the field of environmental auditing and obtaining recognition of the profession of environmental auditing.

The Institute publishes a quarterly newsletter called *Working Papers*.

For more information contact:

The Environmental Auditing Institute
P.O. Box 23686
L'Enfant Plaza Station
Washington, DC 20026-3686

## DEFINING ENVIRONMENTAL AUDITS

Precisely defining environmental audits is a difficult exercise. In part, this is due to the evolving nature of the concept, its rather recent appearance as a formal management tool, and the need to tailor programs to the sponsoring organization. It also has to do with the fact that audit programs are typically designed to meet one or more of the following objectives:

- Assuring compliance with regulations
- Determining liabilities
- Protecting against liabilities for company officials
- Fact-finding for acquisitions and divestitures
- Tracking and reporting of compliance costs
- Transferring information among operating units
- Increasing environmental awareness
- Tracking accountability of managers.

Yet, within these broad boundaries, definitions can be framed. The EPA defines environmental auditing as "a systematic, documented, periodic and objective review by regulated entities[5] of facility operations and practices

---

[5] "Regulated entities" include private firms and public agencies with facilities subject to environmental regulation. Public agencies can include Federal, state, and local agencies as well as special-purpose organizations such as regional sewage commissions.

related to meeting environmental requirements."[6]

An alternative definition is that audits can be said to verify the existence and use of:

Adequate...Systems...Competently...Applied.

Simply put, an audit program is first and foremost a *verification* program. It is not meant to replace existing environmental management systems at the corporate (*e.g.*, regulatory updating), division (*e.g.*, capital planning for pollution control expenditures), or plant (*e.g.*, NPDES discharge monitoring) levels. Indeed, the program should be designed to verify that these environmental management *systems* do, in fact, *exist* and are *in use*. These systems, whether they are to assure compliance or define liabilities, need to be *adequate*, in that they should acknowledge and respond directly to the regulatory and internal requirements that define compliance or liability.

The systems should also be *applied*, meaning that procedures are not simply "bookshelf exercises," out-of-date and out-of-use before the ink dries. And lastly, the systems must be applied *competently*. All plant managers, environmental coordinators, and unit operators must have an awareness of environmental compliance, and conduct their responsibilities accordingly.

## ADVANTAGES AND DISADVANTAGES

Like most anything in life, audit programs can be characterized by both advantages and disadvantages. On the positive side, audits can result in a number of significant benefits, including:

■ Better compliance

---

[6] Taken from U.S. EPA's final Environmental Auditing Policy Statement.

- Fewer surprises
- Fewer fines and suits
- Better public image with the community and regulators
- Potential cost savings
- Improved information transfer
- Increased environmental awareness.

However, these benefits can be offset by some real and potential costs (which will be discussed in more detail in other chapters in this book):

- The commitment of resources to run the program
- Temporary disruption of plant operations
- Increased ammunition for regulators
- Increased liability where one is unable to respond to audit recommendations involving significant capital expenditures.

The last two disadvantages raise a special issue associated with environmental audits. Even where programs are operated under legal protections (*e.g.*, attorney-client privilege) as discussed in another chapter, there is a chance that audit reports could be "discovered" in a legal proceeding. Thus, it is vital that management understand that if a program is initiated, the company must be serious about fixing the problems that are identified in the reports. Otherwise, the report could be most incriminating in a court case or administrative proceeding.

Notwithstanding these drawbacks, most firms, faced with the question of whether or not to undertake an audit program, have opted to do so. The general theory is that in this day of increased litigation and possible criminal suits, it is better to know your liabilities than to remain oblivious to them. As stated some time ago by a former U.S. EPA General Counsel:

"Management ignorance is no defense!"[7]

---

[7] R.V. Zener, In *Environment Reporter*, Bureau of National Affairs, Current Developments, June 26, 1981.

# TRENDS IN ENVIRONMENTAL AUDITING

In the past few years, a number of interesting trends have surfaced within the environmental auditing discipline. These include:

- **A Push for Standards.** With the development of several professional organizations, as described earlier in this chapter, we have seen a push by some segments of the auditing profession towards the development of standards and possibly national and international registration/certification. As discussed in a later chapter, standards such as ISO 14000, BS 7750, the European Eco-Management and Audit Scheme and the like are proliferating. The thought is that standards will help to improve the quality of both audits and auditors and to more clearly define the acceptable environmental audit—presently a nebulous concept. Given the varying types of audits and the multiple objectives a program can be designed to achieve, the development of standards will be a challenging exercise. Nonetheless, as shown in Figure 1.9, the U.S. EPA, the U.S. Department of Justice, the British Standards Institute (BSI) and the International Standardization Organization (ISO) have developed standards and guidelines defining an acceptable environmental audit program.

**Figure 1.9**
**Guidelines for Audit Programs**

## USEPA's Elements of Effective Environmental Audit Programs

1. Top Management Support
2. Independence from Audited Activities
3. Adequate Staffing & Training
4. Explicit Objectives, Scope, Resources & Frequency
5. Process That Collects & Analyzes Information
6. Process to Develop & Distribute Reports
7. Quality Assurance Process

## USDOJ Sentencing Guidelines for Environmental Auditing

1. Sufficient Authority, Resources, Personnel
2. Frequent Auditing
3. Independence from Line Organization
4. All Applicable Requirements
5. Surprise Audits when Necessary
6. Follow-up Countermeasures
7. Continuous Self-Monitoring
8. Reporting Without Retribution
9. Tracking Response Actions

**Figure 1.9** *(cont'd)*

## ISO 14010 Guidelines for EMS Auditing

1. Clearly Defined & Communicated Scope & Objectives

2. Auditor Independence

3. Due Professional Care

4. Quality Assurance

5. Systematic Procedures

6. Appropriate Audit Criteria

7. Sufficient Audit Evidence

8. Written Audit Report

9. Qualified Auditors

## BS7750 Environmental Audits Requirements

1. A Written Plan & Audit Protocol

2. Definition of Areas to be Audited

3. Audit Frequency Based on Risk

4. Assignment of Responsibilities

5. Independence & Expertise of Auditors

6. Audit Findings Reported

7. Objective Approach

8. Reports Submitted to Senior Management

9. Self-Audits & Public Disclosure Encouraged

- **The Use of Computers.** Computers are now being used in a variety of ways in the support of audit programs. They help to maintain on-line regulatory databases; they store audit reports, action plans, and facility schedules; and portables are increasingly used in the field to develop debriefing documents and draft reports. In some cases, permit and other regulatory information, such as training records, are being filed centrally on computers and accessed as part of the audit's pre-visit activities.

- **An Emphasis on Management Audits.** Through the environmental audit, many companies have found that inadequate attention to management systems has been central to the most significant compliance problems identified at operating facilities. Two examples are in the areas of maintenance and emergency response. Lack of an adequate maintenance program can turn a "gold-plated" pollution control facility into a continuing compliance headache. And the lack of a "working" emergency response system can turn a small spill or release into a major remediation project. Thus, many companies are now focusing their audit efforts on assuring that environmental management systems are in place and functioning.

- **An Integration of Compliance Management Programs.** Because of overlapping requirements (*e.g.*, Material Safety Data Sheets, emergency response) and the desire to avoid overwhelming facilities with "the audit of the month," many companies with separate programs are reviewing the possibility of combining environmental and health and safety compliance and audit programs. In some companies, this will require a re-engineering of the organization, because the health and safety program "grew up" in the Human Resources Department while the environmental program, which arrived much later, often had a life of its own. As shown in Figure 1.10, achieving the appropriate balance might be

## Figure 1.10
## EPA *vs.* OSHA: Two Comparisons

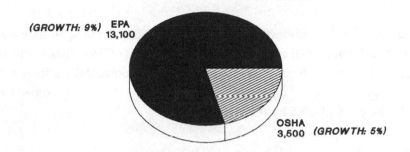

**Pages of Regulations
in the 1994 CFR**

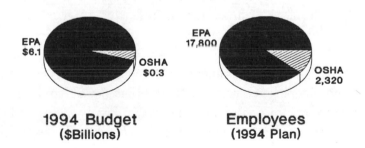

**1994 Budget**
($Billions)

**Employees**
(1994 Plan)

something of a challenge since the U.S. EPA has almost four times as many pages of regulations, twenty times the budget and almost eight times the staff, compared to the Occupational Safety and Health Administration.

While "doing more with less" today may be an understandable attitude and even a worthwhile goal, any fundamental relaxation of compliance efforts by companies presents an unjustifiable risk. Environmental auditing is a sound way to prevent damaging situations from occurring. Its value should not be neglected by management.

# 2

# GOVERNMENT PERSPECTIVE[1]

*"The environmentalists beat up on it. Congress beats up on it. The Administration usually ignores it. In a curious way, the strongest supporters of a forceful EPA are the industries it regulates. They want government to set reasonable standards, and they want the public to know they are being enforced."[2]*

## INTRODUCTION

"Know thyself," the admonition of the oracle of Apollo at Delphi, is particularly sage advice regarding environmental affairs. Apparently, this wisdom is being followed in the regulated community. Eighty-five percent of the corporate respondents to a 1995 Price Waterhouse survey had environmental auditing programs.

Many sound business reasons support having an auditing program. Auditing programs can:

- Provide assurances to management that control systems are functioning
- Decrease the entity's operating and financial risks
- Provide evidence of meeting customers' environmental expectations

---

[1] Note: This chapter was written by Brian P. Riedel and reflects the views of the author and not necessarily those of the U.S. Environmental Protection Agency or any other government entity.

[2] William D. Ruckelshaus, Quoted in "Here Comes the Big New Cleanup," *Fortune Magazine*, November 21, 1988, p. 110.

- Help maintain good public relations
- Satisfy investor criteria and improve access to capital
- Help secure insurance at reasonable cost
- Enhance an entity's image and market share
- Provide the basis for meeting vendor certification criteria
- Demonstrate reasonable care
- Facilitate conservation of input materials and energy
- Improve relations with government.

The vast majority of respondents to the 1995 Price Waterhouse survey cited the following reasons for performing environmental audits:

- Problems can be identified internally and corrected before they are discovered by an agency (96%)
- Assurance can be provided to management that control systems are functioning (91%)
- To improve our company's overall environmental program, and make it more proactive (94%)
- To decrease the company's operating and financial risks (88%)

An auditing program is also one of many tools needed to develop and maintain an effective environmental management program. But identification of the deficiencies in compliance and management is only the first step. An organization must be willing to correct violations and other deficiencies and root causes in order to achieve improved results. If an organization is unwilling to act upon the discovered deficiencies, the audits may become a source of damaging evidence of liability.

However, an *effective* environmental auditing program is a very important piece of the overall environmental management puzzle.[3]

---

[3] For additional reading on broader environmental management practices see: Tapio L. Kuusinen, "Elements of Integrated Industrial Environmental Management," speech presented at Federal Facilities Environmental Program Management: Tools and Techniques for the 1990s, Bethesda, MD, June 1989 available from Regulatory Innovations Staff, EPA (PM-223), 401 M Street SW, Washington, DC 20460; *Environmental Management Review*, quarterly

Auditing can serve as a quality assurance check to help improve the performance of basic environmental management programs by assessing whether necessary control systems and management practices are in place, functioning, and adequate. Environmental audits evaluate—but do not substitute for—direct compliance activities such as obtaining permits, installing controls, monitoring compliance, reporting violations, and keeping records. Although they cannot replace regulatory inspections, audits can supplement conventional federal, state and local government oversight and point toward more efficient allocation of enforcement resources.

The EPA has defined environmental auditing as a systematic, documented, periodic, and objective review of facility operations and practices related to meeting environmental requirements.[4] In Europe, the International Chamber of Commerce has adopted essentially the same definition in its position paper on environmental auditing.[5] Whether domestic or international, an environmental audit can be designed to accomplish any of a number of objectives, including: verifying compliance with environmental requirements; evaluating the effectiveness of environmental management systems already in place; and assessing risks from regulated and unregulated materials, wastes and practices.

Environmental auditing has been developed for sound business reasons, particularly as a means to help manage pollution control affirmatively over time instead of reacting to crises. Auditing can improve facility performance *before* the regulatory inspector arrives. More importantly, it can help indicate effective solutions to common environmental problems, focus facility managers' attention on current and upcoming

---

journal published by Government Institutes, Inc., Rockville, MD; and Friedman.

[4] U.S. Environmental Protection Agency, "Environmental Auditing Policy Statement," 51 *Federal Register* 25004, July 9, 1986.

[5] International Chamber of Commerce, Commission on Environment, 38 Cours Albert ler, 75008 Paris, France, Document No. 210/285 Rev. 2, December 7, 1988.

regulatory requirements, and generate protocols and checklists that help facilities manage themselves better. Auditing also can result in improved management of risks, since auditors frequently identify environmental liabilities that go beyond regulatory compliance.

In 1983, EPA began promoting environmental auditing through a less structured approach labeled "endorsement, analysis and assistance." The agency endorsed auditing at workshops and conferences; analyzed the attributes and benefits of effective audit programs; and assisted those interested in pursuing specific auditing approaches, especially federal agencies.

## EPA POLICY ENCOURAGES USE OF ENVIRONMENTAL AUDITING

In general, EPA encourages sound environmental management practices to help improve environmental results. In particular, implementation of an environmental auditing program can result in better identification, resolution, and avoidance of environmental problems. EPA policy encourages auditing whether done by an independent internal group, by third parties, or by some combination of both. (Larger organizations generally have greater resources to devote to an internal audit team, while smaller ones might be more likely to use outside auditors.)

Although environmental laws do not require auditing, ultimate responsibility for facility environmental performance lies with top corporate managers, who are responsible for taking all necessary steps to ensure compliance with environmental requirements. This creates a strong incentive to use reasonable means, such as environmental auditing, to secure reliable information about facility compliance status. In other words, it's better to manage problems before they surface than to constantly react after the fact and face legal liabilities that could have been avoided.

EPA does not seek to dictate or interfere with the environmental management practices of private or public organizations in its policy or elsewhere. Neither does EPA intend to mandate auditing, though in certain instances EPA will seek to include provisions for environmental auditing in settlement agreements. Environmental auditing systems have been widely adopted on a voluntary basis in the past. Because audit quality depends to a large degree upon genuine management commitment to the program and its objectives, auditing should remain a voluntary activity.

The U.S. Environmental Protection Agency (EPA) recognizes that environmental auditing—and sound environmental management generally—can provide potentially powerful tools toward protection of public health and the environment. In encouraging the use of these tools, EPA has announced the 1986 "Environmental Auditing Policy Statement" (1986 auditing policy) and the 1995 "Incentives for Self-Policing: Discovery, Disclosure, Correction, and Prevention of Violations" (1995 or final self-policing policy). In addition to making statements generally encouraging the use of environmental auditing, the 1986 policy sets forth the widely accepted core criteria for conducting environmental audits. These criteria or elements have provided an important guide for entities wishing to develop and implement auditing programs. In contrast, the 1995 self-policing policy is an enforcement policy. This policy provides penalty mitigation and a criminal safe harbor for entities that discover and promptly disclose and correct environmental violations as long as other safeguards are met.

## EPA's 1986 AUDITING POLICY

On July 9, 1986, EPA published its environmental auditing policy statement. Perhaps the most important contribution of this policy is the listing of the essential elements of effective auditing programs.

## ELEMENTS OF EFFECTIVE AUDIT PROGRAMS

In addition to addressing audit related policy issues, EPA's policy statement also sets forth the essential elements of effective auditing programs. In doing so, the agency hopes to accomplish several objectives. A general description of elements in effective, mature audit programs may help those starting audit programs. Regulatory agencies may also use them when negotiating environmental auditing provisions for consent decrees. Finally, they may help guide states and localities considering their own auditing initiatives.

Private sector environmental audits of facilities have been conducted for several years and have taken a variety of forms, in part to accommodate unique organizational structures and circumstances. Nevertheless, effective environmental audits appear to have certain discernible characteristics in common with other kinds of audits, for which standards have been documented extensively. EPA, when setting standards for environmental audits, drew heavily on two of these documents: *Compendium of Audit Standards*[6] and *Standards for the Professional Practice of Internal Auditing.*[7] EPA's standards also reflect findings from agency analyses conducted over several years and comments from corporate environmental audit managers, especially those active in the Environmental Auditing Roundtable.

An effective environmental auditing system will likely include the following general elements:

- *Explicit top management support for environmental auditing and commitment to follow-up on audit findings.* Such support could be demonstrated by sanctioning the role of the audit program in a written management policy statement on environmental protection, buttressed with specific, tangible implementing actions.

- *An environmental auditing function independent of audited activities.* The status or organizational focus of environmental auditors should be sufficient to ensure objective and unobstructed inquiry, observation, and testing. Auditor objectivity should not be impaired by personal relationships, financial or other conflicts of interest, interference with free inquiry or judgment, or fear of potential retribution.

---

[6] Walter Willborn, American Society for Quality Control, 1983.

[7] The Institute of Internal Auditors, Inc., 1974.

- *Adequate team staffing and auditor training.* Environmental auditors should possess or have ready access to the knowledge, skills, and disciplines needed to accomplish audit objectives. Each individual auditor should comply with professional standards of conduct. Auditors, whether full-time or part-time, should maintain their technical and analytical competence through continuing education and training.

- *Explicit audit program objectives, scope, resources and frequency.* At a minimum, audit objectives should include assessing compliance with applicable environmental laws and evaluating the adequacy of internal compliance systems to carry out assigned responsibilities. Explicit written audit procedures should generally be used. Auditors should be provided with all internal policies, and federal, state, and local permits and regulations pertinent to the facility, as well as checklists or protocols addressing specific features that must be evaluated.

- *A process that collects, analyzes, interprets and documents information sufficient to achieve audit objectives.* Information should be collected before and during an on-site visit regarding matters related to audit objectives and scope. This information should be sufficient, reliable, relevant and useful enough to provide a sound basis for audit findings and recommendations.

## EPA'S 1995 FINAL SELF-POLICING POLICY

On December 22, 1995, EPA announced the "Incentives for Self-Policing: Discovery, Disclosure, Correction and Prevention of Violations," (final auditing or self-policing policy).[8] Under the new policy, the agency will greatly reduce civil penalties and limit liability for criminal prosecution for regulated entities that meet the policy's conditions

---

[8] The policy appeared in the *Federal Register* on December 22, 1995 (60 *FR* 66706).

for discovery, disclosure and correction.[9] The final audit policy represents a refinement of the "Voluntary Environmental Self-Policing and Self-Disclosure Interim Policy Statement" (interim auditing policy) announced on April 3, 1995.[10]

## Policy Incentives: Full and 75 Percent Gravity Mitigation of Civil Penalties; No Criminal Referral to DOJ

Under the policy, EPA will not seek gravity-based[11] civil penalties for violations that are discovered through a compliance management system (CMS) or an environmental audit and that are promptly disclosed and expeditiously corrected, provided the other policy conditions are met. Where violations are discovered by means other than a CMS or an audit, but are promptly disclosed and expeditiously corrected, EPA will reduce gravity-based penalties by seventy-five percent provided the other policy conditions are met. The agency will generally not recommend to the Department of Justice (DOJ) that criminal charges be brought against entities that meet all of the policy conditions. Finally, the policy restates EPA's policy and practice of not routinely requesting environmental audit reports.

### Safeguards

While the final self-policing policy contains significant incentives for encouraging discovery, disclosure and correction of violations, it also contains very important safeguards to deter irresponsible behavior and protect the public and the environment. For example, the policy requires

---

[9] A copy of the policy appears in Appendix B.

[10] 60 *FR* 16875, April 3, 1995.

[11] The "gravity" component of a penalty represents the "seriousness" or "punitive" portion of penalties. The other major part of a penalty, the economic benefit component, represents the economic advantage a violator gains through its noncompliance.

entities to take steps to prevent recurrence of the violation and to remediate any harm caused by the violation. In addition, the policy does not apply to violations which resulted in serious actual harm or may have presented an imminent and substantial endangerment to human health or the environment. Moreover, entities are not eligible for relief under the policy for repeated violations. The policy does not apply to individual criminal acts or corporate criminal acts arising from conscious disregard or willful blindness to violations. Finally, EPA retains its discretion to collect any economic benefit gained from noncompliance in order to preserve a "level playing field" for entities that invest in timely compliance.[12]

## Incentives and Behavior

The final self-policing policy provides additional incentives for entities to utilize the critical compliance tools of environmental auditing and compliance management systems. These incentives add to the many existing business reasons for entities to develop and maintain environmental auditing and compliance management systems. A 1995 Price Waterhouse survey on environmental auditing practices showed that ninety percent of corporate respondents that audit began doing so to find and correct violations before they were found by government inspectors.

---

[12] Under the final self-policing policy, EPA may waive the entire penalty for violations which, in EPA's opinion, do not merit any penalty due to the insignificant amount of any economic benefit.

Some environmental statutes require EPA, in assessing penalties, to consider the economic benefit a violator gains from noncompliance. *See, e.g.*, CWA §309(g), CAA §113(e), and SDWA §1423(c). EPA's longstanding policy has been to collect significant economic benefit gained from noncompliance. *See* A Framework for Statute-Specific Approaches to [Civil] Penalty Assessments, EPA General Enforcement Policy #GM-22, February 16, 1984; *see also* the approximately 24 EPA media and program-specific penalty and enforcement response policies. The reason for collecting economic benefit is to preserve a level playing field for entities that make the timely investment in compliance. Recovery of economic benefit can be likened to the IRS requirement of paying interest or fees on taxes paid late.

Other reasons for auditing cited by respondents included lower exposure to potential third party liability and lower insurance costs.

In 1986, EPA announced that it was the agency's policy to encourage environmental auditing as a means to help achieve and maintain regulatory compliance.[13] Toward that end, the 1986 policy sets forth the basic elements of effective environmental auditing programs.

As memorialized in the 1995 final self-policing policy, EPA's policy toward encouraging the use of compliance tools such as auditing and management systems had evolved into providing penalty incentives and a safe harbor from criminal prosecution. It is important to recognize that this evolution is likely to continue as organizations develop more effective tools to manage the environmental aspects and impacts of their activities, services and products. Environmental management system (EMS) standards such as ISO 14001 and supporting standards hold promise as a means of improving environmental performance. EPA is exploring possible incentives for encouraging the use of such standards insofar as the incentives do not jeopardize protection of human health and the environment.[14]

## Open and Inclusive Process Utilized to Develop the Policy

In May of 1994, the Administrator asked the Office of Enforcement and Compliance Assurance (OECA) to determine whether additional incentives are needed to encourage entities to disclose and correct violations uncovered during environmental audits.[15] Over the next eighteen months, stakeholders representing a spectrum of interests actively participated in the process to develop the final policy. This process included a Washington D.C. public meeting in July 1994, a San Francisco

---

[13] Environmental Auditing Policy Statement, July 9, 1986 (51 FR 25004). *See* earlier discussion.

[14] *See* Brian Riedel's chapter in *The ISO Handbook*, CEEM, Inc., publication expected in early 1996.

[15] *See* June 20, 1994 *FR* notice (59 *FR* 31914).

stakeholder dialogue in January 1995, Chicago stakeholder dialogues in June 1995, and a Washington, D.C. stakeholder dialogue in September 1995. The American Bar Association Section on Natural Resources and Environmental and Energy Law (ABA-SONREEL) hosted the Chicago and Washington, D.C. stakeholder dialogues.

Stakeholders represented state attorneys general and environmental commissioners, district attorneys, industry and trade groups, public interest and citizen groups, and professional auditing groups. In addition, EPA established and maintained the Auditing Policy Docket which makes publicly available hundreds of comments, letters, and documents relating to environmental auditing.[16] The process by which the final policy was developed was a very important part of gaining support for the policy.

## POLICY CONDITIONS

1. **Entity must discover the violation through an environmental audit or compliance management system to obtain full gravity penalty mitigation and criminal safe harbor.**

The final self-policing policy provides full mitigation of gravity-based civil penalties and a criminal safe harbor for entities that discover violations through an environmental audit or system reflecting due diligence (CMS), provided the other policy conditions are met. Note that entities that do not discover the violations through an audit or CMS; *i.e.*, "random discovery," would still obtain seventy-five percent gravity mitigation as long as the other conditions are met.

The final policy defines an *environmental audit* the same way as it is defined in the 1986 auditing policy: "a systematic, documented, periodic and objective review by regulated entities of facility operations and practices related to meeting environmental requirements." Note that this definition covers several types of environmental audits including risk audits and EMS audits as well as compliance audits.

---

[16] The Auditing Policy Docket is accessible by calling (202) 260-7548 and referencing docket number C-94-01.

With respect to due diligence systems (CMSs), the final self-policing policy provides relief to entities that discover violations through an "objective, documented, systematic procedure or practice reflecting the regulated entity's due diligence in preventing, detecting, and correcting violations," provided the other conditions are met. *Due diligence* is defined as systematic efforts meeting criteria based on the 1991 U.S. Sentencing Commission Sentencing Guidelines.[17] The Sentencing Guidelines have had an enormous impact in encouraging the development and implementation of CMSs in the United States.

The due diligence criteria in the self-policing policy include the following:

- the development of compliance policies, standards and procedures to meet regulatory requirements;
- allocation of responsibility to oversee conformance with these policies, standards and procedures;
- mechanisms including monitoring and auditing of compliance and the CMS to assure the policies, standards and procedures are being carried out;
- training to communicate the standards and procedures;
- employee incentives to perform in accordance with the compliance policies, standards and procedures; and
- procedures for the prompt and appropriate correction of violations including program modifications needed to prevent future violations.

The inclusion of CMS or due diligence systems in the final policy represents a very positive and significant revision to the interim auditing policy. Stakeholder written and oral comments indicated that ongoing, comprehensive, and systematic efforts to prevent, detect, and correct violations should be rewarded at least as much as environmental auditing.

---

[17] United States Sentencing Commission Guidelines Manual, Chapter 8—Sentencing of Organizations, Part A—General Application Principles (effective November 1, 1991).

The difference between a compliance audit and a CMS can be likened to the difference between a snapshot and a video.

It is also very significant that EPA may require, as a condition for penalty mitigation, that a description of an entity's CMS be made publicly available. This may entail submission of the CMS description to a national electronic docket. This type of public disclosure has the potential to push the state-of-art in CMS development and encourage benchmarking among suppliers and competitors. The public availability of CMS descriptions can also provide valuable information for insurers, financial markets, investors and lenders—providing the basis for "quasi" market-based incentives.

2. **The policy applies to all violations except those discovered through mandated monitoring or sampling requirements, (e.g. CEM, DMRs).**

Another significant revision made to the interim auditing policy is the elimination of the distinction between violations that are required to be reported and violations that are not required to be reported. The interim policy only applied to *voluntary disclosures,* defined as not required by law to be disclosed. Depending on how this "voluntariness" modifier is interpreted, virtually all violations can be construed as "required to be reported." For example, entities which manage hazardous wastes must "notify" EPA under RCRA. Another example is the failure to obtain an applicable permit. One can argue that a requirement to "notify" or to "apply" for a permit is tantamount to a reporting requirement. Moreover, there does not appear to be a nexus between whether a violation is required to be reported and protection of the health and the environment: many egregious violations are not required to be reported, and many truly "paperwork" violations are required to be reported.

In order to provide maximum opportunity to encourage compliance, and to do so without sacrificing the integrity of critical reporting systems, the policy provides relief on all violations except those discovered through mandated monitoring or sampling requirements, provided that other policy conditions are met. Examples of violations not covered by the policy include emissions violations detected through continuous emissions monitoring, violations of NPDES discharge permits detected through required monitoring or sampling, or violations discovered through a compliance audit required to be performed by the terms of a consent order or settlement agreement.

### 3. Entity promptly discloses the violation in writing to EPA.

Under the policy, the entity must fully disclose in writing to EPA that a violation has occurred or may have occurred, within ten days after discovery. The inclusion of the "may have occurred" language recognizes that in situations where the entity is unsure whether a violation has occurred, it is best for the entity to disclose the potential violation to EPA for a definitive determination. EPA may accept disclosures more than ten days after discovery if more time is needed to make a compliance determination of a complex violation and circumstances do not present a serious threat.

### 4. The entity must disclose the violation prior to imminent discovery by the government.

The entity must identify and disclose the violation before the government has discovered or will discover the violation. Thus, the entity must disclose the violation prior to: commencement of a government inspection or investigation, issuance of an information request, notice of citizen suit, filing of a third-party complaint, or reporting by a whistle-blower.

### 5. The entity must expeditiously correct the violation and remedy harm.

The entity must correct the violation expeditiously and within sixty days certify correction and take appropriate measures to remedy any harm caused by the violation. If more than sixty days is needed to correct the violation, the entity must notify EPA before the sixty-day period has passed. Where appropriate, EPA may require a written agreement, order, or decree to satisfy requirements for correction, remediation, or prevention measures especially where such measures are complex or lengthy.

**6. The entity must agree to take steps to prevent recurrence of the violation.**

The entity's efforts to prevent recurrence of the violation may involve modifying its environmental auditing program or compliance management system.

**7. The violation has not occurred at the same facility within the past three years or is not part of a pattern of violations at the parent company within the past five years.**

The policy does not apply to repeat violators. EPA has established "bright lines" to determine when repeat violators should not be eligible for relief under the policy. Under the policy, the same or closely-related violation has not occurred at the same facility within the past three years or is not part of a pattern of violations at the facility's parent organization within the past five years. This policy exclusion provides entities with a continuing incentive to prevent violations and avoids the unfairness of granting policy relief repeatedly for the same or similar violation.

**8. The violation is not one which has resulted in serious actual harm or may have presented an imminent and substantial endangerment, or violates the specific terms of an order or agreement.**

The policy does not apply to violations which resulted in serious actual harm or may have presented an imminent and substantial endangerment to human health or the environment. Extending coverage of the policy to such violations would undermine deterrence and reward entities for delinquent management of their environmental activities. The policy also does not apply to violations of the specific terms of any administrative or judicial order or consent or plea agreement. This is necessary to preserve incentives to comply with the orders or agreements.

### 9. The entity must cooperate with EPA.

At a minimum, the entity must provide whatever information is necessary and requested by EPA to investigate the violation and any noncompliance problems and environmental consequences related to the violation.

## Relationship to State Laws and Regulations

EPA will work with states to encourage their adoption of policies that reflect the incentives and conditions outlined in the final self-policing policy. EPA remains opposed to environmental audit privileges that would provide a cloak of secrecy over evidence of environmental violations and that contradict the public's right to know. EPA also remains opposed to blanket immunities or amnesty for violations that reflect criminal conduct, present serious threats or actual harm to health or the environment, allow noncomplying entities to gain an economic advantage over their competitors, or reflect a repeated failure to comply with federal law. EPA restates its pledge to work with states to address any provisions of state audit privilege or penalty immunity laws that are inconsistent with the policy and that may prevent a timely and appropriate response to significant environmental violations.

## Applicability and Effective Date

The final self-policing policy supersedes any inconsistent provisions of EPA's media and program-specific penalty and enforcement response policies. The final policy will operate in conjunction with other EPA enforcement policies to the extent they are not inconsistent with the policy. However, an entity may not receive additional penalty mitigation for having met the same or similar conditions under other enforcement policies, nor will the final policy apply to violations which have received penalty mitigation under other enforcement policies. The policy is intended to be utilized for settlement of administrative and judicial enforcement actions, not for pleading purposes. The policy became effective January 22, 1996, but may be applied at EPA's discretion to the settlement of enforcement actions begun before then.

## Tracking of Cases under the Policy

EPA plans to carefully track cases handled under the final self-policing policy and make information about these cases publicly available. This will provide interpretative guidance, help ensure that the policy will be applied consistently, and instill confidence in the policy.

## Policy Redresses Industry Concerns

The policy redresses three general problem areas perceived by some in the regulated community. First, EPA's penalty response to discovered and self-disclosed violations under its approximately twenty-four media and program-specific enforcement policies did not provide the consistency and certainty of enforcement response which regulated industry sought. As outlined above, the final self-policing policy establishes a multi-media, consistent, and certain enforcement response for those entities that discover, disclose, correct environmental violations and meet the other safeguards under the policy.

Second, some in the regulated community perceived that EPA's enforcement policies placed entities that proactively seek to identify noncompliance in no better position—or in a worse position—than entities that have not made the efforts to identify noncompliance. That is, disclosure of violations uncovered through a compliance audit or CMS and the existence of the audit report itself increased an entity's exposure to criminal and civil liability. The discovery of violations could trigger additional reporting requirements and potential criminal liability for knowingly or intentionally failing to report. The audit report could be used as a roadmap to investigate and prosecute violations revealed in the report. These concerns are sometimes collectively referred to as the "seek and ye shall be fined" phenomenon.

EPA believed that perceptions surrounding these concerns were not supported by the facts. For example, EPA and DOJ were not able to identify a single federal or state criminal prosecution of a regulated entity for violations uncovered through an audit and self-disclosed before an independent government investigation was underway. Moreover, a 1995 Price Waterhouse survey indicates that EPA policies and practices have not discouraged auditing: environmental auditing is prevalent among large corporations (eighty-five percent of respondents have auditing programs), the respondents that do not audit generally do not perceive any need to audit, and concern about confidentiality of audit information is one of least important factors in their decisions not to audit. In fact, corporate respondents indicated that EPA activity contributed to their decision to audit: ninety percent of the corporate respondents that do audit stated that they do so in order to find and correct violations before the violations are found by government inspectors.

Nonetheless, through the self-policing policy, the regulated community will have a criminal safe harbor and civil penalty mitigation to further encourage environmental auditing and the development and implementation of systems to prevent, detect, and correct environmental violations. However, entities that are not prepared to promptly disclose and expeditiously correct discovered violations are not protected by the policy and may be running a substantial risk. Overall, the self-policing policy removes perceived disincentives to discover, disclose, and correct environmental violations.

In addition, the final self-policing policy restates EPA's practice and policy since 1986 of not requesting or using an environmental audit report to initiate a civil or criminal investigation of the entity. Thus, consistent with past practice, EPA will not request audit reports during routine inspections. If EPA has reason independent of the audit report to believe that a violation has occurred, however, EPA may seek the relevant information needed to determine liability or the extent of any harm. This policy restatement should further reassure the regulated community that investigators and prosecutors will not abuse their discretion regarding the request for and use of audit reports.

Finally, industry was concerned that criminal acts of "rogue employees" could inculpate the corporation under the general law of agency, and corporate officers under the "responsible corporate officer doctrine," where the corporation and individual corporate officers are not otherwise culpable. The final self-policing policy makes it clear that as long as all of the conditions are satisfied—systematic discovery, prompt disclosure, expeditious correction, no serious actual harm, etc.—EPA will not recommend to DOJ that criminal charges be brought against the entity. This criminal safe harbor is available as long as the violation does not involve a prevalent philosophy or practice to condone or conceal the violations, or a conscious, high-level, corporate or managerial involvement in or willful blindness to the violations. EPA reserves the right to recommend prosecution for the criminal acts of individuals. EPA has not referred a criminal case for prosecution of corporate officers, nor has DOJ criminally prosecuted corporate officers, solely on the basis of the corporate officer's position in the company.

## Privileges and Immunities

In an effort to address some of the perceived concerns regarding government and third party use of audit information, some in the regulated community have turned to state and federal legislators. Since June 1993, fourteen states have enacted legislation to create evidentiary privileges for environmental audits. Some detractors have referred to these laws as "environmental secrecy acts." As the final self-policing policy indicates, EPA opposes audit privileges that encourage secrecy, burden criminal and civil prosecutors, and spawn litigation. Only two states have passed privilege statutes since the agency adopted its interim auditing policy in March 1995.

Nine states (including eight of the "privilege states") have enacted penalty immunity or amnesty provisions for self-disclosed violations discovered through an audit. These penalty immunity provisions vary widely in terms of the extent of exceptions for criminal behavior, serious harm or threats of harm, recovery of economic benefit, and repeat violations. Some detractors have referred to these provisions as "confess and forgive" laws. As the policy indicates, EPA opposes blanket immunities that do not provide these exceptions.

Federal audit privilege and penalty immunity bills have been introduced in both houses of Congress.

## CONCLUSION

The EPA final self-policing policy provides compelling incentives for entities to discover, disclose, and correct environmental violations. These incentives include seventy-five percent or 100 percent mitigation of gravity-based penalties and a safe harbor from criminal prosecution. Beyond systematic discovery, prompt disclosure and expeditious correction of violations, beneficiaries of the policy must satisfy other safeguards to the public and environment, such as remediation of harm, no serious harm or imminent and substantial endangerment, and measures to prevent recurrence. The policy encourages industry efforts to implement an effective "systems" approach toward managing its environmental responsibilities.

# 3

# LEGAL ISSUES[1]

The purpose of this chapter is to explore two important legal issues that must be addressed when considering or conducting an environmental audit: (1) confidentiality of audit-generated data and (2) liability flowing from the performance of an audit to the corporation, its officers and employees, and to the auditors.

## CONFIDENTIALITY

Environmental audits may produce information or opinions that, if revealed, could be harmful to the organization undertaking the audit or to individuals within that organization. For example, audit reports may contain very sensitive data, analysis, and/or recommendations relating to the organization's progress or lack of progress in obtaining full compliance with the law. Such sensitive information could be damaging to the company if it became known to government attorneys who are bringing enforcement actions, or to private plaintiffs who have filed or are considering filing lawsuits for personal injuries or environmental damages.

Certain underlying facts never can be protected from disclosure, and none of the protections discussed herein guarantee confidentiality. Furthermore, those in control of an audit may decide that they are relatively unconcerned about future disclosure. If, however, nondisclosure is potentially important, then, *from the beginning*, the audit must be designed and implemented with an eye toward meeting the requisites of those protections which are available. The early involvement of legal counsel is important in this regard.

---

[1] Note: This chapter was written by Courtney M. Price and Jennifer L. Karas.

The first step in protecting data is to decide what really needs protection. Generally, it is neither possible nor desirable to maintain the confidentiality of all audit-generated information. Therefore, from the outset, auditors should seek legal protection only for those documents for which there is a legitimate claim and need.

## Attorney-Client Privilege

The attorney-client privilege, which protects communications between lawyer and client, is particularly significant in the context of environmental audits. The rationale for the privilege is the assumption that it is more desirable to risk an occasional miscarriage of justice than to inhibit a client's right to obtain effective legal representation. Four elements are necessary to establish this privilege.

First, the privilege applies only to communications between the client and his attorney. It does not protect underlying facts that may have been disclosed to an attorney. For example, the fact of a facility's violation of an environmental permit cannot be protected from disclosure to a third party simply by communicating the fact of the violation to an attorney. However, recommendations or analysis relating to a violation discovered during the course of an environmental audit and communicated by an environmental consultant to the attorney who has been retained by the facility for the purpose of giving legal advice on environmental compliance can be protected, under certain circumstances, by the attorney-client privilege.

This privilege may extend to communications among a wide range of corporate officials and counsel. Since the 1981 decision in *Upjohn Co. v. United States*[2], the definition of "client" in the corporate context is not limited to the most senior employees, but includes middle and lower level employees whose communications to counsel are necessary in order for counsel to render legal advice and are made pursuant to an investigation that management has ordered counsel to undertake.

---

[2] *Upjohn Co. v. United States*, 449 U.S. 383 (1981)

The second element of the attorney-client privilege is that the communications must be made for the purpose of obtaining legal advice. Communications that are for the purpose of obtaining business, technical, or other nonlegal advice are not covered by the privilege. Therefore, a letter retaining outside counsel to assist in undertaking an environmental audit should specify that the purpose of the environmental audit is to obtain legal advice. If in-house attorneys oversee the audit, special care must be taken to try to distinguish their general management role from their role as legal advisers for the purposes of the audit.

The following case aptly illustrates the difficulty of protecting an environmental investigation that is supervised by in-house counsel. The case of *Ohio v. CECOS International, Inc.*[3] dealt with an investigative booklet compiled by CECOS employees in the face of potential criminal prosecution for the alleged pumping of contaminated rainwater into a stream bordering CECOS' hazardous waste disposal landfill in Williamsburg, Ohio. CECOS claimed that the booklet was protected from discovery either by the attorney-client privilege or attorney work-product rule.

The Ohio court reviewed the claims and then held that the booklet was not protected, characterizing it as an internal management report which was not prepared in anticipation of litigation. The court noted that most of the CECOS employees involved in the decision to compile the booklet were not attorneys. Although CECOS' general counsel was involved in the directive to compile the book, the court found that the booklet was not a product of attorney work, but rather a factfinding effort by management preliminary to anticipation of litigation. The court also noted that the booklet was not covered by attorney-client privilege because the corporation failed to demonstrate that the preparation of the booklet was a professional legal activity of the general counsel undertaken for the purpose of providing legal advice.

Another area of particular concern in the context of environmental auditing relates to the retention of technical consultants to assist in the

---

[3] *Ohio v. CECOS International, Inc.*, No. 85-CR-5290C-85-CR-52653C (Ohio C.P. April 23, 1986)

audit. Although the attorney-client privilege requires that communications be made to an attorney for the purpose the obtaining legal advice, in practice, communications to non-lawyers who are assisting counsel in providing legal advice also may be protected by the privilege. In order to maximize the availability of the privilege for this kind of communication, it is recommended that the attorney, rather than the client, retain the technical consultant, supervise his work, and receive the reports of the consultant directly.

In *Olen Properties Corp. v. Sheldahl, Inc.*[4], a California court held that where a party demonstrates that environmental audit materials were generated to assist attorneys in evaluating compliance with environmental laws and regulations, information contained in those materials is covered by the attorney-client privilege and need not be produced.

The third element of the privilege is confidentiality, that is, the communication for which protection is sought must remain confidential. This element becomes particularly important in the context of corporate disclosures. Documents circulated indiscriminately within a corporation may lose the privilege protection. Counsel overseeing the audit should limit dissemination as much as practicable to those immediately concerned with the results of the audit and see that confidential documents are controlled carefully. For example, "privileged and confidential" should be stamped on all documents for which protection will be sought. All persons who participate in the audit should be educated as to the importance of maintaining the confidentiality of certain documents.

The fourth element of the privilege is the absence of waiver. If the holder of the privilege intentionally discloses the communication for which protection is sought, the privilege will be deemed waived. Even unintentional disclosures may destroy the privilege. Thus, counsel overseeing the audit should set up procedures early in the process to minimize the chance of disclosure.

---

[4] *Olen Properties Corp. v. Sheldahl, Inc.*, 1994 U.S. Dist. LEXIS 7125.

An organization undertaking an audit should consider taking the following steps to maximize the chances that certain environmental auditing documents will be protected by the attorney-client privilege:

- An attorney should be involved from the outset in the design and implementation of the audit and should analyze the process for purposes of strengthening the privilege claim.

- Beginning with the letter retaining counsel to assist in performing the audit, all documents should reflect the fact that the purpose of the undertaking is to obtain legal advice.

- Documentation should reflect that: (1) information necessary to perform the audit is known only to those people who are communicating with the attorney or those consultants hired by him; and (2) employees communicating with the attorney have been advised that the purpose of the communication is to enable the provision of legal advice.

- At the outset, procedures should be established to maintain the confidentiality of communications and prevent intentional or unintentional waiver of the privilege.

## Work-Product Rule

In addition to the attorney-client privilege, environmental audit documents also may be protected by the work-product rule, which was formulated first in the case of *Hickman v. Taylor*[5], and is codified in Federal Rules of Civil Procedure 26(b)(3). This rule provides qualified protection for information or material assembled or prepared by or for an attorney in anticipation of litigation or in preparation for trial. Although the rule accords strong protection to the opinion work-product which reflects an attorney's thought processes, discovery can be compelled if the

---

[5] *Hickman v. Taylor*, 329 U.S. 495 (1947).

adversary can show "substantial need" and "undue hardship" in obtaining the information from other sources.[6] Prior to beginning an audit, the elements necessary to establish the protection of the work-product rule should be examined carefully and certain management steps taken to strengthen a potential claim if it is anticipated that this protection will be sought.

First, the materials for which protection is sought must be prepared in anticipation of litigation. This is a threshold requirement. Litigation need not be on-going when the documents in question are prepared, but it must be more than a remote possibility. If it is expected that the protection of the work-product rule will be sought for environmental audit documents, then early in the process, the client's management should communicate in writing to counsel what litigation is anticipated and why. Counsel also should ensure that the audit documents reflect the anticipation of litigation or preparation for trial.

Second, the materials for which protection is sought must be documents and other tangible things. Facts simply known to an attorney are not protected by the work-product rule. Nor will the rule prevent discovery of the existence or location of the documents in question. Basically, the rule only protects an attorney's "mental impressions, conclusions, opinions, or legal theories."

An example of protected material in the context of an environmental audit would be an attorney's opinion on the interpretation of a regulation that the audited facility may have violated.

A final requirement is that materials must be prepared "by or for another party or by or for that other party's representative (including his attorney, consultant, surety, indemnitor insurer, or agent)."[7] This breadth of protection is especially relevant in the context of an environmental audit where technical consultants may be retained to assist the attorney in preparation for litigation.

---

[6] Fed. R. Civ. P. 26 (b)(3)

[7] Fed. R. Civ. P. 26 (b) (3).

As with the attorney-client privilege, the legal protection afforded by the rule may be waived in certain circumstances. For example, disclosure of environmental audit material to third parties without regard for confidentiality would constitute a waiver that would destroy the protection.

In light of the above, counsel should take the following steps if the protection offered by the work-product rule will be sought for environmental audit documents:

- Carefully research the law relating to the work-product rule in the applicable jurisdiction in order to ensure that the necessary requirements can be met.

- Document the reasons for anticipating litigation.

- Take the necessary steps to ensure that the protection of the rule is not waived.

Attorneys and clients should be aware that the circumstances in which this protection is available to environmental audit documents is significantly limited by the requirement that the materials, in fact, be prepared in anticipation of litigation or in preparation for trial.

## Self Evaluative Privilege

The final mechanism for protecting materials generated during an environmental audit stems from the public interest in encouraging companies to undertake self-evaluation. Some courts have given qualified protection to documents relating to self-analysis and certain attempts to correct problems internally. The seminal case in this area is *Bredice v. Doctors Hospital, Inc.*,[8] which involved a malpractice suit. In that case—without a showing of good cause and exceptional necessity by

---

[8] *Bredice v. Doctors Hospital, Inc.*, 50 F.R.D. 299 (D.C.D.C. 1970) aff'd mem., 479 F.2d 920 (D.C. Cir. 1973).

plaintiff)—the court sustained the defendant hospital's refusal to produce minutes and reports recording medical staff reviews of patient treatment protocols, which were conducted with the understanding that they would be confidential. The court noted that without confidentiality constructive self-criticism would not occur, and that such criticism would promote public health interests.

In a more recent case, *Reichhold Chemicals v. Textron,*[9] a Florida district court extended a qualified "self-critical analysis" privilege to information contained in environmental audit materials prepared after the fact of contamination "for the purpose of candid self-evaluation and analysis of the cause and effect of past pollution and Reichhold's possible role, as well as other's, in contributing to pollution at the site."[10]

The court based its decision on the "strong public interest in promoting the voluntary identification and remediation of industrial pollution." The court indicated, however, that the privilege could be overcome if the defendants could demonstrate extraordinary circumstances or special need.

In any discussion of this potential protection, it must be emphasized that the case law in this area is still developing. As such, the law of the relevant jurisdiction should be reviewed to determine the status of self-evaluation protection. Also, it is advisable to carefully document the fact that the audit is being undertaken for self-evaluation purposes.

## State Action on Environmental Audit Privilege

In addition to protections developed from case law, several states have enacted environmental privilege laws. These states include: Arkansas, Colorado, Illinois, Indiana, Kentucky, Idaho, Oregon, Utah, Virginia, and Wyoming. Meanwhile, legislatures in a number of other states are considering similar legislation.

---

[9] *Reichhold Chemicals v. Textron*, 157 F.R.D. 522 (1994).

[10] *Id.* at 527.

Oregon, in 1993, was the first state to adopt an environmental audit privilege. Its statute provides that audit reports are privileged and are not admissible in any civil or criminal action unless (1) the privilege is waived, or (2) a court finds that the privilege was asserted for a fraudulent purpose, the material is not covered by the privilege, or the materials show violations the company has failed to disclose and/or correct. In a criminal action, the privilege may be lost if the court finds there is a compelling need to review the information.

The Oregon statute, however, does not provide companies with immunity from prosecution for disclosing and correcting violations. Colorado's environmental audit privilege law *does* provide companies with privilege and immunity from prosecution when the company reports and corrects violations. Specifically, in order to qualify for immunity, the company must: (1) disclose the violation; (2) promptly initiate efforts to correct the violation; (3) correct the violation within 2 years after disclosure; and (4) cooperate with the regulatory agency. Immunity will not be granted if a company is found to have committed serious violations constituting a pattern of continuous or repeated violations of environmental laws, rules, regulations, permit conditions, settlement agreements, or consent orders.

EPA has criticized states that have adopted laws that grant privilege to environmental audits or give immunity to violations that are discovered through self-evaluation and voluntarily reported. Specifically, EPA stated that "it will scrutinize enforcement more closely in states with audit privilege and/or penalty immunity laws. . . [and the agency] may find it necessary to increase federal enforcement" in those states.

In summary, there are several sources of protection for the confidentiality of environmental audit reports. The laws of the particular jurisdiction should be examined before initiating the audit. Finally, the following is a checklist of actions that should be considered prior to undertaking an environmental audit:

- Senior management directs that the audit be undertaken.

- The audit is conducted through counsel.

- A memorandum is communicated from senior management to legal counsel directing that:
  - The audit be conducted.
  - The audit be undertaken by counsel *in his legal capacity*. (Legal capacity may be easier to establish through outside counsel.)
  - All information be held confidential.
  - The purpose of the audit be for obtaining legal advice, anticipating litigation (asserting the basis for this assumption) and/or self-evaluation.
  - All notes are be logged in the attorney's bound journal only.

- A similar memorandum is directed from counsel to consultants, if they are used.

- Counsel directly retains consultants.

- All written communications are labeled "privileged and confidential" and "do not duplicate."

- All written communications are kept in separate, secured files under the control of counsel.

- Distribution of the report is on a need-to-know basis only.

- An oral audit report is considered only in especially sensitive situations for maximum protection of information.

## POTENTIAL LIABILITY ARISING OUT OF ENVIRONMENTAL AUDITS

### Liability for Violations Discovered

There is be no guarantee that violations discovered during an environmental audit will not result in enforcement actions against the corporation and/or individuals within it. If, however, the corporation and its attorneys promptly respond by disclosing the information and taking action to correct the violations, it is unlikely that the information generated as a result of an environmental audit will result in the imposition of significant penalties on the company or its officers or employees.

The Environmental Protection Agency (EPA) recently released new interim guidelines that seek to promote self-policing and voluntary reporting of violations by providing incentives to regulated entities in the form of reduced civil penalties and limited criminal referrals. Also, several bills are pending in Congress that may provide even greater protection from public disclosure for environmental audit data than is contained in the new guidelines.

Generally, there are two components to an environmental civil penalty—the gravity portion and the economic benefit portion. The new guidelines preserve the EPA's absolute right to recover economic benefit in all circumstances, but permit the EPA to eliminate or reduce the gravity portion of the violation. According to the new guidelines, as long as a violation does not involve either criminal conduct or serious actual harm to human health or the environment, the EPA will eliminate the entire gravity component of a penalty, provided that the following seven conditions are met:

(1) The violation is discovered by the regulated entity through a voluntary self-evaluation.

(2) As soon as it is discovered, the violation is reported to all appropriate agencies— before the commencement of a federal, state, or local agency inspection, investigation or information request, notice of a citizen suit, legal complaint by a third party, or the regulated entity's knowledge that the discovery of the violation by a regulatory agency or third party was imminent.

(3) The violation is corrected, either within sixty days of discovery or as expeditiously as possible.

(4) If the violation created or might have created an imminent and substantial endangerment to human health or the environment, the regulated entity must have expeditiously remedied the condition.

(5) Any environmental harm is remedied and its recurrence is prevented.

(6) The report of the violation does not indicate that the regulated entity has failed to take appropriate steps to avoid repeat or recurring violations.

(7) The regulated entity must cooperate with the EPA, which may entail providing documents and access to witnesses. In addition, the EPA may request the execution of a written agreement, administrative consent order, or judicial consent decree, documenting the terms of the above conditions.

The new guidelines also permit EPA to reduce up to seventy-five percent of the unadjusted gravity component of a penalty if (a) most of the conditions are met, (b) all of the conditions are met and the violation resulted in imminent and substantial endangerment but not serious actual harm, or (c) all of the conditions are met and the violation involved the disclosure of criminal conduct.

Under the new guidelines, the EPA will not recommend that criminal charges be brought against regulated entities if all seven conditions described above are met, and also if the violation does not involve (a) a prevalent corporate management philosophy or practice that conceals or condones environmental violations, (b) high level corporate officials' or managers' conscious involvement in or willful blindness to the violation, or (c) serious actual harm to human health or the environment.

## Who is Liable for Violations Discovered

The EPA's interim guidelines do not apply to criminal acts of individual managers or employees. Recently, there has been a trend on the part of the EPA to bring criminal actions against mid-level managers of facilities. For example, in *U.S. v. Weitzenhoff*,[11] the manager and assistant manager of a sewage plant were convicted for Clean Water Act violations relating to discharges beyond that allowed in the plant's NPDES Permit. Under the new guidelines, these types of suits would continue to be brought.

As for the liability of the auditors themselves, outside counsel and the consultants retained by them who are privy to information from an environmental audit performed for the purpose of providing legal advice are not, in most circumstances, required to report or act on that information pursuant to environmental rules and regulations. In fact, in accordance with the demands of the attorney-client relationship, they can be precluded from revealing client confidences.

To the extent that the attorney or environmental auditor is a corporate employee or manager, however, the duty of reporting could be very different. For example, pursuant to Superfund, any person in charge of a facility must report the release of a hazardous substance as soon as he has knowledge of it, or he could be subject to significant penalties for noncompliance. Therefore, corporate officers and employees should be familiar with specific statutory and regulatory duties and obligations and carefully consider them as they plan, staff and implement an environmental audit. Also, management should provide guidelines to employees on how to report problems.

## Liability for Improperly Performed Audit

Actions conceivably could be brought against an auditor for failure to adequately perform the audit. Possible legal theories for such an action include tort (*i.e.*, professional malpractice); contract (*i.e.*, an implied standard of care in performing the review); or SEC violation (*i.e.*, sufficient connection with a materially misleading misrepresentation or omission). Although this is not an area in which case law is fully developed, the standard of conduct required for those persons performing

---

[11] *U.S. v. Weitzenhoff*, 35 F.3d 1275 (9th Cir. 1993), *cert. denied*, 115 S.Ct. 939 (1995).

an audit will probably be based on reasonable care, good faith, and lack of fraud or collusion. "Reasonable care" requires that auditors use that degree of knowledge, skill, and judgment usually possessed by members of the profession. In determining the standard for the profession, a court would probably look to textbooks on auditing, government documents describing adequate programs (*e.g.*, EPA's recent interim policy statement and other policy statements on environmental auditing) and recorded public statements of experts in the field (*e.g.*, literature distributed at seminars, etc.). A court could hold an auditor to a higher standard if the industry standard did not represent "due care."

An auditor can minimize his liability by having counsel assist in the preparation of the contract by which the auditor is retained. Furthermore, auditors should carefully document the purpose of the audit and the actual steps and activities undertaken to accomplish that purpose.

## CONCLUSION

Throughout this chapter, there are references to the need for counsel, and it is this author's opinion that a lawyer's participation is necessary for an audit to be properly planned, staffed and implemented. Furthermore, an attorney can be the major figure in the performance of an audit. However, the lawyer need not be dominant in order to be useful. In fact, some people prefer to work with non-lawyers in activities such as audits.

However, whether the audit is directed by outside consultants or in-house technicians, it is critical, from the outset, to have legal advice on the legal issues discussed herein.

The development of sound environmental management practices is a necessity for all regulated entities, and environmental auditing is the first step in that development. With the aid of counsel, an audit program can be planned, staffed, and implemented in a way that maximizes confidentiality and minimizes potential liability to a point at which the corporation and its officers and employees can feel secure in the knowledge that they are not taking unreasonable risk.

# 4

# ELEMENTS OF A
# SUCCESSFUL PROGRAM

---

*"The key value is to prevent problems from erupting into crises before management can act."*[1]

---

## PRINCIPLES OF AN AUDIT PROGRAM

The principles of an environmental audit program are well known, but may be usefully summarized in the following five points. The audit team, comprising knowledgeable professionals (either in-house staff, third parties, or both), carries out such duties as:

- Understanding and ascertaining maintenance of schedules and records with respect to all operations having environmental compliance requirements

- Inspecting facilities, equipment, and personnel performance to evaluate adherence to institutional standards

- Submitting written status reports to appropriate senior management

---

[1] Throdahl, M.D., Former Senior Vice President, Monsanto Company, in "Environmental Audits Cut Compliance Risks," *Chemical Week*, May 28, 1980.

- Explaining deviations from the norm and recommending corrective action

- Operating independently of all audited functions, at a peer level with their management.

As these points make clear, the essence of an environmental audit program is to provide assurance to top-level company management and company stockholders that all relevant regulatory requirements are being met in accordance with the company's operating philosophy.

In order to accomplish this goal, an audit program must contain certain key elements around which an individualized structure can be developed. These elements and the relationship between them is shown in Figure 4.1. An audit program must be planned carefully and have the appropriate supportive tools and staff. Audits must be conducted on a continuous basis with additional sampling used only when and where necessary. Finally, the results must be evaluated and turned into solutions and corrective actions. Each of these elements is important to the successful implementation of the program. Accordingly, this chapter presents a detailed discussion of each of the elements within the context of the overall program.

**Figure 4.1.**
**Elements of an Audit Program**

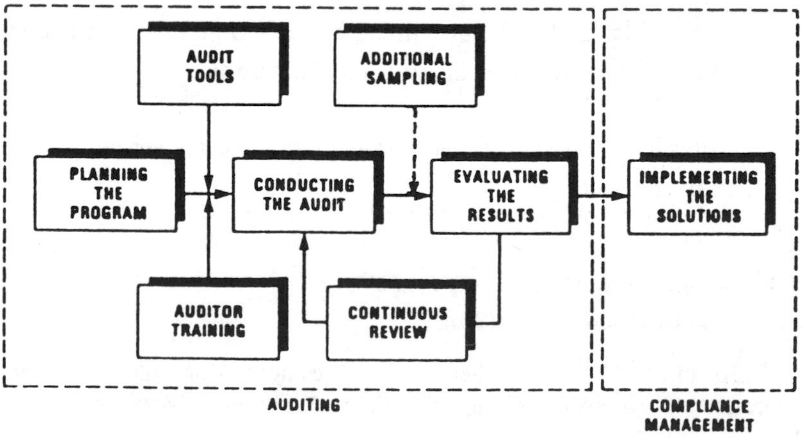

# PLANNING THE PROGRAM

An important first step is planning a program. Among the decisions to be made are: what is the objective to be achieved, who will be involved, how often will audits be conducted, what laws and regulations will be covered, and whether information will be kept confidential. As indicated by the list below, the issues to be resolved are numerous and often complex:

- Program Objectives
    - Assurance of compliance
    - Management of liabilities
    - Accountability of management
    - Tracking of compliance costs

- Organization of the Program
    - Corporate control with full-time corporate auditors
    - Corporate control with part-time auditors selected from divisions
    - A small corporate oversight group with delegation of the audit function to the operating divisions
    - A small corporate oversight group with the use of external, independent auditors

- Legal Issues
    - Written vs. oral reports
    - Retained documentation
    - Corporate counsel involvement
    - Protection from discovery
    - Report format and watchwords

- Program Scope and Coverage
    - Frequency of audits
    - All regulations
    - All plants, random sample, or directed sample
    - Past practices

- - Vendor audits
- - Waste contractor audits

- ■ Process Issues
  - - Pre-visit questionnaire
  - - Number of auditors
  - - Duration of audits
  - - Pre-visit, on-site, post-visit procedures
  - - Sampling procedures
  - - Interview technique
  - - Inspection procedures
  - - Use of portable computers

- ■ Auditor Selection and Training
  - - Skills required
  - - Attorney's role
  - - Training procedures
  - - Full-time vs. part-time auditor

- ■ Management Issues
  - - Expectations of senior management
  - - Measures of successful performance
  - - Unannounced visits
  - - Reporting protocols
  - - Supportive vs. combative style

- ■ Supporting Tools
  - - Pre-visit questionnaire
  - - Compliance checklists
  - - Regulatory updates
  - - Audit reports
  - - Guidance/procedures manual
  - - Follow-up reports
  - - Computer support

Although this book and other sources can provide guidance on what might be considered the "standard" approach to the issues listed above, each company considering a program must decide for itself the optimal program structure. There is no one best answer.

Most companies will develop audit guidance manuals to set the framework for the program and to ensure the consistency, efficiency, and completeness of all audits. An outline of an example audit manual is presented at the end of this chapter. As implied by the outline, audit manuals typically address essential programmatic issues and include a variety of audit tools, including pre-visit questionnaires and audit checklists or protocols. The manuals are designed to communicate the structure of the program to both the auditors and those audited, in order to remove any mystery from the program.

The next few sections discuss the keys to implementing a successful program. They touch upon many of the issues listed above and provide some guidance on preferred approaches. These are the central points which should be addressed in the program's guidance manual.

**Figure 4.2**
**Why Audit? Industry's Response**

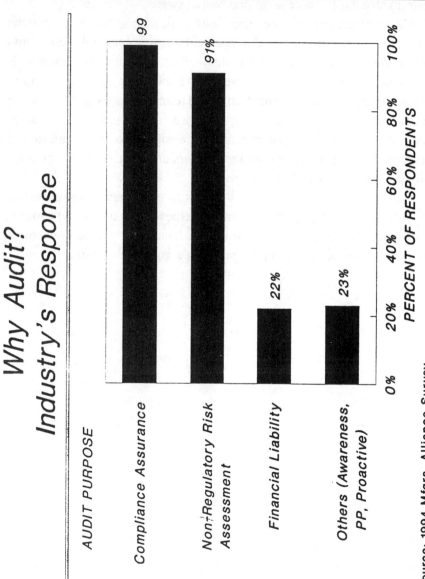

Why Audit?
Industry's Response

AUDIT PURPOSE

Compliance Assurance — 99

Non-Regulatory Risk Assessment — 91%

Financial Liability — 22%

Others (Awareness, PP, Proactive) — 23%

0%   20%   40%   60%   80%   100%
PERCENT OF RESPONDENTS

Source: 1994 Mfgrs. Alliance Survey

**Program Objectives**

An audit program can achieve a variety of objectives. Historically, in most companies, compliance assurance has been the principal objective, with other, secondary objectives evolving along the way. This is confirmed by a 1994 survey of the manufacturing industry, which showed that ninety-nine percent of programs surveyed stated that compliance assurance was one of the main program goals. As shown in Figure 4.2, non-regulatory risk assessment followed closely behind at ninety-one percent.

Yet, it is important to note that companies will frame a program to meet individual needs, and these can vary significantly. One company's programs includes the following five objectives:

■ Provide assurance of adequate EH&S performance and continuous improvement to the board of directors

■ Ensure division and facility compliance with federal, state and local EH&S regulations and corporate policies and procedures

■ Ensure division management accountability for correcting compliance deficiencies identified on audits

■ Increase EH&S awareness and continuous learning by audited personnel and auditors

■ Transfer EH&S technological and administrative innovations across the corporation.

Note that the compliance assurance objective is one of the five listed. But the additional objectives are quite intriguing. For example, the program is also designed to provide assurance and confirmation of continuous improvement of environmental performance to the board of directors. This objective is very consistent with the European Eco-Management and Audit Scheme discussed elsewhere in this book.

Further, the program is designed to hold division management, not the audit program manager, accountable for correcting deficiencies. This applies the compliance pressure to the appropriate organization. And note that it does not hold line management accountable for an excellent rating on any given audit. What it does ask of line management is that, where problems are identified, they be corrected promptly.

Further, the program is designed to be a training tool for both rotating auditors and those who are audited. This is a very common objective for auditing programs and is easily achieved. In probably every instance, an auditor and an auditee can walk away from the audit and say "I learned something from that exercise."

And last, the program is expected to communicate "best practices" throughout the corporation. Auditors are in a great position to do this as they review practices in different businesses and geographical regions. Before the advent of communications tools such as Lotus Notes®, this kind of communication was quite difficult.

As mentioned above, these are only a few of the objectives an environmental audit program can achieve. Individual organizations must decide for themselves which objectives are appropriate. Whatever is decided, the objectives should be stated in such a way that clear performance measures can be established. One must be able to periodically evaluate the program's performance to determine where improvements need to be made.

## Roles and Responsibilities

One of the early keys to developing a successful program, regardless of the strategy selected, is commitment. Top management's role at this stage is critical. Before any audits are conducted, management must develop and communicate a policy that supports the *concept* of an audit program. To be successful the policy must portray the program as a positive move towards helping facility managers enhance compliance and reduce liabilities. Senior management should dismiss explicitly the notion that the program is designed to "check up" on facility managers and operators. In fact, under a properly designed audit program, facility staff have as much, if not more, to gain than corporate managers.

In addition, early in the development of the program, corporate management should designate a senior executive as audit program director. This director should be senior enough to enhance the program's credibility.

The specific structure of the audit program will vary, depending upon the individual organization. Probably the most prevalent, though not the only, approach is to appoint a corporate audit program director and to assign a small group to this individual. This group might include a regulatory specialist and one or more, or no, audit team leaders. The auditors and/or team leaders could come from other corporate groups and division and plant EH&S staffs and would conduct a few audits per year while retaining full-time EH&S responsibilities in their respective organizations. Quite often such audit teams are supplemented by third parties (*e.g.*, consultants) who can provide specialized expertise (*e.g.*, process safety) or an added independence to the program. This "rotating auditor" approach requires significant "buy-in" from line management, which is achieved when there is indeed top management support for the program.

An example of an organizational structure for an audit program is shown in Figure 4.3. Note that, in this case, the audit program manager is tied closely to the board of directors through the company's executive management council. Also, the law department provides significant input but is not directly in control of the program. The audit program manager has organized an "audit core team" which consists of three or four key EH&S staff from both corporate and division/plant groups. This network of individuals sets the policy for the program. The law and financial audit departments also participate on the core team. Qualified auditors can come from anywhere in the organization, but must successfully complete (*i.e.*, pass an examination) an internal training program.

However the audit program is structured, roles and responsibilities should be defined for each of the participant categories. These categories include:

- Senior or executive management
- The audit program manager and staff
- The legal department

**Figure 4.3.**
**Sample Audit Program Organization**

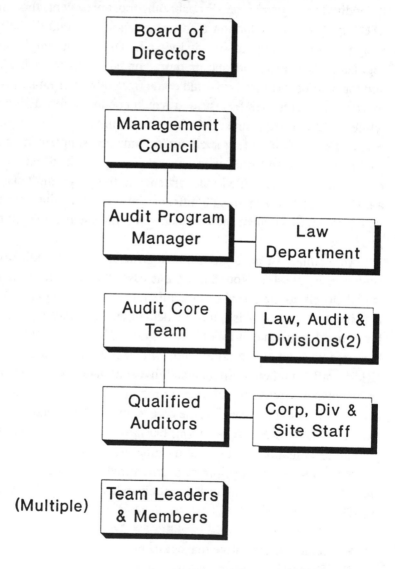

Program Organization

- Division management
- Facility management and staff
- Audit team leaders
- Auditors.

A set of roles and responsibilities for the audit program manager, for example, might include:

- Determining, on an annual basis, the sites to be audited, consistent with the corporate audit frequency policy

- Maintaining and updating program materials including the program guidance manual, the pre-audit questionnaire, and the audit protocols

- Maintaining the auditor training program, including updating the qualified auditor database

- Reviewing all draft and final audit reports to ensure that reports meet quality standards and are submitted on time

- Tracking the status of corrective action plans and formally "closing out" audits when all findings are corrected

- Providing the audit teams, on an as needed basis, with country or state regulatory information

- Conducting an annual evaluation of the program

- Periodically reporting on program status to the board of directors

- With the advice of counsel, managing the program files consistent with the corporation's established records retention policy.

Roles and responsibilities for other participants can be established in a similar fashion. As with the program objectives, the individual participants' responsibilities should be framed so that performance can be measured.

## Legal Protections[2]

The legal protection issue is one of great concern to most companies and, therefore, each organization must decide on the approach that best fits its philosophy. And the best approach is not always invoking maximum legal protections for the program. As shown below, not all companies invoke formal attorney-client privilege protections for audit reports and documents. The results of a recent survey of manufacturing companies suggests that only fifty percent of the surveyed organizations use attorney-client privilege for all documents.

### Audit Confidentiality? Industry's Response[3]

| Management Issue | Response |
|---|---|
| Treat Audit Documents as Confidential | 70% |
| Use Attorney-Client Privilege Veil for ALL Documents | 50% |
| Use Attorney-Client Privilege Veil for SOME Documents | 20% |

[2] An entire chapter in the book is devoted to legal issues associated with audit programs. This section discusses the management consequences of various legal protection approaches.

[3] Taken from a 1994 Manufacturers' Alliance Survey.

Why is there such disparity in approaches? Well, most companies do want to protect their audit reports but some are not sure that it is worth the trouble of invoking formal protection privileges. Use of the privilege puts the legal department, or outside counsel, at the center of the program, not in a review and advisory role. Such extensive legal involvement has often meant a more expensive and elaborate program, and perhaps more importantly, can slow down the report distribution process substantially. Companies that do not invoke the privilege believe that the risk of disclosure is not improved substantially by attorney-client privilege, since the underlying facts are not protected in any event. Moreover, any protections gained are viewed as more than offset by the increased bureaucracy of the program and the inability to produce and distribute reports expeditiously. New state protection legislation might alter how companies view this issue. (*See* Chapter 2 for a discussion of this legislation.)

## Scope and Coverage

Any audit program must have some bounds under which it operates. Which facilities are audited, what they are audited against and what type of audits will be conducted are among the decisions that must be made.

Quite often one of the most difficult tasks for an audit program manager is determining the inventory of facilities subject to the program. With all the mergers, acquisitions, and divestitures taking place in the U.S. and overseas, the structure of any organization can change very quickly. In general, audit programs will be responsible for auditing the following facility types:

- Manufacturing facilities owned and operated
- Manufacturing facilities operated but on leased property
- Joint ventures with majority ownership and operational responsibility
- Joint ventures with minority ownership and operational responsibility
- Warehouses and distribution centers owned and operated
- Administrative and office buildings owned and operated

■ Real estate owned.

For other facility types (such as joint ventures with majority or minority ownership and no operational responsibility, or leased warehouses) the company will attempt to influence the partner(s) to participate in the audit program or ensure that the partner(s) has an equivalent program.

Most audit programs are designed to cover major operating facilities. Other types of facilities that have to be addressed, however, are acquisitions and divestitures, commercial waste disposal contractors and captive toll manufacturers. These are all risk situations in need of some kind of independent review. In some companies, these reviews are considered a formal part of the audit program, and in other companies they are addressed separately but coordinated with the corporate audit program manager.

The second audit issue relates to regulatory coverage. Program managers must decide what regulations are to be addressed. Will environmental, health and safety issues be combined into one integrated audit program? There are as many examples of companies with separate environmental audit programs as there are with integrated EH&S audit programs. Other regulatory issues must be addressed as well. Certain laws such as the Toxic Substances Control Act (TSCA) are not easily auditable at the facility level. Certain programs created to respond to TSCA requirements such as Pre-Manufacture Notification exist more at the business unit level. Companies sometimes develop focused or target audits to address these issues thoroughly but independently of the facility audit program. Similarly, the Chemical Diversion and Trafficking Act (CDTA) is not really an environmental act but can be audited quite easily at the facility level. As a consequence, CDTA requirements are sometimes "piggy-backed" onto the environmental audit program.

Once the company has decided on the regulatory scope of the program, it can then decide on the types of audits that can best meet the program's objectives. This may result in the creation of specialized audits, designed to address multi-organizational issues, as well as the more traditional facility audits. Such specialized audits will be discussed in later chapters.

## Facility Audit Schedules

When first faced with the "audit or not" question, many corporate environmental departments wonder where the significant, additional resources will come from. In an economic environment that is shouting "do more with less," some initial planning and priority setting can help the compliance manager meet the often conflicting objectives of: (1) increased environmental compliance, and (2) reduction of the resources committed to that goal. Fortunately, there are some steps managers can take to better understand the operations within the company that pose the greater risks and to design an audit program that addresses those risks within the context of resource constraints.

Setting facility audit schedules based on a sense of risk can do much to assure a cost-effective program. As shown in Figure 4.4, audit frequency varies considerably among companies. Unfortunately, there is no general guidance on what is an acceptable frequency. The U.S. EPA's landmark 1986 Policy Statement, British Standard 7750 and the ISO 14000 Guidelines are notably silent on the issue. However, the Eco-Management and Audit Scheme (EMAS) in Europe has established audit frequencies as follows:

- One year for activities with a high environmental impact
- Two years for activities with a moderate environmental impact
- Three years for activities with a low environmental impact.

Notwithstanding the EMAS-prescribed frequencies, many companies that have mature audit programs have settled in on a three year frequency for major facilities. This is after each facility has been audited at least once or twice.

**Figure 4.4**
**Frequency of Audits? Industry's Response**

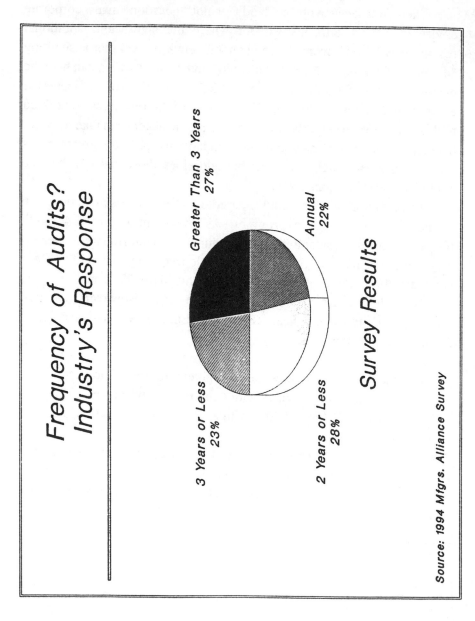

*Frequency of Audits?*
*Industry's Response*

Greater Than 3 Years
27%

Annual
22%

3 Years or Less
23%

2 Years or Less
28%

*Survey Results*

*Source: 1994 Mfgrs. Alliance Survey*

There are several ways to establish a frequency for auditing. One of those is shown in Figure 4.5. This method establishes seven factors that can be used to set a site-audit frequency for each of three major categories. Some judgment is required in using this method because an overall categorization results from an evaluation of how often a site can be placed in one of the categories. That is, a given site may have three factors in Category I, two in Category II and two in Category III. The relative weight of each placement in terms of an overall classification then has to be determined.

## Figure 4.5
## Site-Audit Frequency

| Site Characteristics | Category I (Every 2 Years) | Category II (Every 3 Years) | Category III (Every 4 Years) |
|---|---|---|---|
| Size and Type | Major manufacturing, mining or processing | Minor manufacturing, mining or processing | Warehouses, real estate, administrative buildings |
| Air Emissions | Major source of air toxics or significant emissions; multiple permits | Moderate emissions; some air permits | No sources require air permits |
| Community Relations | Major documented problems with the community | Periodic formal complaints | No or isolated complaints |
| Hazardous Materials Releases | Has 3 or more §313 chemicals | Has 1 or 2 §313 chemicals | Has no §313 chemicals. |
| Hazardous Waste | Large quantity generator | Small quantity generator | Conditionally exempt small quantity generator |
| Wastewater | Operates on-site treatment or pre-treatment plant | Discharges process wastewater to POTW | Discharges only sanitary wastewater or no discharges |
| Spill Potential | On-site bulk petroleum or hazardous substances storage of > 50,000 gallons | On-site bulk petroleum or hazardous substances storage of 1000 to 50,000 gallons | On-site bulk petroleum or hazardous substances storage of < 1000 gallons |

There are, of course, other approaches to risk ranking facilities. Generally, site risk factors can be viewed as two-fold: inherent and external. First, there are indeed inherent risks of operation, which can involve the materials handled, the age of the facility, and the complexity of the process. These risks are important but perhaps more controllable[4] than the second class of external risks that may include the company's compliance history, the community and environmental setting, and the state agency's regulatory stringency.

If one views these two classes of risk in concert, as in Figure 4.6, a facility-by-facility risk evaluation can be conducted. We can find fairly large facilities, such as Facility G, that pose high risks, and efforts can be undertaken to reduce both inherent and external risks to move this facility into either a relatively safe or controllable situation. Such efforts might include increasing measures to reduce noncompliance or investigating the possibility of materials substitution. For another facility, such as Facility E, which poses only modest inherent risk, but is in so unstable an external environment that it is vulnerable to unwanted surprises, a public relations or compliance improvement program can be developed that will move the facility to the "relatively safe" category.

In much the same way, all facilities or units can be evaluated for their relative risk potential. Corrective action programs can then be fine tuned to address the nature and extent of the risk with the most cost-effective solution. One of these solutions could be the development of an environmental audit program which uses the material and facility risk assessment techniques discussed previously as priority setting tools. As shown in Figure 4.7, the scope, frequency, and resource commitment can be assigned based on an estimate of risks posed by the facility and the materials handled at the facility. In this way, resources are committed cost effectively. That is, the number of auditors and the facilities being audited at any one time are minimized.

---

[4] In this context, controllable means those items that are under the purview of management to change; including the substitution of materials, modifying processes, and upgrading process units.

## Figure 4.6
## Assessing Risk in a Multi-Plant or Multi-Unit Setting

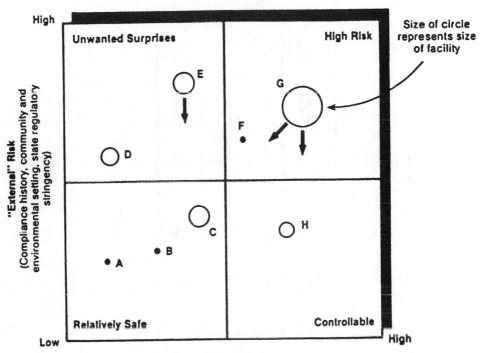

## Figure 4.7
## Sample Site Visit Schedule

### ELEMENTS OF A PROGRAM

| Site | Description | Site Classification* | | | | Overall Rating & Class | Principal Scope | Frequency of Audit | Duration of Audit | No. of Auditors |
| | | Incident History | Materials Handled | Process Complexity | Environ Sensitivity | | | | | |
|---|---|---|---|---|---|---|---|---|---|---|
| I | Large AG-CHEM Processing Facility | 4 | 3 | 3 | 3 | 13 A | RCRA/FIFRA | Semi-Annual | One Week | 4 |
| II | Machine Tool Plant | 3 | 3 | 2 | 4 | 12 B | CWA | Annual | 2-3 Days | 3 |
| III | Drum Storage Facility | 1 | 2 | 1 | 3 | 7 C | RCRA | Every 3 Yrs | Half Day | 1 |
| IV | Truck Terminal | 0 | 1 | 1 | 1 | 3 D | All | Every 5 Yrs | Half Day | 1 |

*Sites are ranked from 0-4 (4 being highest potential risk) in each of four categories. Overall rating is determined by total score as follows
Score 1-4 (D), Score 5-8 (C), Score 9-12 (B), Score 13-16 (A)

Figure 4.7 can also be used as a resource planning tool. If the company's inventory of auditable facilities is established, and then a frequency, audit duration and team size is assigned to each facility, the manpower loading for field audits can be determined for any given year. Further, if the number of field hours are increased by fifty percent or so to account for audit preparation and report writing, the result should indicate full cost accounting for the program, except for management and administration time.

Compiling this information on a spreadsheet will allow the program manager to manipulate critical factors, such as audit frequency, to determine the financial or budgetary impacts of increasing or decreasing the frequency.

## Auditor Selection and Training

Thus far, in the brief history of environmental auditing, companies have tended to design their audit function using one of four fundamental organizational approaches. The first of these, appropriate for larger companies, is to set up a corporate audit function. Under this option, full-time auditors are hired or transferred permanently into a corporate staff unit. This staff function alone is responsible for conducting audits company-wide. This particular strategy can be found in companies where management responsibilities are quite centralized.

The second approach, also most appropriate for large companies, involves establishing audit functions at the division or subsidiary levels. Here, the auditors from one division assess the facilities of other divisions and vice versa. Each audit team report the results of its assessments to corporate headquarters. This strategy is typically used within the operating divisions.

A third tactic is to have each plant audit itself using common guidelines. This approach amounts to self-reporting and is used by some larger companies as well as by one-plant corporations.

Finally, some companies use outside contractors or consultants to (1) develop and manage their audit programs, (2) conduct individual facility audits, or (3) complete periodic third-party reviews of their in-house facility audits. Quite naturally, each of the above approaches has its advantages and disadvantages. Choosing from among them is a difficult exercise with no clear choice emerging.

Independent of the chosen organizational style of an environmental audit program are the questions of who and how many staff should make up a given audit team. Since the assessment team is the heart of each company's program, the success of the program depends in large part on the skill, judgment, and perception of the team members.

Typically, the size of a given assessment team will depend upon the size of the facility, the complexity of the environmental issues, and the time that has elapsed since the last assessment. Experience has shown that three to five people are needed for a week to review complex facilities, with an equal amount of additional time off-site to prepare, write the final report, and respond to inquiries. On the other hand, certain facilities, such as storage facilities or warehouses, can be reviewed by one person in half a day.

Team composition will, to a large extent, be dictated by the type of facility, the environmental issues, and the size of the team. In formulating the team, care should be given to balancing the skills of the team members to include knowledge of:

- The audit process
- The applicable environmental regulations
- Corporate policies
- The individual facility processing the waste treatment operations.

Because of the range of knowledge required, it is not surprising that companies will select from among a number of disciplines, including operations managers, engineers, scientists, attorneys, and accountants (because of their experience with financial audits). However, it is safe to say that the majority of audit teams consist of an operations or environmental manager as team leader who is supported by engineers and scientists familiar with environmental issues. Where a sensitive legal situation dictates, an attorney may provide on- or off-site counsel directly to the audit team.

## Audit Procedures

Establishing procedures for the program is obviously an essential element in the planning process. In fact, an entire chapter of this book is dedicated to this issue. Here, it suffices to say that procedures should be carefully documented so that there is a common understanding among all participants in the program.

One of the key planning issues is defining the time line for individual audits. An example site environmental audit process timeline is shown in Figure 4.8. The timeline shows the entire audit process from initial notification to the site by the audit team leader to submission of the corrective action plan by the site management. All of the interim steps are shown as well. This depiction can be useful in explaining how the process works to the uninitiated.

## Audit Reports and Documentation

Audit programs generate a significant number of documents, many of which can pose significant liabilities to the organization. Thus, document management and retention are critical issues for the program manager. Each company must decide for itself the appropriate document retention procedures. Listed in Figure 4.9 are the classes of documents that are typically generated. Listed also are the functions that are commonly responsible for generation of the documents, the document due dates, the likely recipients of the documents and their ultimate disposition or destruction. The figure presents document management approaches typically used in industry; other approaches than those proposed are, of course, acceptable. Efficient ways of handling this process are discussed elsewhere in this book; but it is important to note here that proper document management is one of the most critical aspects of any audit program.

**Figure 4.8**
**Example Site Environmental Audit Process Timeline**

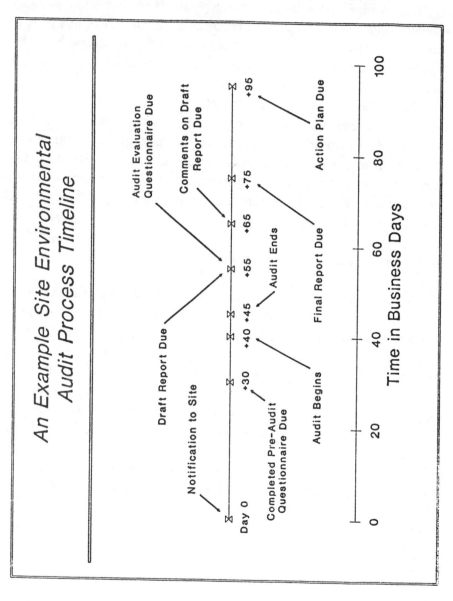

**Figure 4.9**
**Schedule and Disposition for Audit Documentation**

| Document | Responsible Party | Timetable and Disposition |
|---|---|---|
| Pre-audit Questionnaire | Site EH&S Coordinator | Returned completed to team leader 10 business days prior to the audit; retained until the next audit |
| Working Papers (*i.e.*, Field Notes) | Individual Auditors | Returned to audit team leader once draft report is issued; destroyed once final report is issued |
| Draft Findings | Audit Team Leader | Left at the site if draft report is not left at the site; destroyed once draft report is issued |
| Draft Reports | Audit Team Leader | Left at the site (preferred) or within 10 business days of the end of the audit; distributed to corporate EH&S and law divisions, the site, and business management; destroyed once final report is issued |
| Audit Evaluation Questionnaire | Site Management | Due within 10 business days after the audit; distributed to audit program director; retained until annual program review |

| Comments on Draft | Site/Business EH&S Coordinator, Law, Corporate EH&S | Due within 10 business days of receipt of the draft report from the audit team leader; destroyed once final report is issued |
| --- | --- | --- |
| Final Report | Audit Team Leader | Due within 10 business days of due date for receipt of comments; distributed to corporate EH&S and law divisions, plant manager, operations director, business president and site envl. Coordinator; retained until next final report is issued |
| Corrective action plan (CAP) | Plant Manager | Due within 20 business days of final report date; distributed to corp. EH&S & law divisions; site EH&S coordinator; business management; retained until audit is closed out |
| CAP Status Reports | Site EH&S Coordinator | Due quarterly for all open audits; due on 3/1, 6/1 9/1 and 12/1; distributed to corporate EH&S and law divisions and business management; retained until audit is closed out |

## Audit Program Management & Evaluation

Any management program should be evaluated on a periodic basis. Each year, the audit program manager should report to senior management on the successes and failures of the program. The submittals, concise reports on the status of the program, should contain the following information at a minimum:

■ Number of audits completed during the time period and number planned for the next time period

■ Highlights of liabilities most affecting the corporation

■ Trends in the types of noncompliance items to identify potential corporate-wide issues

■ Statistics on business success rates in meeting corrective action plan schedules

■ Results of formal feedback from those who are audited

■ Development and implementation of an auditor training program

■ Timeliness and quality of audit reports

■ Periodic information exchange reports focusing on key learnings and root-cause analysis

■ Annual update of the program guidance manual and audit protocols.

The program evaluation can be done internally or by a third party, and should be completed annually.

## Audit Support Tools

The principal audit program tools are the auditors' checklists or protocols. These are the actual working documents for the auditors. The checklists pose questions relating to:

(1) compliance with federal and state regulatory requirements (the "paper" audit)

(2) effective organizational controls (the "management" audit), and

(3) proper on-site and off-site unit operations (the "technical" audit).

These three areas are reviewed to determine whether: (1) specific regulatory requirements are known and complied with; (2) an organization is in place which can monitor compliance, respond to upsets, or emergencies, and anticipate regulatory changes; and (3) compliance procedures are carried out by unit operators.

Protocols should be developed for key compliance areas and should integrate any federal, state, or local requirements influencing the facility. Specific protocols should be developed for at least the following compliance areas, if they are applicable to the facility:

- Wastewater discharges/pretreatment/stormwater
- Drinking water
- Air emissions
- Solid and hazardous wastes
- Above and underground storage tanks
- PCB handling/disposal
- Pesticide use
- Hazardous materials management
- Asbestos handling/disposal
- Emergency response and spill control
- Community right-to-know
- Worker right-to-know
- Industrial hygiene
- Worker safety
- Process safety
- Product stewardship.

These are considered the major EH&S areas affecting most industrial facilities and should be viewed as a baseline group of protocols. Other protocols can be added to the program as conditions change and new regulations emerge.

The following elements should be part of a good audit protocol:

- It should provide the auditor with an abstract or overview of the key federal, state, and local regulations and requirements relevant to each compliance area at the facility being audited.

■ It should highlight those key definitions that help the auditor conduct the audit without having to refer to the full text copies of the applicable regulations. The importance of definitions in verifying regulatory compliance areas cannot be over-emphasized.

■ It should include all the items strictly required by federal, state, and local statutory laws and their associated regulations.

■ It should include inspection items that permit evaluation of the plant's vulnerability to "common law" environmental problems (*i.e.*, noise, odor, nuisances, community impacts).

■ It should provide an opportunity to verify that internal company environmental management procedures are understood and adhered to by plant staff.

■ It should give the auditor specific action steps so that compliance can be properly evaluated. Use of action verbs such as "examine," "inspect," "calculate," etc., should be used in the protocol so that the auditor understands clearly what must be done.

■ The protocol should provide a space in which comments can be written by the auditor. These notes are then readily available when writing the audit report. This also discourages the use of loose papers and notes that could be misplaced.

# Figure 4.10
## Sample Audit Checklist

### Hazardous Waste

| Requirement | Question | Results | Working Paper Reference |
|---|---|---|---|
| **WASTE IDENTIFICATION** | | | |
| 4-3 Sites which generate solid wastes must determine if the wastes are hazardous wastes. (40 CFR 262.11)<br><br>NOTE: Conditionally Exempt Small Quantity Generators are not subject to regulation under 40 CFR Parts 262 - 266, 268 or 270 or notification requirements of Section 3010 of RCRA. Provided they meet certain requirements (see 40 CFR 261.5) they need only comply with requirements for identification of hazardous waste as indicated in question 4-3 above(40 CFR 262.11). | • Sites should have a hazardous waste management program (GMP).<br><br>• Determine if there is a master list of the types and quantities of hazardous wastes generated, treated and disposed at the site.<br><br>• Verify that the list is complete and updated regularly.<br><br>• Discuss with staff how wastes generated on the site were identified and classified.<br><br>• Determine if the site followed Federal or State criteria for identifying the specific listing or characteristic of hazardous waste, whichever is more stringent (e.g., waste oil is sometimes considered a hazardous waste by state regulations but is exempt under most sections of RCRA). | | |
| 4-4 Each facility should maintain an up-to-date waste emissions inventory that identifies and characterizes each waste that is produced (Colgate-Palmolive BMP 5.II). | • Review facility's waste emissions inventory. Ensure that the inventory identifies the following for each waste:<br>  - Waste Category<br>  - Source or Point of Generation<br>  - Estimated Annual Volume (preferable in metric tons/year)<br>  - In-plant Handling and Storage Procedures<br>  - Waste Disposal Practices<br><br>• Ensure that it is current - should be updated at least annually | | |

Within these guidelines, individual approaches can vary. One example of a particular approach is provided in the Appendix. As illustrated by Figure 4.10 (a sample page taken from an audit protocol) each protocol contains inspection procedures (worksheets) composed of written requirements or guidelines, which indicate possible compliance problems, as well as practices, conditions, and situations that could indicate potential problems. They are intended to focus the auditor's attention on the key compliance questions and issues that should be investigated during the audit. Instructions are provided to direct the auditor to the appropriate action, references, or activity that corresponds to the specific requirement or guideline.

The first column is a statement of a requirement. This may be a direct citation of strict regulatory requirement, or it may be a requirement that is considered a good management practice to maintain compliance, although not specifically mandated by any regulation.

The next column gives instructions to the auditor to help conduct the audit. These instructions are intended to be specific action items that should be accomplished by the auditor. Some of the instructions may be simple documentation checks which take only a few minutes, while others may require physical inspection of a facility. The final column on the worksheet is for the convenience of the auditor so that notations or comments may be documented to provide a permanent record for preparing the final audit report. There is also a column to provide a working paper reference. Typically, additional audit questions are included on the worksheets which reflect the substantive requirements of state/local regulations pertinent to individual facilities.

Remember that the inspection procedures are designed as an aid and should not be considered exhaustive. The auditor's judgment should continue to play a role in determining the focus and extent of further investigation.

The questionnaire presented in Figure 4.10 is very detailed and leaves little leeway for individual auditors. The advantage of this type of approach is that the review will be very comprehensive, even where inexperienced auditors are used. The disadvantage is that where auditors are, in fact, more experienced, the more detailed approach hinders flexibility. In these cases, broader questionnaires can be developed.

One could logically argue that a guide used by government inspectors to audit plant operations could provide an appropriate tool from which to develop an internal audit manual. While this is by no means be a "fail-safe" approach toward assuring compliance, it does provide an internal system of checks consistent with external Federal inspections.[5] Two guides for designing an appropriate questionnaire are the *Multi-Media Inspection Manual*, used by U.S. EPA staff when they inspect industrial facilities and the *Audit Protocols*, used by U.S. EPA staff when they inspect their own facilities, principally laboratories.

The development of an appropriate set of audit protocols is, at a minimum, a challenging task. For example, only three of a virtually limitless number of checklist options have been discussed here. As suggested by Figure 4.11, prior to the selection of a particular style, the manual developer should think long and hard about the specific objectives of the overall corporate audit program, the demands a particular style may place on plant staff and their willingness to accommodate these demands, and the capabilities of the staff who will be designated, either temporarily or permanently, as auditors.

Before developing protocols for an organization, the following steps should be taken:

- **Decide on the scope of the audits.** Which laws and regulations will be reviewed? Will the checklists include state and local requirements?

- **Review your facilities against these standards.** Do you have any NPDES permits or are all plants discharged to municipal systems? Do you have any RCRA disposal facilities? Not all regulations will need checklists if you have no facilities regulated by the standards. A checklist designed to assist in this process is provided at the end of this chapter.

---

[5]However, it is important to realize that state and local regulations often are the governing requirements, and are frequently more stringent than federal regulations.

## Figure 4.11
## Tradeoffs in Questionnaire Development

**UNFORTUNATELY, NO ONE DESIGN FITS ALL CASES . . .**

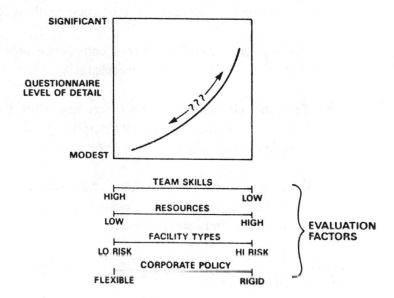

■ **Decide on checklist format.** Use the following sources to choose or develop your style of checklist:

  - EPA/NEIC reports
  - EPA protocols
  - Government Institutes/EPA inspection manuals
  - Published literature
  - Forums such as the Environmental Auditing Roundtable.

■ **Develop one checklist.** Pick a compliance area (wastewater discharges) and develop the checklist.

■ **Test the checklist.** Before proceeding with the remaining checklists, test the first one. It's surprising sometimes how the logic developed in the office is found wanting in the field.

■ **Develop and test the remaining checklists.** Complete and test the remaining checklists, looking for a balance between completeness and manageability.

The end result of this exercise should be checklists that meet the specific organizational objectives.

A key component of any audit is determining the disposition of all waste streams generated at the site. Audit teams should consider developing a table similar to Figure 4.12. This approach will help site management and the auditors to better understand not only the number of waste streams generated, but how and where they are disposed of. Often site managers are either unaware of a low-profile waste stream, or they know the stream and know the contractor that disposes of it, but are unaware of how it is disposed of. In these days of increased restrictions on land disposal options, it is crucial that a generator know exactly what happens to his waste once it leaves the site.

**Figure 4.12**
**Sample Waste Disposition**

## Waste Disposition Summary
### (Complete for ALL Waste Streams)

| Waste Type | Generation Area | Waste Class | Quantity (Per Month) | Transporter | Disposal Company | Disposal Technique |
|---|---|---|---|---|---|---|
| **Example:** TCE | Machine Shop | RCRA Hazardous | 50 Gal. | ABC Co. | XYZ Co. | Solvent Recycling |

Page ☐ of ☐

Reproduce as required

# A KEY PROGRAM ISSUE: CORPORATE STANDARDS AND GUIDELINES

One of the most useful aids for any audit program is a sound set of corporate environmental standards and guidelines. Generally, there are numerous environmental management standards that can be developed to enable a company to more effectively manage its environmentally related activities. Such standards might involve requiring secondary containment for all drum and tank storage, and transferring hazardous materials and wastes, regardless of local regulatory requirements.

Development of a common set of corporate expectations gives the audit team a firm foundation for making risk-based recommendations. For example, in a case where there are no regulatory or corporate requirements for secondary containment of hazardous materials, auditors might receive significant "push-back" from the site if they were to recommend such corrective action, even in cases where the risk clearly justifies construction of containment. Several environmental management standards or procedures that could be developed in any organization are discussed below. Certainly many more procedures could be developed as well. Some will relate to administrative requirements and others might relate to minimally acceptable physical facilities.

## Development of Waste Discharge Inventories

Many companies require that their sites develop air emissions inventories, water balances, and waste inventories. Defining a baseline of discharges assists greatly in measuring environmental improvement and pollution prevention successes. In the United States, documentation of these inventories is almost mandatory due to the Emergency Planning and Community Right-to-Know Act (EPCRA), the Clean Air Act (CAA,) and other recent legislation.

The inventories also are valuable tools for audit teams as they provide a map or guide for the review. This is particularly true for air emissions audits, which can be quite a challenge for the auditor if the site has never completed an inventory. For instance, at large chemical plants, the number of stacks, vents, and other release points can number in the tens of thousands if fugitive sources are included. In fact, on many an audit where there is no documented site air emissions inventory, this lack of an inventory becomes the first finding of the audit team, since it is difficult for the site to determine regulatory applicability if there is no general understanding of site-wide quality and quantity of air releases.

## Secondary Containment for Hazardous Materials and Wastes

It is sometimes surprising how little regulation there is worldwide regarding secondary containment of both liquid hazardous wastes and materials. Even in the United States, there is no federal requirement that hazardous waste accumulation points incorporate secondary containment. However, storage of these materials can pose substantial risk to the organization if there is a release to soils, surface waters or groundwater.

Accordingly, many multinational companies have developed global secondary containment standards for hazardous waste, and material containers and tanks. These standards are fairly straightforward and address both containment integrity (*e.g.*, no cracks or open sumps) and capacity (*e.g.*, the area must hold the contents of the largest container or ten percent of the entire contained volume, whichever is greater, plus an allowance for rainfall if there is no overhead protection). If there is valved drainage associated with the containment area, the standards also require that the valves be closed and locked under normal conditions and that there be a formal, documented procedure for release of accumulated rainwater.

Such corporate standards and audits go hand in glove. They minimize overall corporate risks and provide the legitimacy an audit team requires when making recommendations to install secondary containment measures. Using the "it's a good management practice" argument is often not viewed as compelling by the site management.

## Internal Reporting of Environmentally Related Incidents

During the course of the year, facilities will experience violations, excursions, spills or other environmental incidents. These occurrences must be brought to the attention of corporate headquarters on a timely basis. The objective of the internal reporting procedure is to provide a mechanism for the prompt reporting of these incidents as well as to define responsibilities throughout the company for the reports. Figure 4.13 contains examples of incident reporting forms which can be used to document environmentally related incidents.

The information, when received by headquarters, is typically compiled into a database so that the total liability of the corporation can be readily determined. This policy is not geared toward the periodic reporting requirements of regulatory agencies as stipulated in permits and other regulations. Rather, this procedure should emphasize the proper reporting of a violation from the point at which it happens, up through the chain of command to the corporate headquarters officer responsible for environmental affairs.

All facilities need to develop standard forms, and distribute them along with instructions for completing them and submitting them to headquarters when a violation occurs. Large companies commonly develop a management information system (preferably computer-based) to store records of the reported violations at corporate headquarters. This system gives corporate headquarters the "real time" information it needs to effectively manage the environmental affairs of the company and helps auditors better understand the environmental management issues at the site.

# Figure 4.13
# Model Regulatory Compliance System

ENVIRONMENTAL INCIDENT REPORTING FORMAT

| LOCATION OF INCIDENT | DATE | TIME | REPORTED BY (NAME, DEPT. & TITLE) |
|---|---|---|---|
| 1. | 2. | 3. | 4. |

DESCRIBE EXACT INCIDENT LOCATION AND TYPE:

| BUILDING & AREA | | DEPARTMENT | SUPERVISOR IN CHARGE AT TIME OF INCIDENT |
|---|---|---|---|
| 5. | | 6. | 7. |

| AREA AFFECTED:<br>____ IMMEDIATE<br>____ SURROUNDING AREA<br>____ OUTSIDE NIH<br>8. | DID A PERSONAL INJURY OCCUR?<br>____ YES ____ NO<br>9. | WAS THERE PROPERTY DAMAGE?<br>____ YES ____ NO<br>10. | PROBABILITY OF RECURRENCE<br>__ HIGH<br>__ MEDIUM<br>11.__ LOW |
|---|---|---|---|

WAS A COMPLAINT RECEIVED: ___ YES ___ NO

| WHO MADE THE COMPLAINT? IF YES: ___ EMPLOYEE<br>___ GOV'T. AGENCY<br>___ NEIGHBOR<br>___ OTHER<br>12. | NAME, ADDRESS AND AFFILIATION OF COMPLAINANT:<br>13. |
|---|---|

RELEASED MATERIAL:

| ___ GAS<br>___ LIQUID<br>14. ___ SOLID | NAME OF MATERIAL: | QUANTITY RELEASED: (LBS, GAL)<br>15. | INCIDENT DURA-TION (MIN.)<br>16. |
|---|---|---|---|

DID IT ENTER A SEWER OR OUTFALL?__YES __NO DID IT ENTER ATMOSPHERE? ___ YES ___NO

| IF YES: | SEWER OR OUTFALL NUMBER:<br>17. | IF YES: | STACK IDENTIFICATION:<br>18. |
|---|---|---|---|

WAS IT AN EXCURSION FROM A REGULATORY REQUIREMENT?

| IF YES: | NAME PERMIT AND REGULATORY AGENCY<br>19. | NATURE OF EXCURSION:<br>20. |
|---|---|---|

SIGNATURE       DATE

| FOR ENVIRONMENTAL PROTECTION BRANCH USE ONLY. PLEASE DO NOT WRITE IN THIS SPACE.<br>RECEIVED BY: _____DATE:_____TIME:_____<br>FOLLOWUP: |
|---|

**Figure 4.13** *(cont'd)*

ENVIRONMENTAL INCIDENT

FOLLOWUP REPORT

Incident #_____

| DESCRIBE HOW INCIDENT OCCURRED: |
| --- |
| |
| RESULTS OF THE INCIDENT (DEGREE OR AMOUNT OF VARIATION FROM PERMIT OR REG-ULATION, AND OTHER IMPORTANT CONSEQUENCES: |
| |
| WHAT ACTS, FAILURE TO ACT AND/OR CONDITIONS CONTRIBUTED DIRECTLY TO THE INCIDENT? |
| |
| BASIC REASONS FOR THE ACTS OR CONDITIONS ABOVE: |
| |
| IMMEDIATE CORRECTIVE ACTION TAKEN: |
| |
| FUTURE ACTION NECESSARY TO PREVENT RECURRENCE: |
| |

## Environmental Recordkeeping/Records Retention Procedures

The objective of this management procedure is to specify environmentally the related data, records, reports, and files that should be maintained by the company. Also, the procedure should identify responsibilities for maintenance and custody of environmental records, including minimum retention periods.

Environmentally related data, reports, records, and files should be interpreted broadly to include all materials strictly required by statute, rule, or regulation, as well as those written materials useful and necessary for the company to manage its environmental compliance program but not required by law.

In some cases, joint custody of particular records should be established between departments that share responsibility for compliance activities. Some records and reports must be maintained for specified periods of time as stipulated in various statues, rules, and regulations. Many records, however, have no statutory requirements for retention, and, therefore, good management practice considerations should be used to establish the retention periods.

Records are normally maintained in a central location. Maintaining them in this manner will aid in retrieval of environmental compliance-related information, and will enhance the site's ability to respond to internal audits and external regulatory inspections.

## External Regulatory Inspection Procedures

The objective of this management procedure is to specify procedures to be followed when compliance inspections are conducted by external regulatory agencies. Most federal and state agencies have broad statutory authority to conduct inspections at industrial facilities.

Companies should develop a procedure that includes steps for dealing with these visits. Figure 4.14, found at the end of this chapter, contains an example of a management procedure that addresses the conduct of regulatory inspection procedures. The procedure deals with items that should be covered prior to, during, and after the inspection. Plant management should have clear guidance on such things as: (1) limits of authority of inspectors, (2) objective and scope of the inspection, (3) plant personnel who will accompany the inspector, (4) and legal counsel participation.

During the inspection, plant personnel should have guidance on how they should respond to deficiencies found, what notes they should take during the inspection, and what their general attitude should be in dealing with inspectors.

After the inspection, plant personnel should have instructions from corporate headquarters on preparing memoranda for the record and developing an action plan for resolving identified deficiencies.

The common elements of a compliance inspection procedure can be grouped into five procedural categories: (1) pre-inspection preparation, (2) entry of inspector, (3) opening conference, (4) physical inspection, and (5) closing conference. These categories will be discussed at length in various chapters of this book.

## ADDITIONAL SAMPLING

There is no real consensus on whether or not to conduct sampling as part of an audit and most companies do not do so. These companies typically focus on verifying compliance in this area through a review of the site's sampling and analysis quality assurance/quality control (QA/QC) program. If that program is found to be strong, then duplicate (split) sampling is considered unnecessary.

However, some companies do collect split samples and send one set of samples to an outside, independent laboratory to further verify their QA/QC program.

In a third approach, the audit team conducts analyses during the audit with pH probes and the like to determine whether the stream is in compliance that day. This approach is used in only limited cases, and is not generally believed to be an effective means of verification.

## CONTINUOUS REVIEW

Auditing is not something that is done once and forgotten for ten years. Noncompliance can surface quickly. Therefore, audits should be conducted on a regular basis. As the audit program evolves into an integral part of operations, senior management should periodically review the program to assure that it is continuing to meet its originally stated objectives. And, in fact, some companies use outside firms periodically to conduct audits as a mid-course review on their program.

## EVALUATING THE RESULTS AND IMPLEMENTING THE SOLUTIONS

These two elements must both be included for a successful program. Corporations are most liable where compliance problems have been identified, yet no solutions are planned. These solutions must be documented, even where management initially can only commit to a phased response for a needed major capital expenditure. It is vital that senior management have a system in place to assure that where violations have been noted, the facility operator is, in fact, on the way towards achieving compliance. History indicates that the corporation and its executives are at great risk where violations have been identified and no plan is in place to remedy the situation.

In recent years there has been a particular trend to: (1) assure that corrective-action plans are, in fact, just that and (2) to provide formal briefings for senior management (including the board of directors) on the status of the audit program. Automated tracking systems have been put in place in most companies. These systems produce monthly status reports for each facility's action plan and, subsequently, produce statistics (similar to safety program reporting) for senior management on those items completed ahead of, on, or behind schedule. This kind of reporting to senior management has done much to increase the internal credibility of the audit program; after all, senior management can no longer escape the civil and criminal liability associated with noncompliance.

## Figure 4.14
## Environmental Audit Program—Draft Guidance Manual

**Proposed Outline**

I. **INTRODUCTION**

o   Managing Environmental Compliance
o   Overview of Environmental Auditing
o   Purpose of the Manual

II. **STATEMENT OF PURPOSE**

o   Letter from Management
o   Environmental Policy Statement
o   Environmental Auditing Policy Statement
o   Program Objectives

III. **ROLES AND RESPONSIBILITIES**

o   The Audit Organization
o   Legal Department
o   Corporate Environmental Management
o   Groups/Divisions
o   Plants
o   Health and Safety

IV. **SCOPE OF THE PROGRAM**

o   Facility Compliance
o   On-Site Contractors
o   Real Estate Transactions
o   Waste Disposal Contractors
o   Contract Work on Suppliers' Sites

V. **AUDIT COVERAGE**

o   Compliance Areas

-Air Emissions               -Wastewater
-Hazardous Wastes            -Stormwater

|  |  |
|---|---|
| -PCBs | -Drinking Water |
| -Solid Wastes | -Hazardous Materials |
| -SARA/Prop. 65 | -Noise |
| -Toxic Substances | -Asbestos |
| -Electromagnetic Pollution | -Pathological Waste |
| -Low-Level Radioactive Waste | -Explosives |
| -Pesticides | |

o   Federal, State, and Local Requirements
o   Corporate Policies
o   Good Management Practices
o   Regulatory Updating

## VI.  AUDIT PROCEDURES

o   Prior to the Audit

-Team Formation
-Team Responsibilities
-Regulatory Research
-Pre-visit Questionnaire
-Site Agenda and Notification
-Checklist Refinement and Updating

o   During the Audit

-Opening Conference
-File Reviews
-Inspections
-Interviews
-Closing Conference

o   Following the Audit

-Preparation and Distribution of Draft Report
-Preparation and Distribution of Final Report
-Action Plans
-Follow-Up
-Records Retention

## VII.  LEGAL PROTECTIONS

o   Imminent Hazards Notification
o   Report Protections
o   Legal Review of:
       -Checklists
       -Pre-visit Questionnaires
       -Reports

## VIII.    FACILITY AUDIT SCHEDULES

- o  Priority Setting
- o  Schedules

## IX.  AUDIT TEAM SELECTION

- o  Philosophy
- o  Process
- o  Assignments

## X.  AUDIT REPORT STRUCTURE AND DISTRIBUTION

- o  Report Outlines
- o  Report Responsibilities
- o  Report Distribution
- o  Report Quality Assurance

## XI.    MANAGING THE PROGRAM

- o  Setting Goals
- o  Program Evaluation
- o  Managing Information
    - -Checklists
    - -Regulatory Updates
    - -Working Papers
    - -Reports
    - -Internal Data (e.g., Permits)
    - -Action Plans
    - -Status Reports
- o  Reporting to Management

## APPENDIX

- A.  Pre-Visit Questionnaire
- B.  Compliance Checklists
- C.  Auditor's Checklist

ASSESSING THE NEED FOR ENVIRONMENTAL AUDIT PROTOCOLS
AT INDUSTRIAL MANUFACTURING FACILITIES

(CHECK ALL FACTORS THAT COULD APPLY TO FACILITIES
TO BE INCLUDED IN YOUR PROGRAM)

| ASSESSMENT CATEGORY | YES | NO |
|---|---|---|

**I. General**

- **Types of facilities**

  - Major Manufacturing Facilities ☐ ☐
  - Laboratories ☐ ☐
  - Distribution Centers/ ☐ ☐
    Warehouses

- **Environment**

  - Remote ☐ ☐
  - Industrial/Commercial ☐ ☐
  - Residential ☐ ☐

- **Locations**

  - U.S. only ☐ ☐
  - Number of states — ☐
  - International ☐ ☐

- - - - - - - - - - - - - - - - - - - - - - - - - - - - - - - - - - - - - - - -

**II. Air Emissions**

- Process vents/stacks ☐ ☐

- Boiler(s) ☐ ☐

- Degreasers ☐ ☐

- VOC bulk storage ☐ ☐

- Air pollution control equipment ☐ ☐

- Asbestos insulation to be removed ☐ ☐

- - - - - - - - - - - - - - - - - - - - - - - - - - - - - - - - - - - - - - - -

---

III. Water Supply

- Municipal ☐ ☐
- Own wells ☐ ☐

---

IV. Wastewater/Stormwater

- Septic tanks ☐ ☐
- NPDES stormwater discharges ☐ ☐
- Unpermitted storm sewers/ditches ☐ ☐
- NPDES wastewater discharges ☐ ☐
- Pretreatment/sewers ☐ ☐
- Use of unlined impoundments/basins ☐ ☐

---

V. Solid Waste

- Trash/refuse ☐ ☐
- Asbestos in place ☐ ☐
- Recyclables ☐ ☐
- Sludges ☐ ☐
- Used oils ☐ ☐
- Solvent/oil contaminated rags ☐ ☐
- On-site landfills ☐ ☐

---

VI.  Hazardous Waste

- Large quantity generator      ☐ ☐
- Small quantity generator      ☐ ☐
- Part B RCRA Permits or Interim
  Status

  - Treatment      ☐ ☐
  - Storage      ☐ ☐

  - On-site disposal
    - incineration   ☐ ☐
    - lagoon   ☐ ☐
    - landfill   ☐ ☐
    - other treatment   ☐ ☐

- Off-site disposal      ☐ ☐

-------------------------------------------------------------

VII. Oil/Hazardous Materials

- Underground storage tanks
  - Heating oils   ☐ ☐
  - Other oils   ☐ ☐
  - Hazardous materials   ☐ ☐

- Bulk storage
  - Oils   ☐ ☐
  - Hazardous materials   ☐ ☐

- Drum Storage (>10 drums)
  - Oils   ☐ ☐
  - Hazardous Materials   ☐ ☐

- Gas cylinders
  - For welding   ☐ ☐
  - For other purposes   ☐ ☐

-------------------------------------------------------------

VIII.PCB Contaminated Equipment

- In-service      ☐ ☐

- Out-of service      ☐ ☐

- In storage      ☐ ☐

-------------------------------------------------------------

---

IX. Pesticides

- Manufacturer ☐ ☐

- Self-application to sites ☐ ☐

- Contractor application to sites ☐ ☐

---

X. Miscellaneous

- TSCA applicability/
  premanufacture notifications ☐ ☐

- SARA Title III applicability ☐ ☐

- Pathological/Infectious wastes ☐ ☐

- Radioactive wastes ☐ ☐

- OSHA Hazard Communication Standard ☐ ☐

---

XI. Special Issues?

---

XII. Which factors that were checked "Yes" do you not want
     included as part of your environmental audit program?

# 5

# IMPACT OF INTERNATIONAL STANDARDS ON ENVIRONMENTAL AUDIT PROGRAMS[1]

What do General Motors and Ben and Jerry's Homemade Ice Cream have in common? Both companies have endorsed the CERES Principles, a model corporate code of environmental conduct developed by the Coalition for Environmentally Responsible Economies (CERES). This "model code" generally goes well beyond what is now typically required of industry to maintain compliance with already stringent requirements established by regulatory agencies.

The CERES Principles are not the only game in town. Many other environmental initiatives have surfaced both in the U.S. and abroad over the past few years. Expectations for corporations are rising even as these initiatives compete for recognition and acceptance. Notable disasters such as Bhopal and the Exxon *Valdez* accidents, as well as more chronic environmental issues such as deforestation and depletion of the ozone layer, have done much to heighten the world's concerns over the environment by generating support for establishing a "global-level playing field" that protects the environment while allowing for sustainable

---

[1] Adapted from an article originally published in the Summer 1994 issue of *Total Quality Environmental Management*, Executive Enterprises, Inc., New York, NY.

development. It is this concept that is really considered to be the underlying driver for all of the increased pressure on transnational companies to rethink their performance. Sustainable development in its basic form is the process of meeting the needs of the present without interfering with future generations' ability to meet their needs.

Presently, there are seven primary sets of initiatives that will likely shape the future of corporate environmental performance expectations and environmental auditing programs. These are:

- The CERES Principles
- International Standards Organization (ISO) Environmental Standards
- European Community Eco-Audit Program
- International Chamber of Commerce Business Charter for Sustainable Development of Environmental Management Principles
- Responsible Care®
- British Standards Institution (BSI) Standard 7750
- U.S. EPA's Environmental Leadership Program.

This chapter discusses these seven sets of initiatives and provides some thoughts on what companies are doing presently to prepare for the future. And a challenging future it will be.

## SEVEN INITIATIVES

The grouping of the initiatives into a set of seven is, admittedly, somewhat arbitrary. For example, the ISO standards themselves consist of seven sets of pending individual standards. The dissimilarity and fragmentation of the initiatives will be the major challenge facing those who wish to comply with the progressive requirements, although there has been some historical coordination among the originating organizations. This dilemma is similar to that of facility emergency response planning in the U.S., where separate plans, with separate requirements, can be required under the Clean Air Act, the Clean Water Act, the Resource Conservation and Recovery Act, the Emergency Response and Community Right-to-Know Act, and the Occupational Safety and Health Act. How then does one assure that a site's emergency response planning efforts will respond to individual requirements while resulting in a program that will indeed work if there is an emergency? One could ask the same question about the environmental management initiatives under discussion.

## The CERES Principles

CERES is a nonprofit membership organization comprised of leading social investors, major environmental groups, public pension funds, labor organizations, and public interest groups. Together this coalition "represents more than ten million people and over *$150 billion* in invested assets."[2] Thus the power it wields in making investment choices can be significant.

CERES began as a result of the Exxon *Valdez* incident that occurred in March 1989. In fact, the principles were originally called the Valdez Principles. As shown in Figure 5.1, the consequences of the *Valdez* incident on Exxon's stock price were measurable, resulting in roughly a ten-dollar-per-share shortfall a little over a year after the incident when compared to five major Exxon competitors. Investor strategies likely had a significant bearing on this shortfall.

---

[2] Information provided in: *A Healthy Economy & a Healthy Environment*, undated brochure, CERES, Boston, MA.

**Figure 5.1**
**Consequences of the Exxon Valdez**

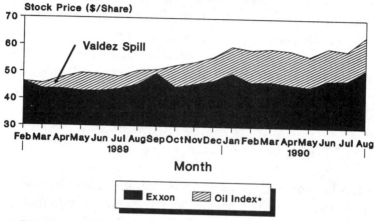

# Exxon Stock Growth
## Did Valdez Hurt???

•Normalized average of 6 Exxon
Competitors

*CERES Principles...*

## Notable Signatories (1995)

- Ben & Jerry's Homemade
- Domino's Pizza
- Louisville Sewer District
- The Sun Company∗
- The Timberland Company
- General Motors∗
- H.B. Fuller∗
- U.S. Trust Co. of Boston
- County of Westchester, NY
- Polaroid∗

*∗ - Fortune 500 Company*

The ten CERES Principles, listed below, demand that "companies that endorse the Principles pledge to go voluntarily beyond the requirements of the law."[3] These Principles are as follows:

- Protection of the Biosphere
- Sustainable Use of Natural Resources
- Reduction and Disposal of Wastes
- Energy Conservation
- Risk Reduction
- Safe Products and Services
- Environmental Restoration
- Informing the Public
- Management Commitment
- Audits and Reports.

One noteworthy Principle is the Management Commitment, which requires that a company's board of directors and chief executive officer be fully informed about pertinent environmental issues. Directly or indirectly, this has resulted in many cases where companies now have appointed environmental experts on their boards of directors. This trend is discussed in more detail later in the chapter.

One of the most problematic of the CERES Principles for signatories is that of the audits and annual CERES Report. "The CERES Report represents the first attempt to produce a comprehensive and accessible environmental reporting format for corporations. Signatories to the CERES Principles complete the Report on an annual basis, and in it disclose information vital to assessing the real environmental impacts of

---

[3] *Ibid.*

their corporations."[4] Not surprisingly, reporting to the public on environmental issues is troubling to many companies. However, it is a trend that can be found in several of the initiatives (*e.g.*, the European Eco-Audit Program), all of which have been preceded in the U.S. by the public reporting requirements of federal Community Right-to-Know legislation passed in 1986.

Figure 5.1 shows that as of early 1995, only a handful of well-known companies had endorsed the CERES Principles, with General Motors, the Sun Company, H.B. Fuller, and Polaroid being the only Fortune 500 companies in the group. But the endorsement language must be read quite carefully—GM and Sun, in their statements, stop short of *adopting* the principles, an original requirement of the CERES coalition. For example, Sun's statement is as follows: "In February 1993, Sun Company, Inc. became the first Fortune 500 firm to endorse the health, environmental and safety principles of CERES. In addition, CERES has recognized Sun's Principles of Health, Environment and Safety as being consistent with their own."[5] Sun's CEO has been heard on radio and television talking about his company's commitment to the environment and the CERES Principles—perhaps the most pro-active effort to date by a Fortune 500 company in getting its message to the public.

Similarly, GM's endorsement statement as reported by *The New York Times* was as follows: "GM said on February 3, 1994 that it would uphold the CERES Principles. In return for support of the Principles, CERES endorsed GM's own principles as consistent with the goals of the CERES

---

[4] *Ibid.*

[5] Taken from an undated Sun Company brochure.

Principles."[6] The skeptic might wonder what is really being said here. Nonetheless, the recent endorsement by two Fortune 500 companies has likely provided an impetus for the endorsement or adoption of the Principles by others.

## International Standards Organization (ISO) Environmental Standards

The standards that probably have the most potential long-term impact, and yet are among those that are the least developed, are those being crafted by the International Standards Organization (ISO). ISO has a long history of developing quality standards (*i.e.*, the ISO 9000 series) that are firmly entrenched in Europe and are becoming entrenched in the remainder of the developed world, including the U.S. Becoming a registered ISO 9000 company is almost a mandatory requirement. In time, a facility seeking ISO 9000 quality registration may be required to have or be seeking ISO environmental registration as well.

Early in 1993, the ISO constituted a new technical committee (TC 207) charged with developing international standards in the field of environmental tools and systems. ISO/TC 207 was the result of a two-year study by ISO's Strategic Advisory Group on the Environment (SAGE), which concluded that international environmental management standards would:

- Promote a common approach to environmental management, similar to quality management and the ISO 9000 series
- Enhance organizations' ability to attain and measure improvements in environmental performance

---

[6] *The New York Times*, February 4, 1994.

- Facilitate trade and remove trade barriers.

In mid-1993, several ISO/TC 207 subcommittees were established to address particular aspects of environmental management systems. These subcommittees and the countries designated to lead the effort are:

- Environmental Management Systems (United Kingdom)
- Environmental Auditing (The Netherlands)
- Environmental Labeling (Australia)
- Environmental Performance Evaluation (U.S.)
- Life-Cycle Assessment (France)
- Terms and Definitions (Norway)
- Environmental Aspects in Product Standards (Germany)

It is very likely that official ISO 14000 Standards related to these efforts will be finalized in early- to mid-1996, particularly with regard to the "fast-tracked" standards of environmental auditing and management systems.

The environmental auditing standards/guidelines are really comprised of three separate guidelines, which are as follows:

- ISO 14010 - General Principles of Environmental Management Systems (EMS) Auditing
- ISO 14011 - EMS Auditing Procedures
- ISO 14012 - Qualification Criteria for Environmental Auditors

The Committee Draft Standards are effective as of mid-1995. It should be noted that the ISO 14000 auditing guidelines are just that: guidelines. This is to distinguish them from the ISO 14001 Environmental Management Systems (EMS) specification, which carries with it the "shall" descriptor. Also, realize that the ISO 14000 Standards are only environmental standards, as opposed to health and safety standards, and will only be required if the global marketplace calls for them. In other words, ISO 9000 Standards have been adopted principally because customers of multinational companies have demanded it or because the manufacturers have viewed their adoption as providing a competitive advantage. The same will likely be true of the ISO 14000 Standards.

## The European Community Eco-Audit Program

The European Community Eco-Management and Audit Scheme (EMAS) was adopted in mid-1993 and became effective in April 1995. It is designed to "promote continuous improvement in the environmental performance of industry."[7] This "voluntary" regulation affects only European Union (EU) members, and if a company operating in one of these countries wishes to participate, only individual sites can be registered. In order to participate, the following requirements must be met:

- An environmental policy must be adopted consistent with requirements listed in the EMAS regulations.
- An initial site review must be conducted by the company.
- An environmental management system must be developed consistent with requirements listed in the EMAS regulations.

---

[7] "Council Regulation (EEC) No. 1836/93 of 29 June 1993 allowing voluntary participation by companies in the industrial sector in a Community eco-management and audit scheme," *Official Journal of the European Communities*, No. L 168/1, July 10, 1993, p. 2.

- Independent, ISO 10011-consistent audits must be conducted at least every three years, using internal, but site-independent, staff or third parties.
- An environmental statement must be prepared annually.
- The environmental statement must be verified by an independent third party who was not involved in the audit.
- The environmental statement is to be released to the public.

Arguably, the most controversial provision of the EMAS regulations is the last requirement: release of the environmental statement to the public. The reason is that the statement is likely to contain sensitive information. The statement is to have the following organization:

- Overview of site activities
- Assessment of significant environmental issues
- Summary data on emissions and releases
- Other environmental performance data
- Company environmental policy and overview of environmental management systems
- Deadline for issuance of next environmental statement
- Name of verifier.

There are additional controversies associated with the EMAS regulations, such as setting accreditation standards for the third-party verifiers. However, the apparent momentum is such that major companies operating facilities in the EU will be compelled to comply. Facilities that do participate will be allowed to use the EMAS logo on their stationery but not on their products or in their advertisements.

## ICC Business Charter's Environmental Management Principles

In early 1993, the Global Environmental Management Initiative (GEMI) published its Environmental Self-Assessment Program (ESAP), a direct consequence of the establishment of the ICC's principles. GEMI is a group of twenty-three leading multinational companies dedicated to fostering environmental excellence in businesses worldwide. Member companies include AT&T, Dow, DuPont, Kodak, Procter & Gamble, and Union Carbide, among others.

The ESAP is a self-assessment tool designed to evaluate whether a company is operating consistent with sustainable development objectives. That is, the "ESAP is designed to measure and improve corporate environmental management performance over time, with a focus on corporate-level policy, systems, and performance measurement programs. This program can assist a company, regardless of its size, business sector, or geographic scope of operations, in evaluating its environmental management performance relative to the sixteen principles of the International Chamber of Commerce (ICC) Business Charter."[8] These sixteen principles are:

- Corporate Priority
- Integrated Management
- Process of Improvement
- Employee Education
- Prior Assessment
- Products and Services
- Customer Advice

---

[8] *Environmental Self-Assessment Program*, April 1993, First Edition, Global Environmental Management Initiative (GEMI), Washington, DC.

- Facilities and Operations
- Research
- Precautionary Approach
- Contractors and Suppliers
- Emergency Preparedness
- Transfer of Technology
- Contributing to the Common Effort
- Openness to Concerns
- Compliance and Reporting

The ESAP allows a company to actually score its performance in each of these categories and to assess performance over time, as well as allowing the company to compare performance against four absolute standards:

- Level 1 - Compliance
- Level 2 - Systems Development and Implementation
- Level 3 - Integration into General Business Functions
- Level 4 - Total Quality Approach

The ultimate goal is to assure the highest level of performance (*i.e.*, Level 4), which assumes that integrated environmental management systems are applied to operations globally and evaluated continually for improvement opportunities.

Of course, the GEMI organization is neither a standard-setting body nor a regulatory authority and, as such, cannot impose its expectations on industry unilaterally. However, peer pressure and many companies' public commitment to sustainable development have resulted in early adoption of the ESAP as a tool used to assure that a company is meeting its environmental objectives.

## Responsible Care®

The chemical industry worldwide has historically been under tremendous scrutiny related to its environmental, health and safety practices. In an effort to better manage its environmental liabilities, the industry developed the Responsible Care® Program a few years ago. In the U.S., this has been spearheaded by the Chemical Manufacturers Association (CMA), which includes in its membership all major U.S. chemical companies. Presently, the adoption of the Responsible Care® Codes of Practice is a requirement for membership in the CMA. The six codes are:

- Community Awareness and Emergency Response
- Process Safety
- Product Safety
- Employee Safety
- Distribution
- Pollution Prevention.

Each of the codes prescribes the management systems that should be expected in a company in order to meet the objectives of a sound program. Each year, member companies are required to evaluate or audit how far along they are towards full implementation of an effective program.

The Responsible Care® Program is a good indicator of how and why expectations can become fragmented and confusing. The Responsible Care® concept was initially developed by the Canadian Chemical Producers' Association (CCPA) and modified by CMA for use in the U.S.

The CCPA codes include:

- Community Awareness and Emergency Response
- Research and Development
- Manufacturing
- Transportation
- Distribution
- Hazardous Waste Management

While the U.S. and Canadian codes are similar, they are not identical. Variations have typically occurred in standard-setting programs, either due to the variation in geo-politics of the individual country or because, frankly, most organizations wish to place their own particular signature on developing trends.

## British Standards Institution Standard BS 7750

Standard-setting organizations in individual countries have been quite active in the environmental management systems arena. Probably the best known initiative to date is the British Standards Institution Standard BS 7750, effective March 16, 1992. The standard "is designed to enable any organization to establish an effective management system, as a foundation for both sound environmental performance and participation in 'environmental auditing' schemes."[9] BS 7750 specifies the elements of an effective environmental management system, intended to apply to all types and sizes of organizations. These elements include:

---

[9] "Specification for Environmental Management Systems," British Standards Institution, BS 7750, 1992, p. 2.

- Environmental management system
- Environmental policy
- Organization and personnel
- Environmental effects
- Environmental objectives and targets
- Environmental management program
- Environmental management manual and documentation
- Operational control
- Environmental management records
- Environmental management audits
- Environmental management reviews.

The standard was developed to complement the EMAS regulation and "in particular to specify the requirements for an environmental management system as a foundation for [EMAS] registration."[10] Notably, BS 7750 does not itself enumerate specific environmental performance criteria. The draft standard has recently undergone a two-year field test, and a final standard was published in March of 1994.

Other environmental management system initiatives similar to BS 7750 have surfaced in countries such as France, Ireland, Canada, South Africa, and the U.S. Of particular note is the forty-chapter "Agenda for the 21st Century" (Agenda 21). This document was the result of the June 1992 United Nations Conference in Rio de Janeiro, which was attended by representatives from countries all over the world. Agenda 21 contains nonbinding recommendations for business and industry on protection of the global environment, thus embracing the principle of sustainable development.

---

[10] *Ibid*, Annex C.

## U.S. EPA's Environmental Leadership Program

A Federal Register notice from June 21, 1994 requested proposals for Environmental Leadership Program (ELP) pilot projects and outlined the criteria that facilities must address to be considered for participation. The pilot projects would explore ways that EPA and states might encourage facilities to develop innovative auditing and compliance programs and reduce the risk of noncompliance through pollution prevention practices. In addition, the projects will help EPA design a full-scale leadership program and determine if implementing such a program can help improve environmental compliance. The ELP is similar to OSHA's Voluntary Protection Program (VPP), which has been in place for years. The long-term benefits of program participation could include fewer regulatory agency inspections and mitigation of fines and enforcement actions should a problem occur.

Facilities applying to the pilot projects must describe their existing or proposed environmental management and auditing programs, their systems for resolving issues raised by these programs in a timely manner, and their systems for evaluating and adjusting these programs on a regular basis. EPA is currently evaluating its environmental auditing policy, and intends to base any decision to reinforce or change existing policy on empirical data. The ELP pilot projects may generate useful data on auditing measures and methodologies, and may permit experimentation with different incentives. Facilities applying to the ELP must demonstrate a willingness to disclose the results of their audits.

# CORPORATE RESPONSES

In many ways, the future expectations for environmental excellence can be overwhelming to corporate executives. Nonetheless, the likelihood that all of this will simply go away is quite remote. Thus corporations have already begun to change the way they operate, anticipating an increasingly challenging future. Four rather interesting developments are:

- Appointment of environmental experts to corporate boards
- Voluntary public reporting
- Third-party program evaluations and certifications
- Benchmarking studies.

Each of these is discussed briefly below.

## Appointment of Environmental Experts to Corporate Boards

As previously mentioned, investor groups have requested that U.S. companies adopt certain principles to help ensure sound environmental management. One of the principles asks that companies appoint members of environmental organizations, or at a minimum environmental experts, to company boards. In fact, this is happening.

A listing of several companies and the environmental experts on their boards is shown in Figure 5.2.[11] It is a diverse group of companies and a diverse group of board members, including several previous administrators of the U.S. EPA. Moreover, more than fifty company executives are also on the boards of national environmental organizations; Union Carbide, for example, has its directors on the boards of the World Wildlife Fund, World Resources Institute, and the Natural Resources Defense Council.

---

[11] The figure first appeared in: Cahill, L.B. and S. P. Engelman, "Bolstering the Board's Environmental Focus," *Directors and Boards Magazine*, Vol. 18, No. 1, Fall 1993.

## Figure 5.2
## Environmentalists on Boards of Directors

| | Environmentalists on Boards | |
|---|---|---|
| **Company** | **Environmental Director** | **Affiliation** |
| Ashland Oil Inc. | Patrick Noonan | President, Conservation Fund |
| Atlantic Richfield Co. | Frank Boren | Conservation Fellow, World Wildlife Fund/ Conservation Fund |
| Baxter International Inc. | James Ebert | Director, Chesapeake Bay Institute |
| Chevron Corp. | Bruce Smart | Senior Counselor, World Resources Institute |
| Dexter Corp. | Jean-François Saglio | Former Director, French Administration of Environment Protection |
| Du Pont Co. | William Reilly | Former Administrator, U.S. Environmental Protection Agency |
| Exxon Corp. | John Steele | Senior Scientist, Woods Hole Oceanographic Institution |
| Metaclad Corp. | Alan John Borner | Founding Director, National Association for Environmental Management |
| Monsanto Co. | William Ruckelshaus | Former Administrator, EPA |
| Niagara Mohawk Power Corp. | Bonnie Guiton | President and CEO, Earth Conservation Corps |
| Union Carbide Corp. | Russell Train | Chairman, World Wildlife Fund; Former Administrator, EPA |
| Waste Management Inc. | Kathryn Fuller | President, World Wildlife Fund and Conservation Foundation (presently on leave from the board) |
| Weyerhaueser Co. | William Ruckelshaus | Former Administrator, EPA |

*Sources: Directors & Boards; Directorship; The Wall Street Journal*

## Voluntary Public Reporting

Public reporting of environmental issues is something that creates great angst among chief executives in major corporations. In the U.S., however, with the passage of the Emergency Planning and Community Right-to-Know Act (EPCRA) in 1986, public reporting is here and here to stay. It is now quite simple to identify and rank the top ten companies that release EPCRA-listed toxic chemicals, as is done in the accompanying Figure 5.3. This is one top-ten list that companies will attempt to avoid.

While these toxic release reports are mandated by regulation, many companies have now concluded that, since this freely available information will be reported to the public by independent environmental organizations and newspapers, it might be better to produce voluntary comparable information, along with any positive developments, in an annual environmental report. As a result, annual environmental reports are becoming commonplace among progressive companies. Over seventy companies in North America, Europe, and Japan have produced free-standing environmental reports.[12] These reports typically are being produced in response to new regulations; emerging business requirements, including the need to report environmental liabilities accurately; and changing public expectations.

---

[12] Deloitte Touche Tohmatsu International, The International Institute for Sustainable Development, and Sustainability, *Coming Clean: Corporate Environmental Reporting,* 1993.

## Figure 5.3
## Toxics Release Rankings

*1993 Toxics Release Inventory...*
# State Rankings

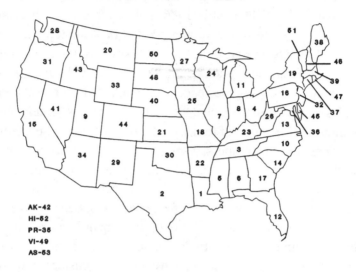

AK-42
HI-52
PR-35
VI-49
AS-53

*1993 Toxics Release Reporting...*
# The Top Ten Chemicals

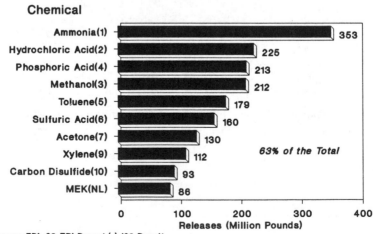

Chemical

| Chemical | Releases (Million Pounds) |
|---|---|
| Ammonia(1) | 353 |
| Hydrochloric Acid(2) | 225 |
| Phosphoric Acid(4) | 213 |
| Methanol(3) | 212 |
| Toluene(5) | 179 |
| Sulfuric Acid(6) | 160 |
| Acetone(7) | 130 |
| Xylene(9) | 112 |
| Carbon Disulfide(10) | 93 |
| MEK(NL) | 86 |

*63% of the Total*

Source:EPA 93 TRI Report;( )-'92 Result

**Figure 5.3** *(cont'd)*

*1993 Toxics Release Reporting...*

## The Top Ten Companies

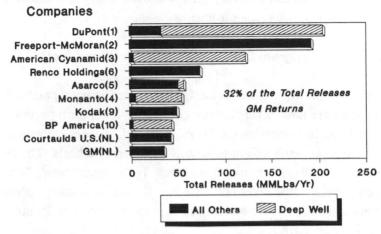

Source:EPA '93 TRI Report;( )-'92 Result

*Toxics Release Reporting...*

## Trends in Releases 1988-1993

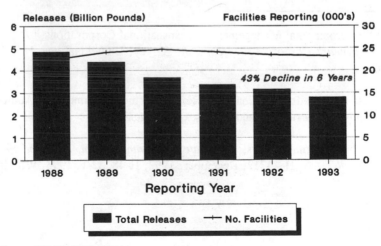

Source: EPA '93 TRI Report

Moreover, a corporate environmental survey conducted by the United Nations in 1993 showed that about forty-seven percent of transnational corporations have a separate environmental section in their annual reports.[13] However, the content and substance of these public environmental disclosures vary greatly and are likely to focus only on the company's positive accomplishments.

## Third-Party Program Evaluations and Certifications[14]

Consistent with good Total Quality Management practices, many companies are now having outside consultants (*i.e.*, third parties) or their internal audit departments (*i.e.*, second parties) evaluate their environmental management programs on a periodic basis. The annual or biennial studies help the companies both meet their "continuous improvement" objectives and answer to the increasing pressure by stakeholders to provide assurances that the company is identifying and remedying its EH&S issues.

And within these reports, environmental audit programs are occasionally, but increasingly, addressed. For example, DuPont in its 1992 and 1993 *Corporate Environmentalism Reports*[15] provided the

---

[13] "Environmental Management in Transnational Corporations," *Report on the Benchmark Corporate Environmental Survey*, United Nations, New York, 1993.

[14] Also discussed in the Chapter: "Benchmarking Environmental Audit Programs: Best Practices and Biggest Challenges."

[15] DuPont External Affairs, *Corporate Environmentalism: 1992 and 1993 Progress Reports,* H-44712, December, 1992, H-48082-1, 1993, Wilmington, DE.

executive summary, along with DuPont's response, from a third-party evaluation of the company's corporate environmental audit program. Eastman Kodak produced a similar summary of a third-party evaluation in its 1994 environmental report. And finally, Hoechst Celanese Corporation included an environmental auditor's opinion in its 1994 environmental, health and safety performance report. A copy of this opinion is included as Figure 5.4.

Further, companies are now conducting evaluations of their entire environmental management systems. This trend has developed as the draft ISO Environmental Standards are beginning to evolve and be distributed. Many companies wish to be well prepared to receive an ISO certification as soon as the standards are officially released. The 14001, 14010, 14011, and 14012 (Environmental Auditing Guidelines) have been published as Draft International Standards and are expected to be finalized in the fall of 1996. Since organizational and procedural changes can take some time to implement, preparing over a year in advance is not inappropriate.

## Benchmarking Studies

And finally, one might ask how environmental expectations should be managed appropriately in the future. Companies have important strategic choices: they can be leading edge, middle-of-the-pack, or follower companies. Depending on a variety of factors, each of these strategies could be appropriate. One of the key challenges, however, is defining at any given time what each means. The rules and expectations are changing constantly; leading edge this year could be follower next year.

# Figure 5.4
## Third-Party Assessment of the Hoechst Celanese Corporation

*June 6, 1995*

Environmental Resources Management, Inc. (ERM) conducted an assessment of Hoechst Celanese Corporation's (HCC's) Environmental, Health & Safety (EH&S) Audit Program managed by the corporate Environmental, Health & Safety Affairs (EHSA) Department. ERM evaluated the elements and performance of the Program in order to render an independent opinion about its effectiveness in achieving improved EH&S performance throughout the company. The assessment was conducted during May 1995 and included a review of Program documentation, interviews with Program directors and staff, and selected interviews of division and site representatives who have been subject to the evaluations.

The Program was evaluated against (1) audit program criteria developed by the U.S. Environmental Protection Agency in its 1986 Environmental Auditing Policy Statement and by the U.S. Sentencing Commission in its Proposed Sentencing Guidelines, (2) HCC's internal EH&S Audit Program guidance and policy as principally articulated by the Program Director and Audit Managers and in available Program documents, and (3) generally accepted audit practices existing in comparable companies.

ERM reviewed the scope and elements of the Program, the procedures utilized, the resources applied to implement the Program and the degree and quality of management commitment. Based on the information made available to ERM by HCC, ERM has concluded that HCC's Program is generally consistent with, and in some cases exceeds, expectations of the established criteria. In our opinion, the Program provides competent, reliable and objective information to management about the status of the company's EH&S compliance programs and performance. Further, HCC's management is responsive in correcting deficiencies when they are identified by the Program.

During 1994, HCC contracted with an independent reengineering firm in order to fully evaluate the Program. That evaluation of the Program resulted in a number of strategic improvements, many of which have been implemented. A number of the Program's elements, such as the mechanisms used to communicate results with senior management, the experience of the auditors, the on-site audit process, including the development of a draft report prior to leaving the site, the integration of the Compliance Assurance Letter process with the audit program and the audit tracking system, including the use of follow-up verification audits, are quite advanced when compared to practices in other companies. A select number of more tactical issues, including the need for the updating and formalization of certain audit procedures and protocols, the need to develop program performance metrics and the need to prioritize audit findings, were identified during the ERM assessment as areas still needing improvement. Management has been informed of these issues and is currently taking steps to respond to them.

*Environmental Resources Management, Inc.*

**Figure 5.5**
**Environmental Audit Benchmarking Study**

*Environmental Audit Benchmarking Study*
## Classifying Findings by Significance

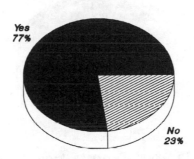

### Do You Classify Findings?

*Environmental Audit Benchmarking Study*
## Use of Attorney-Client Privilege

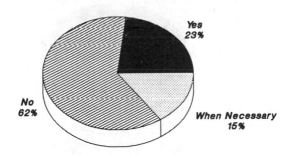

### Do You Utilize Attorney-Client
### Privilege Protections?

**Figure 5.5** *(cont'd)*

Environmental Audit Benchmarking Study
## Auditor's Summary Opinion

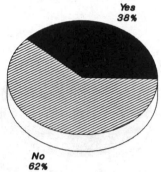

Do Your Reports Include an
Auditor's Summary Opinion?

Environmental Audit Benchmarking Study
## Organizational Program Reporting

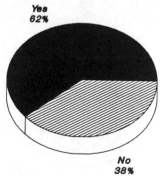

Does Program Manager Report to
a Corporate VP or Higher?

Conducting regular benchmarks of comparable companies can help to better define the situation. As an example, Figure 5.5 displays the results of an analysis of four audit program elements taken from a survey of fifteen companies. Note that the entire programs were not benchmarked, only certain elements were, such as the classification of findings, the presence of an auditor's summary opinion in the audit report, the use of attorney-client privilege, and organizational program reporting. Conducting simple benchmarking studies such as this can help audit program managers make better decisions about the direction they should take and can also help them understand the possible consequences of those decisions.

In summary, stakeholder expectations of corporations are increasing at a rapid rate. Environmental management initiatives are exploding worldwide, resulting in a confusing and challenging geopolitical setting that is not likely to change any time in the near future. Progressive companies are taking action now to respond to present and future demands. They have gotten the message that was delivered at the February 1990 World Economic Forum, where 650 industry and government leaders ranked the environment as the number one challenge for business. Sustainable development is no longer a fashionable cliché. The corporate performance expectations and emerging initiatives discussed in this chapter have made it a reality.

---

*Notes: Any discussion of a Company's individual environmental audit program approaches in this chapter is based on information provided to the public at large through technical papers, presentations and the like. Any discussion of a commercially available audit product does not imply an endorsement of that product.*

# 6

# POTENTIAL EFFECT OF ISO 14000 STANDARDS ON ENVIRONMENTAL AUDIT TRAINING IN THE UNITED STATES[1]

## INTRODUCTION

Training has been an important component of the environmental audit profession since its outset some fifteen to twenty years ago. Training and education have taken on a variety of forms—from audit texts and articles, to formal courses, to video tapes, to on-the-job training programs. These efforts have helped to advance the sophistication of environmental auditing over the past few years.

Now, with the advent of the universal ISO 14012 Environmental Auditor Qualification Guidelines[2], the auditing profession in the U.S. and around the world will probably be transformed in the near future. In fact, many organizations are now re-engineering their programs in anticipation

---

[1] Originally published in the Spring 1995 Issue of *Total Quality Environmental Management*, John Wiley & Sons, Inc., New York, NY.

[2] Revised Committee Draft, ISO/CD 14012: "Guidelines for environmental auditing—Qualification criteria for environmental auditors," 4 October 1994. The 14012 Guidelines are part of the larger ISO 14000 Environmental Management standard-setting scheme, under which guidelines will also be developed for environmental management systems, environmental performance, life-cycle analysis and the like. These standards are "close cousins" of the ISO 9000 Quality Standards, which have been widely adopted by industry worldwide.

of the emerging guidelines. The expected transformation applies to training and education programs as well.

This chapter suggests some of the ways that environmental audit training is likely to be reshaped as a result of ISO 14012. People who offer training courses and seminars would be wise to anticipate the changes and modify their programs accordingly—a quick response could provide a competitive advantage. Moreover, corporate audit program managers should conduct critical evaluations of commercial offerings to assure that the programs' contents are consistent with ISO 14012 auditor certification expectations.

## ENVIRONMENTAL AUDIT TRAINING IN THE U.S.

Environmental audit training in the U.S. has become something of an industry in and of itself. Videotapes, books, and articles abound, as well as enumerable training courses and seminars. The courses come in a variety of forms, as presented below. However, there is often uncertainty over whether an individual course teaches environmental auditing or property transfer environmental liability assessments. This is an important distinction and, therefore, "buyer beware." ISO 14012 guidelines expect that environmental auditing will be the subject of the training.

The following are descriptions of different types of training programs:

- **University certificate programs.** Several universities provide certificate programs in environmental auditing. For example, the University of California at Irvine's Extension Program offers an extensive certificate program in environmental auditing. The certificate is awarded to those who complete at least 180 course hours with a grade of "C" or better in each course. On the other hand, the University of Texas offers a one-time, two-day seminar in environmental auditing. This seminar is not part of a larger program, and no formal certification is offered.

- **Professional organizations' certificate programs (with examinations).** Several professional organizations offer two- to three-day certificate programs that typically include examinations. These programs are offered by the National Association of Safety and Health Professionals, the National Registry of Environmental Professionals, and the Environmental Assessment Association, among others.

- **Commercial, open programs.** Organizations such as Government Institutes, Federal Publications, and Executive Enterprises have offered environmental audit courses for over ten years. There are now some relatively new entrants in this business, such as the Environmental and Occupational Health Sciences Institute. These programs range from two to three days and are held in hotels. Instructors are likely to be from consulting firms, law firms, and/or industry. There are no prerequisites and generally no examinations. Continuing Education Units (CEUs) and certificates of completion are awarded to participants.

- **In-house, internal, and third-party programs.** Some of the best training programs are the in-house programs developed by large companies. These programs last from three to five days and, more importantly, usually involve a "mock audit" as part of the curriculum. This mock audit allows the participants to apply the techniques they have been taught in the classroom; hence, these programs are often held at or near a plant site. Instructors are likely to be experienced in-house auditors and/or consultants.

- **Consulting firms.** Many consulting firms teach environmental auditing skills and techniques. These programs are usually taught at a client's site, although in some cases the training takes place at the offices of the consultant. Certificates are often awarded for completion, but their distribution is generally not dependent on the results of a rigorous examination.

One of the most important things to remember about the current status of auditor certification in the U.S. is that as of early 1995, there is no single, generally accepted "certification" or "registration" program for environmental auditors in the marketplace. That is, there is nothing comparable to the PE, CSP, or CIH. At least, not yet.

One must therefore be very careful in evaluating an individual training program and should review the true meaning of any certification. The ISO 14012 guidelines presently set the requirements for Environmental Management Systems (EMS) auditing. In the future, training organizations may want to certify auditors that meet these EMS auditing guidelines, as well as other types of audits (*e.g.*, compliance audits) that the guidelines currently do not address.

An important development that might affect training organizations in the U.S. is the recent initiative of the Environmental Auditors Registration Association (EARA), a United Kingdom-based "independent, non-profit making organisation representing the interests of both the providers and recipients of services relating to environmental auditing and environmental management systems."[3] The EARA has both an auditor certification and a training course accreditation program, under which it is now licensing groups in other countries to conduct auditor certifications and training program accreditations using the EARA model. The first licensee is the Asian Pacific Institute for Environmental Assessment. The members include Singapore, Malaysia, the Philippines, and Borneo. As of early 1995, the EARA has had minimal impact in the U.S., although the EARA has registered environmental auditors who are based in the U.S. The EARA's impact could expand if the U.S. does not develop some form of national response to the ISO 14000 guidelines.

## ISO 14012 GUIDELINES

The ISO 14012 guidelines provide guidance on qualification criteria for environmental auditors and lead environmental auditors, including education, experience, and training criteria. Among the criteria, the guidelines require auditors to complete formal and on-the-job training in order to carry out environmental audits.

---

[3] Environmental Auditors Registration Association brochure, undated.

The ISO 14012 guidelines require formal training to address: environmental science and technology; technical and environmental aspects of facility operations; relevant requirements of environmental laws, regulations, and related documents; environmental management systems and standards against which audits may be performed; and audit procedures, processes, and techniques.

In addition, the current draft of the guidelines calls for on-the-job training, including a minimum of twenty workdays of auditing and four audits which occur within a period of not more than three consecutive years. This training should include involvement in the entire audit process under the supervision and guidance of a lead auditor.

Like most ISO 14000 standards, there has been an attempt in ISO 14012 to achieve harmonization with the in-place ISO 9000 standards. Some of the consequences of this philosophy are discussed below.

## U.S. IMPLEMENTATION OF THE ISO 9000 STANDARDS: A STANDING PRECEDENT

The ISO 9000 registration process has been in effect for some time now in the U.S. Since many of ISO 14000's standards are being modeled after the ISO 9000 quality standards, it would be helpful to understand how that process works in this country. As shown in Figure 6.1, the Registrar Accreditation Board (RAB) has taken the leadership position in the U.S. ISO 9000 registration, accreditation and certification process. RAB, a nonprofit organization, is an affiliate of the American Society of Quality Control (ASQC), which formed RAB in 1989 as a separate, self-supporting organization.

The RAB performs several functions. "RAB accredits third-party organizations, known as registrars, using criteria based on internationally recognized standards and guides. Registrars, in turn, audit and register suppliers also using international guides and standards. The ultimate intent of RAB's accreditation process is to assure purchasers that their suppliers have implemented proper quality systems as defined by the ISO 9000 standards. To support the registrar accreditation program, RAB operates the U.S. program for quality systems auditor certification and an accreditation program for auditor training courses."[4]

Many of the terms used in the ISO 9000 process are quite similar and easily confused. The highlights of the process are as follows:

- The RAB accredits registrars
- Registrars register companies (*i.e.*, suppliers) to ISO 9000
- The RAB accredits training programs
- Accredited training program providers train auditors
- The RAB certifies auditors.

A similar, but probably not identical, approach for registration, accreditation, and certification will be developed under the ISO 14000 standards. With respect to the impact on environmental audit training programs, in particular, note that under ISO 9000, trainers do not certify auditors, but the RAB does. This is partly because the qualification criteria include an evaluation against education, experience, training, and personal attributes and skills, not simply a person's ability to successfully complete a training program.

---

[4] The source for both the text and the figure is an undated document entitled *Registrar Accreditation Board (RAB) Overview*, provided by the RAB.

## Figure 6.1
## Accreditation and Registration Process

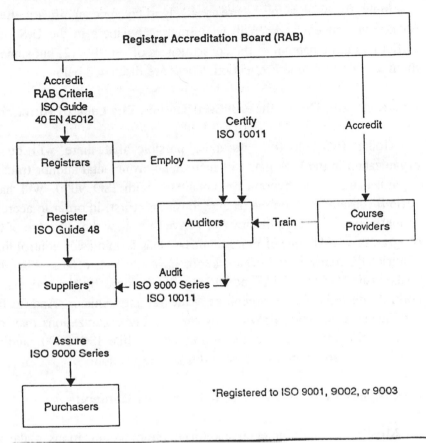

# POTENTIAL EFFECTS

In this dynamic setting, there are likely to be some significant effects on the current environmental audit training business in the U.S. It is difficult to anticipate all of the consequences of ISO 14012, but some of them are apparent upon reflection. These are discussed below.

## Accreditation, Oversight and Certification: The Lack of Autonomy

Under ISO 14000, it is quite possible that there will be an organization in the U.S. that will accredit environmental auditor training organizations. This hypothesis, consistent with ISO 9000, will have various consequences on the training business. First, in order to accredit trainers, oversight organizations will have to review the structure of the programs and audit one or more seminars. This is an outside control that, historically, many of the organizations have not had to face. Also, the costs associated with these activities will be borne by the trainers. If nothing else, this will raise the price of the training seminars. And finally, certifications currently provided by the training organizations may not have significant impact on the profession if, like ISO 9000, auditor certifications are made by the accrediting organization.

## Length and Type of Courses: A Variety of Demands

Most U.S. environmental auditor training organizations make no distinction among the various levels of experience that trainees may have. However, under ISO 9000, there are different training requirements for associate auditors, auditors, and lead auditors. For example, the RAB grants accreditation for two types of courses: a thirty-six hour course meets the requirements for its auditor certification program, while a sixteen hour course, along with an ASQC Certified Quality Auditor (CQA) certificate, also meets those requirements.

Similarly, ISO 14012 guidelines classify individuals as auditors or lead auditors, depending upon their education, experience, and training. Thus it is likely that environmental auditor trainers will have to acknowledge the varying degrees of experience and qualifications of their customers and design programs accordingly, especially in the first few years of ISO 14000 implementation, when many individuals will wish to be "grandfathered" with minimal training.

Both ISO 9000 and ISO 14000 also call for refresher training in order to assure that auditors maintain their skills and currency with evolving audit methodology improvements. This means that training program providers should provide refresher training programs. EARA recommends that this training accomplish the following:

- Ensure that auditors' knowledge of environmental management systems standards and requirements is current
- Ensure that auditors' knowledge of environmental laws and regulations, auditing processes, procedures, and techniques is current
- Ensure that auditors' experience in executing an audit is current.[5]

Meeting these requirements can also be accomplished through on-the-job training, but assurances to the marketplace will be less clear under that scenario.

### Class Size: Too Small to be Economical?

Under ISO 9000, the RAB has issued the guidance document entitled "Requirements for Accreditation of an Auditor Training Course (Rev. 1-930421)." These requirements apply to ISO 9000 training providers in the U.S. The RAB requires that the number of students does not exceed twenty and that there must be at least two instructors for each course. However, there is a possibility of equivalent requirements existing under ISO 14000, regardless of who accredits trainers. This class-size constraint will affect the economics of many current programs, which have no

---

[5] Bacon, Ruth A., "The Environmental Auditors Registration Association (EARA) Scheme," prepared for the ISO SCRAG meeting on September 27, 1994, Baltimore, MD.

limitations on class size and, in fact, achieve acceptable profitability *only* when class sizes *contain more than twenty students*.

One should also note that, at the same time course sizes may be limited, course costs will be increasing due to accreditation fees. For example, U.S. ISO 9000 course providers wanting to receive accreditation from RAB are expected to pay between $10,000 to $15,000 in one-time application fees and up-front program audits and evaluations, as well as $4,000 to $6,000 in annual fees. Similarly, EARA, under ISO 14000, requires an application fee of about $1,500 and annual fees of about $15 per student, or $300 for each twenty-person program.

## Course Content: The Shift to Management Systems

In the past few years, most audit programs in major companies have changed their focus towards evaluating management systems and away from more conventional detailed compliance reviews. In actual practice, however, this re-emphasis has been difficult to achieve fully. The reasons are varied. First, as evidenced by Figure 6.2, evaluating management systems requires a "paradigm shift" that many auditors have found difficult to achieve. It implies that the auditor look at issues almost at right angles from what he or she is used to in conducting compliance audits. Second, many auditors are scientists and engineers, and they find it difficult to conduct an evaluation that they feel is more in line with what an MBA might do in an organization study. And third, conducting management systems audits means that one is attempting to identify root causes: asking the why's, not just the what's. This type of analysis takes substantially more time and typically is not as "clean" as a compliance audit.

ISO 14012 does, in fact, require training programs to formally address management systems audits; therefore, audit training programs in the future will have to emphasize the techniques used in management systems evaluations. This may be as difficult for the instructors as it is for the students.

**Figure 6.2**
**Evaluating Management Shifts**

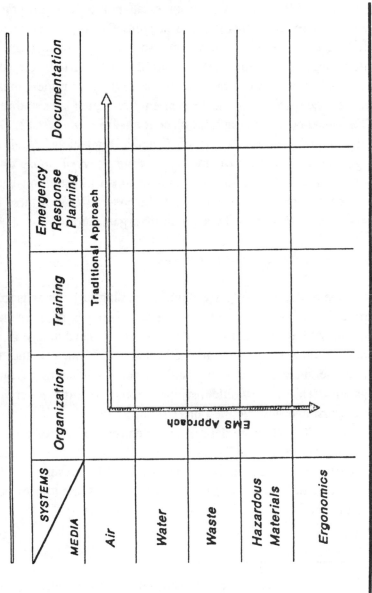

## Course Content: Teaching Relevant Regulatory Requirements

As stated above, ISO 14012 requires that a training program address relevant requirements of environmental laws, regulations, and related documents. Most current environmental audit training programs address regulatory requirements in some fashion. This can be an overwhelming challenge in the U.S., however, so the U.S. delegation to ISO is attempting to modify the current draft of ISO 14012 to better acknowledge this challenge. For example, does this regulatory review mean that the trainer needs to discuss all federal environmental laws and regulations? This would be a daunting task, since there are now over 12,000 pages in Title 40 of the U.S. Code of Federal Regulations. And what state regulations should be covered—none, a sampling, or those from all fifty states? Also, if the auditors are going to conduct international audits, how are regulations in other countries to be covered? All of these issues will need to be resolved by the accrediting organization.

## Course Content: Field Exercises

One of the most effective auditor training techniques is to conduct a "mock audit" as part of a program. This, of course, requires that the class be able to visit a location that might have a regulated unit (*e.g.*, hazardous waste accumulation point) and audit that unit. Typically, the site will set up some situations of noncompliance or breakdowns of management systems. The class would review (altered) records, interview (actor) operators, and observe the (altered) unit.

The ISO 14012 guidelines are silent on the use of mock audits in training. However, EARA Foundation accredited courses "must include at least one practical exercise/site visit. ...[The presumption is that] the training organization running the courses will be able to provide or secure with the cooperation of a local company, a site visit."[6]

---

[6] Environmental Auditors Registration Association, undated brochures.

Responding to this requirement or suggestion will be easier for some training organizations than for others. The commercial programs held in hotels will find it difficult to do anything but conduct simulations of the mock audit using slide photographs. On the other hand, industrial on-site programs typically incorporate this technique already.

## Course Examinations: Evaluating Personal Attributes

Under ISO 14000, examinations will be expected at the conclusion of the training courses. If expectations are comparable to ISO 9000, the exams will last two hours. This provision is not currently found in many auditor training courses, but it could easily be incorporated.

However, the ISO 14012 guidelines expect auditors to meet not only technical qualification criteria but "interpersonal attribute" criteria as well. Namely, the ISO 14012 guidelines suggest that auditors should possess attributes and skills that include, but are not limited to:

- Competence in clearly and fluently expressing concepts and ideas, both orally and in writing
- Interpersonal skills conducive to the effective and efficient performance of the audit, such as diplomacy, tact, and the ability to listen
- Ability to maintain independence and objectivity sufficient to permit the accomplishment of auditor responsibilities
- Skills of personal organization necessary for the effective and efficient performance of the audit
- Ability to reach sound judgments based on objective evidence.

It may be that the trainer will not be expected to conduct this "softer" evaluation. Under ISO 9000, it is done by the auditor accreditation organization—the RAB. But if course providers are expected to evaluate participants against these criteria, the situation could become problematic. Many of the criteria are quite subjective, and written exams are not a good measure of an individual's capabilities.

Consequently, the trainers will likely be expected to place participants in role-playing situations to evaluate their interpersonal skills—a difficult task for both trainer and participant. For instance, what if the trainer finds the participant lacking and does not award certification of successful course completion? What then? Will there be an appeals process? There is under ISO 9000.

Additionally, some auditors with less than perfect personal attributes (*e.g.*, they are dogged in their verification efforts but not very personable in how they go about it) can be valued team members on certain types of audits when coupled with the right partner (such as in a good cop, bad cop scenario). These individuals may be unnecessarily dropped from a training course.

### Documentation: New Expectations

In the future, training organizations will have to be considerably more rigorous about maintaining records. The course syllabus, course manuals, results of examinations and the like will have to be available to auditors from the accrediting organization. Yes, the auditor trainers will be audited; this is quite the turn of events.

## CONCLUSIONS

Environmental auditor training in the U.S. will probably change considerably over the next couple of years, due to the finalization of ISO 14012 guidelines. Programs will need to meet conformance criteria to be recognized as accredited trainers under ISO, even though the current version of ISO 14012 only addresses certification and accreditation for conducting and training for environmental management systems audits. In the future, there may be additional requirements set for other types of audits as well. For example, the door remains open for the development of requirements for compliance audits. Training providers should watch the evolving developments closely; there should be excellent opportunities under a national environmental auditor training accreditation program.

# 7

# A REVIEW OF SOME TYPICAL PROGRAMS

---

*"Our plant managers like the reviews;*
*they give them some security."*[1]

---

Although most environmental audit programs exhibit the same basic structure, individual program approaches are usually tailor-made to meet the objectives and constraints of the sponsoring organization. Thus, one can find distinctly different programs in the same basic industry. Over time, one can even find a dramatically changed program in the same company; the changes can occur due to reorganization, mid-course program evaluations, or, in these days of mergers and acquisitions, a major restructuring of business lines.

A survey of over twenty programs indicated that the extent of fine-tuning varies, but is significant in some cases. The survey, which was a combination of literature review and phone interviews, evaluated the programs against the following twelve factors:

- Program overview and scope
  - Program name
  - Start date
  - Purpose
  - Program organization
  - Program scope

---

[1] Callahan, Edward W., Vice President, Environmental Affairs, Allied Corporation, in *Chemical Week*, May 28, 1980.

- Program methodology
  - Audit methodology
  - Reporting findings
  - Follow-up mechanisms

- Program operations
  - Audit staffing
  - Audit duration
  - Number of plants per year audited
  - Frequency of audits.

Summaries of the individual programs can be found attached in the Appendix, but company names are not listed in order to protect the confidential nature of some of the information. Descriptions of the companies' sizes and types of business are also provided. Presented in this chapter is an assessment of the survey for each of the evaluation factors; the highlights are presented in Figure 7.1.

## PROGRAM OVERVIEW AND SCOPE

### 1. Program Name

While the majority of companies use the word *audit* to describe the nature of the program, several companies use surrogate words, such as *assessment*, *surveillance*, and *systems review*. Some of the reasons given for using surrogates are that (1) audits can connote a more rigorous approach (vis-a-vis financial audits) than is typically the case; (2) the word audit (meaning "to examine, verify, or correct accounts, records, or claims") has a "gotcha" flavor to it; and (3) environmental reviews are more than the paperwork exercises that financial audits imply.

### 2. Start Date

Estimating the program start date was difficult because most companies have had informal programs for some time, and the formal programs have had several stages of development. Notwithstanding these constraints, most companies estimate that their program, roughly as it exists today, started anywhere between 1978 and 1988—the beginning of which coincidentally spans the time when SEC required Allied (1977), U.S. Steel (1979), and Occidental (1980) to complete costs-of-compliance audits, and the first set of RCRA hazardous waste regulations was promulgated (1980). Later in the 1980s, programs were re-engineered to take advantage of the lessons learned.

## Figure 7.1
## Summary of the Review of Twenty Environmental Audit Programs

| Evaluation Factor | Most Common Response | Range of Other Responses |
| --- | --- | --- |
| Program Overview and Scope | | |
| Program Name | Audit | Surveillance |
| Start Date | 1985 | 1976–1988 |
| Purpose | Compliance with regulations and company standards | Compliance with good management practices; awareness; technology transfer |
| Organization | Corporate environmental audit department with 2–4 full-time auditors | Most programs placed in corporate with 1–30 staff; some programs delegated to divisions with corporate oversight |

| Program Scope | U.S.; environmental separate from H&S a little more than half the time; multi-media reviews; recent emphasis on management systems and global roll-out | Special issues such as medical and product stewardship sometimes included; some focused, single-media reviews |
|---|---|---|
| **Program Methodology** | | |
| Audit Methodology | Standard pre-audit, on-site, and post audit activities; use of computers for in-field report writing | Surprise audits when needed; infrequent physical sampling; use of spill response drills; community participation on some audits |
| Reporting | Reports due within 30 days; brief exception reports; recommendations included; findings classified by significance; legal review initially and as needed thereafter; periodic summary reports to management | Highly variable; verbal reporting only in unique cases |
| Follow-up | Formal plant response required within 60 days of audit; formal calendar-based quarterly tracking | Informal response in some cases; "red tag" plant shut down procedure in one case; follow-up verification audits conducted |

| Program Operations | | |
|---|---|---|
| Audit Staffing | 3 technical staff | 1–8 staff (some legal) |
| Audit Duration | 3 days | 1–15 days |
| Plants/Year Audited | 20 | 5–80 |
| Frequency of Audits | Every two years; slowly moving to once every three years | 1–5 year cycle |

## 3. Purpose

More often than not, the stated purpose of the program is to attain and maintain compliance with federal, state, and local regulations. Most (but not all) companies also state explicitly the objective of meeting good management practices (GMPs) and corporate policies and procedures. In the past few years, there has been widespread development of corporate environmental standards and guidelines, which developed in part to take the guesswork and judgment out of GMPs. This has helped to improve the rigor of environmental audits, especially in third-world countries where there are few environmental regulations. While other benefits are recognized by most companies (*e.g.*, technology transfer, increased awareness), they are typically seen as the "icing on the cake."

Recently, several companies have re-designed their programs to emphasize reviews of management systems. This focus is an attempt to remedy the underlying causes of noncompliance and is consistent with emerging ISO 14000 environmental management systems standards.

## 4. Program Organization

A certain amount of independence is seen as an essential element of a program, so almost every company has established a corporate audit function that is responsible for the program. This function is usually housed in the corporate environmental (health and safety) department. Other options include a function in the financial department (where the financial auditing function is located) or in the legal department (ostensibly to provide as much protection from report discovery as possible).

The size of a corporate audit staff varies considerably. In some companies, one or two staff provide an oversight function, drawing on a pool of division and plant staff and consultants to conduct the audits. In other cases, a small group (two to three) of corporate staff conduct the audits but with help from division and staff people. Finally, some companies have a corporate staff of as many as thirty whose responsibility it is to run the program and conduct the audits (without assistance). In all cases, however, audit teams are comprised of staff who are organizationally independent of the plant audited.

## 5. Program Scope

So far, most companies have focused on their U.S. facilities, although there is an increasing trend to broaden the program to all facilities worldwide. The Bhopal tragedy has done much to accelerate this trend. One company has developed checklists and protocols in Spanish in its effort to audit South American plants. Several companies have left the implementation of the audit program overseas to regional environmental coordinators; audit teams are selected from trained staff within the particular region. This approach can minimize cultural and language problems. Still other companies conduct all their overseas audits using U.S. staff.

Companies are divided fairly evenly as to whether their program should be strictly environmental or should be broader to include other areas, such as hazard communication and product stewardship. However, those companies that have strictly environmental audit programs usually have companion or separate health and safety audit programs as well.

The great majority of companies do perform multimedia audits (*i.e.*, water, solid and hazardous wastes, PCBs, and pesticides), but as the programs have evolved and facilities have been audited several times, there has been a tendency to reduce the scope of the audit to those media that present the greatest liability at that plant site.

# PROGRAM METHODOLOGY

## 6. Audit Methodology

There has evolved over time a "standard" audit methodology that most companies in this survey used. It involves the three phases shown in Figure 7.2. With only a few exceptions, this approach is used by companies with existing audit programs. Some of the more significant exceptions include:

- Surprise audits—used by some companies where situations warrant them

- Effluent/emissions sampling—some audit teams conduct these independently

- Ranking of facilities based on a scoring system—some companies use this

- Vendor audits—normally the plant's responsibility, but occasionally the audit team will do this as well

- Emergency response drills during an audit—their use is becoming more prevalent

- Community participation on selected audits—used by some companies such as DuPont

Although the use of detailed checklists and procedures is pretty much universal, the format for the checklists is typically unique to the company. The main components of the checklists include the basic protocol, topical outline, detailed audit guide, yes/no questionnaire, open-ended questionnaire, and a scored questionnaire/rating sheet.

How the checklists are used varies as well. Some auditors use them religiously as they review, inspect, and interview. Other auditors, perhaps those with more experience, use them as references at the beginning and end of each day, to assure themselves that they are not overlooking anything. Still others use the checklists as the source for a one-page list (per medium) of important items to review during the audit. Each of these approaches has merit; their use must depend on the skill and preferences of the auditors and the audit team leader.

A final point concerns the use of computers. Many companies are now outfitting their audit teams with laptop computers. The objective is to develop both a debriefing document and a draft report before the audit team leaves the site. The more preformatted the audit report, the easier it is to accomplish. One company goes so far as to have the audit team telecommunicate (via modem) the debriefing document and draft report to the corporate audit staff prior to releasing it to the site management.

**Figure 7.2**
**Standard Environmental Audit Methodology**

| Phase | Steps |
|---|---|
| I. Pre-Audit | - Organize the audit team<br>- Assign responsibilities<br>- Notify the site<br>- Review relevant regulations<br>- Review the last audit, protocols, and pre-audit questionnaire |
| II. Audit | - In-brief site management<br>- Meet with environmental staff<br>- Participate in orientation tour<br>- Review records and management systems<br>- Interview site personnel<br>- Inspect facilities<br>- Brief site daily<br>- Debrief site management |
| III. Post-Audit | - Draft audit report<br>- Respond to reviews<br>- Submit final audit report<br>- Enter report findings into follow-up tracking system<br>- Close-out audit |

## 7. Reporting Findings

Most audit reports are brief (five to twenty pages), but some can be as long as fifty pages. Increasingly, audit teams attempt to draft a preliminary report before leaving the site. When developed, these are used to debrief site management. At a minimum, most companies strive to develop a formal list of findings and prepare them so they can be inserted into the audit report.

There is no clear consensus on whether audit reports should discuss good performance as well as out-of-compliance issues. Some companies believe that good performance should be incorporated because the report, as a stand-alone document, will provide a more accurate picture of total compliance. And these companies are also concerned that the audit reports are legally "discoverable" and could wind up in the hands of the public. The alternative philosophy is that audit reports should not be political, lengthy, or self-congratulatory documents, but should instead be straightforward, factual summaries of compliance items in need of attention. For the most part, audit programs use the exception-report format.

There is also no consensus on whether audit reports should contain recommendations on how to fix the stated problem. Some companies include recommendations because they believe the process and report are "cleaner" that way; for every identified problem, there is a solution. This approach avoids the situation where the audit report is "discovered" without a companion piece stating the company's planned efforts to remedy the problems.

The alternative approach is for no recommendations to be included in the audit report because this responsibility is thought to be that of the plant manager. The underlying philosophy is that the auditors are responsible for simply identifying noncompliance and should not dictate to the plant manager their particular suggestions on how to resolve the problem. One company resolved the "recommendations" dilemma by having the auditors state only their findings in the report but leaving room (after each finding, for inclusion at a later date) for the plant manager's proposed solution. Hence, auditors do not dictate to the plant staff, but each audit report is a complete document with recommendations.

Finally, the companies are split once again on whether proforma legal review of draft audit reports is required. Some companies have their legal department review all audit reports, while others have legal review only when the audit team leader deems it necessary.

## 8. Follow-Up Mechanisms

Detailed follow-up systems have evolved rapidly and are discussed elsewhere in this book. Most companies, at a minimum, require a formal response from the plant within thirty to sixty days. This response would consist of a detailed action plan for remedying all identified problems. The trend presently is to load all action plans onto a computer system that generates monthly "tickler" reports, which in turn report on the status of the action plan items. One company goes so far as to "score" each plant on its action item completion rate and submit those scores to senior management. One company has a "red tag" procedure, whereby a unit can be shut down if follow-up is deemed non-responsive. Although used sparingly, this procedure has been used successfully in the past. Presently, the mere threat of its use seems sufficient leverage to obtain adequate responses.

# PROGRAM OPERATIONS

## 9. Audit Staffing

Audits are viewed by most companies as principally a technical exercise with regulatory overtones. Thus audit teams consist almost exclusively of experienced technical staff (*e.g.*, engineers, scientists, environmental professionals). In some instances, lawyers are part of the audit team, but this is more the exception than the rule. On average, two to three staff make up the audit team, but the range includes anywhere from one to eight, depending upon the audit objectives and the size and complexity of the facility.

## 10. Audit Duration

On average, audits will take from three to four days, but the range includes anywhere from one day to three weeks; again, this depends upon the audit objectives and the size and complexity of the facility.

## 11. Number of Plants Per Year Audited

The number of plants per year that are audited varies considerably. One company may audit as few as six facilities, but these audits include checks on hazard communication and product stewardship. On the other hand, another company may audit as many as seventy-five facilities per year, but this company does strictly environmental audits and has ten full-time corporate audit staff members. Of course, the number of facilities audited annually also has a lot to do with the size of the organization, so it is difficult to create a measure of appropriateness.

## 12. Frequency of Audits

Generally, the goal for companies is to audit major production facilities once every two years and to audit minor facilities (*e.g.*, warehouses, distribution centers) once every three to five years. However, there is a trend in companies with mature audit programs to extend the frequency of audits from once every two years to once every three years.

# CONCLUSION

It is clear that within the overall concept of environmental auditing, each company must design a program to meet its own identified needs and objectives. This program must be structured so that it is consistent with available resources. This "simple" exercise of matching objectives with available resources is, in large part, the reason for the variety of programs encountered today.

# 8

# BENCHMARKING ENVIRONMENTAL AUDIT PROGRAMS: BEST PRACTICES AND BIGGEST CHALLENGES[1]

A few years ago, I was leading an audit of a coal mine in Wyoming. In preparing for the upcoming closing conference with the president of the company, I mentioned to their environmental, health and safety (EH&S) director that several programs were operating quite well and, therefore, we would have no findings in those areas. I said casually, "If it ain't broke, don't fix it." I thought the EH&S director would have a coronary on the spot. It seemed that the president was a big supporter of the then new concept of Total Quality Management (TQM) and had lectured his staff incessantly on the need for continuous improvement. My director friend suggested that I would be thrown out on my ear if I made the same casual comment in the closing conference. So went my first exposure to TQM.

Since that time, TQM has become an important concept in learning how to manage environmental, health and safety audit programs more effectively. An especially useful TQM tool has been competitive benchmarking. Companies are using benchmarking studies to identify "best practices" that could be incorporated into their programs. In conducting benchmarking studies, evaluators often also identify the biggest common challenges facing audit program managers.

This chapter discusses these best practices and biggest challenges associated with environmental audit programs. The conclusions are based on a number of benchmarking studies and third-party evaluations of

---

[1] Originally published in the Summer 1994 Issue of *Total Quality Environmental Management*, Executive Enterprises, Inc., New York, NY.

corporate audit programs. As the sources are necessarily limited to the author's own experiences, there are no doubt many other best practices, in particular, that are not discussed in this chapter. These will surface over time.

## OVERVIEW OF BENCHMARKING

The concept of competitive benchmarking has received considerable attention in the literature of the past few years.[2] As defined by the Xerox Corporation in its Leadership Through Quality Program, *competitive benchmarking* is "the continuous process of measuring company products, services and practices against the toughest competitors or those companies recognized as leaders." The steps used in a benchmarking study are relatively straightforward and are shown in Figure 8.1. It is not the intent of this chapter to provide a discourse on each of the steps involved in the benchmarking process; that is better handled through other sources. However, based on previous environmental audit program benchmarking experiences, five steps were found to be key when conducting a successful study. Those five steps are discussed below.

### 1. Precisely Define the Scope

One can benchmark any and all components of an audit program. For example, a chemical company that was in the midst of a reorganization wished to determine the best reporting relationship for its audit program. Thus their goal in a benchmarking study was to determine just that: among a dozen targeted companies, to whom did the corporate audit program manager typically report?

More broad-based studies can help as well. However, there are many components to an audit program, and it is probably best to define and analyze only those that are most crucial (*e.g.*, use of attorney-client privilege, frequency of audits, follow-up systems).

---

[2] For example, see: Leibfried, K.H.J. and C.J. McNair, *Benchmarking: A Tool for Continuous Improvement,* The Coopers & Lybrand Performance Solutions Series, HarperCollins Publishers, Inc., New York, NY, 1992.

**Figure 8.1**
**Total Quality Management—**
**The Benchmarking Process**

*Research, Interviews, Observation, Verification & Analysis*

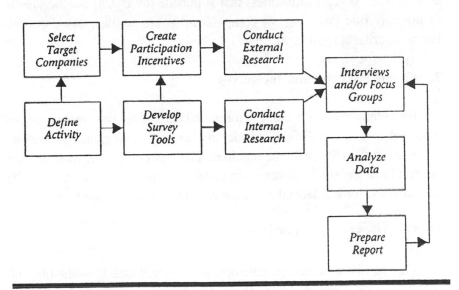

## 2. Select Target Companies Using a Variety of Techniques

One can select among companies in similar businesses, industry in general, or companies with known "best-in-class" audit programs. Any of these approaches would suffice, depending on the objectives of the study. An example of one type of technique that can be used is shown in Figure 8.1. In this figure, fourteen companies are evaluated against one another using three criteria: company size, return on equity, and the percent of the company's sales that are in service businesses. Other criteria can be used as well (*e.g.*, percent of business that is outside the U.S.), but the idea is to possibly find groupings of companies that are similar, based on key financial criteria against which to benchmark.

## 3. Create Participation Incentives

Benchmarking has become a common business analysis technique, so quite often it is not difficult to identify willing participants. However, some incentive will usually be required. This can be a report summarizing the results of the study, although in order for a participant to receive the full analysis, some financial participation in the study is usually expected.

## 4. Develop Measurable Criteria

This can be a difficult challenge, especially if one is evaluating the "softer" components of a program. However, developing measurable criteria in advance is crucial to comparing results from disparate companies. These criteria might include the following:

- Frequency of audits for major facilities
- Type of report, if any, left with site staff at the close of the field audit
- Draft and final audit report types (*e.g.*, exception reporting only) and schedules
- Frequency and type of follow-up system
- Use of legal protections
- Frequency and type of reporting to management

- Organizational levels between the audit program manager and the chief executive officer
- Budget for audit program per unit of company sales.

Each of the above criteria is generally quite measurable and requires short cryptic responses. The evaluator can then make broad conclusions based on the results (*e.g.*, fifty percent of the benchmarked companies use attorney-client privilege protections to protect their audit reports).

### 5. Utilize Focus-Group Sessions

Bringing together the participants of a benchmarking study for a day can be an extremely useful exercise. It can help to assure that individuals are not discussing apples and oranges when addressing complex program issues. The technique can also help to identify subtle nuances in programs that otherwise might not surface during a one-on-one interview. If participants have difficulty assembling in one location, teleconferencing can be a suitable substitute.

## BEST PRACTICES

In benchmarking audit programs over the past few years, a number of "best practices" have surfaced. Not all of these might be applied effectively to a given audit program, but using them selectively should result in an improved program that meets the ever-increasing expectations of stakeholders. This section describes a number of those practices.

### 1. Reports to Management

Reporting health and safety statistics to executive management has been a common practice in U.S. industry for many years. More recently, overall environmental, health and safety performance is being reported not only to executives, but to the public as well. Annual environmental reports are becoming commonplace among progressive companies.[3] And within these reports, environmental audit programs are occasionally, but increasingly, addressed. For example, DuPont in its 1992 *Corporate Environmentalism Report*[4] provided the executive summary, along with DuPont's response, from a third-party evaluation of the company's corporate environmental audit program.

For environmental audit programs to be successful, some form of reporting to the company's executives is paramount. In a few companies, like Hoechst Celanese, the chief executive has shown enough interest that he reviews every report, and calls are made to line managers where it is perceived that individual issues are not being resolved quickly enough. This demonstrates the top management commitment that the U.S. Environmental Protection Agency calls for in its Environmental Auditing Policy Statement.[5] More commonly, senior management might receive a quarterly or semi-annual briefing on issues identified by the audits. These issues might be noncompliance problems that cut across the corporation and are in need of a systemic remedy; statistics among business units on timely resolution of findings; or instances where the audit program saved the company money through, for example, the avoidance of fines or substitution of less toxic materials.

## 2.  Relationship to Compensation

In order to assure support for an environmental audit program, many companies will factor audit results into the bonus equation used for plant and/or environmental managers. This usually gets the attention of the

---

[3] For example, see: Deloitte Touche Tohmatsu International, The International Institute for Sustainable Development, and Sustainability, *Coming Clean: Corporate Environmental Reporting*, 1993.

[4] DuPont External Affairs, *Corporate Environmentalism: 1992 Progress Report*, H-44712, December, 1992, Wilmington, DE.

[5] U.S. Environmental Protection Agency, *Environmental Auditing Policy Statement*, FR25003, Vol. 51, No. 131, July 9, 1986.

individuals principally responsible for remedying problems; however, some caveats are in order.

First, managers should be held more accountable for fixing identified problems than for the results of the first audit at a given facility. This first audit usually sets the baseline and should not be a pejorative exercise. How well the site staff respond after the audit is really what should be evaluated.

Second, one has to be careful of getting into the trap of numerically scoring the results of an audit in order to apportion compensation. For one thing, it is difficult to compare facilities and their audit performance. There are many factors beyond the plant manager's control, such as the regulatory stringency of the state in which the plant is located; the type, age, and size of the facility; the toxicity of the materials used; and the nature of the property (*e.g.*, the presence of wetlands). Also, using a scoring system tied to compensation can heighten the tension of an audit because the plant staff will typically be more argumentative about the number and significance of findings.

## 3. Use of Spill Drills

One of the most important EH&S compliance areas to audit is emergency response. A good response program can save both money and lives. Moreover, emergency response is now an integral component of regulations, promulgated under a variety of statutes:

- Spill Prevention Control and Countermeasure Plans under the Clean Water Act (CWA)
- Contingency Plans under the Resource Conservation and Recovery Act (RCRA)
- Hazardous Waste Operations Emergency Response Planning (HAZWOPER) under the Occupational Safety and Health Act (OSHA)
- Hazard Identification and Release Reporting under the Emergency Planning and Community Right-to-Know Act (EPCRA)

■ Risk Management Planning under the Clean Air Act (CAA) Amendments.

This has made developing an effective emergency response program a complex exercise. The more progressive audit programs will not only verify that the emergency response program meets regulatory requirements but, more importantly, that the program will also work when needed. To provide these assurances, firms like Safety Kleen actually conduct "spill drills" during their audits, and many of these take place in the early hours of the morning. As a suitable but less direct alternative, some audit programs will review the results of spill drills conducted during the year. Either way, emergency response programs should include drills, and audit programs should assess the results of those drills.

## 4. Red Tag Shutdowns

This was quite an interesting approach used by a west coast aerospace company. Essentially, their auditors carried with them red equipment tags that could be attached to an individual unit to shut the operation down if observed non-compliance issues seemed to warrant it. Obviously, the technique proved to be very powerful—it was used once and only once. Subsequently, the mere threat of its use provided sufficient leverage to obtain adequate responses.

## 5. Community Participation

In 1992, DuPont initiated a demonstration program in which a community member participates in a select number of audits each year. This program appears to be unique among U.S. corporations. The rationale is as follows: "As one element of our efforts to understand and respond to community concerns, drive improved environmental performance, build trust, and continue to achieve this consent, we will consider community participation in each corporate environmental review."[6] Selected audit team members must come from the local Community Advisory Panel (CAP), and their participation does not typically extend to employee interviews because such participation "might inhibit the free flow of information essential for an environmental review and compromise employee rights."[7] The demonstration program continued through 1994.

## 6. Next Site Participation

One way to reduce the anxiety of being audited is to have a site environmental manager participate in an audit, as a team member, directly prior to the audit of his or her site. Some companies use this technique to help the individual better understand the process and prepare effectively for the audit of his or her site.

## 7. Use of Portable Computers

Use of portable computers is becoming almost mandatory for audits. Computers are used in a variety of ways. Some companies, such as BFGoodrich, use commercially available automated checklists in a Windows® environment to directly insert findings generated from a review of the checklists into a report skeleton written in WordPerfect®. The checklists also contain regulatory digests so that the multi-volume Code of Federal Regulations does not have to be carried to every audit. In general, computers are used for three purposes on audits: (1) access to an on-line, CD ROM or floppy-disk-based regulatory database, (2) generation of reports prior to leaving the site, and (3) use of automated checklists in the field. The ability to leave a draft report with the site, in particular, is especially valuable in that it helps to maintain the momentum of the audit. Computers are also helpful for keeping in touch with the home office through electronic mail.

---

[6] *DuPont Environmental Auditing: Corporate Guidelines*, Appendix B, Community Participation in Corporate Environmental Reviews, June 4, 1992.

[7] *Ibid.*

## 8. Assessment of Ancillary Operations

Most audits rightfully focus on the site's line operations. However, the better programs address certain ancillary operations as well. These include

- **Off-site hazardous waste treatment, storage, and disposal (TSD) facilities.** This is an especially important area to review. Audit teams typically are not expected to visit off-site TSD facilities. At a minimum, however, the audit team should check to assure that only corporate-approved facilities are being used by the site, and/or the facilities have been visited at a frequency consistent with corporate guidance.

- **Purchasing.** Purchasing staff should be interviewed to determine if they use any environmental guidelines in procuring materials. For example, one company's policy requires the purchasing of materials in 55-gallon drums on an exception basis only. Other companies have a similar policy for chlorinated solvents. The two relevant audit questions are: Is purchasing required to meet any environmental guidelines, and if so, are these guidelines being followed consistently?

- **Maintenance.** This function is almost always audited; however, the depth of the audit can at times be quite shallow. Key environmental issues to address include the use and disposal of maintenance chemicals. The safety of the maintenance equipment (*e.g.*, grinders, lathes, drill presses) should also be reviewed. An area often overlooked is the storage and application of pesticides.

- **On-site contractors.** Temporary and permanent on-site contractors, such as asbestos removal companies, should be reviewed by the audit team. Their contracts should also be reviewed to verify that appropriate EH&S provisions are included and that they are being followed.

- **Nearby warehouses used by the site.** Off-site owned or leased warehouses can create liabilities for companies and, therefore, should be audited on occasion. If the warehouse is nearby, many audit programs will include it in the site audit.

- **Local publicly owned treatment works (POTWs).** Quite often a site will discharge some or all of its wastewater into a local sewer system. Visiting the local POTW can provide the audit team with insight into the municipality's view of the company's compliance status and the likelihood of forthcoming changes in the effluent standards or surcharges.

Addressing these ancillary areas does add time to the audit. However, many companies believe that this is time well spent.

## 9. Use of Verification Audits

One of the biggest challenges associated with audit programs is assuring that problems are corrected in a timely fashion. Companies typically set up sophisticated databases to track the status of corrective actions. How data are entered into these databases, however, is the real challenge. Typically, sites send quarterly status reports to a central location and the data are entered there. This approach works reasonably well except in the instance where site management's perception of a "fix" diverges from that of the audit team.

One interesting approach, used by Hoechst Celanese, is to conduct verification audits of a select number (*e.g.*, ten to twenty percent) of sites with outstanding corrective actions. The audits would involve only one or two auditors on site for one or two days. Their sole objective is to review the status of completed and outstanding corrective actions from the previous audit report. These audits help to "verify" the accuracy of the quarterly data being submitted and assist in resolving any problems the site might be having with interpretation of a finding. They also help to "keep the responses honest" because sites recognize that there is some possibility each year that they will be the recipient of a verification audit.

## 10. Site-Satisfaction Questionnaire

One way for the audit program manager to assure that the program is achieving its objectives is to have the site staff being audited complete a questionnaire evaluating the audit team's performance. This questionnaire is typically given to the site management at the close of the audit, and it is mailed back to the audit program manager, not the team leader. Survey topics include the competency and reasonableness of the team, the adequacy of the interpersonal skills of the team, the depth of the audit, the perceived value of the audit, and any improvements that could be made.

Companies like Hoechst Celanese keep extensive statistics on the returned questionnaires, which are used to make adjustments to the program. It should be noted, however, that the completed questionnaires must be reviewed very carefully. One has to be careful if the site responses are overly positive. While this might be a good result, it could also mean that the audit team was quite lenient in its dealings with the site management. Therefore, site management should respect the audit team but not necessarily be pleased with the results of the audit.

## 11. Periodic Third-Party Evaluations

Consistent with good TQM practices, many companies are now having outside consultants (*i.e.*, third parties) or their internal audit departments (*i.e.*, second parties) evaluate their environmental audit programs on a periodic basis. The annual or biennial studies help the companies meet their "continuous improvement" objectives and help relieve the increasing pressure by stakeholders to provide assurances that the company is identifying and remedying its EH&S issues.

Third-party evaluations conducted in the U.S. usually include a review of program documentation, including audit reports and corrective action plans, interviews with key program participants, and observation of a select number of audits. Programs typically are compared with the Elements of Effective Environmental Auditing Programs provided as an

Appendix in U.S. EPA's 1986 Auditing Policy Statement,[8] as well as with other companies' programs, and internally developed standards and policies.

In Europe, third-party reviews should be much more prevalent and formalized because the European Community's Eco-Management and Audit Scheme (EMAS) became effective in April 1995. This program requires site environmental statements and audits to be validated through an external environmental verifier. Although the EMAS is voluntary, companies like DuPont have committed to implementing the scheme for all facilities in the European Community.

## 12. Development of a Program Newsletter

There should be no secrets about the objectives, implementation, and operation of an environmental audit program. Most programs have guidance manuals that describe the workings of the audit process. The manuals provide a needed program constancy, should the audit program manager be reassigned. Rohm & Haas took this one step further by developing a program newsletter in 1993. This attractive single-sheet, two-sided document is produced quarterly and includes, among other items, remarks from managers who have been audited, a discussion of exceptional EH&S programs identified on audits, profiles of auditors, upcoming audit training seminars for those who might want to become part of the auditor pool, and fun EH&S facts. The newsletter can do much to advertise the program in a very positive light and to communicate important information, such as the exceptional EH&S programs, which might otherwise go unrecognized.

There are surely other best practices among audit programs that have not been identified in this chapter. If one is interested in identifying other techniques that are being used, these could be identified, surprisingly enough, through a benchmarking study.

---

[8] *Op.Cit.*, EPA Policy Statement, pp. 25008–25010.

# BIGGEST CHALLENGES

Benchmarking studies and third-party evaluations identify not only best practices, but usually the biggest challenges as well. Listed below are a few items that seem to recur time and again as deficiencies in environmental audit programs.

## 1. The Program Manual

Putting together an environmental audit program manual can be an onerous task and quite often, as a result, it is not done formally. Some programs will have no written guidance document whatsoever, while others might have a presentation package describing the program that is used in the opening conferences on audits. The lack of a program manual does not necessarily mean that a given program is not operating well; however, a manual does help document and communicate the program's objectives and procedures. Further, it is difficult to assess the relative successes or failures of a program without some written guidelines against which this evaluation can be made. And consistent with TQM principles, all programs should be evaluated on a periodic basis. Finally, a manual helps to guide the program during the transition that would occur if the program manager were to vacate the position.

## 2. Protocol Updating

The problem with audit checklists or protocols is that they are typically out-of-date once they are reproduced. This poses a problem because auditors who rely heavily on the checklists, and not a fundamental knowledge of the regulations, may not be evaluating a site based on the most recently applicable requirements. Many audit program managers do find it difficult to allocate the resources or the time to update protocols more than once every couple of years. An update, at least annually, is consistent with good audit practices.

Use of commercially available automated checklists[9] can help to avoid the problem of outdated program documents. These checklists are updated automatically by the company providing the product. The initial investment in the checklists may seem high, but this update service can be very cost-effective.

## 3. State Regulatory Review

It is imperative that audit teams independently evaluate a site's compliance against both federal and state requirements. The federal government establishes only a regulatory floor from which state agencies can and do develop more stringent requirements, although these state requirements are not always addressed appropriately on audits. Because many audit checklists and protocols emphasize only the federal requirements, too often the auditors rely on the site EH&S manager's knowledge of the applicable state requirements, and this is clearly not an independent assessment. There are now several regulatory databases available that allow for an independent review of state regulations prior to an audit.[10] For example, Rohm & Haas uses these sources to actually develop a state checklist prior to any audit. State requirements must be addressed and addressed independently.

## 4. Misleading Closing Conferences

The closing conference is one of the most difficult elements of the audit process. There is a strong tendency to "sugar coat" the findings in

---

[9] As of early 1994, there were two principal automated checklist offerings: Audit Master® by Utilicom, Inc., Rochester, NY and CompQuest Pro+® by Semcor, Inc., Mount Laurel, NJ. Each provides semi-annual updates as part of the maintenance contract.

[10] Two on-line regulatory databases that cover both federal and state requirements are Earthlaw by Infodata Systems Inc., Falls Church, VA and the Computer-Aided Environmental Legislative Data System (CELDS) by the U.S. Army Corps of Engineers through the University of Illinois, Champaign, IL.

order to keep from antagonizing site management and to make the meeting go smoothly. All too many closing conferences begin with an extended discourse on how wonderful the site staff are and how well the site is operated. I observed one audit, in particular, where the team leader opened with the statement that the large chemical plant "was found to be in compliance," whereupon the plant manager left the meeting, leaving the EH&S manager to deal with the forty or so findings that the audit team subsequently raised. Audit team leaders must have the fortitude to make sure that plant management gets the right message. If there are significant findings, they should be raised early in the meeting and given their rightful emphasis.

## 5.   Timeliness and Quality of the Report

Considering the emphasis that most companies place on the audit report, one would think that there would be few problems in this area. However, this is not the case. With established programs, there are generally more problems with late reports than anything else. Late reports can destroy the momentum gained by the audit and can create liabilities for the organization. Quality is always an issue with any program; yet, after a few reports have been developed, there are typically adequate models to follow. A good technique that is used by many companies is to develop an audit program writing manual, which provides general guidance and samples of acceptable findings and complete reports. Eastman Kodak has developed one of these manuals and gone one step further. The company has a database containing hundreds of findings, taken from previous reports, that auditors have access to in the field. Finally, any audit training that is conducted as part of the program should stress the written report as much as anything else.

## 6.   Insufficient Follow-up

This is a chronic problem in most any program. Audit reports are developed; corrective action plans follow; and then many systems break down. There are no systematic assurances that findings are being corrected in a timely fashion. This, of course, can create major liabilities for the organization. Development of a sound database, which allows periodic (*e.g.*, quarterly) tracking of corrective action status, can help to alleviate this problem. Verification audits, as practiced by Hoechst Celanese and discussed earlier in this chapter, are another useful technique. One should not wait until the next scheduled audit, which might be three or more years later, to verify completion of corrective actions; although the closure of findings from a previous audit should be a formal part of any audit.

---

*Notes: Any discussion of a Company's individual environmental audit program approaches in this chapter is based on information provided to the public at large through technical papers, presentations, and the like. Any discussion of a commercially available audit product does not imply an endorsement of that product.*

# 9

# ENVIRONMENTAL AUDITOR QUALIFICATIONS: GREAT EXPECTATIONS[1]

There has been much discussion lately in the profession about certifying environmental health and safety auditors based on certain performance criteria. But how does one evaluate for such important measures as "physical stamina"?

This chapter discusses the qualifications that environmental auditors should have if they are to lead or conduct a quality audit. It becomes evident after reviewing these expectations that *certifying* auditors through a traditional written examination might ignore some of the most important skills an auditor should possess. Before certifying organizations progress too far down the "tight restrictions," written-exam road, they should determine how these other skills are to be assessed. For instance, no amount of "book learning" can overcome the drawbacks of a poor or combative interviewer, or even of an exhausted auditor. On the other hand, some years ago an exceptional auditor with a masters degree in regional planning could not become a registered environmental assessor in California because she did not have a degree in engineering, science or law. Thankfully, that restriction has since been relaxed, but these kinds of arbitrary restrictions are not appropriate. Environmental auditing is a demanding profession that requires a whole range of skills. This chapter will explore those skills and attributes that are necessary for good auditing.

---

[1] Excerpted from an article originally published in the Spring 1994 Issue of *Total Quality Environmental Management*, Executive Enterprises, Inc., NY, NY.

## CORE SKILLS

There are certain "core" skills that every auditor should have. Some of these can be easily learned; others are inherent in an individual's makeup.

### Learned Skills

First, the learned skills—a working knowledge of the regulations, a familiarity with the facilities being audited and, in today's world, computer literacy.

Auditors should have a working knowledge of the applicable regulations. This knowledge can certainly be acquired through courses, textbooks and the review and application of audit checklists or protocols. This last way of learning about regulations brings up an important point. Regulations do not have to be, and more importantly, cannot be committed to memory. As of 1992, there were *over 11,000 pages* of federal environmental regulations in Title 40 of the *Code of Federal Regulations* and another *3,000 pages* of health and safety regulations in Title 29. Therefore, the term "working knowledge" is just that. Auditors should be familiar enough with the regulations to be able to use a protocol effectively. And the protocols need to be detailed enough to support the auditor's efforts.

There is a secondary issue related to regulatory knowledge. Auditors too often limit themselves to their very defined area of expertise. It is important that auditors also become comfortable reviewing areas *related* to those in which they are expert. This makes them more valuable and flexible as team members and allows them to "pinch hit" should one of the team members become indisposed during an audit.

The auditor should also be familiar with the facilities being audited and should be comfortable in a plant setting. This does not mean, however, that to be effective every auditor of a chemical plant must be a chemical engineer with twenty years of plant experience. In fact, a good, solid, smart professional with an unabashed, natural curiosity and some familiarity with the facility being audited is probably your best candidate to do a quality audit. Further, there is no one particular educational background that is best suited for auditing. Engineers, scientists, lawyers, managers and, yes, even financial auditors can bring valuable skills to the table.

More recently, auditors are finding that computer literacy is also a necessary core skill, and for several reasons. First, it has become routinely expected that a working draft report will be prepared on a laptop computer before the team leaves the site. Also, some companies are now using automated checklists and/or protocols loaded onto laptop computers. Further, computerized regulatory databases are providing additional field support to audit teams. And lastly, plant data files (*e.g.*, training records, discharge monitoring reports) are now more frequently automated. Thus, it is presently difficult to get by on an audit without some familiarity with computers.

## Inherent Skills

All of the above skills can be acquired with a little effort and dedication. However, there are other core skills that are more difficult to attain unless there are existing, fundamental "building blocks" within the auditor's personality. The two that come to mind are interpersonal traits and physical stamina.

Good interpersonal skills go a long way in conducting quality audits. In fact, of all the core skills required, these are clearly the most important. Such skills include being able to interview people effectively, having a high degree of curiosity, adjusting smoothly to changes in schedule, responding professionally to challenges made to your verbal and written statements, working well under pressure, and generally keeping a cool head when everybody around you is panicking.

The "bottom line" is that an auditor must be both a good communicator and an excellent listener. This takes sincerity, patience and, at times, a great sense of humor.

While certainly these skills can be learned through experience, many individuals will never attain them to a sufficient level to work effectively as an environmental auditor. Remember that audits, unlike regulatory agency inspections, are meant to be a supportive function and, therefore, individuals who are inherently brusque, volatile, argumentative, and overly egocentric will not be successful.

It may sound ludicrous to place a strong emphasis on physical stamina as an important trait for auditors. However, any of us who have participated in audits know how important this can be. Consider the pace of an audit. Auditors are often traveling on the weekends or late at night. They climb towers and buildings all day which can be exhausting. They are under constant pressure to perform and to assure that they are diligent and thorough in the investigation.

And the most taxing tasks of all are the daily late-night team meetings to discuss observations and findings. When a draft report is to be prepared prior to leaving the site—a general trend these days—the night before the day of the closing conference can become the early hours of the morning if the data-gathering phase has not been completed sufficiently early. Attempting to craft an articulate, accurate, and precise finding at 1:00 am is no simple feat.

## OBSERVATIONS IN THE FIELD

Now, how do these skills and attributes actually contribute to or hinder the quality of an audit? As an audit team leader or program manager, what should you be looking for, or watching out for, in an auditor. Examples of the worst and best behaviors are discussed below.

### The Worst Attributes

These are behaviors that are commonly exhibited but should be avoided.

**Insufficient Records Review.** Too many auditors want to do a field inspection immediately upon beginning an audit. But records must be reviewed first to determine the applicable requirements. This problem is often a result of a lack of familiarity or comfort with the regulations. An auditor should take the time early to determine the requirements.

**Too Much Records Review.** This often occurs when auditors are intimidated by the size or complexity of the site and just don't know where to begin. They will bury themselves in the records and not come up for air. A better approach would be to first get a "windshield" tour of the facility, and then cut off small chunks of the operation and visit appropriate locations in a modular fashion.

**Identification of Symptoms.** A very common problem arises when the auditor becomes too dependent on the checklist: he or she focuses on symptoms instead of causes. Auditors add value when they address underlying causes. They should take some time to think through what really might be happening at the site to cause the identified problems.

**Jumping to Conclusions.** This is the opposite of the problem discussed above. Some auditors have a tendency to draw broad conclusions before all the evidence is in. Auditors can sometimes let their egos get in the way of doing a thorough analysis and verifying their findings. They've seen it all before and they simply *know* what is causing the problem. This results in statements such as "the hazardous waste management system at the site is deficient," which is not a very articulate or helpful finding. Auditors need to take a step back and make sure that there is evidence to support the conclusion.

**Poor Time Management.** There are auditors who never seem to finish on time. This often occurs when they are given more than one compliance area to cover and spend 90 percent of their time on the area where they feel the most comfortable. Make sure that auditors attack compliance areas in parallel, not in series. It is risky to leave one area for the last day of the audit.

**"In My State" Syndrome.** Nothing annoys site staff more than having auditors preach to them about how things are done in their state. Auditors should research the appropriate state regulations for the site, audit the site against those regulations and leave the preaching for Sunday. They should of course tactfully report on what is being done at their site if it might be a helpful suggestion.

**Too Easy/Too Tough.** Auditors must strike a fine balance between being supportive and providing an honest assessment of the site's performance. Over the years, auditing programs do have a tendency to swing back and forth between "good cop/bad cop" scenarios. When the plants scream that the program is becoming punitive, then the "white hats" go on and the reports become so vague that management can't tell if there's a problem or not. There is no easy solution here, nor should there be. Auditors should not, however, pull any punches in the closing conference or report. Site managers need to understand all of the ramifications of their actions.

## The Best Attributes

What is it that makes an individual a good auditor. Communication skills certainly count, but other characteristics contribute as well.

**A Good Even Disposition.** Volatile personalities do not make good auditors. This is a very stressful occupation and requires a level head and an ability to adjust to new people and constant change.

**Flexibility.** Audits never go quite as planned. An auditor needs to be flexible to adjust to changing dates, schedules, situations and the like.

**Natural Curiosity.** Mentioned previously, this is an important trait. Some would say that the better descriptor would be a "healthy skepticism." In either case a natural inquisitiveness is important.

**High Energy Level.** There is no time to relax on an audit. Days typically start at 7:00 am with a breakfast meeting and end at 9:00 pm with a discussion of findings. This can be very taxing, and an auditor has to enter into the process understanding that.

**Poise Under Fire.** Auditors are constantly challenged by site staff during an audit. It is simply the nature of the process. Auditors must be able to handle this with professionalism. This is especially true during the closing conference when those doing the challenging are more likely to be senior staff.

**Under Control Ego.** A healthy ego is probably an asset for an auditor, who must present his findings confidently. But a "know it all" attitude is not helpful.

## MAKING GOOD THINGS HAPPEN

How does one assure that auditors do indeed have the appropriate skills? Most firms accomplish this through a variety of techniques.

First, there are certain individuals that can be eliminated because of their personalities. This does not necessarily mean that they are not good performers, just that they might not be suited to be auditors.

Second, most firms require that auditors attend a formal training program. These programs can be tailored to address those issues most in need of attention. In fact, some companies have both basic and refresher seminars. The best basic programs usually include a "mock audit" of an actual plant so that auditor candidates can get a true feel for the experience. Such programs feature simulated opening and closing conferences, and actual interviews, records reviews and facility inspections. Improving communication skills through role playing and group exercises is usually an essential element of the program.

The refresher programs can focus on historical problems experienced in the field by the audit teams. These problems can be identified using several techniques. For example, audit team leaders can and do critique both the process and the auditors. And they should be confident enough to do this during the audit, as well as after it is complete. On-the-spot *constructive* feedback is one of the best ways to improve performance. Further, as part of a quality assurance program, some companies will have an oversight auditor participate on a select number of audits during the year to evaluate the process. And lastly, site feedback questionnaires are used to identify problems from a "customer" perspective.

## SUMMARY

Attaining and maintaining good auditor skills is a challenging and never-ending task. To be effective, auditors need more than just a knowledge of the regulations. The challenge is to assure that auditors are trained properly and receive continuous feedback. Third-party certifying organizations must assure that auditors they certify have the full arsenal of skills necessary to conduct a quality audit.

# 10

# TRAINING AUDITORS

---

*"An environmental assessor is an individual who, through academic training, occupational experience, and reputation, is qualified to objectively conduct one or more aspects of an environmental assessment."*[1]

---

## THE NEED

Auditor training is an essential element of any audit program. Sending poorly trained staff to conduct audits at operating facilities can create technical, legal, and organizational problems, particularly in cases where companies or institutions are attempting to establish the credibility of a corporate or headquarters audit program. Moreover, some states, such as California, have developed formal training criteria for the voluntary registration of environmental auditors. That is, the criteria assume only trained individuals will be registered.

There is also a continuing groundswell within the environmental auditing profession that some independent certification program should be developed. Although support for this approach is mixed at best, it is likely that it will be a reality once ISO 14012 guidelines are adopted formally.[2] Thus, there are externally driven, as well as inherently logical, reasons for assuring that an organization's auditors are formally trained.

---

[1] California State Bill S.B. 1875 enacting the Environmental Quality Assessment Act of 1986.

[2] *See* the Chapter that discusses how ISO 14000 is likely to affect auditor training programs in the U.S. for further information on this topic.

Fortunately, with a modest investment of effort up front, management can feel comfortable knowing that the auditing staff are at least trained in the fundamentals of manufacturing operations, regulatory requirements, interviewing and inspection techniques, and perhaps most importantly, dealing effectively with plant management.

Choosing the best training approach is a decision that must be made by management for their particular organization. There is no one solution that fits each and every case. However, the environmental manager can go through the following analytical process to define the training needs of the organization. The steps include:[3]

- Diagnosing the situation
  - What is the status quo?
  - What discrepancy exists between the current behavior of the trainees and what is desired?
  - What are the realistic goals for training?
  - What methods are needed to achieve the instructional objectives?

- Developing the instructional plan
  - Title?
  - Trainer(s)?
  - Sponsoring organization?
  - Participants?
  - Dates and times?
  - Places?
  - Overall training objectives?
  - Training plan?
    - Time period
    - Goal
    - Method
    - Materials needed

---

[3] Taken from Friedman, P.G., and E.A. Yarbrough, *Training Strategies from Start to Finish*, Prentice-Hall, Inc., Englewood Cliffs, N.J., 1985.

- Evaluation procedure?

■ Implementing and monitoring
  - Frequency?
  - Updating?
  - Annual evaluations?

The remainder of this chapter attempts to provide some insight into how management might go about answering these programmatic questions. The chapter addresses issues such as who should do the training, who should be trained, where the training should take place, and what techniques should be used. Several suggestions for ensuring that the training is effective are included, along with a training "starter kit."

## SELECTING THE TRAINERS

There are three basic choices involved in selecting the trainers for an audit program. These are:

■ Using experienced in-house auditors to train junior staff.

■ Sending inexperienced staff to publicly available courses on environmental auditing.

■ Hiring consultants to come in and train staff in groups.

The first option can be done formally or it can be provided on the-job. That is, junior staff can be teamed with more senior staff on actual audits and as they gain experience, they can be given additional responsibilities. Using the in-house option has the advantage of giving the auditors company-specific training. However, it can be taxing on the lead auditors. It also does not expose trainees to how other companies handle the variety of issues that auditing presents, and thus may stifle program advances.

There are several publicly available courses that can be used as training vehicles for auditors—in fact, this book is used as the text for one of those courses. The course approach has merit because it exposes auditors to other organizations' programs and to the wide variety of strategic and operational approaches used by these organizations. However, in order to cram a considerable amount of material into one or two days, these courses cannot tailor the program to an individual company's needs, and typically cannot include field exercises as part of the agenda.

A third approach is to contract with consultants experienced in environmental auditing. The consultant can conduct a one- to two-day training seminar that not only highlights the broad issues but is also tailored to meet specific program requirements. The consultant can also take photographs in advance of typical plant situations to add realism to the training. This seminar can be held at a plant site or, in larger companies, during an annual environmental management meeting, and might include several field exercises as well as classroom instruction. As worthwhile as this approach is, it is also probably the most expensive approach and would be beneficial only if there were a sufficient number of auditors to train on an initial and ongoing basis.

## SELECTING THE TRAINEES

Obviously, those staff members who will be conducting the audits should be trained prior to their first audit. In addition, however, refresher training of all auditors should be considered. This ongoing training can be accomplished through conventional courses or through participation in organizations such as the Environmental Auditing Roundtable. Management should also consider including those *to be audited* (*e.g.*, plant managers, environmental coordinators) in the training programs so that they are aware of the objectives of the environmental auditing program.

# SELECTING THE SETTING

With instructional courses and on-the-job training, management has little control over the setting in which the training is conducted. That setting is either in the cities where the courses are offered or in the plants where the audits are conducted. However, when a formal in-house classroom approach is selected, there are some choices:

- At headquarters
- At a plant site
- At a training center.

Each of these options has its advantages and disadvantages. The headquarters location is usually more accessible to most staff than a particular plant site (unless they are coincident). Moreover, there are often periodic meetings of the one kind or another at headquarters locations that could provide a forum for a training program. However, a plant site offers a "real world" setting for incorporating field exercises into the training program (*e.g.*, conducting a mock audit of the hazardous waste storage area). Some companies have regional training centers, and these provide effective locations for a program, particularly if they are close to a plant site. One of the real advantages of a training center, regional or otherwise, is the resources it offers in the way of rooms, audio-visual equipment, and other amenities.

# SELECTING THE TECHNIQUES

A variety of teaching techniques are available for structuring an environmental audit-training program. Those that are most appropriate are outlined below.

## Lecture

Probably the most commonly used technique, the lecture, can be useful in explaining the fundamentals of environmental auditing. Lectures can be supported by visuals to make them more interesting. In particular, the use of photographs/slides to portray typical situations an auditor might find in a plant can be especially informative. The biggest risk with lectures is overusing them; that is, the communication is only one-way. Lectures need to be supported by more interactive techniques in order to make the entire training experience valuable.

## Problem Solving

We have each learned (some of us the hard way) that there is nothing like problem solving to help us better understand the subject at hand. The use of problem solving to better understand environmental auditing is no exception to this rule.

Several examples of problem-solving exercises are included in Appendix G under the category of "auditing techniques." These exercises simulate actual situations an auditor might experience in the field, including situations that might occur while he is conducting audits of wastewater discharges, air emissions, oil spill control, and PCBs. These exercises are not only meant to instruct staff on how to conduct audits of each of these compliance areas, but to present to the trainee the possibility that a variety of unusual circumstances can arise during an audit. This latter lesson is illustrated particularly in the "Dealing with Regulatory Agencies and Handling Sensitive Situations" exercises. Finally, a field exercise is included in the Appendix that involves conducting a mock audit of a hazardous waste storage area. This can be used if the training takes place at a plant or can be incorporated as a homework assignment.

## Role Playing

Role playing can be a valuable and enjoyable supporting technique in any training program. It allows program participants to more directly experience situations that occur during audits.

Two role-playing exercises that can be used to strengthen auditing skills are presented in the Appendix. The first is an exercise in which an audit team leader is asked to debrief a plant manager at the conclusion of an audit. This role playing is free-form; that is, no scripts are used. The auditor is asked to present certain findings to the plant manager, and the manager is free to react in any way he or she deems appropriate.

The second role-playing exercise is a scripted one in which two auditors, each with an entirely different and purposely exaggerated approach, interview a plant engineer on the issue of hazardous waste management.

The training approach for both exercises is to have training volunteers role play through the situations first. Then the performance of the "auditors" is evaluated. The evaluations should be done in a constructive way, and should come from the class participants, with the process facilitated by an experienced trainer.

## Keys to Making It Work

Developing and implementing a successful training program is not easy. Some of the keys to making a success of the program are:

- **Be Prepared.** Adequate preparation is essential. The program planner needs to outline the objectives, develop the agenda, select the trainers and audience, select the methods, and develop the training tools—in advance. Don't be afraid to "dry-run" the materials. If there are field exercises, survey the facilities before the training session. Attend to details; make sure audio-visual equipment is available, that it works, and that spare parts (light bulbs) are available.

- **Make it Fun.** Don't rely too heavily on lectures. Use interactive techniques such as in-class problem solving, role playing, and field exercises.

- **Evaluate the Program.** Always prepare an evaluation sheet to assess the program. Training programs should not be static. Improvements should be made constantly. Feedback is essential.

- **Leave the Trainees With a "To Do" List.** In many training programs, the trainee receives a barrage of materials and can easily become overwhelmed by the volume and complexity of the issues. It's important to sum up the training by focusing on the most important points a trainee should take from the training. Examples could include:

- Develop a checklist for yourself on what you need to accomplish during the pre-audit phase.

- Develop a list of the materials you need to take with you on any audit.

- Spend at least four hours a week reviewing the environmental literature in order to keep up on regulatory trends.

These exercises can help trainees identify the most important techniques to be learned from the training.

# 11

# ENVIRONMENTAL AUDIT TRAINING
# IN THE FAR EAST[1]

With many auditors in the environmental profession doing more international work these days, I thought it might be helpful to describe the highlights of a recent two-and-a-half week experience in Singapore and Japan.

There is a certain heightened anxiety in doing any assignment overseas. And frankly, if you aren't just a bit anxious over teaching different cultures about environmental auditing, particularly about issues such as how to conduct an effective closing conference or prepare an adequate audit report, you just haven't thought the challenge through. It is just not the same as taking a short plane ride in the U.S. to conduct a seminar, knowing that the class shares a common background and that you are only a short trip from home or office. If you have forgotten something on a U.S. assignment, you know you can have it shipped or faxed rather easily. This is not so easy overseas. On the other hand, communications overseas are much easier now than they used to be. For this 1995 trip, a great deal could be accomplished via phone and fax.

You also will want to have certain bulky items, such as course manuals, translated and shipped ahead of time—this should be done far enough in advance so that you can confirm that the materials have reached their destination. Care should be taken in such places as Singapore; there can be heavy censorship. Although the content of the training materials, as such, should not be offensive, customs review of such things as videotapes can hold them up for a considerable time. I was able to send certain non-critical course materials in advance to both Singapore and Japan by an express delivery service and, much to my amazement, they did arrive—but three days after I expected them. This experience was

---

[1] This chapter is an account of Mr. Cahill's experiences as recorded in his travel journal during his trip to the Far East.

quite similar to one I had in the early 1970s when my noise analysis equipment, which was crucial to completing my work, was tied up in French customs for three days.

And lastly, how do you pack for two weeks so that you only need one carry-on bag? It was difficult to do for this trip since Singapore, only eighty miles north of the Equator, is a casual-dress business environment with average daily maximum and minimum temperatures in April of 88° F. and 75° F., respectively, and Japan is a formal-dress business environment with average daily maximum and minimum temperatures of 62° F. and 46° F., respectively, for the part of the country that I was going to visit. Two very different types of clothing were required.

This trip over took thirty hours from Philadelphia to Singapore, via San Francisco and then Tokyo, on three separate jets. Local Singapore time was two hours different from U.S. Eastern Time, so jet lag was brutal. It was five nights before I slept through a complete night.

## FIRST STOP: SINGAPORE

Singapore turned out to be quite a surprise. It is a very rich but small (2.7 million people and 225 square miles) independent city-state just off the southern coast of Malaysia. It is about half the size of Los Angeles with about 80 percent of the population. The population is seventy-six percent Chinese, fifteen percent Malaysian and seven percent Indian. The official business language is English. The currency is the Singaporean dollar, worth about three-quarters of a U.S. dollar.

Traveling around in Singapore presented its challenges. In the great British tradition, they, of course, drive on the left side of the road. It takes a while to get used to sitting in the left front passenger seat with no steering wheel in front of you. Singapore also has a number of interesting laws, such as no spitting and no jay walking.

Communicating with home by phone was incredibly easy with the new direct dial systems, and the quality of the connections made you feel like the person on the other end was sitting in the next room. Water and food were not a problem; the mainly Chinese food is excellent and there are fifty-seven McDonalds in Singapore.

On this trip, I was traveling with a colleague and co-trainer, Darwin Wika of DuPont, who had lived in Singapore for three and a half years, while on assignment. I must admit, as a result of this and the Anglicized environment, I had a relatively easy time adjusting on this leg of the trip.

My assignment while in each of the countries was to conduct a week-long environmental audit training program. The first two days would be classroom training and the last three would be an actual audit of the plant site, aided by the recently trained class. This was a difficult two-fold objective, complicated further by the need to produce a draft, typed audit report by the end of each week.

There were about fifteen people in the Singapore class, who came from Korea, Taiwan, New Zealand and Singapore. All spoke at least some English, and although we did have difficulties with language, the interaction was excellent.

Good interaction overseas is not always the case, even when the class members are supposed to understand English. Even the most fluent are sometimes timid in using a second language in public forums, such as training classes. I am reminded of a training seminar I gave in Europe a few years ago. An attendee raised his hand early in the training and I became excited because I thought we might actually have an interactive session. He said quite seriously: "Mr. Cahill, could you please hurry up and slow down?" When I asked that he clarify his request, I found out that what he meant was I was speaking too quickly for him, so I should slow down my speech, but, at the same time, he wanted me to get through all the material, so could I hurry it up. Upon reflection, I found it quite an interesting and fair request.

The training session in Singapore went quite well, although we did face the usual problems. It seems that the site was undergoing an ISO 9000 audit at the same time as our visit. Although this did constrain us somewhat, it demonstrated to the class that they must learn to be flexible. On the positive side, when given classroom assignments the students attacked the problems with tremendous energy. They had a wonderful sense of humor and were also quite prompt, sometimes arriving as much as a half hour ahead of time. This trait is greatly appreciated by trainers trying to live up to a pre-set agenda.

The mock field audit worked especially well, although things become complicated when just-trained people are unleashed on a site. As a rule, I prefer to have enough trained coaches from the client's staff so that each field team has someone who can guide them through the investigative process. There were three coaches, including myself, in this program. In Singapore this turned out to be enough "seasoned" coaches. In Japan, during the second week, this became more problematic, because these same coaches were Americans and many of the team members were not very fluent in English. Forcing the students to conduct the audit in English constrained the process too much, so the mock audit was conducted mostly in Japanese. This allowed for only an occasional review of progress by the coaches.

For the most part, the course materials worked well. I brought over 300 35mm slides, including many facility photographs portraying typical non-compliance situations. These included pictures of visible emissions, spills, lack of secondary containment, rusting drums, etc. Without exception, the students preferred these images over word slides describing what one should focus on during an environmental audit.

The course manual was also helpful as it contained, among other things, numerous class exercises. These exercises, or case studies, allowed for a break from lecturing for both the instructor and the students and can really add value to the training experience by reinforcing the lecture points. Harvard Business School knew what it was doing when it decided years ago to use the case-study approach as its principal teaching tool. Course evaluations received back from students often say that the best part of the training was the interaction with the other students during the exercises. But these are intelligent people, who, with a little advance work on my part, basically teach themselves through use of the exercises.

We conducted mock opening and closing conferences with the plant manager. These mock conferences worked extremely well in exposing the students to the difficult questions that they might encounter on an audit. The plant manager was fed a number of "loaded" questions, including:

- I won't be around for the closing conference. Is that okay?

- We just got word there's to be an agency inspection this week. Can you postpone the audit?

- How would you rank my facility against the others you've audited?

- If I fix some of the deficiencies today, can they be removed from the report?

- If I fix all of the items you've identified, and later have an agency inspection, can you guarantee we'll get a "clean bill of health"?

This plant manager played his role so well that the students could not tell the planted questions from the real ones. And this did complicate the situation. Recall that the training program had two objectives—to train internal staff and to actually conduct an audit of the site. With some effort, we did work through this issue.

The training ended with the presentation of a draft report to plant management in a closing conference. The three teams were given forms to use in writing and classifying their findings and recommendations. Because the class generally followed directions in completing the form (one form per finding), it was easy to incorporate the findings into a report skeleton that had been prepared in advance. It did make for a long Thursday night for the person responsible for completing and critiquing the report, but the advantages of being done with the draft report before leaving a site half way around the world greatly outweighed the extra effort one had to make.

One interesting and legitimate accountability issue was raised by the plant manager in the closing conference. It seems that the wastewater treatment plant had exceeded its daily flow limits about fifty percent of the time over the previous month. The team presented this as a serious finding. The plant manager's question, then, was should he shut the entire facility down if he had every reason to believe that tomorrow the treatment plant would once again exceed its permit limits? We suggested to the class that the most appropriate response was for the audit team to simply state that, if the plant were allowed to operate under these circumstances, yet another noncompliance incident would occur; it is the plant management's responsibility to decide on the relative risks of the various alternative actions and to make a business decision based on their assessment of these risks. Beyond tomorrow, however, something should be done by plant management in the short term to mitigate, if not eliminate the problem (*e.g.*, reporting of the exceedances, revising the permit, installing a retention pond or tank, examining possible infiltration or cross-connections in the process sewer system).

We finished up the closing conference by discussing the next steps in the process, including finalizing the report and corrective action plans. The plant manager was not elated that we had found some real issues, but that is the nature of the audit process.

## THE TRANSITION: SINGAPORE TO JAPAN

The flight from Singapore to Tokyo was six hours, a relatively short hop compared to the trip from the U.S.

Many of the things that we hear about Japan are indeed true. Sushi is quite popular, so get used to raw fish. One does sit on the floor for formal, traditional dinners. Shoes are removed upon entering a residence or restaurant. Also, it is very impolite for you to refill your own beer or wine glass. This is done by your dinner companion. And, of course, it is your responsibility to keep their glass full as well. Unless you are careful, this can make for a challenging and foggy evening.

The Japanese take business card exchanges very seriously. You should have your business cards ready. The Japanese are able to present theirs to you with the same speed that Wyatt Earp drew his six shooter. And, it *is* a presentation. The cards are presented with two hands with the words facing the recipient. You then read the card, acknowledge the individual, and place the card in a pocket of importance, which is not your pants' hip pocket. You return the favor by presenting your card in a similar fashion. Sliding cards across a meeting table in Japan is quite the insult.

I did notice that the Japanese seemed to be a very forgiving people when an American makes a cultural mistake, and you *will* make some. As long as you make an effort to comply with the customs, and to enjoy them, the Japanese will forgive your mistakes and will, at times, find them quite humorous.

## SECOND STOP: JAPAN

We conducted the training in Japan similar to the way we did in Singapore: two days of lecture (in a local hotel this time) and two and a half days of a mock audit at a site. There was an extra challenge in Japan, however. Because of language problems with some, but not all, of the class, all program materials had to be translated into Japanese. This posed an interesting quandary when using word-slide overhead transparencies. As an instructor, should you put up the English version, which has the same page number as the Japanese version in the class manuals, and is therefore trackable by the class? This method allows the instructor to better recall the issues he needs to address or highlight. Or should you put up the Japanese version, which is better for the class, but can be confusing to the instructor? We opted to use a mixed approach depending on the subject matter.

All of our lectures had to be translated into Japanese as we went along. This was done for each slide, so that there was first the lecturer's discussion and then a translation of that discussion. This made the presentations almost twice as long as normal. It did curtail discussion of some of the audit "war stories" that instructors use to enliven presentations. But, in some ways there were unforeseen advantages in having the translations. The most significant of these was that the one to two minute respite that occurred during the translation allowed the American instructor to think about anything important that was missed and/or to prepare for the next slide. This made for a very organized and thoughtful presentation.

Observing and participating in the translation process was very interesting for many reasons. Two are worth particular mention. First, our instructors became very sympathetic towards the translator, who was a senior staffer for the Asia-Pacific environmental group of the client organization. We, the instructors, would be relaxing about half the time. The translator had to not only do the translating, but had to pay attention during the English presentation so that he could, in fact, translate. Thus, the translator was the only person in the room who had to be focused 100 percent of the time. Our translator was still enthusiastic but very tired after two days.

Second, it turned out that many in the class understood English quite well. We found this out when many of the students would laugh at our funny remarks before the translation even occurred. Once I began to understand this, I would periodically ask one of the students in English if the translations were accurate. He would always tell me the same story. They were mostly accurate but sometimes the translator would add some of his own comments and sometimes he would delete some of my comments—a consequence of having a professional environmental staffer as the translator. Actually, however, this was a plus. Our translator was confident enough to ad lib on occasion because he had attended all of the lectures in Singapore the week before. This was an invaluable preparation step that considerably improved the quality of the Japanese program.

I must say that one of the things that was marvelous and, at times, almost disconcerting, was the punctuality of the Japanese. As the scheduled start time of 9:00 am approached on the first day of the training, all twenty-five students were at their places, with course materials in front of them and hands folded, waiting for us to begin. This was different from the U.S. or Europe, where for most classes you have to call people into the room. Also, in the U.S. you often have to play little mind games with the class to get them to return from breaks on time. One of these, for example, is giving the class a thirteen and a half minute break to reinforce the idea of punctuality. This game playing is totally unnecessary in Japan.

I did have a difficult time remembering students' names in Japan. I do require name tents, filled out with bold magic-markers, for all courses that I teach. This helps me remember the names. Name badges just do not work because they cannot be read from the front of a class. Even with the name tents, however, name recollection was quite a challenge. The Japanese go by their last names followed by the salutation *san*—so I was known as Cahill-san and Darwin was known as Wika-san. A sample roll call of the course attendees was: Yoshimura-san, Watanabe-san, Inuizawa-san, Takekawa-san. While not overly difficult, it was still a challenge for me.

The value of class exercises or case studies was something we were concerned about in Japan. Many of the class exercises that we typically use require considerable interplay between the instructors and the class. In Japan, the two-way translation process would impede a free-flowing conversation. Instead, we redesigned the exercises to incorporate more visual report backs. For example, in the U.S. (and in Singapore) after a lecture on air emissions auditing, I would typically break the class up in to groups of five or six students and ask them in an open-ended way how they would go about doing an air emissions audit at a particular site. A case study would be given to them with specific questions related to this hypothetical site. In Japan, we reworked the questions to be more closed-ended so that the answers could be easily posted on a flip chart, easel, or white board. So for example, instead of the open-ended, free-wheeling approach, the case study might pose specific questions like:

- How many air sources would you review?
- Would you go up on (1) all roofs, (2) some roofs, or (3) no roofs?
- Would you talk directly to an asbestos contractor working at the site?

Each of the four teams would then post the answers on the board, which was initially turned away from the class. Then, once the exercise was over for the teams, the instructor would discuss his evaluation of the results with the entire class. This minimized the amount of needed translation without significantly compromising the value of the case studies. We designed one case study on the first day in this way and took a vote in class on the value of the approach. There was an overwhelmingly positive response, so we spent that evening redesigning the remainder of the case studies. We completed four more in this fashion on the second day.

The mock audit at the site went well, although most of it was conducted in Japanese. Each of the three American coaches joined one of the three audit teams (air, water, or waste). One of the coaches was the Asia-Pacific Safety, Health and Environmental Director for the client organization (a U.S. citizen) so his involvement helped him get to know the Japanese staff better and it was more cost-effective than sending a third person from the U.S. There was at least one student on each team who was bilingual, so we three coaches were able to check progress every now and again.

Although the Japanese were disciplined and organized, it appeared that they were not as probing as a comparable team of U.S. auditors. Americans are suspicious about everything, but in the Japanese culture it is not proper to be suspicious in a public setting. As a result, the verification process was probably not as rigorous as we might expect in this country.

We had an interesting time prioritizing the findings of the mock audit—both the commendable activities and the deficiencies. We had each of the teams list on a white board, in Japanese, their top three or four "good and bad" findings. We had a quick verbal translation and then a vote by the entire group. Each person was allowed to vote three times for his or her choice of priority findings. It was remarkable how quickly this stratified the results, and before we knew it, we had our executive summary.

After the fact gathering and analysis, our goal was to prepare a draft report on my laptop computer, print it out and distribute it during the closing conference on Friday. Again, as in Singapore, we used a prescribed form for each of the findings and the teams were quite responsive in completing it accurately and fully. We had prepared the introduction and background material in advance, so it was simply a matter of inserting the twenty or so findings into the report on Thursday evening. Our translator took the forms filled out in Japanese, translated them "on-the-fly," and inserted a rough English version into the report. I then edited this version for technical correctness and language usage. It turned out that the site had a brand new Macintosh Power PC, which was great because I was using a PowerBook 520. With all the right adapters and consistent software, this part of the process flowed smoothly. In my experience around the world on these training audits, this process does not always work as well, although you can always seem to jury-rig some kind of computer support.

We did however commit one of the cardinal sins of auditing on the mock audit. We allowed the site to schedule the group dinner on Thursday evening, the night on which the audit team normally puts together the draft report or draft set of findings. Adding a three-hour formal dinner onto the schedule for this night makes for one long, long evening. Every site environmental manager should know this.

Our last responsibility was to conduct the closing conference. I introduced the process and summarized the results, being sure to emphasize that we had a dual mission—to train people and to conduct an audit that would be as thorough as possible, given the time constraints. Each of the teams presented their findings in Japanese to the twenty or so site people attending the meeting. We concluded our meeting, ate our last sushi meal and gracefully departed.

# 12

# INFORMATION MANAGEMENT

---

*"Environmental managers are discovering that manual methods of data management are no longer adequate."*[1]

---

Ensuring environmental compliance has become a paperwork nightmare. Companies typically need to generate and track a variety of documents such as permit applications, manifests, monthly and annual operating reports, exception reports, audit action item reports, and the like. In addition, they need to manage the incoming information stream generated by federal, state, and local regulatory agencies in their continuing efforts to promulgate standards, regulations, and guidance.

In response to this ever-increasing flood of paper, corporate environmental auditing staffs are developing in house, or purchasing from vendors, computerized data management systems. Generally, for audit programs these systems fall into four classes:

- Federal, state and international regulatory databases
- Automated audit checklists
- Environmental property information databases
- Program management software.

This chapter addresses the first two classes of systems: regulatory databases and automated checklists. Environmental property information

---

[1] Raybourn, R. and W. Rappaport, "Computerizing Environmental Information: What to Look for in Systems," in *Pollution Engineering,* January, 1985.

databases and program management software are discussed in other chapters of this book. For the two classes presented in this chapter, the discussion focuses on the factors that should be considered when evaluating any given package and/or approach.

Because the environmental software field is so dynamic, this chapter does not evaluate individual products or on-line services. For detailed reviews and comparisons of current products, the reader can refer to environmental software buyers guides that can be found in periodicals such as *The National Environmental Journal* and *Environmental Solutions*.

## REGULATORY DATABASES

Current information on most federal and state regulations is easily accessed on a variety of "user friendly" computerized databases. The data is either provided on-line (*e.g.*, EarthLaw, Lexis/Nexis) or on CD-ROM (*e.g.*, Enflex, BNA Environment Reporter, RegScan, FastRegs). More and more auditors are using these systems, both in preparation for an audit and during the on-site activities. This on-site use means, of course, that the audit team must have access to a computer during the audit and that computer must have either a modem to connect with an on-line service or a CD-ROM reader. The regulatory databases are especially important in identifying applicable state regulations as most audit protocols will address the federal regulations only.

Automated regulatory databases should incorporate the following characteristics:

- **Coverage.** All major federal environmental, health and safety regulations should be a part of the system. This includes regulations on the environment (Title 40 of the Code of Federal Regulations), health and safety (Title 29) and hazardous materials transportation (Title 49). In addition, most advanced products now include the environmental, health and safety regulations for all fifty states. As discussed above, this state coverage is essential. And finally, some of the products also offer international coverage (*e.g.*, the European Union, individual countries). As audit programs go global, this added feature is becoming more important.

- **Currency.** The systems should be relatively current; updating should occur at least every six months, if not more frequently. More frequent updating at the federal or national level should be expected, as in the U.S. the *Federal Register* provides an expedited means of doing so. Updating of individual state regulations is more problematic since many states are not as systematized in codifying their regulations.

- **Ease of Use.** These systems should be, and for the most part are, user friendly. Moreover, they should be friendly to both the casual and power user. The Windows© environment has done much to improve the usability of several of the systems, such as Enflex.

- **Flexibility.** Systems should be designed for single-user applications, local-area networks, wide-area networks and the evolving groupware software. A user should be able to access the database from other software such as Lotus Notes® and Microsoft™ Windows. The provider should be willing to arrange for a corporate license that responds to these varying needs without breaking the corporate bank.

- **Powerful Search Capabilities.** The programs should incorporate some kind of search engine (*e.g.*, Fulcrum Technologies' SearchServer™) that allows for easy and powerful searching in a variety of ways, including searches by key word, acronym, citation number, and subject. The user should be allowed to search across multiple jurisdictions simultaneously.

- **Notation of Changes.** Some of the databases have incorporated a feature that marks text to show the user exactly what has been added, deleted, or amended since the last update. This is especially useful for auditors who need to keep abreast of changing requirements in an efficient manner.

- **On-Line Help.** The database should have on-line help that is easy to access and simple to understand.

- **Speed and More Speed.** With tens of thousands of pages of regulations now on the books, the database should work very quickly in its searches.

- **Ease of Exporting or Printing.** Although this certainly is the computer age, it is still sometimes helpful to be able to print hard copy or otherwise output the regulations that are being researched. The database should have a printer and/or export function that will allow regulations to be printed in a high-quality format (including tables) and/or to be directly inserted into reports.

Although computerized regulatory databases are powerful tools for auditors, sometimes it is still helpful to actually talk to a fellow human being for interpretations of requirements. The U.S. EPA offers several hotlines that can be called for these regulatory interpretations. These include:

| | |
|---|---|
| Asbestos: | (202) 554-1404 |
| FIFRA: | (800) 858-7378 |
| Radon: | (202) 475-9605 |
| RCRA/CERCLA: | (800) 424-9346 |
| SARA: | (800) 535-0202 |
| SDWA: | (800) 426-4791 |
| TSCA: | (202) 554-1404 |

These numbers can be called anonymously. The lines are usually staffed by contractors, not EPA employees and, further, the advice is not the "official voice" of the agency. For particularly complex issues, it is probably a good idea to call the applicable hotline twice and determine whether the interpretations are consistent.

## AUTOMATED CHECKLISTS

Auditors' checklists or protocols are a key component of any audit program. The checklists typically contain hundreds, if not thousands, of questions related to applicable government regulations, company standards and guidelines, and good management practices. Historically, checklists have been fairly large hard-copy documents, making it difficult and costly for companies to both initially develop and then routinely update these checklists. Because of this and other difficulties associated with maintaining audit checklists, several software development companies have developed automated checklists (*e.g.*, Dakota Software's Audit Master™ and Semcor's CompQuest Pro+). These checklists can be loaded onto a portable computer and used as the principal audit protocol tool in the field. The automated checklists can be very useful but do not come without some disadvantages. The pluses and minuses of the checklists are discussed below.

### Advantages

The automated checklists offer a number of key advantages.

- **Regulatory Coverage.** The checklists typically cover the major federal environmental, health and safety regulations that would be of concern to most any organization. In some cases, state requirements are addressed as well.

- **Regulatory Searches.** The automated checklists are generally contained in a linked database that provides the actual regulations that drive the audit requirement. The user can quickly access these regulations by using a "hot key." The user has access to a detailed interpretation of the requirement without having to refer to a separate document, system, or regulatory database.

- **Currency.** One of the real advantages of the automated checklists is that they are kept current by the supplier. Most checklists are updated every three–six months, incorporating new regulatory requirements. Updated computer disks or CD-ROMs are routinely distributed to customers.

- **Consistency.** In providing each auditor with a comprehensive set of audit protocols, the company is likely to achieve more consistent auditing over time.

- **Training.** The checklists can be an invaluable training device. They contain a wealth of information on regulatory requirements, and particular areas of interest can be accessed quickly. The checklists are quite often used as part of site self-assessment programs in which site environmental coordinators select one or more compliance areas to audit every other month or so.

- **Report Generators.** Most of the automated checklists contain report generators, which can automatically incorporate findings through "red-flag" triggers. That is, the audit questions are structured so that certain answers will trigger a finding, and these are imported into the report generator automatically. Some advanced users have been able to use this powerful red-flag tool in a Windows environment and have transferred the findings to a more traditional word processing software (Word for Windows or WordPerfect©).

## Disadvantages

The automated checklists bring with them a number of challenges that must be overcome.

- **State Coverage.** Not all the automated checklists contain the environmental regulatory requirements at the state level. This can be a significant drawback because ensuring adequate state coverage is a key component of any environmental audit. The programs do allow for incorporating additional questions in each module, which can be questions addressing corporate standards and guidelines and/or state requirements. Adding these components to the checklists will increase their value, but updating them will burden the purchaser of the product.

- **Small Site Applicability.** In order to be comprehensive, the automated checklists can contain several thousand questions. This comprehensiveness can be a problem when auditing a small manufacturing site or laboratory. An auditor does not usually have the time to review several thousand questions on a one- or two-day audit of these smaller facilities. As a consequence, the programs often provide for an initial screening of the areas in need of review and also have the ability to follow up with a more focused audit. Nonetheless, auditors sometimes feel more comfortable with a paper checklist so that screening and culling can be accomplished more quickly and effectively.

- **Multiple-User Costs.** Like any software product, the commercial automated checklists have become much more cost-competitive over the past few years. However, if a large corporation adopts one of these packages as a universal audit tool, the costs can become significant because each user requires a separate license. The software companies will work with purchasers of multiple systems to develop a corporate license that can reduce significantly the cost per user.

- **The Occasional User.** Automated checklists work best when auditors use them often enough to become very familiar with their capabilities. The packages are more of a challenge when auditors use them only occasionally. Such occasional usage is very common with audit programs that use part-time auditors from divisions and plants. It is difficult to become facile with a software package when one uses it only once or twice a year as many of these part-time auditors do.

- **Interference with the Audit.** One of the most often-communicated problems associated with the automated checklists is that they interfere with the normal audit process. Sometimes auditors become so engrossed with interacting with and completing the checklists that they do not spend sufficient time interviewing staff, reviewing records, or inspecting the facility.

In sum, automated checklists are definitely worth exploring, especially for programs where self-audits will be expected. There are some risks in adopting this approach, however. Careful planning can do much to mitigate these risks.

## EVALUATING AUDITING SOFTWARE SYSTEMS

Choosing from among the numerous[2] commercially available environmental auditing software systems is not easy. Some systems can be purchased outright for less than $1,000. Others, if they were to be purchased and used company-wide, might cost as much as $100,000 or more for a corporate license. Yet, these more expensive systems do provide for significantly more flexibility and use throughout the corporation.

To select from among the options offered, in general, potential purchasers should first develop program specifications and requirements and then evaluate the offerings against these specifications. For example, a buyer might want an audit program management system to meet some or all of the following requirements. That is, in a perfect world, the system should be able to:

- Be operated on the company's hardware configuration and be accessible from portable computers with a model
- Allow for widespread access to environmental regulatory databases

---

[2] The Annual Environmental Software Buyers Guide in the July/August 1995 issue of the *National Environmental Journal* listed seventeen (17) separate audit software packages and seventy-four (74) separate environmental regulatory databases.

- Load and sort company-designed or commercial federal and state compliance checklists
- Allow easy updating of the checklists
- Load (at times by modem from the field) and search audit reports
- Produce tabular audit-findings reports
- Produce periodic "incomplete action items" reports, across all facilities
- Code and retrieve findings by class (*e.g.*, air emissions), location, year, organization, and severity (*e.g.*, low, medium, and high risk)
- Calculate simple statistics (*e.g.*, number of significant air compliance issues in California) across all audits
- Produce graphics-oriented management reports, using data across all audits
- Provide for selective data security.

Once the above design specifications are further refined, staff need to evaluate any offering against the following criteria:

- Is the program truly pre-existing, or will the buyer be paying for its design or major modifications?
- Can the system readily meet the design specifications?
- Is the system flexible?
- Is it user friendly?
- What type of reports can be generated?
- Are search capabilities broad enough?
- Is documentation complete and clear?
- Can the system handle text (*e.g.*, checklists) as well as numeric data?
- Will it have on-screen help?
- Can data be transferred among modules?
- Are presentation graphics available?
- Is updating/upgrading the program design easy or a major undertaking?

It is possible that an otherwise attractive system may not fulfill all these requirements in its current form. However, the question really is: can the system be modified readily to take into account the needs of the company? If so, many of these software packages can add significantly to the auditor's tool set.

# 13

# USING GROUPWARE TO MANAGE THE EH&S AUDIT PROGRAM DOCUMENTATION PROCESS[1]

## IMPORTANCE OF SOUND DOCUMENT MANAGEMENT

Environmental, health and safety (EH&S) audits are used by the regulated community to help assure continued compliance with laws and regulations, company policies and procedures, and good management practices. One of the biggest challenges for any program is to ensure effective management, control, and distribution of the documents generated by the audit.

Several critically sensitive documents are generated as a result of each audit. These include: (1) draft and final audit reports prepared by the audit team, which document the deficiencies or findings at the site or within the audited organization or system, (2) a plan generated by the audited entity, which defines the measures needed to correct the deficiencies, and (3) a report that continuously tracks the status of all proposed corrective actions for all audits. It is generally the responsibility of the corporate (or headquarters) EH&S audit program manager to monitor and analyze the data and information generated by these audit reports and plans.

One of the more important challenges for an EH&S audit program manager is to provide documented assurances to the corporation that the sites' corrective actions are being completed on time and consistent with the findings of the audit teams. It has been stated many times by attorneys and regulators that one is better off not implementing an audit program if

---

[1] Originally published in the Summer 1995 *Total Quality Environmental Management*, John Wiley & Sons, Inc., New York, NY.

the corporation cannot assure that identified deficiencies are corrected in a timely fashion. In sum, organizations need to reach *verifiable* closure on all audit findings.

A second critical documentation issue for the audit program manager is determining whether audit reports and documents can be protected from discovery in a legal proceeding and, if not, deciding what should be done to minimize liabilities to the corporation posed by a potential release of these sensitive documents. In 1986, the U.S. Environmental Protection Agency stated only that they "will not routinely request audit reports."[2] In the meantime, at least "four states—Colorado, Kentucky, Indiana, and Oregon—have passed laws enacting environmental audit privileges of varying breadth and depth. Other states have similar legislation pending."[3] These state-level initiatives providing protections for audit reports have not been met with open arms by the EPA. As a consequence, the EPA has been holding public hearings to determine what might be the best approach for the Agency in dealing with audit report privilege protections. As of early 1995, what the Agency will conclude is anybody's guess.

These and other documentation issues demand that audit program managers treat records management, retention, and destruction very seriously. Historically, this has been an arduous task since most programs principally used paper filing systems. In the past few years, automation has made some breakthroughs, particularly in the area of tracking audit reports and the status of corrective action plans. New software technology—that is, groupware—if used properly can revolutionize the management of audit program documentation.

---

[2] U.S. Environmental Protection Agency, "Environmental Auditing Policy Statement; Notice," *Federal Register,* Vol. 51, No. 131, July 9, 1986.

[3] Ronald, David, "The Case Against an Environmental Audit Privilege," *National Environmental Enforcement Journal,* National Association of Attorneys General, Vol. 9, No. 8, September 1994.

# TRADITIONAL RESPONSES TO AUDIT DOCUMENT MANAGEMENT

The audit report documentation process can be quite complex and involved in many organizations. As shown in Figure 13.1, the process typically involves staff at the corporate and site level as well as the audit team.

Typically, as a first step, the audit team will develop a draft list of findings and draft and final audit reports. These receive extensive reviews by site staff, line management, corporate environmental affairs staff and corporate legal staff. In some companies, these reviews are conducted simultaneously, which accelerates the process, but at the risk of the audit team receiving conflicting comments.

**Figure 13.1**
**Audit Reporting Process Organizational Responsibilities**

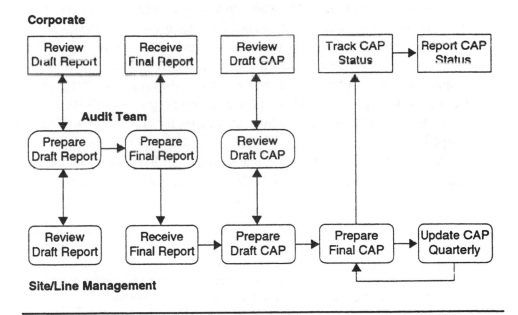

In other companies, corporate environmental affairs and legal staff must receive copies of the draft reports first (even draft findings sometimes), and comments must be incorporated before the site or line management receive their copies. This helps to protect the auditors and the corporation, but the process often bogs down substantially, with the site at times not receiving draft reports for several months.

There are similar challenges associated with the corrective action planning process. The site will develop a draft corrective action plan (CAP) and submit it to the audit team leader and corporate audit program manager for review. The legal department will typically receive a copy as well. Comments will be incorporated and a final CAP distributed. The corporate audit program manager will be responsible for tracking the progress of each CAP, and this is usually accomplished through the development and distribution of quarterly status reports by the sites.

In an organization with scores of sites, one can imagine how quickly the document management process can get out of control. At any given time, there are a variety of document types for each site, in various stages of completion, winding their way through the organization. As an example, take a company with twenty sites, each on a two year audit cycle. After six years, the organization will have developed sixty draft audit reports, sixty final audit reports, sixty draft CAPs, sixty final CAPs, as many as 480 quarterly status reports, and numerous other program documents. It adds up. Is it any wonder that most programs can't seem to meet their established scheduling standards for report and CAP completion?[4]

As a result of these challenges, many companies have increased their reliance on portable computers in the field. Field use of computers allows

---

[4] Report timeliness and quality and insufficient follow-up have been identified previously as two of the top six deficiencies typically observed in environmental audit programs. See: Cahill, L.B., "Benchmarking Environmental Audit Programs: Best Practices and Biggest Challenges," *Total Quality Environmental Management*, Executive Enterprises, Inc., New York, NY, Vol. 3, No. 4, Summer 1994.

audit teams to develop the skeleton of a report prior to visiting the site. This can accelerate the report-generation process.

Companies also have instituted semi-automated audit tracking systems. These systems, relying on commercially available databases, such as Paradox®, monitor the generation of audit reports and codify individual findings and their associated corrective actions.

Using these systems, audit program managers can monitor the progress of draft and final audit reports and periodically (usually quarterly) review the status of all outstanding corrective actions. Audit reports, where all corrective actions are completed will be formally closed out, and the program manager will "red flag" those planned actions that are behind schedule.

Examples of the types of data that are tracked are shown in Figure 13.2. This demonstrates that many factors associated with each finding and corrective action must be tracked.

For the corporate periodic tracking to be effective, the audited entities must submit a status report of some kind, which is usually entered manually into the database, which is then updated. Examples of the types of output typically generated from the submitted reports are discussed later in this article.

In addition, the data are also reviewed to determine the progress the site has made toward completing the corrective actions by the original target date. In fact, timely completion of corrective actions is probably the most critical performance factor for the site that was audited, even more so than other factors, such as the number of deficiencies observed by the audit team.

Unfortunately, many of the current audit automation systems in use are cumbersome and not easily maintained and updated. For example, audit reports will be completed using a word processing program, tracking data are maintained on a relational database, and performance charts are developed using a charting program. Although many of these programs can be used relatively seamlessly, the linkages are often manual. That is, a clerk re-enters the data from the audit report into a database and, subsequently, enters key elements of the database into the charting program. Not surprisingly, without a true champion, the databases, quarterly reporting, and tracking often fall by the wayside.

**Figure 13.2**
**Audit Reporting Process**
**Critical Data Elements for Tracking**

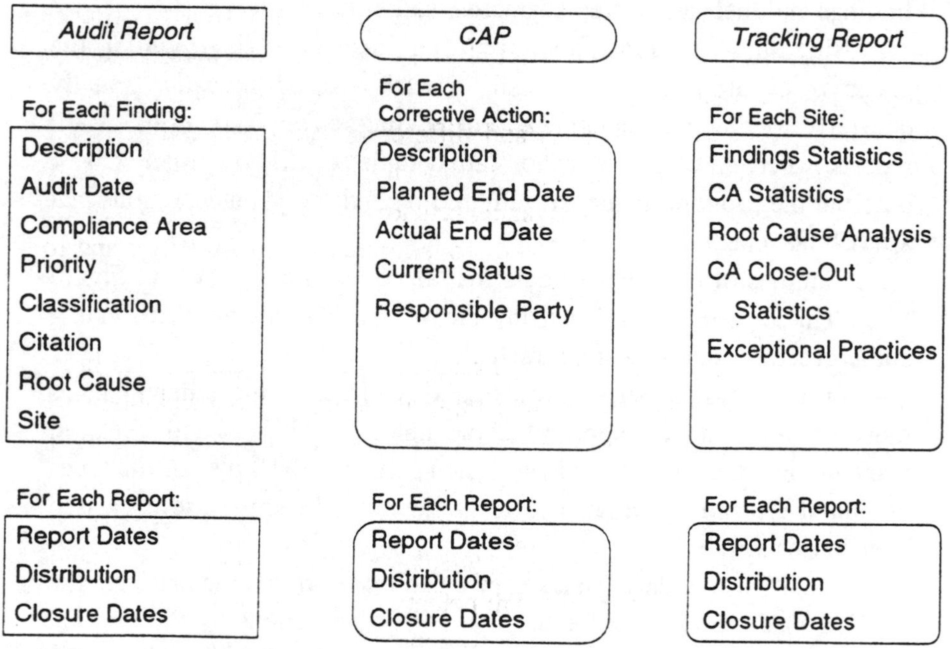

| Audit Report | CAP | Tracking Report |
|---|---|---|
| **For Each Finding:** | **For Each Corrective Action:** | **For Each Site:** |
| Description | Description | Findings Statistics |
| Audit Date | Planned End Date | CA Statistics |
| Compliance Area | Actual End Date | Root Cause Analysis |
| Priority | Current Status | CA Close-Out |
| Classification | Responsible Party | Statistics |
| Citation | | Exceptional Practices |
| Root Cause | | |
| Site | | |
| **For Each Report:** | **For Each Report:** | **For Each Report:** |
| Report Dates | Report Dates | Report Dates |
| Distribution | Distribution | Distribution |
| Closure Dates | Closure Dates | Closure Dates |

# THE GROUPWARE SOLUTION

Groupware such as Lotus Notes® can make the audit tracking process much more manageable and effective. "Groupware allows groups of users to work together by sharing information across a network. The information can include files created in the groupware, files from other software packages, comments attached to those files, and a variety of other file formats and types of information. You can also combine different types of information and different file formats into a single document or database."[5] As relayed by the inventor of Lotus Notes®, Ray Ozzie, in a recent *Fortune* magazine article[6], the intention of groupware is to "enable people in business to collaborate with one another and to share knowledge or expertise unbounded by factors such as distance or time zone differences."

"One of the special powers of Lotus Notes®, in particular, is in its replication capability. Performing replication passes your changes to a document to the shared database, and/or receives other user's changes from the shared database."[7] Replication routines assure that common files are periodically updated globally.

Conceptually, groupware could be a very powerful tool for managing an audit program. If taken to its full potential, groupware could allow for a *paperless* audit documentation system. Let's explore some of the possibilities.

Under the groupware scenario, findings from audits can be loaded directly onto the system network during the audit, allowing for real-time review of the audit reports by the site and other interested parties (*e.g.*,

---

[5] Sim, Allen W., *Lotus Notes® for Novices—A Guide for the Perplexed*, Ballantine Books, New York, NY, 1994, p. 2.

[6] Kirkpatrick, David, "Why Microsoft Can't Stop Lotus Notes," *Fortune*, Time Inc., New York, NY, December 12, 1994, p. 142.

[7] Sim, p. 161.

corporate legal). This would allow the legal department to review a draft audit report the evening before the closing conference and to have any proposed changes incorporated into the report prior to the meeting. The same could be done by the corporate audit program manager, and all comments could be incorporated and the file replicated so that all recipients have the same version. In fact, where all relevant parties can have this access to the report findings, there may be no need for a written report, although a printed set of findings might be useful for the closing conference at the site.

Further, the corrective action plan can be quite easily loaded onto the groupware system and keyed to the audit findings and report. Screens can be designed so that for each audit, individual findings with their proposed corrective actions can be shown together. Real time adjustments can be made here as well.

One of the more powerful ways to use groupware is in the continuous updating of corrective action plans. There really would no longer be a need for formal, quarterly status reports. Updating can be done at the site level since actions are completed through direct input into the network as opposed to transmitting status reports by paper or electronic mail to be manually entered into a database.

It might be prudent to exercise certain controls or security measures over the corrective-action updating process, because the corporation may not want to allow uncontrolled modification of the corrective action plans. That is, only certain individuals might be given authority to update the plans. Establishing read-only files for most users of the audit reports and corrective action plans might be an appropriate measure. This is quite easy to accomplish with groupware.

And finally, groupware allows the audit program manager to easily develop and update the periodic tracking reports and distribute the reports directly and electronically to the interested parties.

Under groupware, audit program output reports can take a variety of forms. Graphical presentations can be developed from the fundamental tracking tool, the audit finding. For Lotus Notes®, in particular, this capability has been enhanced by the recent release of Lotus Notes ViP®, a visual programmer that allows developers to build custom applications based on the Notes environment.

## Figure 13.3
## Important Program Characteristics

**Newark Plant Summary of Audit Findings**
**(By Type and Significance)**

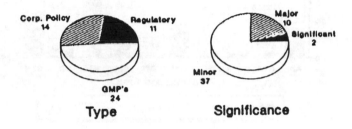

Type                    Significance

**Corporate Audit Findings...**
**Root Cause Summary**

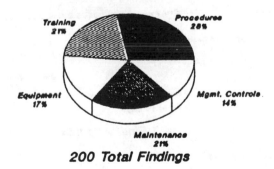

**200 Total Findings**

## Figure 13.3 *(cont'd)*

*Corporate Audit Findings...*
### Summary by Compliance Area

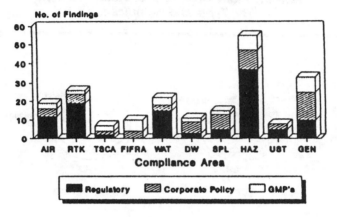

*Corporate Audit Reports...*
### Review of Final Report Timeliness

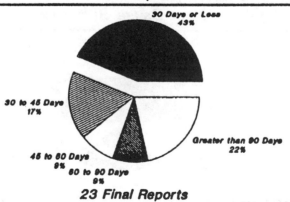

As shown in Figure 13.3, certain important program characteristics such as root causes, report timeliness and findings type and significance can be developed and analyzed. The audit program manager should decide in advance which characteristics are appropriate to analyze and display for his or her organization. Some companies will statistically analyze each plant's performance, others will assess only a division's performance, while still others will "roll-up" the data to show only overall corporate performance. These strategic views can help the audit program manager better understand the relative strengths and weaknesses of the organization's EH&S program.

## CONCLUSION

According to *Fortune*, as of early 1995, some 4000 companies have purchased Lotus Notes®, the premier entry in an emerging field of software known as groupware. For many companies making the commitment, installation of the software has meant significant capital expenditures since wide area and local area networks and client-server hardware had to be installed to make effective use of Notes' capabilities nationwide and worldwide. A few companies are in the early stages of developing environmental, health and safety management capabilities. At first blush, the EH&S audit document management process appears to be an ideal application for groupware. It is vitally important that audit tracking systems are put in place and operate effectively. Groupware can help to achieve this objective.

# 14

# MANAGING AND CRITIQUING AN AUDIT PROGRAM

---

*"A company must evaluate a number of general considerations in devising a rational and cost-effective environmental audit program tailored to its unique needs."*[1]

---

Assuring regulatory compliance is a complex challenge for today's industrial manager. Shaping environmental audit programs to help meet this challenge is no less a challenge in and of itself. Programs must be well thought out and receive the support of senior managers.

This chapter is directed at those managers who will be given the complex chore of developing and/or managing an environmental audit program within a corporate structure. The chapter first presents an approach for developing a program, consisting of eight steps, beginning with a policy setting and planning exercise and ending with full implementation of the program. Using that as a starting-off point, the remainder of the chapter discusses a variety of management techniques that can be used to control, assess, and evaluate an ongoing audit program. The chapter closes with twenty tips for achieving a successful audit.

---

[1] W.N. Hall, "Environmental Audits—A Corporate Response to Bhopal," in *Environmental Forum*, August, 1985.

# BEGINNING A PROGRAM

As one is immersed in the complex and pliable concept of environmental auditing, one can become overwhelmed with the choices that need to be made in setting up a comprehensive yet workable program. It is often useful to attack the problem one step at a time. Presented below is an outline of one step-by-step approach that could be used in creating a successful audit program.

## Develop an Environmental Audit Policy and Procedure

Successful environmental audit programs uniformly receive strong support from senior management. The explicit backing of the program by senior managers creates a supportive attitude throughout all levels of a company. Thus, quite often the first step in any program is the development and distribution of a policy statement, signed by a senior corporate executive, describing and supporting the initiation of a formal environmental auditing program. This policy statement sets the tone of the program.

The policy statement should include a set of program procedures, which is a natural follow-up to the policy. These procedures deal with key planning and structural issues, such as those suggested below. The statement should be responsive to the criteria selected by the organization, as appropriate. These criteria might include the U.S. EPA Environmental Auditing Policy, the U.S. DOJ Sentencing Guidelines, ISO 14000 Auditing Guidelines, the European Union Eco-Management and Audit Scheme and/or BS7750. Lastly, the policy statement should include a tentative site-visit timetable for the planned implementation of the program.

| KEY AUDIT PROGRAM PLANNING ISSUES | |
|---|---|
| **Type** | **Issues** |
| Policy | • Objective of the program <br> • Expectations of senior management <br> • Measures of successful performance |
| Management | • Corporate vs. divisional control <br> • Internal vs. external auditors <br> • Unannounced visits <br> • Reporting protocols |
| Legal | • Written oral reports <br> • Retained documentation <br> • Corporate counsel involvement |
| Coverage | • Frequency of audits <br> • All regulations <br> • All plants, random sample, directed sample <br> • Past practices |

The successful environmental audit program should also respond to any Total Quality Management principles established by the organization. Some that come to mind include:

**Focus on Customers.** Define the customers of the program. They include corporate management, line management, EH&S management and site staff. In some cases, shareholders are included as well. Make sure that key documents, such as the facility audit reports and third-party certifications, respond appropriately to those customers.

**Improve Constantly.** The program should be evaluated annually to determine if it has met pre-established objectives. Moreover, a new set of objectives should be established each year, which should correspond to program improvements. It is especially important that line management perceives that the program is adding value each year and not simply becoming "more picky" on each subsequent audit.

**Drive Out Fear.** Although it is difficult to remove the "performance evaluation" flavor of an environmental audit completely, the program should focus on management systems and site procedures and performance. Depersonalizing the program can drive out the fear at the site level.

**Remove Staff Barriers.** The audit program can help the sites understand that environmental compliance is not simply the responsibility of EH&S managers. Line management has a clear role in achieving compliance. With line and staff organizations working together, more can be accomplished.

**Eliminate Slogans and Quotas.** Audit programs should be careful about incorporating broad, unachievable objectives, such as, "compliance will be improved." The audit program is only part of that equation. One should also be careful to put quotas on the number of audits that need to be conducted each year, without some risk categorization of the sites being audited.

**De-emphasize Mass Inspections.** There is a common tendency for organizations to "count up" the number of findings on each audit and compare one site against another. This should be avoided since each site is different and not generally comparable. One comparison that can be made is how a site performs on audits over time. Are deficiencies being corrected? Is the site performing better over time?

**Institute Training.** Most companies have a requirement that auditors must be formally trained prior to their first audit and that experienced auditors receive refresher training and periodic updates on relevant regulatory issues. This is a good practice and consistent with ISO 14012 expectations.

**Don't Buy On Price Alone.** In reviewing ancillary operations during an audit, make sure that the site is purchasing services and equipment that aren't always the low-cost providers. Many companies' sites have been compromised environmentally by low-bid, on-site contractors.

## Set Up the Organization

The second step on the agenda is deciding on how to organize the audit program. For larger, multi-division companies, there are a variety of options, including:

- A corporate audit function with full-time corporate auditors
- A corporate audit function with part-time auditors selected from divisions
- A small corporate oversight group with delegation of the audit function to the operating division
- A small corporate oversight group with self-auditing conducted by plant staff
- A small corporate oversight group with the use of external, independent auditors.

Even in smaller companies, similar organizational issues arise. Beyond the choices listed above, the types of individuals who should be auditors must be decided (*e.g.*, engineers vs. attorneys), as well as deciding how corporate counsel is to be involved; whether auditors should be full-time, full-time rotation, or part-time; and who is to have ultimate responsibility *and* authority for running the program.

Unfortunately, there is no standard organizational solution for the management of an audit program. Each company must investigate the strengths and weaknesses of the variety of options open to it and select accordingly.

## Develop Tools

The third part of program planning process is the early development of the audit tools. Many of the tools and techniques commonly used in industry have been discussed elsewhere in this book. These include:

- Statement of policy
- Pre-visit and audit evaluation questionnaires
- Site visit and opening/closing conference protocols
- Audit checklists and protocols
- Audit report, corrective action plan and status report formats.

Each of these tools should be developed in a standard and compatible way to ensure consistency within and among individual audits. Eventually, they can be included in an all-encompassing audit manual that will serve as a training device, as well as the audit working document.

## Train Staff

The fourth step in the process is auditor training. A certain amount of auditor training will be required early in the program. The extent and nature of the training will vary depending upon the backgrounds of the individual auditors (*e.g.*, corporate environmental staff *vs.* plant environmental engineers *vs.* corporate attorneys). Additionally, the frequency of recurring training will vary depending upon the organizational choice selected earlier. That is, the training of full-time auditors is pretty much a one-time effort, while the use of full-time rotational or part-time auditors will require training on a regular basis.

Understandably, auditors must be trained technically to be able to deal with very complex compliance management requirements. But perhaps more importantly, they must also develop skills in conducting probing interviews with personnel having a wide variety of educational and employment backgrounds. Auditors must walk a fine line between being supportive to plant supervisors in their management of environmental affairs and being assertive in their probing for areas of known or suspected noncompliance. It is a task requiring both diplomacy and firmness.

A final point on training—and this is an area where outside help may be useful. Audit consultants can bring a large number of experiences to a training exercise and can provide the expertise or program start-up. These may be otherwise unavailable to the organization.

## Test the Program

With the organization, tools, and trained auditors in place, the fifth step is to test the program. Testing involves conducting one or more practice audits at plant sites. In this light, the audit program manager may want to test the auditors' abilities to conduct an investigation commensurate with the potential liabilities associated with the range of company sites. Therefore, he may want to select one large, complex facility and one small, fairly risk-free facility (*e.g.*, warehouse, distribution center) to audit. The intensity and duration of the investigations for these facilities should vary considerably and, therefore, testing the audit method at each site would be a useful exercise.

## Set Review Schedule

With the program fully tested, the sixth step is to set a schedule for reviewing all of the company's sites. To provide some rationale for determining how often sites are to be audited, for how long, and by how many auditors, many companies rank sites based on risk-potential criteria. An example of how this is typically done was shown in Chapter 4. In the example, each of four sites was first classified based on risk potential. The resulting classifications were then used to determine frequency of audit, duration of audit, and audit team size. This type of exercise can be completed for all facilities to help assure that audit resources are applied effectively and efficiently.

## Conduct Audits

With all preparatory tools in place and a schedule set for facility audits, the seventh step is to actually conduct the audit using the tools and timetable developed. The approach used to conduct individual site audits is discussed in detail in Chapter 15.

## Implement Full Program

The central purpose of an environmental audit program is to assure continued compliance with regulatory requirements. Audits are not one-time efforts but must be followed up on a regular basis. As has been stated, where instances of noncompliance have been identified, these must be corrected within a reasonable length of time so that corporate vulnerability is not exacerbated. Thus, in implementing the full program, most companies have designed a follow-up mechanism to monitor subsequent corrections made to eliminate noncompliance. One approach has been to develop brief "exception reports" or "action item reports" as a result of each and every audit. These reports would then be entered into a computer, together with a program that would produce monthly status reports on completion of action items. As corrective actions are taken at sites, these are noted. In addition, semi-annual or annual reports are generated that summarize key compliance management statistics such as: items of noncompliance per facility type; type and cost of corrective measures; lag time between identification of noncompliance and implementation of corrective action; and so forth.

# MANAGING A PROGRAM

Too often the value of a management technique, such as an environmental audit program, is lost to an organization because of execution. This can happen even in cases where the program planning exercise has been extremely successful. Presented below are some management techniques that can be useful in keeping a program on track.

## Assessment Tools

Audit programs often are implemented without any *a priori* consideration given to how program success might be measured. This is not a good management practice. Truly successful programs will typically incorporate several monitoring and evaluation techniques. Most of these use the reduction over time of noncompliance items as a yardstick.

The formal incorporation of compliance measures into the performance standards of plant and division managers is a technique sometimes used. Since community complaints, excursions, notices of violation (NOVs) and fines are, in fact, measurable events, management can be held accountable. In some observed cases, plant managers have maintained graphs on their office walls depicting historical compliance at the facility. As demonstrated by the chart in Figure 14.1, these graphs can depict the trend in both overall and individual media environmental compliance.

Incorporating these techniques into the management system, and evaluating managers based on successful execution, will help to ensure that risks and liabilities are understood and handled appropriately. It establishes a clear message that environmental management is taken seriously at high levels within the organization.

**Figure 14.1**
**Compliance Monitoring**

## Plant Manager Tracking System

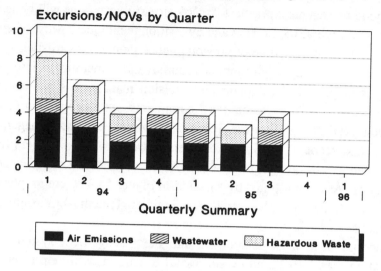

Excursions/NOVs by Quarter

## Quarterly Report to Management

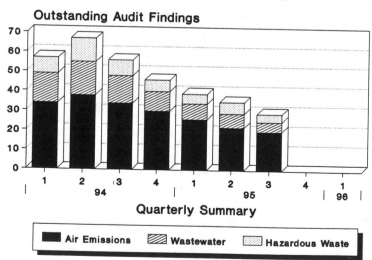

Outstanding Audit Findings

## Management Reports

Top management's role in assuring environmental compliance is crucial. Constructive environmental policies and procedures must be developed *and* communicated. Recent history is filled with examples of employees illegally dumping wastes and falsifying compliance records, all done "in the best interests of the company." Senior management must demonstrate a serious commitment to assuring compliance, and surely there is enough reason to do so.

In addition to being involved in the setting of policies and procedures, senior management also must be apprised of the successes and failures that occur during program operation. In organizations where the reporting of audit results does not go beyond the corporate environmental manager, this manager has, in some cases, found himself in a troubling situation. He can indeed lack the institutional leverage needed to convince line managers—often his peers—that expenditures for achieving environmental compliance should be budgeted on par with normal operating expenditures.

One way to ensure management awareness and commitment is to design an audit program that formally requires reports to be delivered to senior management. Precedence for this can be found in corporate safety programs in which summary safety statistics are routinely reported to levels as high as the Board of Directors. Environmental reports could (and should) be a part of this process.

A key to making the "reports to management" idea work is to design a report that provides useful information efficiently. Most senior executives are not in a position to read all audit reports, and even if they did, they would typically be unable to reach any strategic conclusions without some supporting trend or snap-shot analyses.

Typically, then, reports to management should be brief tabular and graphical presentations, transmitting the maximum amount of information in the minimum number of pages. Assuming that these reports would probably be quarterly, they might contain the following information:

- Number of audits completed in the quarter (by organizational unit)
- Highlights of liabilities most affecting the firm (*e.g.*, new or pending underground storage regulations)
- Trends in noncompliance items identified (*See* Figure 14.1)
- Statistics on success rates in meeting "action item" schedules
- Specific instances where audit program findings and recommendations have resulted in significant cost savings (*e.g.*, major potential fines avoided or substitution of nonhazardous materials for hazardous materials).

Presenting this information in a concise, digestible format will aid the environmental audit program manager in his efforts to establish and maintain corporate credibility for his program.

## Staff Management

The role of an environmental auditor is a difficult one. Even under the auspices of a supportive program, much of what an auditor does involves evaluating the performance of the company's environmental and operational staff. Constantly "dinging" facilities (and their management) in written audit reports can become a fatiguing exercise (although some people, for better or worse, seem born to it). Also, in most situations, extensive travel is required. And while this seems glamorous at first, regularly spending weeks in Nowhere, U.S.A. can add to the fatigue.

In order to maintain a motivating environment for auditors, the program manager should initiate some efforts to counter the downsides of auditing. These could include:

- **Rotating Staff.** Most companies rotate audit staff on a regular basis. This helps to keep auditors fresh and enthusiastic and will also help to spread environmental awareness among a large group of individuals.

- **Establishing Career-Progression Tracks.** Supplementing the need for rotation is the need to convince staff that the environmental auditing slot is a logical step in one's career and not a "dead end" or "holding pattern." One company, in fact, has a standing pool of corporate "fast trackers" who, as part of their individual development plans, will serve as environmental auditors for a period of time. The goal is to create a sense of environmental awareness in the managers of the future.

- **Maintaining the Technical Edge.** Keeping up with environmental trends is an overwhelming task. It seems that every month one significant federal environmental regulation is either proposed or promulgated. Audit staff need to do more than read the *Federal Register* (a task to be reserved for strong hearts and sharp eyes) to keep abreast of developments. Participation in regular seminars (*e.g.*, the Environmental Auditing Roundtable) will maintain a cutting edge in staff, put auditors in an environment with auditors from other companies, and provide a perquisite that can enhance morale.

There are, of course, many other traditional techniques managers can use to motivate staff (*e.g.*, good salary increases for good performance), and they should indeed be used, for over time an audit program will only be as strong as its people.

## Quality Assurance

One of the key challenges with any program is to ensure it meets or exceeds its stated objectives. Assuring a quality program requires regular performance reviews. One way to accomplish this is to have an annual self-evaluation. The evaluation can be completed as part of the year-end planning and scheduling for next year's site visits, which is done by most companies. Thus, the planning exercise should not only be a look forward but a look backward as well.

- What have we accomplished?
- Did we meet our objectives?
- Where did we exceed our plan?
- Where did we come up short? Why?
- What changes can we make to improve our program?

In addition to the self-evaluation, programs also benefit from occasional third-party evaluations. For a modest expenditure to an outside firm, a company could conduct an audit of their audit program. This audit could include requesting that the third party observe an actual audit or review audit reports for substance, format, and language.

Lastly, the program manager is faced with a very difficult QA challenge in order to keep current with changing environmental regulations. Most programs use some form of audit checklist drawn from current standards and regulations. If no effort is made to keep these checklists updated, they will quickly be viewed as irrelevant, and the program's viability will be jeopardized. As discussed earlier in another Chapter, some mechanism for regulatory updating must be chosen *and used*.

## Maintaining Awareness

It is frequently surprising how much of a job—any job—is selling. We all must sell ideas, approaches, plans, strategies, and the like to our bosses, our peers, and even our staff. Strong-arming goes only so far. Running a successful audit program is no different. Senior management and line managers must be convinced of its value. And perhaps most importantly, plant operators, through the audit program, must understand the environmental consequences of their actions. They are where "the rubber meets the road." Techniques companies have used to make this happen include, for example, the posting of important notices and procedures in the workspace of unit operators. Too often emergency response procedures, spill procedures, and target effluent limitations are found as blueprints on the shelves in office buildings but not in the operating environment where they would be the most useful. Also, regular training and orientation of both new and seasoned employees is an important awareness technique for management. This should include not simply classroom exercises but practice drills for spills and upsets. Finally, advertising good environmental practices in the way safety performance has been traditionally rewarded will help improve the visibility and credibility of the environmental audit program.

## TWENTY TIPS FOR ACHIEVING A SUCCESSFUL AUDIT

When all is said and done, there is no such thing as a perfect environmental audit. However, if one keeps the following twenty tips in mind, the audit will be the best it can be.

1. Do not ignore or underestimate the need to prepare for and plan the audit and its logistics.

2. Develop and maintain a "living" agenda throughout the audit.

3. You will never finish, so manage your time wisely. It is better to address all areas to some extent than to ignore one or two areas entirely.

4. Keep a balance between records review, interviews, and observation. Review records first.

5. Things will happen on the audit. Be flexible.

6. Always be on time and be courteous, even when the site staff are not.

7. Remember that audits will always be considered a performance evaluation by site staff. Be empathetic, not sympathetic.

8. Use sound judgment. In spite of the existence of governmental regulations and company standards and guidelines, considerable judgment is still required of the auditor.

9. Remember that every country is different; every state or region is different. Local standards apply, with specific exceptions determined by corporate standards and guidelines.

10. Learn and apply the audit protocols, but don't forget to use your common sense and natural curiosity.

11. Try to observe things (*e.g.*, spill drills, wastewater sampling, chemicals loading and unloading, waste pickups) as they happen.

12. Look at ancillary operations (*e.g.*, maintenance, laboratories, on-site contractors).

13. First observe, then articulate, then write—in that order.

14. Perform true and complete root cause analysis only where the risks dictate it. This process can be difficult and time consuming. It is the site's responsibility to identify underlying causes and to effect long-term and permanent solutions.

15. Remember that writing is the hardest part; developing "bullet-proof" findings is an elusive goal. Findings are verified statements of fact; leave the speculation and editorializing for other endeavors. Avoid conjecture and hearsay.

16. Write the findings as you go. Get a draft report or draft set of findings done before you leave. Use a standard form to compile individual findings. Use a portable computer and report "skeleton findings" to expedite the process.

17. Prepare well for the opening and closing conferences. Use standard presentation tools. Listen attentively to questions, and

respond directly. If you don't know the answer, admit it and commit to gathering the information needed to respond.

18. Make sure the site hears the real story in the closing conference. There will be a tendency to sugar-coat the findings.

19. Plan to be done when you leave the site. Minimize issues requiring future analysis.

20. Try to enjoy yourself. It's hard work but personally and professionally rewarding.

This chapter has been designed to assist company managers and staff in developing and running audit programs to monitor compliance at their plant sites—be it one site or fifty. It is our hope that the concepts and approaches presented herein will guide the development of efficient and effective programs that will minimize both the liability exposure of corporate managers and the pollution exposure of the public at large.

# PART II

# CONDUCTING THE AUDIT

# 15

# CONDUCTING THE
# ENVIRONMENTAL AUDIT

---

*"Management ignorance is no defense."[1]*

---

## INTRODUCTION

The core of any audit program lies in the actual completion of individual facility audits. Audits must be conducted efficiently, consistently, and comprehensively by competent auditors who must elicit the cooperation of plant managers and staff. This can be a tall order for uniquely complex facilities whose basic day-to-day operations fully tax on-site staff resources.

Accordingly, this and the following chapter discuss the detailed steps and procedures that are typically undertaken in the course of an environmental audit. It should be noted, however, that every facility is different, and this chapter can only address those elements of an audit that are common to most facilities and programs. That is, some of what is presented here may not be applicable to certain programs and, conversely, there may be some environmental compliance issues that are not included here that are of concern to your operations.

Nevertheless, these next two chapters will provide insights and perspectives into the conduct of an environmental audit from the "typical" plant viewpoint. The following key questions will be addressed:

---

[1] Robert V. Zener, Former EPA General Counsel, in *Environment Reporter*, Current Developments, June 26, 1981.

- How should I prepare for the audit?
- What records and documentation should I review?
- What part of the plant should I inspect and what compliance issues are key?
- Who should I interview and how should I do it?
- What should be done after the audit is completed?

There are three major phases to an environmental audit. They are: (1) pre-audit activities, (2) on-site activities, and (3) post-audit activities. Because of the distinct procedural differences in each of these phases, the remainder of this chapter addresses each separately. The next chapter discusses specific examples of inspection items and compliance problems that one might observe at a facility.

# PRE-AUDIT ACTIVITIES

There are several key activities that should be completed before an audit team arrives on-site. They include:

- Team selection and formation
- Completion and review of the pre-audit questionnaire
- Review of relevant regulations
- Audit scope definition and establishment of team responsibilities
- Review of audit protocols
- Development of detailed agenda

A discussion of each of these activities is presented below.

## Team Selection and Formation

The audit program manager selects the audit team leader and together they typically select the remainder of the team members. Team members should have complementary and overlapping skills, if at all possible. As discussed elsewhere in this book, auditors should be prepared for any circumstance in the field. Auditors can become ill, especially overseas, or otherwise indisposed. One is sometimes asked to pitch in and cover the areas that one of the team is not addressing.

Thought should also be given to paring up relatively inexperienced auditors with suitable mentors; not every audit should get the "A" team, otherwise the experienced auditor pool will never expand.

In the U.S., it is helpful to have one or more auditors on the team from the same state but independent of the site. This aids in conducting an independent review of state regulatory applicability.

Some companies assign to the audit team the environmental coordinator from the next site to be audited. This gives that person a good idea of the process and will help them prepare better for the upcoming visit.

One should also consider language when creating an audit team. Even in North America this issue can come into play. Having an auditor who can speak French on an audit in Quebec or an auditor who can speak Spanish on an audit in Mexico or Puerto Rico can be very advantageous.

And, for better or worse, cultural issues should be considered, at least to some extent. On one recent third-party review of a corporate audit program, one of the criticisms of the audits was that "corporate keeps sending these Yankees with attitudes down here to audit us. We don't appreciate it." Cultural diversity on audit teams can be a wonderful thing in order to be able to audit a wide geographic and culturally diverse area.

Finally, once the audit team is selected, there should be a team meeting prior to the audit. This can be handled in a variety of ways. A face-to-face meeting is preferred but not always possible. A teleconference or video conference is an acceptable substitute. And lastly, the team can schedule a meeting for the evening before the audit. This approach should be used sparingly because it might be too late at this time to make any last minute adjustments. And there is always the risk that one or more auditors' flights will be late and the meeting will never take place.

## Completion and Review of Pre-Audit Questionnaire

The purpose of the pre-audit questionnaire is two-fold. First, it enables the audit team to become familiar with the general environment-related activities and operations before they arrive on-site. Secondly, it serves as a timely alert to the facility manager so that he may better prepare himself and his staff for the audit.

An important element of a good pre-audit questionnaire should be its brevity. Normally ten to twenty pages is sufficient. The pre-audit questionnaire should ask only for general information on the facility. Its primary purpose is to enable the audit team to be sufficiently aware of plant environmental activities so that relevant regulations and requirements can be obtained and reviewed. An example of a pre-audit questionnaire is shown at the end of this chapter.

Several important lessons on pre-audit questionnaires are worth mentioning. If the audit team wants the questionnaire completed, make sure it is simple and straightforward (*e.g.*, the questionnaire should not require the site to provide reams of performance data that can be reviewed better when the audit team arrives on-site) and make sure it is distributed to the site well in advance. And most importantly, each member of the team must review the questionnaire prior to the on-site activities. There is nothing more inconsiderate than to require the site to complete a questionnaire and then realize it has not been reviewed by the team.

## Review of Relevant Regulations

There are numerous regulatory requirements that are administered by federal, state, and local authorities. Figure 15.1 is a summary of key compliance areas and the regulatory authorities that typically have some jurisdiction over these areas. As shown, each level of government will have a major influence on some compliance areas and have little or no effect at all on others. In some cases, each level of government will influence a compliance area.

Once the pre-audit questionnaire has been returned to the audit team, a thorough review of the relevant regulations affecting the facility should be undertaken. There are several methods that can be used to obtain the regulations. Direct contact with regulatory agencies asking for copies of the regulations is an appropriate technique. This is particularly useful on the local level for such items as sewer ordinances, county/city effluent limitations, air emission regulations promulgated by air quality management districts, and flammable/combustible storage and handling regulations issued by local health or fire departments.

There are also several computerized regulatory database systems available that provide fast and cost-effective retrieval of federal and state regulations through the use of key words. More information on these systems is presented in the chapter on information management.

The Appendix lists some of the key federal and state regulations that typically should be reviewed prior to the audit, if they apply to the facility. The regulations should be reviewed and digested down to a series of specific audit elements for use by the audit team. This was discussed previously in Chapter 4 under Audit Protocols.

## Definition of Audit Scope and Establishment of Team Responsibilities

An important pre-audit activity is to define clearly the audit scope and audit team responsibilities. It is not necessary that all areas be audited during the site visit if it is felt that time would be better spent concentrating on a few compliance areas that represent a higher exposure. An environmental, health and safety audit scope could consist of any or all categories shown in Figure 15.2.

Once the team decides on the areas that will be the subject of the audit, each team member should be given a specific assignment and responsibility to serve as the auditor for that compliance area. The auditor can then prepare to be fully ready to start his or her portions of the audit efficiently once the team arrives on-site.

**Figure 15.1**
**Key Compliance Areas and Regulatory Authority Jurisdiction**

| Compliance Area | Federal | State | Local |
|---|---|---|---|
| Air Emissions | Secondary | Primary | Primary |
| Wastewater Discharges | Secondary | Primary | Secondary |
| PCBs | Primary | Secondary | --- |
| Pesticides | Secondary | Primary | --- |
| Hazardous Waste | Primary | Secondary | --- |
| Hazardous Materials | Secondary | --- | Primary |
| Hazard Communication | Primary | Secondary | --- |
| Community Right-to-Know | Primary | Secondary | Primary |
| Underground Storage Tanks | Primary | Secondary | Primary |
| Solid Waste | Secondary | Secondary | Primary |

**Figure 15.2**
**Typical Compliance Areas on an Environmental,**
**Health and Safety Audit**

| Compliance Area | Typical Audited Elements |
|---|---|
| Air Emissions | Fuel burners<br>Incinerators<br>VOC sources<br>Hazardous air emissions (NESHAPS)<br>Fugitive emissions<br>Vehicle emissions |
| Wastewater Discharges | Direct discharges (NPDES)<br>Indirect discharges (POTW)<br>Stormwater discharges<br>Treatment plant operations<br>Certification/licenses |
| PCBs | In-use equipment inspection<br>Temporary storage<br>Permanent storage facility<br>Disposal |
| Hazardous Waste | Generator requirements<br>TSD requirements<br>Manifests/records<br>Off-site disposal |
| Oil Spill Control | SPCC Plans<br>Spill reporting<br>Training |
| Pesticides | Application certification<br>Storage and handling<br>Disposal |

| | |
|---|---|
| Drinking Water | Water use permits<br>Sampling/analysis<br>Reporting to agencies |
| Hazardous Materials | Flammable/combustible storage and handling<br>Acid storage<br>Reportable quantity discharges |
| Solid Waste | Permitted facilities<br>Monitoring for segregation<br>Proper disposal |
| Underground Storage Tanks | Registration of tanks<br>Leak detection<br>Monitoring<br>Reporting |
| Hazard Communication | Training<br>MSDS<br>Labeling |
| Health and Safety | Ventilation systems<br>Medical surveillance<br>Exposure monitoring<br>Training<br>Recordkeeping |
| Community Right-to-Know | Inventory monitoring<br>Reporting of releases<br>Emergency response planning |

In some cases, it may be appropriate to have two members of the team audit the more complex compliance areas. This should be decided by the team before the audit commences.

Other issues that should be discussed by the audit team before the on-site activities start are whether staff organization and company management systems, procedures, and policies are going to be reviewed during the audit. For instance, is this audit going to look past the "strict regulatory requirements" and evaluate the attitude and effectiveness of the plant environmental staff? A should the team review potential liabilities under common law that could create exposure for the plant? Will the team assess the need for improved environmental management procedures at the facility? These items, which go beyond the traditional regulatory requirements, are typical of the questions the audit team should discuss during their pre-audit activities.

## Development of Detailed Audit Agenda

The success of the on-site audit will depend largely on how well the audit team has prepared itself and the facility staff. A good technique is to develop a daily agenda that describes the activities planned for each day of the audit. A typical daily agenda for an environmental audit is shown in Figure 15.3. The agenda allows the facility manager to alert key facility staff when they will be required to talk with the auditors and, generally, the audit topics that will be covered.

## Review of Audit Protocols

Audit protocols can be used with some discretion. More experienced auditors will often use the checklists as reference documents, reviewing them in detail each morning and evening, to assure themselves that they are not overlooking any aspect of compliance. These auditors sometimes develop one-page reminder sheets, which cover the major compliance categories for each area. Other, perhaps less-experienced, auditors will use the checklists religiously during the course of the inspections and interviews. The advantages of the first approach are the increased flexibility for the auditor and his ability to interview staff without the intimidating presence of a checklist. Although these advantages are typically lost in the second approach, use of the more rigorous second approach is less likely to result in the auditor overlooking noncompliance items.

## Figure 15.3
## Itinerary for Typical Audit

---

**Day 1--Monday a.m.**

The audit team meets with host facility manager and environmental manager to make introductions and review purpose of the audit. Refine schedules for interviews, tours, and special requirements.

The environmental coordinator makes a brief presentation to audit team, highlighting major facilities and features of the facility.

There is a general tour of the facility to provide audit team with orientation of overall facility. This is a brief "ride through" of key areas on the facility.

**Day 1--Monday p.m.**

The team conducts a detailed audit of *wastewater discharge* compliance. Physically inspect discharge points, treatment facilities, NPDES permits, monitoring records, and other related data. Conduct interviews with key facility staff.

The team conducts a detailed audit of *air emissions* compliance. Physically inspect fuel burning facilities, incinerators, sources of VOC emissions, and other air emission sources. Review air monitoring records, inspection records, permits, and licenses. Conduct interviews with key facility staff.

The team meets at the end of the day, without site environmental staff, to discuss interim findings and general observations.

**Day 2--Tuesday a.m.**

The audit team briefs the site environmental staff on interim findings and they plan the daily agenda.

Conduct audit of *hazardous waste management* compliance. Physically inspect hazardous waste generation points (shops, process areas, storage yards, warehouses). Inspect manifests, waste analysis records, RCRA permits, groundwater monitoring, and inspection plans/schedules. Conduct interviews with key facility staff. (Note: Depending on facility size and hazardous waste activities, this audit could take a full day or more.)

**Day 2--Tuesday p.m.**

Conduct audit of *hazardous materials management* compliance. Physically inspect hazardous waste chemical storage areas and warehouses. Review spill contingency plans, spill reporting procedures, and incident reports. Conduct interviews with key staff.

Conduct audit of *fuel/oil storage and spill response* compliance. Physically inspect oil storage areas, tank farms, and drum storage areas. Review oil spill contingency plans, spill reports, and SPCC Plans. Conduct interviews with key staff.

The team meets at the end of the day, without site environmental staff, to discuss interim findings and general observations.

**Day 3--Wednesday a.m.**

The audit team briefs the site environmental staff on interim findings and they plan the daily agenda.

The team conducts an audit of *PCB handling/storage* compliance. Physically inspect areas where transformers/capacitors are maintained. Review records of testing for PCB content, inspection logs, annual reports, and disposal records. Conduct interviews with key staff.

The team conducts an audit of *pesticides handling/storage* compliance. Physically inspect pesticide storage areas. Review records of pesticide applications, certifications for applicators, and disposal records. Conduct interviews with key staff.

**Day 3--Wednesday p.m.**

The team conducts an audit of *water supply/distribution* compliance. Physically inspect well facilities and pump house treatment facilities. Review records of water quality monitoring, water withdrawal, distribution system inspection records, and conduct interviews with key staff.

The team conducts an audit of *solid waste handling/disposal* compliance. Physically inspect areas of facility with asbestos insulation. Inspect handling/disposal procedures. Review records of disposal of asbestos. Conduct interviews with key staff.

The team meets at the end of the day, without site environmental staff, to discuss interim findings and general observations.

## Day 4--Thursday a.m.

The audit team briefs the site environmental staff on interim findings and they plan the daily agenda.

The team conducts a general inspection of areas just outside facility boundaries, especially residential areas. Look for potential "common law" environmental nuisance problems (emissions, noise, odor, etc.).

The team continues audits as before.

## Day 4--Thursday p.m.

This period is reserved to conduct audits of any special areas that are uncovered once the team is on-site. It is a period to review the previous days' activities. Additional data or information needs are identified and any unfinished items are completed. Perhaps a second visit to an area on-site is needed or some discussion on the applicability of certain regulations is required. This period is also used by the audit team to develop a preliminary list of audit findings for an information briefing to key staff.

## Day 5--Friday a.m.

The audit team will make an informal briefing to key environmental staff on preliminary findings of audit. The purpose of this briefing is to alert the staff of any significant compliance problems found so that corrective actions may be initiated. The meeting is also used to verify the validity of the findings with site environmental management prior to meeting with the facility manager.

The audit team conducts a formal closing briefing to facility manager.
The audit team departs the facility.

Regardless of the protocol-use techniques selected, the auditor should familiarize himself or herself thoroughly with the protocol prior to the audit. If the auditor chooses the first approach (*i.e.*, using the protocols as reference tools), he or she probably should develop the one-page summary sheets prior to the site visit. It is also important to include compliance items on the protocols that reflect local regulatory requirements (*e.g.*, city, county, region).

## ON-SITE ACTIVITIES

On-site activities during an environmental audit will include the following primary functions:

- Opening conference
- Orientation tour
- Records/documentation review
- Staff interviews
- Physical inspection of facilities
- Daily reviews
- Closing conference
- Audit report.

### Opening Conference

Each audit should begin with an opening conference involving the audit team and site operational and environmental management. This meeting, while it might begin with a welcome by the plant manager, is the responsibility of the audit team leader, who sets the agenda and keeps it running on time. Advice on how to run the opening conference is presented elsewhere in this book. One of the most important things to remember is to—arrive on time. The audit will not proceed smoothly if the audit team is late and the site's management team is sitting and waiting for them to arrive. This happens more than one might imagine.

## Orientation Tour

After the opening conference, the audit team is usually taken on an orientation tour. Some tips for completing this tour are as follows:

1. Give the site escort a time limit for the tour (usually less than one hour) and, for subsequent scheduling purposes, assume you will run over by 15 minutes. Tell the escort what you would like to see. Use the pre-audit questionnaire to help focus the exercise.

2. For environmental (as opposed to health and safety) audit initial orientations, work from the outside of the facility inwards. Focus on the following general areas:

   - The fence line
   - Stormwater & wastewater outfalls
   - Critical environmental areas (*e.g.*, wetlands)
   - On-site disposal & equipment "boneyards"
   - Critical storm drainage areas
   - The waste water treatment plant
   - Major tank farms & materials loading/unloading areas
   - Underground storage tank locations
   - Major liquid materials tank & container areas
   - Solid/hazardous waste tank & container areas
   - Hazardous waste treatment systems (*e.g.*, neutralization, incineration)
   - Used oil & solvent storage areas
   - The powerhouse and electrical substations
   - Maintenance (shops, pesticide storage)
   - Laboratories & medical clinics
   - Process & manufacturing areas

3. If you need to request a vehicle to observe remote areas, do so. Generally, you will be able to return to most plant areas later, but for

remote areas this may not be the case. Make careful initial observations in these situations.

4. Be careful not to spend too much time in the process or manufacturing areas. Auditors generally find these areas personally fascinating but detailed observation of manufacturing does not always provide substantial environmental compliance or performance information.

5. If the audit, because of the size of the site, requires a large team (more than 4), split the team in two and request two escorts. Organize the orientations based on relative areas of interest among the two audit sub-teams.

6. Take notes when on the tour but, other than for remote areas, do not get into too much detail. It will slow down the process. Try to note particular areas you would like the team to re-visit.

7. Bring a one-page plot plan or site map with you to help with site orientation.

8. If feasible, do not retrace your steps so that you see as much of the facility in as short as time as possible. For example, try not to return to the office by the same route you left.

9. Try to have the team stay together as much as possible. There will be a tendency for team members to wander off. This not only slows down the process but the escort often makes significant comments during the tour and team members might not hear them.

10. Arrive at the plant early on the first day of the audit and attempt to drive along the site boundary to obtain a different perspective. This may help during the orientation tour.

## Records/Documentation Review

As noted earlier, a good audit protocol will give the auditor instructions on the records, files, and reports he should request during the audit. And, more importantly, it should help the auditor understand what he should be looking for when he reviews the documentation. Normally, records should be reviewed just before conducting interviews and inspecting facilities. However, often it is useful to review documentation at the same time an interview is being conducted or an operation is being inspected. This will depend on the area being audited and the individual style of the auditors. As a general rule, however, auditors should spend their initial efforts reviewing the records and files associated with the specific compliance area, to gain a better understanding of the key compliance issues at the facility.

Figure 15.4 lists some of the most significant records, files, and reports that should be reviewed by the auditors. They are broken down by the typical compliance areas and operations normally found at industrial facilities.

One of the important things to remember when reviewing documentation is that many compliance requirements are contained within the records. For example, an NPDES permit may have numerous conditions on Best Management Practices (BMPs) listed in the permit. The auditor will need to scan these documents, looking specifically for applicable terms, conditions, or requirements, and then verify that these requirements are being met.

When appropriate, auditors should obtain copies of relevant documentation uncovered during the audit when it is felt that this can aid them in evaluating the compliance status of the facility. Also, it is useful as reference material to write the audit report at the conclusion of the audit.

During the course of the audit, a file of working papers will develop. It is not uncommon for an audit team to leave the facility with a thick file of records, reports, or other documentation. However, the rule-of-thumb is to make copies only of those materials that the audit team needs to perform a complete and thorough evaluation of the facility's compliance status. And finally, be careful about leaving the site with too many documents. This may mean that you really haven't completed your review and it will be difficult to conduct a follow-up verification of apparent deficiencies once back in your office.

**Figure 15.4**
**Environmental Compliance Audit Program**
**Documents to Review**

| *General* | *Pesticide Use* |
|---|---|
| General correspondence file | Application permits |
| Notices of violation | List of certified applicators |
| Consent orders | Training records for applicators |
| Agency inspection reports | List of restricted use pesticides |
| Corporate environmental policies | Soil plan for storage area |
| | Pesticide waste disposal records |
| *Boiler Operations* | Records of pesticide applications |
| | Annual reports |
| Boiler operating permits | Medical tests for applications |
| Opacity records | |
| Daily boiler records | *Drinking Water* |
| Stack emission test records | |
| Boiler operation inspection records | Records on water system repairs, |
| Fuel use reports | maintenance, changes |
| | Permits |
| *Process Air Emissions* | Analytical records |
| | |
| Stack/vent operating certificates | *Oil Operations* |
| Source emission inventory | |
| Records on petroleum storage | Spill plan |
| facilities and vapor control systems | Oil discharge reports/notifications |
| Permits | Oil storage systems inspection rec'ds |
| | Waste oil disposal records |
| | |
| *Hazardous Waste Generation* | *Land Use* |
| | |
| Hazardous waste manifests | Deed restrictions |
| Biennial/annual reports | Wetlands |
| Manifest exception reports | Reclamation requirements |

| | |
|---|---|
| Waste analysis test results | Closure requirements |
| Contingency plans | Post-closure requirements |
| Inspection records | |
| Medical waste records | ***Hazardous Materials*** |
| Training records | |
| Waste minimization plan | Hazardous materials spill plan |
| Treatment/ disposal permit | Hazard communications program plan |
| Superfund involvement | Hazard communications training records |
| | MSDSs |
| ***Wastewater Discharges*** | Community right-to-know reports |
| | Spill/incident reporting |
| Discharge permit | |
| Sludge disposal permit | ***PCBs*** |
| Discharge monitoring reports | |
| Exception reports | Annual PCB reports |
| Laboratory certifications | PCB disposal records/manifests |
| Off-site disposal records | PCB inspections records |
| Operator training certifications | PCB spill reports/notifications |
| Stormwater management records | Testing records of PCB equipment |
| Septic tank maintenance records | |

## Interviews With Facility Staff

Proper interviewing during an environmental audit is perhaps the single most important aspect of the audit process. Yet, too often, lack of attention to the subtleties of interviewing results in less than optimal results. There are a number of interviewing skills and techniques that, if followed, can enhance the outcome of the audit.

### Schedule Interviews Ahead of Time

Popping in on busy superintendents or shop supervisors is a sure way to cause ill feelings and poor audit results. An interview should be scheduled well ahead of time so that the person being interviewed is prepared and ready for the auditor. If possible, the interviewee should be told ahead of time the purpose of the audit, the general types of questions to be asked, and the types of documentation the auditor wants to review. Often the interviewee will suggest that someone else in the shop be available for the interview because the auditor's questions can be better answered by that person. This is to be encouraged because the whole purpose of the interview is to obtain the best evidence available, relative to the compliance status of the facility. If at all possible, at least a one-day advance notice should be given to all interviewees.

## Conduct Interviews in Work Spaces

Conducting interviews in the work space of the interviewee serves several purposes. First of all, the person being interviewed is more relaxed and less defensive than he might be if summoned to the plant manager's office or other similar location for the interview. Secondly, often key records and files are maintained in the work space and will be more readily available to the auditor. Thirdly, the auditor has an opportunity to observe first-hand actual work place conditions, attitudes of plant staff, and other subtleties that would not be observable in the plant manager's office or conference room.

The one disadvantage to conducting interviews in the work place is that distractions and interruptions likely will occur. The interviewee will face phone calls, employees with a problem, and other similar distractions. Also, it is often noisy in many work places and difficult to communicate.

The auditor should make it clear to the interviewee that he would like his attention and would appreciate some uninterrupted time. An opening statement may be appropriate such as, "Hello Mr. Smith, I know you also have a tight schedule, as I do. If you could put your telephone calls on hold and minimize interruptions from your staff, I know I can be finished with these questions in short order and let you get back to work."

## Be Sensitive to the Interviewee's Nervousness and Defensiveness

This may seem obvious, but often is not perceived by auditors. There is a normal negative reaction associated with an audit of any kind regardless of the auditor's plea that "I'm only here to help you . . ." The auditor must continue to reassure the interviewee that the purpose of the audit is to uncover the areas that need more attention, resources, staff training, and the like in order to comply with the law. It is not an audit of the interviewee, but rather an evaluation of the things that are needed to help him or her do his or her job better and more efficiently.

Some people will view the auditor as a threat to their job security. The auditor should be sensitive to this and ask questions in a constructive, respectful, and unemotional manner. If an interviewee becomes overly negative, sarcastic, or uncooperative, it is better to terminate the interview at that point and discuss the problem with the team's site host. Most people who are interviewed are nervous to begin with and become more so if they sense their responses to the auditor's questions are not the correct ones. Sometimes, they will dig a hole for themselves and continue to try and grope for the right answer. When the interviewee is struggling, the auditor should change the subject rather than probe more deeply. Move on to either other audit questions or perhaps curtail the audit temporarily and talk about some other topic to relax the interviewee.

## Avoid Yes/No Responses to Questions

Questions asked during an audit should be focused on getting the interviewee to open up and discuss compliance issues. Don't ask questions that give the person an easy "yes/no" response. "Do you inspect the PCB storage facility every 30 days as the regulations require?" may result in a simple "yes" because he assumes that's what he is supposed to do. He may not even be aware of any requirement to inspect PCB storage facilities. "Tell me about your procedures to ensure your PCB equipment isn't leaking" is a more open-ended question, which will reveal much more about the interviewee's knowledge of PCB inspection requirements.

In summary, if these suggestions are followed, auditors will obtain more and better information about the conditions at the facility.

Interviewing is a personal, interactive process. Each auditor will have their own unique style and attitude which suit them best. As an aid, a list of recommendations that all auditors should review before conducting interviews is provided below.

1. Call the interviewee and schedule the exact time and place of the interview at least a few hours ahead of time.

2. Upon meeting the interviewee, introduce yourself and explain who you are, where you are from, and why you are auditing the facility.

3. First, relax the interviewee and get him to explain who he is and his roles and responsibilities.

4. Don't become captive to your checklist so that effective communication is hampered.

5. Ask open-ended questions, avoid yes/no type questions.

6. Don't ask questions about compliance items that can be better evaluated by physical inspection or documentation review at a later stage of the audit.

7. Don't preach to or intimidate the interviewee by your superior knowledge of the regulations. Don't quote or give citations unless this information is requested and you think it will help the interviewing process.

8. Ask the interviewee about previous audits/inspections relative to the compliance area you are auditing. It's a good place to start.

9. Solicit input from the interviewee about how the audit should be conducted for maximum effectiveness. Make him a part of the audit process not the subject of it.

10. Be careful not to tell or order the interviewee to do anything during the interview. Remember you are there to obtain information and identify problems, not to solve them during the audit. That will come later.

11. When appropriate, and if it helps the interview to proceed more effectively, general comments make that are instructional in nature as long as they are not construed to be directives from the auditor.

12. At the time that noncompliance items are being identified, don't make the interviewee uncomfortable by your reaction. Continue to assure him that the audit is not intended to be a personal inquisition but a process to identify problems that are typically found at facilities.

13. Be empathetic as you talk to the interviewee but don't cross the line and become too sympathetic to his problems. The interview could drag on longer than it needs to.

14. Be quick to distinguish the "hostile interviewee" who doesn't want to cooperate, from the "nervous interviewee" who is just too uptight about being questioned.

## Physical Inspection of Facilities

There are many facilities, operations, pieces of equipment, and other physical activities that need to be observed or inspected during an audit. In this section some fundamental concepts relating to audit techniques that should be employed during the on-site inspection will be discussed. The next chapter will present more specific examples of inspection items and typical compliance problems that auditors should look for in each of several compliance areas.

### Inspect Remote Areas

Many facilities will have remote areas that in the past or perhaps currently have been used for storage or handling of hazardous materials. It is important that the auditor thoroughly review the facility for these locations, because the plant staff may not realize they represent potential problems. A good example is shown in Exhibit 15.1, a picture all too common at industrial facilities. Trailers like that shown in the picture are favorite hiding places for forgotten drums of hazardous materials. The drums shown stacked up are empty according to the plant staff, but are they? Why does the road seem oily? Do the transformers contain PCBs?

Areas like that shown in the picture may have become part of the landscape to people working at the facility. The auditor's job is to diligently inspect the area to determine if potential problems exist.

The same can be said for remote buildings or buildings that are said to contain no hazardous materials or operations, as shown in Exhibit 15.2. Is this building truly free of hazardous materials? Are any liquid chemicals, including oils, stored near floor drains? What about asbestos? What about floor drains?

## Observations Must Be Timed to Verify Compliance

There are inspections of certain compliance areas that can only be verified adequately if the inspection is timed to occur at the optimal moment. This is normally related to equipment operation inspection items.

For example, air quality regulations in most states usually allow no visible emissions except during periods of start-up, maintenance, or trouble-shooting. During these events, short periods (four to six minutes/hour) of opacity are permitted. If the auditor does not observe the stack shown in Exhibit 15.3 during one of these events, a true verification of compliance may not be obtained. Observing the air emission source only during the period it is running in a steady-state condition may lead the auditor to the conclusion that the facility is always in compliance. A good technique, therefore, is to schedule the inspection of the source shown in Exhibit 15.3 at a period when it is started-up for the day or during a maintenance period.

**Exhibit 15.1**
**Inspecting Remote Areas**

Inspection of Facilities

**Exhibit 15.2**
**Inspecting Buildings**

Inspection of Facilities

## Observe the Facility From the Outside

The facility shown in Exhibit 15.4 may look different to the auditor from the outside looking in than it does from a perspective inside the plant. Nuisance conditions that are sometimes more obvious from a distance include odors, fumes, mists, excessive noise, fugitive dust, and even unknown discharges or runoff under the fence line. With the growing interest and awareness by the public of the community right-to-know legislation and their rights under these regulations, it is more important then ever that industrial facilities use the audits as an opportunity to evaluate and monitor the environmental profile they create in the neighboring community.

## Observe Contingency Plans In Action

Many environmental regulations require industrial facilities to prepare and maintain contingency plans that should include specific response actions in the event of inadvertent spills, discharges, fires, or other unplanned emergencies. Clearly, as part of the records review phase of the audit discussed earlier in this chapter, the auditor conducts an evaluation of the contingency plans for completeness. But what is also perhaps more revealing to observe first hand is how staff at the facility actually respond when an environmental incident occurs.

**Exhibit 15.3**
**Timing Inspections**

**Exhibit 15.4**
**Inspecting the Fence Line**

Inspection of Facilities

Accordingly, it is recommended that a drill or exercise be scheduled during the audit, so that the auditor can observe how the facility personnel react and if the plan is understood by those who are responsible for responding (*see* Exhibit 15.5). A drill or exercise may not be necessary if the auditor happens to be on-site when an actual emergency release or unpermitted discharge or leak occurs. If this happens during the inspection, the auditor should immediately seize this opportunity, stand back and observe how the facility staff carry out the procedures in the contingency plan.

## Observe Sampling/Monitoring Procedures

In some cases, chemical or physical sampling and analysis is included as part of an audit, as in the case of taking split samples of a wastewater discharge to verify NPDES permit compliance. However, the majority of audits being conducted today do not include this activity as part of their audit protocol. The prevailing attitude is that a one-shot grab sample is not a valid technique in evaluating compliance on a daily basis. Most firms seem to focus on auditing proper sampling and analysis methodology during the audit rather than expending resources on actual analysis of samples.

As shown in Exhibit 15.6, the technician is observed by the auditor to verify if proper sampling protocols are followed. Items observed would include: (1) use of clean sterile sampling containers, (2) not exceeding holding times and temperatures, (3) proper chain of custody procedures, and (4) samples sent to certified laboratories.

These fundamental audit techniques are considered to be effective mechanisms to verify compliance when auditors are conducting on-site physical inspections. Obviously, they must be used with judgment depending on the audit objectives, scope, resources available, size of facility, and other related factors.

Following these audit techniques, the next chapter will present detailed inspection items and compliance problems that auditors should look for. But before that, this chapter concludes with the final element of the on-site activities—the audit debriefing.

## Exhibit 15.5
### Emergency Response Drills

Demonstration of Contingency Plan

**Exhibit 15.6**
**Auditing Sampling Procedures**

## Daily Reviews

It is important that the audit team meet with the site environmental staff daily to discuss progress and interim findings. Based on experience, these meetings are better held in the morning rather than the afternoon. This gives the audit team the night before to organize themselves and cross-check the validity of each auditor's findings.

## Conduct Audit Debriefing

A debriefing should take place at the facility being audited before the audit team leaves the site. The debriefing serves several useful purposes. First, it provides an opportunity to review, informally, the audit findings before anything is written in a report. Misinterpretations and mis-understandings can often occur during the audit. The debriefing provides a forum to rectify and clarify any inconsistencies between the audit team and the plant staff.

For example, a required report, license, or certification may have been found to be lacking by the audit team. The debriefing may determine that the facility was recently exempted from this requirement or perhaps the audit team asked the wrong person for the document and, in fact, it is held by someone else at the plant. These are examples of typical misunderstandings that can be cleared up before the official audit report is submitted.

Another purpose for the debriefing is that it provides the audit team an opportunity to identify additional data or information it needs to complete its analysis. Sometimes, records or files are maintained at some location other than the plant (*i.e.*, division headquarters or corporate offices). During the debriefing, the audit team can list the information it wishes to receive, which was unavailable during the audit.

A third reason for the audit debriefing may be the most important. The audit team is likely to uncover activities, operations, or conditions that represent an imminent health or safety hazard to the facility personnel or the public. The debriefing is the time to identify these concerns so that immediate remedial measures can be taken by plant management. Leaking PCB transformers, cracks in containers holding hazardous materials, containers in questionable condition, ignition sources in close proximity to flammable materials, and contaminated wastewaters flowing into stormwater catch basins are but a few of the typical audit findings that require immediately attention by plant management. Waiting for the audit report to come out with these findings before taking remedial action can put the facility in a potentially high-risk situation.

Debriefing can be a critical part of the audit because that is the only time the plant staff will hear first hand about their compliance problems from an independent third party. However, if the auditors have not prepared for and organized the debriefing, the value of its impact may be lost on the staff at the facility. Provided below is a list of items in checklist format that auditors should review before conducting a debriefing.

*Preparing For The Debriefing*

1. Determine time, place, and invitees to debriefing as soon as possible so that all parties are properly informed.

2. Organize findings by compliance areas (air, wastewater, hazardous waste) and prepare a brief handout for people attending the debriefing.

3. Plan debriefing to be relatively short (less than one hour) and focus only on key compliance problems. Leave details to the audit report.

4. Organize findings according to either (1) regulatory compliance findings or (2) risk management findings.

5. Designate within each category each finding as minor, major, or significant depending on degree of risk, health impact, or other similar criteria.

6. Focus findings on the root causes of the compliance problems not the symptoms.

7. Rehearse the debriefing with other audit team members and challenge each other to be sure that the findings are valid.

*Conducting The Debriefing*

8. Introduce the audit team and state the objectives of the audit and the format for the debriefing.

9. Express appreciation to plant management for cooperation and assistance.

10. Try to avoid using specific names of plant staff when presenting negative findings but specify individuals' names when presenting commendable findings.

11. Avoid confrontation any with any plant staff in the debriefing who may disagree with your findings.

12. Avoid comparisons to another plant's performance unless it is helpful to make a point.

13. Resist requests to "re-audit" an area before the audit team departs the site so that findings can be left out of the audit report.

14. Leave the debriefing handout with all participants at the debriefing, summarizing key findings, but stress that they are preliminary findings.

15. Inform management of imminent hazards that need immediate corrective action.

## The Audit Report

Just about every audit results in a written audit report. This topic is so important it is the principal topic of another chapter in this book. In general, most audit programs are now attempting to develop draft audit reports or draft audit findings, at a minimum, before the team leaves the site. The team will usually have a report skeleton loaded onto a portable computer, which is brought to the audit. The introduction and facility background sections can even be written before the on-site activities. Preparing reports on site can make for some long evenings during the week, but this cost is more than offset by the fact that no (or very little) written work is required of the team once they leave the site.

# 16

# TYPICAL COMPLIANCE PROBLEMS FOUND DURING AUDITS

---

*"The emerging lesson in a company that ignores environmental requirements is that the middle of the corporate ladder can be a perilous place to perch. At the lower end...the "Nuremberg Defense" can be an effective escape.... And top management may be too far removed from actual or constructive knowledge of polluting activities."[1]*

---

In this chapter, typical compliance problems and inspection items are presented for each of several areas that may be a part of the scope of an environmental, health and safety audit. This listing is not intended to be all inclusive.

In the first section, for each of a variety of compliance areas, the compliance problems most often uncovered during audits are listed. The second section highlights many of the key inspection items associated with these same compliance areas. Photographs of typical plant facilities illustrate the inspection items, which an auditor should address when on-site.

Finally, at the end of the chapter is an auditor's checklist, which can be used as a reminder for the team leader as he or she prepares for the audit.

---

[1] Last, Michael P., "Toxic Torts Put CEOs at Risk," *Environmental Liability*, July/August 1989.

# COMMON COMPLIANCE PROBLEMS

## PCB Management

- PCBs in hydraulic systems or heat transfer systems not known

- PCB transformers not inspected quarterly and inspection records incomplete

- PCB transformers not registered with local fire department

- No knowledge of PCB content of utility-owned transformer located on property

- No labeling of PCB-free or PCB-contaminated transformers (making it difficult to identify all PCB transformers)

- Combustible materials stored next to PCB transformers

- Out-of-service PCB storage facility not designed properly, including lack of secondary containment, presence of drains and cracked floors, and lack of spill clean-up materials

- PCB-contaminated equipment and large, out-of-service PCB high-voltage capacitors stored outside PCB storage facilities, not stored on pallets, and not inspected weekly

- Liquid PCBs not stored in U.S. DOT-approved containers

- PCB annual reports missing or not prepared in accordance with EPA regulations

- Disposal records/manifests not available or incomplete.

**Wastewater Discharge**

- NPDES permit out of date, or re-application not submitted six months before expiration date

- Proper sampling procedures not followed

- Monitoring equipment not calibrated

- Discharge monitoring reports not submitted on time

- Occasional exceedances of discharge limits

- Excursions not reported immediately to regulatory agencies

- Changes in plant operations or discharges not reflected in permit revisions or renewals

- Inoperative or poorly maintained monitoring equipment

- Unpermitted process wastewaters in shop areas discharging to sewer lines, septic tanks, or streams

- Runoff from hazardous materials storage areas discharging to sewer lines or streams

- Discharges to a publicly-owned treatment works (POTW) contain materials that are fire or explosion hazards, corrosive hazards, or flow obstructions

- Unpermitted septic tanks, leach fields, or other on-site wastewater disposal areas

- Unpermitted stormwater contaminated discharges to storm sewer or streams

- QA/QC procedures not followed by analytical lab; lab not certified.

## Air Emissions

- Air emission points not completely identified, including stacks, vents, wall fans, exhaust ports, incinerators, and fume hoods

- Air emissions inventory not available or incomplete

- Air emissions sources not permitted or exemption letters not available

- Performance testing records not available

- Monitoring equipment not calibrated or maintained properly

- Vapor control systems not installed on required sources

- Submerged fill pipes not installed on gasoline fuel tanks

- Modifications to air sources not reported to regulators

- Air pollution alert and emergency plan not available or incomplete

- Sulfur content of fuel oil not in conformance with regulatory limitations for fuel burners; no certificate of analysis for sulfur content

- Fugitive dust impacting neighborhood areas

- Asbestos abatement or demolition activities not reported to regulatory agencies prior to the event

- Incinerator operations not maintained at correct temperature requirements and charging rates.

## Oil Spill Control

- Spill prevention control and countermeasure (SPCC) plan not signed by a registered professional engineer

- SPCC plan not updated in the past three years and not consistent with oil storage facilities at the plant

- No evidence of spill response training

- Lack of adequate spill control equipment and materials

- Lack of specific procedures for handling used oil

- Appropriate containment measures (dikes, berms) not installed around oil facilities

- Secondary containment structures cracked, drains present, or signage inadequate for volume of oil present

- Valves for secondary containment structures left in the open position

- No procedure for the assessment and disposition of accumulated rainwater in containment structures

- Oil/water separators not installed before stormwater is discharged to stream or sewer

- Oil/water separators not maintained and cleaned periodically for efficient operation

■ Aboveground tanks over certain capacities not leak tested and inspected periodically.

## Hazardous Waste Generation

■ Inadequate waste analysis plan, resulting in hazardous waste being handled as nonhazardous waste

■ Satellite accumulation drums improperly labeled and kept in open condition, venting to the atmosphere

■ Hazardous waste manifests file incomplete; return copies or land disposal restrictions notifications/certifications missing

■ Hazardous waste containers stored longer than 90 days

■ Accumulation start dates and other required information missing or incomplete on hazardous waste containers

■ Open top funnels and missing bungs on hazardous waste drums

■ Grounding clips and wires missing on stored drums of highly flammable hazardous waste

■ Accumulation point not maintained, with improper segregation of incompatible wastes, insufficient aisle space, and unsafe stacking height

■ Accumulation point manager not specifically designated in writing and not properly trained

■ Weekly inspection records of accumulation points not available, incomplete, or missing.

■ Emergency response equipment and material not available or inadequate at the accumulation point.

- Contingency plan not available or not maintained at the accumulation point.

## Community Right-To-Know

- Notification not made to state and local authorities

- List of chemicals stored on-site not reported to local emergency planning committee

- Hazardous chemical inventories (TIER I and TIER II) not submitted on an annual basis

- Toxic chemical release inventory (TRI) form (FORM R) not submitted by July 1 for previous year to local regulatory authority.

- Selected chemicals missing from Tier I, Tier II and TRI inventories.

## Worker Health and Hazard Communication

- Baseline exposure monitoring for hazardous chemicals in the workplace not properly documented for all affected employees

- No documentation available for employee noise exposures

- Periodic noise level measurements not routinely taken

- Written respiratory protection program not documented

- Emergency self-contained breathing apparatus (SCBA) equipment not maintained, tested, or stored properly

- Hazard warning labels not placed on chemical containers in shop areas

- Material identification labels missing on chemical containers in shop areas

- Personal exposure monitoring system not established for exposure to X-rays

- Written hazard communication plan incomplete or out of date, not updated for changes in operations and use and handling of hazardous chemicals in shop areas

- Employee notification of health monitoring results not given in a timely manner

- MSDS sheets not complete for chemicals in the workplace and MSDS files not maintained in an area accessible to employees

- Hazard communication training for employees not conducted or documented.

## Employee Safety

- Records and certificates for personnel trained in first aid and CPR not available for inspection

- Ventilation systems not routinely maintained, air flow rate not regularly tested, and filters not regularly changed

- Machine guards removed by employees or not properly installed

- Overhead cranes and lifting equipment not labeled with tonnage rating

- No safety manual or guidelines promulgated or posted for employees access

- Fire-fighting piping and hoses not hydrostatically tested or physically inspected on an annual basis

- Annual fire extinguisher training for employees not conducted or documented

- All exits not marked with illuminated signs, and exit paths not clearly marked

- Work-related injuries and illnesses records (OSHA 200 LOGS) not maintained properly

- Hard-hat, safety-shoes and safety-glasses procedures not uniformly monitored and enforced by plant management.

## Pesticide Management

- Pesticide applicators not state-certified or properly trained

- Health monitoring for pesticide-handling personnel not conducted or documented

- Pesticide-application records not maintained at the plant

- Pesticide storage facilities not segregated from other non- related spaces

- Concentrated pesticides improperly released to sanitary wastewater or stormwater drains

- Use of banned pesticides

- Pesticide storage not in a dry, well ventilated area designed for two air changes per hour

- Personal protective clothing, such as respirators, masks, gloves, coveralls, and equipment, not available

- Restrictions on application of "restricted use" pesticides not followed.

## Underground Storage Tanks

- Tanks not registered with state regulatory agency

- Monthly monitoring not conducted for new tanks

- Monthly inventory control and tank tightness testing not conducted every five years

- New tanks not installed with proper corrosion protection

- New tanks not constructed with proper spill/overfill protection

- New and existing suction piping not tested every three years

- New and existing pressurized piping not outfitted with automatic flow restructure, automatic shut-off device, continuous alarm system; no annual line testing or monthly monitoring.

## Hazardous Materials Storage

- Inside bulk storage of flammable materials not in compliance with design standards (containment, ventilation, self-closing doors, explosion-proof lighting)

- Flammable/combustible bulk storage inside buildings not meeting stack heights, aisle space, and quantity restrictions

- Outside storage areas for flammable/combustible materials not graded or contained to keep spills away from buildings

- NFPA-approved metal cabinets not used for incidental storage of flammable materials around shops and work areas

- Flammable materials not put in NFPA-approved safety cans when used around shop areas

- Dispensing areas for flammable materials missing drip pans under containers

- Grounding clips missing on containers of highly flammable materials

- Compressed gas cylinders not tied down and stored in segregated area in one-story building

- Spills of reportable quantities of hazardous materials not reported to the National Response Center and Coast Guard.

## Drinking Water

- Water withdrawal permits not obtained from regulatory agencies

- Water treatment plant operators not properly trained or certified, or documentation not available

- Drinking water not monitored for all required parameters (nitrates, fluoride, bacteriological, turbidity, radiological, trihalomethanes)

- Drinking water systems not checked for corrosivity

- Exceedances of primary drinking water standards not reported to state within seven days

- Public notification procedures not followed when maximum contaminant limits (MCLs) are exceeded

- Monthly monitoring reports not sent to state regulatory authorities for bacteriological and turbidity analysis

- Annual monitoring reports not sent to state regulatory authorities for radiological and chemical analysis.

## Solid Waste Disposal

- Solid waste disposed of/buried on-site without permit or approval from local regulatory agency

- Use of off-site disposal facilities that are not properly permitted for solid waste stream

- No knowledge of identity of off-site solid waste facilities being used by haulers

- Solid waste containing asbestos not sealed in leak proof containers and not disposed of at approved disposal facilities

- Solid waste receptacles not meeting design and operational specifications

- No procedures to insure that only nonhazardous waste materials are placed in solid waste receptacles

- Ash residues or sludges not tested for hazardous constituents before being disposed of as a solid waste

- Used oils contaminated with hazardous constituents being disposed of as a non hazardous waste

■ Effective recycling/reclamation procedures not in place for solid waste stream.

## TYPICAL INSPECTION AREAS

On the following pages are photographs of facilities normally inspected during an environmental audit. On the page facing each photograph is a list of the key inspection items the auditor should cover during the audit. These photographs arc only a representative sample of some of the more common inspection areas; many more facilities will be inspected during an actual audit.

## Exhibit 16.1
## In-Service PCB Transformer

**Key Compliance Issue**: Labeling and Inspection of PCB Transformers

**Exhibit 16.1**
**Typical Inspection Items**

---

1. Compare PCB inventory with PCB equipment on the facility for consistency.

2. Inspect PCB transformer locations, look for PCB labels affixed to equipment. They should contain phone numbers/contact for emergency response.

3. Look for presence of drains adjacent to PCB transformers and evidence of leaks from equipment.

4. Inspect any PCB transformer locations that pose an exposure risk to food and feed. Ensure weekly inspections are conducted for these units.

5. Inspect logs to ensure all other PCB transformers are inspected quarterly for leaks.

6. For PCB transformers found to be leaking, ensure cleanup is initiated within 48 hours.

---

**Exhibit 16.2**
**PCB Transformers Out of Service and Awaiting Disposal**

**Key Compliance Issue**: PCB Storage Facility Standards and
Requirements

**Exhibit 16.2**
**Typical Inspection Items**

1. Inspect PCB storage facility for the following:

   - Roof/walls to exclude rainfall
   - Six-inch curb or containment
   - No drains, valves, joints, or openings
   - Floors/curbing continuous and impervious
   - Doors locked/access controlled.

2. Verify that inspection of PCB storage facility is conducted at least every 30 days.

3. Observe presence of cleanup/absorbent materials in storage facility.

4. Observe that all PCB equipment in storage facility is marked with the date it was placed in storage. Disposal must occur within one year.

5. Observe that containers used for storage of PCB liquids do not have removable heads. Typical DOT specs are 5, 5B, and 17C.

6. Observe that PCB equipment found leaking is contained and cleanup is initiated immediately.

7. Observe that hazardous waste labels are affixed to PCB equipment awaiting disposal.

**Exhibit 16.3**
**Wastewater Treatment Facilities**

**Key Compliance Issue**: Proper Operation and Maintenance to Meet
Effluent Limitations

## Exhibit 16.3
## Typical Inspection Items

1. Observe general housekeeping, plant conditions, presence of odors.

2. Observe provision for standby power and look for adequate alarm systems in case of equipment failure.

3. Observe proper operation of all treatment units.

4. Observe spare parts inventory for critical equipment.

5. Inspect O&M manual for consistency with actual plant configuration.

6. Inspect operator and superintendent state certifications.

7. Inspect daily plant operating logs for neatness and clarity.

8. Observe daily sampling procedures at sampling points.

9. Observe effluent bypass points and discuss instances of any hydraulic or organic overloads at the plant.

10. Observe any on-site sludge disposal areas. Look for evidence of disease vectors, surface runoff control, seeding, or other cover measures.

11. Observe posting of safety or occupational hazards and contingency measures.

### Exhibit 16.4
### Stormwater Discharges

**Key Compliance Issue**: Contaminated Stormwater Discharge

## Exhibit 16.4
## Typical Inspection Items

1. Observe outfall for obvious signs of contamination (odors, discoloration, oil sheen).

2. Observe location for possible sources of contamination upstream of discharge point.

3. If permit is required, verify location is as described in permit.

4. If monitoring is required in permit, observe location of sampling points, proper operation of measuring for sampling device.

5. Observe sampling procedure of field technician, looking for properly cleaned sample containers, proper preservation techniques, hold times, and chain-of-custody procedures.

6. Inspect any upstream oil/water separators for proper operation and maintenance.

*Note: Recent EPA regulations require any stormwater discharge in an "industrial area" to be permitted.*

## Exhibit 16.5
## Hazardous Waste Accumulation in Drums

**Key Compliance Issue**: Labeling and Container Standards

## Exhibit 16.5
## Typical Inspection Items

1. Inspect containers awaiting transport off-site for leaks, corrosion, and bulges.

2. Inspect labels on drums for accurate description of waste. Ensure "Hazardous Waste" label is clearly visible.

3. Observe that accumulation start dates are on each container. Ninety days is the maximum period for storage; a permit is required.

4. Observe that containers are tightly sealed. Open funnels are not permitted. Bungs should be closed.

5. Containers of highly flammable waste should be electrically grounded.

6. Inspect accumulation points for presence of:

   - Telephone
   - Fire extinguishers
   - Spill cleanup equipment
   - Warning signs.

7. Drums should be on pallets and adequate aisle space maintained.

8. Observe that any containers holding ignitable or reactive waste are located at least 50 feet from property line.

**Exhibit 16.6**
**Hazardous Waste Storage Facility**

**Key Compliance Issue:** Storage Facility, Security, Emergency, and
Contingency Requirements

## Exhibit 16.6
## Typical Inspection Items

1. Observe facility is secured with locked doors, fenced areas, and is monitored by security guards.

2. Signs with wording "DANGER—UNAUTHORIZED PERSONNEL KEEP OUT" and "HAZARDOUS WASTE STORAGE AREA" are posted at each entrance.

3. Observe that communications equipment (telephone/radio) is present and operating.

4. Inspect safety and emergency equipment for proper operation.

5. Inspect the PSD inspection log. Verify that entries are properly recorded:

   - Date/time of inspection
   - Name of inspector
   - Notation of observances
   - Date/nature of repairs.

6. Inspect containers for tight seals, leaking, bulging, or rusting.

7. Inspect storage area for adequate aisle space.

8. Observe storage area for containment system:

   - Base is free from cracks and gaps, and impervious to spills or leaks.
   - Base is sloped to drain spills or leaks.
   - Containers are elevated to prevent contact with spilled liquids.
   - Capacity of containment system is adequate to contain ten percent of the volume of the container.

**Exhibit 16.7**
**Above-Ground Oil Storage Tank**

**Key Compliance Issue**: Spill Containment and Inclusion in SPCC Plan

## Exhibit 16.7
## Typical Inspection Items

1.  Inspect oil storage facilities and ensure that all facilities are included in SPCC Plan.

2.  Observe if any changes to oil storage/handling facilities have been included in amended SPCC Plan.

3.  Observe that appropriate containment (berms, dikes) are present to prevent discharge of petroleum products.

4.  Inspect condition of dikes for cracks or overflow.

5.  For tanks over 660 gallons, observe that secondary containment for entire contents is provided.

6.  Observe that dikes areas are impervious to spilled contents.

7.  Observe that appropriate cleanup equipment is available at the facility:

    ■ Absorbent materials
    ■ Oil retention brooms
    ■ Sand bags
    ■ Fuel recovery pumps
    ■ Protective gear (boots, gloves, respirators).

## Exhibit 16.8
## Solid Waste Incinerator

**Key Compliance Issue**: Particulate and Visible Emissions

**Exhibit 16.8**
**Typical Inspection Items**

1. Observe that incinerator is on facility emissions inventory.

2. Determine that emissions estimate has been calculated correctly.

3. Observe that permit to operate incinerator is posted and effective.

4. Observe that incinerator has multiple chambers fired by supplemental fuel.

5. Observe that timing/control mechanism for supplemental fuel feed operates properly to ensure complete combustion.

6. Observe visible emissions during combustion for exceedance of emissions limitations.

7. Observe security measures (fence, locks) are in place.

8. Observe area around stack for excess ash.

9. Observe allowable permit limitations for temperature or charging rates are followed during operation.

*Note: Incinerators with capacities greater than 50 tons/day are regulated under EPA's NSPS. Many states regulate incinerators with lower capacities.*

## Exhibit 16.9
## Solvent Metal Cleaner

**Key Compliance Issue**: VOC Emissions

## Exhibit 16.9
## Typical Inspection Items

1. Observe that cover is always in closed position when not handling parts in the cleaner.

2. Provisions for draining parts must be internal to degreaser.

3. Cleaned parts are allowed to drain for 15 seconds or until dripping stops.

4. Operating instructions are posted near the cleaner.

5. Solvent stream does not splash above freeboard level.

6. If solvent vapor pressure is more than 0.6 pounds/psi, or if temperature is above 120°F, check for either a water cover over the solvent, or a freeboard ratio of 0.7 or less.

7. Observe operation of any thermostats or safety switches.

*Note: Freeboard ratio is measured by dividing the height of the freeboard by the width of the tank.*

**Exhibit 16.10**
**Pesticide Storage Facility**

**Key Compliance Issue**: Proper Facility Design Standards and Storage
Requirements

**Exhibit 16.10**
**Typical Inspection Items**

1. Observe that the pesticide storage facility is not co-used by non-related functions.

2. Observe that storage, mixing, and preparation areas are separate from laundry, office, showers, and locker rooms.

3. Observe that facility is dry and well-ventilated with no drains or cracks in floor.

4. Observe that curbing or other containment is provided for mixing areas or decontamination areas.

5. Observe that all rinse or washwaters are contained and do not connect to sanitary or stormwater lines.

6. Inspect mixing area ventilation for six changes of air per hour.

7. Inspect storage area ventilation for two air changes per hour.

8. Observe that personnel protective clothing and equipment is provided:

   ■ Respirators
   ■ Masks
   ■ Gloves
   ■ Safety shoes.

9. Observe that no smoking, food consumption, or drinking is allowed in pesticide storage areas.

10. Observe that pesticide containers are kept closed when not in use, stored upright above the facility floor with all labels plainly visible.

11. Observe that signs which read "DANGER—POISON, PESTICIDE STORAGE" are posted near entries to storage facilities.

## Exhibit 16.11
## Hazardous Materials Dispensing Area
## for Shop Use

**Key Compliance Issue**: Proper Storage Spill Containment Procedures

## Exhibit 16.11
## Typical Inspection Items

1. Observe that drums/containers are not leaking and are tightly sealed when not in use.

2. Drip pans or absorbent materials should be placed under containers.

3. Observe that all containers are clearly marked with contents, and incompatible materials are separated.

4. Observe that dispensing areas are not located near catch basins or storm drains.

5. Observe that highly flammable (Class I) materials are electrically grounded.

6. Observe that approved safety cans are used for transporting and dispensing flammable liquids in quantities of five gallons or less.

7. Observe that chemical dispensing areas are located away from adjacent property lines.

8. Observe that "NO SMOKING" signs are posted near areas where flammable materials are dispensed.

## Exhibit 16.12
## Flammable/Combustible Materials Storage Facility

**Key Compliance Issue:** Facility Design Standards and Specifications

### Exhibit 16.12
### Typical Inspection Items

---

1. Observe that walls meet NFPA Fire Resistant Standards (2 HR) NFPA 251-1969.

2. Observe that wall/floor joints are liquid tight.

3. Observe that electrical wiring and equipment meet NFPA standards (explosion-proof).

4. Makeup air is vented directly to exterior of building.

5. Exhaust air is vented directly to exterior of building.

6. Observe that exterior markings and warning signs are posted on building.

7. Observe that stacked containers are separated by dunnage or pallets, and stacks are no closer than three feet to nearest beam, girder, or sprinkler.

8. Observe that all containers are marked with contents and hazard markings.

9. Observe that at least one 10-BC-rated fire extinguisher is located outside and within 10 feet of opening.

10. Observe that all positive sources of ignition (open flames, cutting, welding, radiant heat) are prevented near the storage facility.

---

## Exhibit 16.13
## Solid Waste Disposal Facility

**Key Compliance Issue**: Proper Segregation and Disposal of Solid Waste

## Exhibit 16.13
## Typical Inspection Items

1. Inspect sample of solid waste receptacles.

2. Observe that solid wastes receptacles are vermin-proof and waterproof, and have functioning lids.

3. Observe that any ash residues or sludges disposed of as a solid waste have been tested for hazardous characteristics.

4. Observe that asbestos wastes are disposed of in leak-proof containers, and are properly labeled. Verify that disposal facility is approved to accept asbestos waste.

5. Inspect plans of solid waste disposal facility. Verify that its operations are in compliance with permit conditions.

6. Observe that proper solid waste disposal procedures are current and posted for all employees.

7. Interview shop personnel for awareness of solid waste/hazardous waste segregation procedures.

**Exhibit 16.14**
**Drinking Water Sampling Point**

**Key Compliance Issue**: Proper Sampling Procedures and Protocols

## Exhibit 16.14
## Typical Inspection Items

1. Observe sampling procedures for the following:

   ■ Containers are clean, sterile.

   ■ Water is allowed to flow freely before sample is taken.

   ■ Record of sample is properly documented, including source of sample, time of collection, purpose of sample, and person collecting sample.

2. Inspect sampling points to verify they are representative of entire water supply system.

3. Observe wellhouse facilities. Look for sources of possible contamination to well casing.

4. Inspect chemical, bacteriological, and other analytical data for exceedances of primary drinking water standards. Discuss causes/sources of exceedances and remedial action taken.

5. Observe state certification of lab.

6. Observe certifications of treatment facility operators.

# AUDITOR'S CHECKLIST

Conducting an environmental audit is a complex exercise. It involves mobilization of a team, conduct of detailed regulatory reviews, coordination with site management, a relatively short on-site visit, and team preparation of a report.

In preparing for an audit, the team leader, in particular, will have numerous details to consider. The following is a checklist of items that will guide the team leader through the process.

### Preparing for the Audit

1. What is the scope of the audit?

   - Location of site (*e.g.*, state, county)
   - Acreage of site
   - Square feet of manufacturing space
   - Number of employees
   - Age of facility
   - Number of tanks (above and below ground)
   - Number of permitted air discharges
   - Presence of wastewater treatment plant
   - RCRA permitted TSD facility.

2. What should be decided beforehand?

   - Audit coverage (RCRA, CWA, etc.)
   - Team members (number, checklist responsibilities)
   - Duration of audit
   - Days of audit
   - Agenda
   - Hotel, cars
   - Handling of working papers (*e.g.*, bound journals)
   - Assigning a team member to every section of the audit report, including introduction and facility overview.

3. What should be done prior to the audit?

   - Establish plant contact
   - Send previsit questionnaire
   - Establish need and desire for:

     - Hard hats
     - Work clothes
     - Safety glasses
     - Work boots
     - Cameras
     - Security requirements

   - Set exact time of arrival
   - Get directions to facility
   - Get hotel recommendations
   - Request office space and telephone
   - Obtain names and phone numbers of state and county/city agencies
   - Make airline reservations, early, and for arrival the night before
   - Make sure plant will be operating during the audit
   - Ask that paperwork be assembled
   - Identify all key site staff
   - Ask if emergency response drill could be scheduled during the audit.

4. What should be reviewed before the audit?

   - Previous audit
   - Previsit questionnaire
   - Checklists
   - State/local regulations
   - New federal regulations
   - Plant background reports.

5.  What problems should be anticipated?

    ■ Didn't guarantee hotel room for late arrival
    ■ Previsit questionnaire not returned
    ■ One team member inexperienced or unprepared
    ■ Last minute request to delay the audit
    ■ One or more team members confused about responsibilities and/or arrangements
    ■ Team leader forgets to make confirming phone call the Friday before the audit week.

## Conducting the Audit

1.  Brief the site management.

    ■ Introductions
    ■ Team member responsibilities
    ■ Identify escort and key interview candidates
    ■ Purpose of the audit; supportive but thorough
    ■ Who initiated the audit
    ■ Audit approach (review of records, interviews, field inspections)
    ■ Program approach (how long for report, who gets it, what do facility staff need to do?)
    ■ Protocol if imminent hazard (*e.g.*, uncontrolled spill) is observed
    ■ Approximate schedule for debriefing
    ■ Who to call with concerns about the audit.

2.  Become oriented to facility.

    ■ Obtain plot plan
    ■ Ask for overview of manufacturing/processing operations
    ■ Ask for orientation tour.

3. Review paperwork.

   - Review important paperwork first (*e.g.*, permits)
   - Ask about making/taking copies
   - Read local newspaper.

4. Interview staff.

   - Interview all key players
   - Be courteous; schedule ahead
   - Ask open-ended questions
   - Interview staff in their workplace
   - Be supportive.

5. Inspect facility.

   - "Leave no stone unturned"
   - Walk the fenceline
   - Walk key roofs
   - Respect (but visit) secured areas
   - Identify all adjacent property uses
   - Visit during evening/night shifts
   - Schedule emergency response drill.

6. Conduct debriefing.

   - Conduct complete dry-run with team first
   - Acknowledge assistance of plant staff
   - Provide overall statement of performance
   - Have each team member present his/her findings
   - Be supportive, precise, thorough, and practical.

7. Complete report.

   - Decide on format and outline
   - Assign every section

- Decide on report manager
- Set schedule for first draft
- Set schedule for final draft and distribution.

8. Anticipate problems.

- Team members spend too much time reviewing records and not enough time interviewing or inspecting operations
- Lukewarm reception by plant staff
- No office available
- Site is more complicated than anticipated
- Paperwork wasn't assembled
- Audit team member addressing compliance areas in sequence and not in parallel; leaving one important area for the last day
- Escort was unaware of agenda or need for his/her significant involvement
- A significant compliance issue is raised by a plant staff member but there is no corroborative evidence
- One team member can't seem to complete his/her report sections on time
- A team member is imprecise in the debriefing, raising questions about the general competence of the audit team.

# CONDUCTING EFFECTIVE OPENING AND CLOSING CONFERENCES[1]

The room was filling slowly for the Friday afternoon meeting. There was a feeling of nervous anticipation among the staff of the large chemical plant located in Texas. Just then, the audit team leader announced that the closing conference would begin. The plant manager and his fifteen staff members took their seats alongside the five-person audit team. The team leader, noticeably on edge himself, began with "We found this site to be in compliance..." and paused for effect. As he paused, his team members looked at each other in masked disbelief. The plant manager did much the same, whispered a comment to his deputy, and left the room, seemingly satisfied that all was well. He returned two hours later, as the meeting was breaking up, and after the audit team had presented some thirty findings of significance resulting from the audit. Due to a misleading opening, the plant manager had missed the true message of the audit—all was *not* well at his facility.

This is a true story and took place in the 1990s on an audit in a Fortune 100 Company with a well-developed environmental (health and safety) audit program. It demonstrates that, if not handled correctly, a closing conference can jeopardize an otherwise excellent effort. Both opening and closing conferences are crucial elements of an effective environmental audit program. In fact, closing conferences, in particular, have been cited in a previous article as one of the six biggest challenges in maintaining an effective audit program.[2] And, as many firms go about

---

[1] Originally published in the fall 1994 issue of *Total Quality Environmental Management,* John Wiley & Sons, Inc., New York, NY.

[2] Cahill, L. B., "Benchmarking Environmental Audit Programs: Best Practices and Biggest Challenges," *Total Quality Environmental Management*, Executive

re-engineering[3] their audit process—which they are in fact doing—focusing on the beginning and end points of the on-site process makes eminent sense.

This chapter discusses a variety of issues that often arise during the preparation and conduct of opening and closing conferences. If these issues are dealt with appropriately, the audit will have a greater potential for success.

## OPENING CONFERENCES

The opening conference, not surprisingly, occurs at the very beginning of the audit. It typically involves the full audit team, the site manager, the site environmental, health and safety (EH&S) manager, and selected site staff. The basic intent is to provide face-to-face introductions, establish the audit objectives, set the agenda for the audit, and to discuss, in a preliminary fashion, issues raised by a review of pre-audit information. The meeting quite often has two phases: the first, shorter phase, involving the plant or site manager, followed by a longer issues-discussion phase during which the site manager may or may not be present.

The opening conference sets the tone for the entire audit. It is a meeting that can be stressful and is filled with potential "land mines" for the audit team. If the following six recommendations are kept in mind by the audit team, many of the mines can be avoided.

### 1. Take Charge of the Meeting

This may seem to be a simple point, but it is often lost on the audit team leader. The opening conference or meeting should be orchestrated by the audit team, not by site management. This can be especially difficult

---

Enterprises, Inc., New York, NY, Vol. 4 No. 3, Summer 1994, pp. 457-467.

[3] For more information see: Hammer, M. and J. Champy, *Reengineering the Corporation—A Manifesto for Business Revolution,* Harper Collins Publishers, Inc., New York, NY, 1993.

with a domineering site manager who is several positions higher than the team leader in the organizational hierarchy.

Certainly, the site manager will most often want to welcome the team, make general introductions, and provide an overview of the site's history and operations. But after this is accomplished, the team leader should take over, based on an agenda that has been discussed previously with site management. This is not to say that site staff don't have a major contribution to make during the meeting—they do indeed. However, the audit team is on-site only for a short while and will have quite an ambitious agenda; therefore, any substantial deflections can jeopardize meeting the objectives of the audit. One must keep the audit on track at all times.

## 2. Be Organized

Once the team leader assumes responsibility for the flow of the meeting, he or she should conduct the meeting in an organized fashion. A predeveloped series of slides and/or handouts works well to keep things moving ahead and to assure that all audits, regardless of the makeup of the team, are conducted consistently. This somewhat formal method is especially useful when discussing the audit objectives, approach, results, and follow-up. Quite often, companies develop the handouts as part of the audit program guidance document so that they are generally available to all auditors. A list of topics that should be covered in the opening conference includes:

- Audit objectives and scope
- Audit "flow" including follow-up and closure
- Identification of key interview candidates and availability (including obtaining relevant organization charts)
- A summary review of the operations at the facility (including obtaining plot plans)
- Identification of important site activities occurring during the week
- A review of the pre-audit information
- Scheduling the daily and closing conferences

- Identification of the audit team workroom and phone protocol
- Identification of site work hours and visitor safety and security protocols
- Identification of computer/printer support
- Discussion of the site escort protocol for visitors.

A couple of points here need to be kept in mind. The meeting should be crisp and to the point. The team's questions and inquiries should clearly reflect that they have done their homework; that the pre-audit information has been read and digested well in advance of the audit. There is no greater insult to the site staff than to have them realize that one or more audit team members has not reviewed (in detail) the pre-audit information that they so painstakingly prepared.

And finally, remember that not all questions will be answered in full during the opening meeting. Don't involve ten or so people in a protracted discussion of an issue that can be resolved much more easily and quickly in a follow-up, one-on-one discussion.

## 3. Discuss Logistics

One of the biggest challenges during the audit week is to manage the logistics of the effort. During the opening conference, the pre-set agenda should be modified as appropriate, and actual names of interview candidates and escorts should be placed in the "time blocks" of the agenda.

Notwithstanding this noble objective, one can be assured that within two hours of the formal initiation of the audit, a change to the agenda will need to be made. A new interview candidate will be identified, another will be called away for a day, or a surprise regulatory agency inspection will occur, justifiably deflecting the attention of the site staff.

The audit team needs to be responsive to the various incidents that might require schedule changes; flexibility and common courtesy are paramount. Several techniques are useful in this regard. First, the audit team should leave some open time near the end of the audit agenda. This can be used to accommodate needed changes. (I have never seen this time go to waste.)

Second, the auditor should address compliance areas in parallel, not in series. In other words, if an individual audit team member is responsible for several areas, such as wastewater, PCBs and spills, some work on each of these areas should begin early in the process, even if it is simply to review paperwork and schedule interviews. This accomplishes two things. It allows for better time management, minimizing the chance that an area will not be addressed at all if completely left to the morning of the last day. And it also avoids the issue of having the key site staff member take a holiday on the last day of the audit, unbeknownst to the auditor.

And third, don't think of the audit as only taking place from 8:00 a.m. to 5:00 p.m. each day. Conflicts in scheduling can sometimes be remedied through breakfast and dinner meetings. After all, what else is there to do far from home in a hotel two miles from the plant?

## 4. Find Out What's Happening at the Site

Far too often, audit teams miss opportunities to observe important activities that might occur during the week of the audit. Examples include:

- A spill drill
- A hazardous waste pickup
- Startup of a boiler
- Monthly wastewater or stormwater sampling
- Weekly or quarterly inspections
- A fuel or chemical delivery.

There is no better verification technique than to actually watch to see if something is done consistent with written procedures. These activities should be observed if at all possible.

I am reminded of an audit interview with the head of a medical research laboratory who was discussing the lab's wonderful waste segregation procedures—each station had both a "black-bag" and a "red-bag" container, which allowed for complete segregation of infectious wastes. As we discussed the procedures, I was looking over his shoulder and noticed a technician at a nearby lab hood seemingly discarding waste with no distinction being made between infectious and non-infectious wastes. Later, when interviewed separately, the technician stated that he thought the different bag colors were simply a result of the medical center's desire to "liven up the place!"

The team may never have uncovered this issue without actually being there to observe the activity as it took place. Take some time in the opening conference to schedule real-time observation of as many activities as possible; written procedures and what managers believe happen are not always consistent with reality.

## 5. Schedule Additional Conferences

Most mature audit programs include not only opening and closing conferences, but daily audit team and site EH&S staff meetings as well. During the opening conference, a tentative time should be set for the closing conference and the daily meetings with the site.

The daily meetings do take time away from the actual audit, but they are crucial in maintaining adequate communications. These meetings allow participants to review findings to date, resolve any mis-understandings early on, and revise the audit agenda as needed. Based on numerous observations, the daily meetings with site staff appear to work better first thing in the morning as opposed to last thing in the afternoon, which is the more traditional approach. That way, the morning meetings occur when people are fresh and not fatigued by the day. People on both sides of the table are often more articulate and more patient. The morning meeting also allows the audit team to have some time alone the evening before to prepare and compare notes.

## 6. Address Important Topics

There are certain special topics that should be addressed during the opening conference. If they are ignored, misunderstandings can arise. These topics are as follows:

- *There will be findings.* Site staff need to be told that an audit will result in findings, no matter how well managed the site is. Stated another way, audit findings in and of themselves are not necessarily reflective of poor management; it is more the number and significance of the findings.

■ *Audits require verification.* Acceptable audit methods require verification; hearsay evidence is generally not good enough. This implies that when a site employee claims, for example, that all staff members have received appropriate training, the auditor typically will not or cannot take that statement at face value; audit records will have to be reviewed. However, unless site staff are made aware of the need for verification, any given individual could be insulted if the auditor seemingly does not believe his or her statements.

■ *All areas must be covered.* Simply because the completed pre-audit questionnaire stated, for example, that there were no underground storage tanks or PCB-containing equipment at the site, does not mean that the audit team can ignore those areas. The person completing the questionnaire might not have considered a septic tank serving the maintenance shop as an "underground storage tank" and/or might not have known that fluorescent light fixtures sometimes have PCB-containing ballasts. These issues can be important and cannot be ignored.

■ *How will "fixed" findings be handled?* It is quite likely that some identified deficiencies will be remedied during the course of the audit. The audit team needs to inform the site staff on how these are to be handled in the report. Site staff typically will request that the fixed findings not be included in the report; however, most audit programs do include them but indicate that they were remedied during the audit. This approach is generally preferred because the findings, as trivial as they might be to the site staff, could sometimes be symptoms of a much larger site issue or even a corporate-wide issue. The larger issues could be lost if the symptoms are not included in the report.

- *How will "significant" issues be handled?* There may be, on occasion, a truly significant issue (*e.g.*, an uncontrolled PCB spill, operating without a permit) identified by the audit team. The team should not wait until the end of the audit before raising these issues with site management. Site management should be told that if a significant issue is uncovered, it will be brought to their attention immediately and the audit may be halted temporarily until a plan of action is implemented.

As can be gathered from the above comments, the "bottom line" for opening conferences is free and open communication.

## CLOSING CONFERENCES

The closing conference is the final on-site activity of the audit process. It typically involves the full audit team, the site manager, the site EH&S manager, and selected site staff; the participant makeup is thus pretty much identical to the opening conference. The basic intent of the meeting is for the audit team to report back to the site on the findings of the audit. The meeting should be crisp and to the point; a forty-five minute goal (not including the inevitable questions and comments) is usually appropriate.

More often than not, the closing conference is directly preceded by a final working meeting with site EH&S staff. This meeting is where most, if not all, of the "differing points of view" are resolved; it is not unlike the daily meetings discussed previously.

The closing conference can be a stressful meeting, especially if the audit team is working right up to the last minute to produce a typed, formal list of findings or a draft report, as many audit teams are directed to do. Keeping the following points in mind can help to reduce, but not eliminate, the stress.

### 1. Take Charge of the Meeting

Having the audit team take control of the closing conference is an even more important issue than for the opening conference, so it's a point worth repeating. The same issues discussed with respect to the opening conference apply here as well. An agenda should be established by the team leader; there should be an allowance for significant input by site staff; and the team should respect the position of the plant manager without being overly deferential.

Further, the correct message must be communicated to plant management, and this can only be done if (1) the team leader is in charge and (2) the plant manager is present. If, in fact, the plant manager is unavoidably absent, the team leader should find some way to communicate the findings to him or her separately, whether by phone or face-to-face at a different time and/or location. This direct communication is essential because no matter how well-intentioned the site EH&S staff is in relaying the information, there will always be a filtering or screening of the findings when transmitted secondhand.

## 2. Be Organized

The audit team must certainly be organized for the closing conference; most often, time with the plant manager, in particular, is limited. It would not hurt to once again review the audit objectives, scope, approach, report format, and follow-up requirements. Using the formal overheads or transparencies developed for the program, as discussed previously, is a useful technique. Although much of this information was transmitted in the opening conference, a second review helps when additional site staff are in attendance for the closing conference and, moreover, because the staff who were in attendance, for better or worse, will not have retained all of the information in the days that have transpired since the opening conference.

A proposed agenda for the meeting might be:

- General reintroductions
- Statement of objectives, scope, and approach
- Overall summary of the audit, highlighting particularly commendable items and significant deficiencies
- Review of findings by individual compliance area
- Discussion of the audit report format and schedule

■ Discussion of the corrective action report format and schedule.

An organizational issue that audit team leaders face is who should present the findings for the individual compliance areas. The audit team leader could make all the presentations himself or herself, or each auditor could discuss his or her individual compliance area. In most cases, it is advisable to have the individual team members present their own findings, even if their presentation skills are not quite as good as the team leader's. Using typed transparencies of the findings is a good support tool, even for experienced presenters, and it should be considered.

Having individual auditors present their own material is advantageous for several reasons. If the auditors know that at the end of the audit they will be standing before a demanding audience and defending their findings, they are more likely to be careful about both the accuracy and precision of each finding. They are also the persons who directly observed the deficiency and should be better able to respond to questions. And lastly, each of these closing conferences is an opportunity to improve presentation skills and, ultimately, to groom future lead auditors.

## 3. Be Appreciative but Don't Bury the Message

It is very difficult to stand up in front of ten to twenty people who might also be fellow employees and colleagues and go into detail about all the things they are doing wrong. Thus, more often than not, closing conferences begin with a speech by the audit team leader, thanking virtually everyone on the site for their help on the audit and praising them for their extensive knowledge and expertise. This is certainly courteous behavior and is appreciated by the site staff, but if the praise becomes too effusive, site management can misinterpret what is being said. Managers will focus on the recognized expertise of their staff and assume the deficiencies or findings are trivial, administrative shortcomings, whether or not this is true. Auditors should not overdo the thanks and acknowledgments; be courteous but be straight and to the point.

## 4. Set Priorities

On audits where there are numerous findings, some prioritization is in order. Many programs distinguish between regulatory, company policy, and "good management practice" findings. Each of these finding types are often classified by level of significance as well.[4] In the closing conference, some similar organizational structure is needed. The site manager should know before leaving the meeting whether there are any so-called "show-stopper" findings. These significant findings should be addressed either at the very beginning or the very end of the meeting.

## 5. Minimize Praise for "Acceptable" Performance

In school, scoring a 95 out of a possible 100 will get you an A+. In regulatory compliance, it could get you a major fine or jail time. People often forget this important distinction. And therefore, audit team leaders need to stress that 100 percent compliance with environmental rules and regulations is the expectation of enforcement agencies. Sites should not receive effusive praise for being "mostly" in compliance. This philosophy is quite consistent with total quality management principles of continuous improvement, no-defects production, and "six sigma" quality control.

## 6. Respond Professionally to Challenges

It is the goal of each audit team to resolve all conflicts with site staff prior to the closing conference. This is done mostly in the daily meetings and the final preparatory meeting with site EH&S staff. However, it is possible that controversies will arise in the closing conference.

The most important thing to remember in the closing conference is that the audit team should respond professionally to any challenge of the findings, if for no other reason than the site staff might be correct in their

---

[4] See Cahill, L. B., "Preparing Quality Audit Reports: Ten Steps (and Some Leaps) to Improve Auditing," *Total Quality Environmental Management*, Executive Enterprises, Inc., New York, NY, Vol. 4, No. 2, Spring 1994, pp. 319–324.

opinion. A few years ago, a member of my audit team was convinced through interviews that the required quarterly inspections of PCB equipment were not being performed at the site being audited. This was brought up as a finding in the closing conference and found to be incorrect; the auditor never interviewed the correct individual—the site electrician. This "error" had to be handled quite delicately.

In dealing with these potential controversies, it is helpful to use the guidance given in the classic 1981 book on negotiation—*Getting to Yes*.[5] In this text, the authors propose that it is best to argue your "underlying interest," not your "position." In the context of an audit, the team leader's position is quite often that "there will be no successful challenges to the team's findings during the closing conference." The position of the site staff might be that "there will be no significant findings resulting from the audit." If both parties enter the meeting with these positions in mind, it could be a difficult session. If, on the other hand, the team leader can convince participants, as well as himself or herself, that the underlying interest of all is to have a fair, thorough, and accurate audit of the site, things might work out a bit better.

## 7. Focus on Root Causes but Avoid Evaluations of Staff Performance

There is now a general trend in auditing to get behind the symptoms and to identify the root causes of the problems. This quite often results in a focus on management and organizational systems. The advantage of this approach is that focusing on root causes will more likely result in long-term fixes.

For example, a label could be missing from a drum containing hazardous waste. If this is reported as a finding, the site's response will most probably be to stick a label on the drum and be done with it. Yet, the audit team focusing on root causes would attempt to identify not only the "what" but also the "why" of the problem (*e.g.*, why was the label

---

[5] Fisher, R. and W. Ury, *Getting to Yes: Negotiating Agreement Without Giving In*, Houghton Mifflin, Boston, MA, 1981.

missing), and there could be half a dozen reasons. Focusing on these why's, such as lack of training, no ownership of the drum, poor labels, and so forth, should result in more appropriate corrective actions.

Yet, as valuable a tool as root cause analysis can be, it can also more directly identify an individual's or group's lack of performance as being at the core of the problem. This is a much more difficult point to raise in the closing conference and should be handled with great care. The old adage "praise in public, admonish in private" should apply. If there is indeed a sense on the part of the audit team that there is a chronic performance problem, the team might want to (1) make absolutely sure it's true (*e.g.*, maybe the individual did not receive the appropriate training) and (2) deal with it outside the public arena of the closing conference.

### 8. Understand How to Handle Repeat Findings

Every audit program should have a policy on how to handle findings that arise repeatedly from audit to audit. Many organizations will escalate the significance of the finding simply because it has not been corrected since the last audit. There is also the issue of how these findings are handled in audit reports. Some programs will specifically identify them as repeat findings, while others prohibit this practice because they are concerned about the potential liabilities of apparent "willful and knowing" noncompliance. Should there be any repeat findings on the audit, the audit team leader should be ready to address these issues consistent with corporate policy during the closing conference.

### 9. Avoid Comparisons

Whether they ask the question or not, site managers are always curious about how their operations measure up with others in the organization. Responding to this question in the closing conference is generally a no-win proposition for the audit team leader. If the team leader states that the site is exceptional by comparison, then any message of significance is lost on the site management. Also, the "about average" or "lower quartile" responses are almost never taken at face value without challenges.

Some programs do require the audit team to score the facility, and this is done in a variety of ways for a variety of reasons. This is occasionally done against an absolute four-point scale, such as:

- In compliance (4.0)
- Generally in compliance (3.0)
- Needs improvement (2.0)
- Substantially out of compliance (1.0).

Even this more simplified scoring can create disputes and arguments, however, given the competitive nature of most organizations.

Notwithstanding the above policies of scoring of some companies, it is best for the audit team not to provide comparative judgements when, in fact, sites cannot be evaluated fairly against one another. Sites handle different materials, use different technologies, are of varying ages, and are located in different physical and regulatory settings (*e.g.*, wetlands, non-attainment areas for air pollutants.)

## 10. Avoid Guarantees

Another wish of the site manager is that the audit will identify any and all problems so that if a regulatory agency inspection were to occur within a short time after the audit, no surprises would result. This guarantee should never be made by the team leader. An audit, especially a systems audit, is a spot check of the facility's operations using statistical sampling techniques. That is, it occurs over a limited period of time, with a limited scope, and not *all* records are necessarily reviewed. It is quite possible that an agency inspector could subsequently review a certain compliance area in considerably more depth. The inspector could, therefore, come across a compliance problem not identified by the audit team.

## 11. Leave Written Findings

It is remarkable to observe the look on the site EH&S manager's face when told by the audit team leader during the closing conference that the team will leave no written record of the findings with the site. There is usually a look of bewilderment and disappointment, and one can sense that the momentum built up during the week of the audit is in jeopardy of being lost.

Some believe that providing a verbal debriefing alone should indeed be sufficient feedback to the site during the audit. After all, the site staff is hearing the essence of the message. These same people would prefer that any written report, be it a draft list of findings or a draft report, receive an independent (legal) review before even a limited distribution to the site and others.

There are at least two problems with this approach. First, momentum can be lost as the site waits anywhere from two weeks to three months for the draft report. The lack of on-site written documentation lessens the likelihood of quick or immediate responses. One could argue that the liability stemming from not addressing an issue in a timely fashion as one awaits the draft report is greater than the liability associated with leaving a draft list of findings with the site, which may need legal "massaging" later. And in point of fact, the audit team leader should stress that if there are to be changes of significance between the draft list of findings and the draft or final report, site staff will be alerted prior to distribution of the report. This is common courtesy.

Second, there is quite often a major difference between the spoken and written word. Time and again people will agree with what the auditor is saying in a closing conference, yet once they see it in written form, they will disagree with both form and substance. This is because there are communications filters between the observation, the verbalization of the observation, and the documentation of the observation. Misinterpretations caused by these filters should be resolved face-to-face in the closing conference, not by "e-mail missiles."

## 12. Discuss the Next Steps

It is important for the audit team leader to close the meeting with a precise discussion of the next steps in the process. These steps relate principally to any follow-up work required of the audit team, schedule and distribution of the draft and final reports, and schedule and distribution of the corrective action plan. The team leader should explain that much is expected of the site in the final phase of the audit process. The site really determines the success of the audit in their dedication to correcting identified deficiencies in a timely manner. If all participants leave the meeting agreeing to the importance of this commitment, the audit has been a success.

---

*Acknowledgment: The author wishes to thank Ray Kane, with whom he has led scores of environmental audit training seminars over the years. Many of the ideas and issues raised in this chapter have been brought up not only on audits but also during these seminars.*

# 18

# PREPARING QUALITY
# AUDIT REPORTS[1]

*"Vigorous writing is concise. A sentence should contain no unnecessary words, a paragraph no unnecessary sentences, for the same reason that a drawing should have no unnecessary lines and a machine no unnecessary parts. This requires not that the writer make all his sentences short, or that he avoid all detail and treat his subject only in outline, but that every word tell."*[2]

*"Writing is easy. All you do is stare at a blank sheet of paper until drops of blood form on your forehead." (Gene Fowler)*

There has been much written on developing effective environmental audit reports. And this is usually done with some trepidation. Authoring a chapter on how to write effective audit reports is a bit like giving a speech on how to be an effective speaker. Anything but perfection will not be tolerated.

Notwithstanding this heart-felt reservation, this chapter examines a variety of issues that need to be considered when developing an audit report. These are issues that seem to never quite go away, no matter how many reports are written or how often the frustrated audit team leader refers an auditor to the classic Strunk and White book, *The Elements of Style*.

---

[1] Parts of this Chapter are adapted from "Preparing Quality Audit Reports: Ten Steps (and Some Leaps) to Improve Auditing," *Total Quality Environmental Management*, Spring 1994, Executive Enterprises, Inc., New York, NY.

[2] William Strunk, Jr. *The Elements of Style*. Third Edition, 1979.

The audit report is the document in which the findings of the audit team are presented with respect to the compliance status of the facility. As such, it is the *key* document of the audit program. The audit report should be a working, supportive document that will assist the facility staff in attaining and maintaining compliance.

Assuring that an audit report is accurate, precise, thorough, and helpful to facility staff is a considerable challenge. It requires proper preparation during the audit, deliberate efforts in the week following the audit, and a consistent follow-up program. This chapter discusses each of these equally important phases.

# FIELD PREPARATION

Preparing well in the field is vital to developing a useful audit report. In most cases, a return trip to the site to gather overlooked information is infeasible. Follow-up phone calls are about the only avenue open to the auditor. There are several things that an auditor can do to help assure that sufficient data has been obtained to write the report.

## Keep the Customer in Mind

One of the key tenets of Total Quality Management, "keeping the customer in mind," is often forgotten in the heat of the battle. Audit reports should be written principally for the site management—those responsible for correcting the deficiencies.

Yes, there are other customers. Senior management should have a sense of whether they have a problem site on their hands after reading the executive summary. And there should indeed be an executive summary, for quite often that is all some reviewers will read.

But the body of the report, the findings, must be written so that the reader both understands the nature of the problem and can readily envision a corrective action. Thus, when the team leader reviews the report, he or she should put his or her plant hat on and ask, for every finding, do I understand the problem well enough to know how to fix it?

If the answer is yes, then the auditor has done the job of serving the customer.

## Look for Underlying Causes

It is the responsibility of the auditor to determine the root cause of the problem. For example, the lack of secondary containment at a hazardous waste storage facility may be due to: (1) an untrained site manager who is unaware of the requirement, (2) the permitted use of a temporary location while a permanent one is being built, (3) a disapproved capital improvement request by division management, or (4) a general uncaring attitude on the part of site management. The recommended corrective action will vary depending upon which of the above is the underlying cause. In a case where site staff are not trained in the requirements of RCRA, simply constructing a dike or berm will not likely avoid the possibility of additional problems arising in the future.

## Organize Daily

An individual auditor can produce hundreds of pages of field notes during the course of the on-site work. How these field notes are recorded should be left to the personal preferences of each auditor. Some will use bound journals, some loose papers, and others loose-leaf or spiral-bound notebooks. Regardless of the technique that is used, it is imperative that the auditor take some time daily to organize his thoughts and findings. Otherwise, the auditor will probably be unable to decipher his own work when it comes time to write the report. Organizing the working papers daily also allows the auditor to determine whether any particular compliance area has been overlooked. This organization task may require allocating some time each evening in a hotel room, but this will be time well spent.

## Bottom-Line Interviews

An auditor might interview scores of people during the on-site effort. And interviewing is probably the most important of the field activities.

During an interview, an auditor is typically very busy; especially where, as is often the case, the auditor is conducting the interview alone. The auditor will be referring to the protocols, asking questions, hopefully listening to the answers, and taking notes. Too often, auditors are thinking about what the next question should be, as opposed to truly listening to the answer. The auditor should not be intimidated by "dead air time" during the course of the interview; take some time to allow your notes to catch up with the conversation.

Quite often auditors question whether a small, pocket tape recorder might not help in documenting interviews. In general, this is not an effective approach. Interviewees quickly clam-up when a recorder is present. If the auditor later wishes to record notes to himself on the tape recorder as a way to document findings, though, this would be acceptable.

Probably the most important thing an auditor can do is to "bottom-line" the interview. In other words, interviews should be scheduled so that time is available after the session to quickly summarize the results of the interview. The benefits of this effort will be realized later when the auditor attempts to write the report using field notes.

Another effective technique is to summarize the results of the interview at the end with the interviewee; that is, repeat back to the individual your perception of the key findings resulting from the conversation. The auditor will often find out that the recap has lost something in the translation, and this gives the interviewee a final opportunity to clarify any misunderstandings.

## Develop an Annotated Outline

The auditor should begin to develop an annotated outline of his or her sections of the report, beginning the very first day. This will help to prepare for the debriefing and will be a check to assure that all areas the auditor is responsible for are covered adequately. This effort will also aid in organizing field notes properly, as discussed above.

## Challenge Each Other

In most cases, the audit team will be required to debrief site management as a way of concluding the field work. It can be almost guaranteed that the audit team's findings will be challenged in this meeting. Therefore, it is important that the team members challenge each other's findings throughout the field work in preparation for the debriefing. This often proves to be a valuable exercise and will result in more precise findings and in double-checking sources of information. Each team member should be aware of the benefits of this internal challenging and should not immediately become defensive if another team member questions a finding.

## Develop a Consistent Debriefing Approach

The debriefing is a vital part of the entire process. It informs facility staff of significant findings and clears up any misunderstandings held by the audit team. If the debriefing is structured properly, it can also provide the basis for the audit report.

Included at the end of this chapter is a proposed format for the debriefing presentation: the Facility Environmental Compliance Status Form. The form summarizes regulatory and procedural deficiencies by compliance area. Significant, major, and minor deficiencies are counted and classified, and a color code is used to help facility managers set priorities among compliance areas. Together with an actual listing of the deficiencies, this approach has been used by the U.S. Air Force and works quite well as a report summary and debriefing handout.

# REPORT PREPARATION

Having properly done everything in the field, the auditor is now prepared to undertake what for many is the most difficult part of the process: writing the report. The audit report typically includes a discussion of the audit process, an overview of the facility, an executive summary, and a presentation of findings and recommendations. The report addresses both regulatory and procedural findings. A regulatory deficiency is one involving federal, state, or local regulations. A procedural deficiency is one involving internal requirements or good management practices. Findings are often classified as significant, major, or minor. Where a finding is corrected prior to the departure of the audit team, the report will contain the finding but will indicate it as corrected.

Audit team members will write those sections of the report that relate to the compliance areas reviewed. The audit team leader is responsible for assigning other sections of the report to himself or other team members (*e.g.*, Introduction, Facility Overview, Executive Summary). Based on past experiences, both good and bad, team members would do well to adhere to the following guidance:

## Organize for Monitoring

Many reports are rambling narratives in which it is difficult to sort background information from the findings. This does little to help the site or to allow for later tracking of corrective actions.

The audit report should be organized so that the findings are broken out, listed, and codified. And it is normally best to list only findings that are deficiencies. It makes little sense to codify a finding that requires no corrective action. If positive findings are listed, there will be a numbering gap in the sequencing of corrective actions that can confuse those responsible for managing the close-out process. Save the positive findings and commendable items for the executive summary.

One company, Hoechst Celanese, assigns each finding of each report a two-field numerical code, which includes the year of the audit and the number of the finding. When coupled with a code for the plant that was audited, this allows for computerized tracking of the closure of all findings. And this closure step is critical to the success of a program. In my independent reviews of audit programs, problems with the management of the findings follow-up and closure system is one of the most commonly observed deficiencies.

## Start Early

When the auditor returns to his office, he always does so with good intentions; the first order of business is to complete the audit report. But most people who have been away from the office for a few days to a week will spend the first couple of days back returning phone calls, reading mail, and catching up with the latest office politics. It is crucial not to let these distractions prevent the auditor from completing the report in a timely fashion. To paraphrase what Yogi Berra might say in this regard, "The more you put it off, the more you'll put it off." Procrastination builds on itself.

As an auditor, you do not want to be in the position of committing to write a report in the field as you undertake yet another audit. Nor do you want to be back at your office with several reports due to team leaders. You can become quite confused in trying to recall exactly what you observed at any given site.

There are some approaches that will help the auditor complete the report quickly. Developing an annotated outline before one leaves the site is a technique that works well, although one of the more interesting approaches is the increasing use of laptop computers in the field. In some cases, draft reports are basically completed before the team leaves the site. The laptops are also helpful in developing very impressive debriefing packages.

In summary, however, each auditor must do what works best for him. For many of us, simply taking an hour to organize our notes for one compliance area and writing the first paragraph is enough to get us going. If we continue to struggle, reviewing a previous audit report should be a source of ideas on how to approach the material.

## Establish a Report Format

Regardless of the report format chosen, it should be adhered to consistently so that subsequent audit reports are prepared in the same manner. This allows comparison of reports among different facilities to be made against common elements. It also allows for easier assessment of how a specific facility is improving its compliance status over time by comparing findings in one report to the findings in the subsequent audit report.

Provided at the end of this chapter is a general outline for an audit report. Although there is no one correct approach, many companies follow the outline presented. The philosophy of the report should be to make it as short as possible without compromising on necessary details. Long reports slow down the process markedly and jeopardize the relevance and currency of the findings. One should aim to produce reports of less than twenty to twenty-five pages if at all possible.

In the outline presented, Section I stipulates the who, what, where, and when information to orient the reader to the administrative aspects of the audit.

Section II is a brief executive summary written for upper management that highlights the key findings and recommendations of the audit report. The aforementioned Facility Environmental Compliance Status form should be included here.

Section III, Audit Findings, should be written with respect to noncompliance items as a first priority. Many companies take the viewpoint that the audit report should be an exception report; that is, the report should only document those items that were found to be deficient. Conditions found in compliance are not documented, but an individual section begins with a phrase along the lines of, "The facility was found to be generally in compliance with hazardous waste regulations with the following exceptions. . ." Other companies take the view that all findings, good and bad, should be documented.

The exception report approach appears to be the most prevalent form of documentation of audit findings by companies with audit programs. When reporting findings, every effort should be made to specify the regulatory citation that applies to the finding. This can be a federal citation (*e.g.*, 40 CFR 761.50), a state citation (*e.g.*, Pennsylvania Department of Environmental Resources Regulation 150.50), or a local regulation (*e.g.*, City of Philadelphia Sewer Ordinance 115).

Section IV, Recommendations, typically stipulates the auditor's suggestions on actions needed to come into compliance. These may range from simple administrative suggestions to recommendations for a capital improvement project (*e.g.*, a treatment plant upgrade). The recommendations may also focus on the need for additional investigation or further analysis before a final solution is chosen to correct the noncompliance condition observed.

One of the significant issues of concern to companies with audit programs is the liability created when recommendations documented in an audit report are not implemented. The only guidance that can be given is to document only recommendations that are known to be implementable by the facility. If there is some doubt that the audit's recommendations will be carried out (due to lack of resources, staffing or funds), then it would be better to recommend that the facility management come up with their own action plan to address the problems.

Specifying a responsible party and a time limit or date for a recommendation is more often left to the site's corrective action plan. The concern of having a recommendation in an audit report, which can't be implemented, is a valid one. Audit report writers should be sensitive to the positions that they potentially are putting facility managers into when they pose their recommendations.

Section V, Supporting Data and Information, is sometimes used to provide certain relevant backup information, such as tables of analytical data, copies of Notices of Violations (NOVs), plot plans, schematics, or photographs. There is great variability among audit programs regarding inclusion of this material in a report. Photographs, in particular, are excluded from many audit reports because they are considered to be potentially damaging and provide only a snapshot of the facility.

The benefits of including supplementary materials in a report should be weighed against the impact the material might have on the reader of the report. A picture may be worth a thousand words, but it might also be worth a thousand words to a plaintiff's attorney in the event the report is used as evidence in a lawsuit. A simple rule of thumb is to include only supplementary information that is absolutely needed to convey the point to the reader. If the point can be made clearly and completely in the body of the report, the use of supplementary material to support the finding should be discouraged.

## Pay Attention to Repeat Findings

The handling of findings that remain uncorrected from the previous audit is quite a challenge for most programs. Many lawyers will argue, rightfully, that labeling them as "repeat" findings in the audit report creates added liabilities for the organization. Yet, for senior managers, it is important to know if documented problems are being corrected. Some would say that not knowing whether problems are being fixed is an even greater liability than documenting repeat findings, and I would tend to agree.

One company, DuPont, has a specific and proactive policy regarding repeat findings. The policy calls for auditors to pay particular attention to problems identified in the previous audit that have not been corrected. In classifying these findings, auditors are asked to consider assigning them a higher risk priority than otherwise would be the case.

## Be Careful of "Good Practices"

Most audit programs classify findings into one of three types: regulatory, company policy, or good management practice (GMP), sometimes also called good engineering practice or observations. In my experience, difficulty often arises in crafting effective GMP findings.

I have observed hour-long discussions (or, more precisely, arguments) between the audit team and site management over the validity of a single GMP finding. The discussion usually centers on the justification for the finding. When the auditor says "that's the way we do it at our site" or "we will accept nothing less than best practice procedures", this usually raises the hair on the back of the site manager's neck.

There is a place for GMPs in an audit report. Regulations or company policies do not always address all practices that might pose a risk to the organization. However, there needs to be solid justification for each GMP finding.

One way to develop that justification is to frame GMP findings in the same way as suggested above for repeat findings. What is the situation? And what is the requirement (*e.g.*, protection of groundwater)? Too often, GMP findings begin with phrases such as, "The site should...", "The site needs to...", and the like. These are *not* findings; they are really "soft" recommendations. And this is the trap that many auditors fall into, crafting GMP findings as recommendations with no supporting justification.

## Set Priorities

Not all findings are created equal. Some are more important than others. Thus, it is usually helpful to categorize findings by significance. Some companies do this by bringing forward the most important findings into the executive summary. In other cases, individual findings are classified by level of significance. This can, of course, create heartburn among the legal profession, but it can be a valuable tool for management.

Some organizations (*e.g.*, the U.S. Air Force) use a scheme that is based on a classification system suggested by the U.S. EPA in its Federal Facilities Compliance Strategy. This results in a three-tiered classification, as follows:

Significant. A problem categorized as significant requires immediate action. It poses, or has a high likelihood of posing, a direct and immediate threat to human health, safety, the environment, or the installation mission. Some administrative issues can be categorized as significant. For example, failure to ensure that hazardous waste is destined for a permitted facility, failure to report when required, and failure to meet a compliance schedule are all significant deficiencies.

Major. A problem categorized as major requires action, but not necessarily immediately. This category of deficiencies usually results in a notice of violation from regulatory agencies. Major deficiencies may pose a future threat to human health, safety, or the environment. Immediate threats, however, must be categorized as significant.

Minor. Minor deficiencies are mostly administrative in nature. They may involve temporary or occasional instances of noncompliance.

The above classification scheme can be helpful because the definitions are based on the U.S. EPA's sense of significance, and it allows the auditor to highlight truly significant findings.

DuPont has taken this scheme and modified it to serve the purposes of its own environmental audit program. DuPont classifies its audit findings as follows:

**Level I: Immediate Action Required.** Situations which could result in substantial risk to the environment, the public, employees, stockholders, customers, the company or its reputation, or in criminal or civil liability for knowing violations.

**Level II: Priority Action Required.** Does not meet the criteria for Level I, but is more than an isolated or occasional situation. Should not continue beyond the short term.

**Level III: Action Required.** Findings may be administrative in nature or involve an isolated or occasional situation.

This classification has been helpful in setting facility action priorities within DuPont. No matter what approach is used, however, there should be some way to establish priorities among findings and to highlight those items that are truly in need of immediate attention.

## Be Clear and Concise

A good audit report should not be an attempt to write the great American novel. Reports should not be gripping, suspenseful, emotional stories. Findings should be written in a simple and straightforward manner based solely on observation, tests, or interviews. Use of general, vague statements will do little to help the reader understand the magnitude and nature of the compliance problems found during the audit. Best put, the writer should put himself in the facility manager's place and ask the question: If I were to read this report, would I know the exact nature of the problem and would I know how to resolve it? Again, as was said before, the auditor needs to be precise and accurate about underlying causes.

The importance of writing clear and precise audit reports and findings has been discussed at length in the classical audit texts. Here, the intent is to reinforce the importance of the topic and to suggest an approach to writing findings that can help achieve a quality product. Most audit findings can be broken up into three components: the situation, the requirement, and the reference. A sample finding written in this format is presented below.

**The Situation:** At the Building 27, ninety-day hazardous waste accumulation point, there was one drum containing waste solvents that had an open-top funnel.

**The Requirement:** RCRA requires that containers holding hazardous waste be closed except when adding or removing waste.

**The Reference:** 40 CFR 265.173(a)

Now, each finding does not have to be organized with the three headings listed as shown. But the intent of the organization is to first, in one to two sentences, precisely describe the *situation* so that the reader could readily go to the location in question (*e.g.*, Building 27) and locate the deficiency (*e.g.*, the drum with the funnel). This helps the customer solve the immediate problem.

By paraphrasing the *requirement*, the auditor is also helping the site staff understand what is needed. Finally, listing the reference allows them to conduct additional research if necessary. With a little more effort, even management systems findings can be handled in this way.

When using numbers to quantify a finding, put the information into its proper perspective. Finding three improperly completed manifests may mean that the facility has a major problem. One could simply report that there were discrepancies in manifest management at the site, leaving the reader to wonder if this is a major problem. However, if 500 manifests were reviewed and the problem was an incorrect abbreviation for drum versus missing return copies, the problem may, in fact, be trivial. This clarifying information should be included in the writeup.

The use of indefinite adjectives should be discouraged or minimized in audit reports. "Very," "some," "significant," "small," "high," and the like are often found throughout audit reports and their meaning and intent are wrongly left to the interpretation of the reader. One tip is to spend the time to count things when it's appropriate. For example, it is often a good idea to count drums at hazardous waste accumulation areas. Then, when it is reported that five drums were stored past the ninety-day deadline, the auditor can state whether it is 5 of 10 or 5 of 200. There is a difference.

The auditor should not attempt to prove a point using sensational language or hyperbole in an audit report. Words such as "dangerous," "negligent," "willful," "criminal," and the like are often over-statements and imprecise and can lead to misinterpretation by the reader.

The auditor should also use proper English and avoid slang and excessive use of acronyms. Sentences, according to writing experts, should be in the range of fifteen to eighteen words average length; that is, short, crisp sentences make the technical information easily digestible by the reader. Where acronyms are necessary, they should be spelled-out the first time they are used, or the auditor should include a table of acronyms in the report. One of the best books discussing writing style is Strunk and White's *Elements of Style*. This short book is an excellent source and is must reading for any auditor.

## De-emphasize Numbers

Total Quality Management principles suggest that numerical quotas should be eliminated. In the context of an audit report, this suggests that one has to be very careful how the *number* of findings is handled within the organization. It seems that no matter what is said or done, line managers and others have a tendency to add up the findings and compare. This is neither a meaningful nor constructive exercise and should be discouraged by senior management.

It is nearly impossible to make effective performance evaluations solely by keeping track of the number of findings. First, individual site situations, even within the same class of facility, differ greatly due to such factors as the surrounding area (*e.g.*, wetlands, non-attainment areas) and the state in which the facility is located (*e.g.*, regulatory stringency).

Second, audit teams differ in composition and makeup. Based on many of my own training experiences, one can send two different audit teams to the same location and have them audit something as straightforward as an accumulation point and wind up with a wide variance in the number of findings.

And third, as audit programs have evolved over the past few years, there has been a movement towards evaluating management systems. This has resulted in the "rolling up" of individual compliance findings into an overall system finding. This makes comparisons over time even more difficult.

In summary, using the number of audit findings as a measure of performance is fraught with problems. As an alternative, some companies have asked audit teams to classify the site overall. An example ranking system, based in part on a scheme sometimes used by financial auditors, might be to classify the site as follows:

- Good
- Qualified
- Needs Improvement
- Unacceptable.

This approach is more compelling, although each of the classifications would have to be defined.

## Use Evidence in the Discussion of Findings

During the course of the audit, evidence will be obtained using one of three methods: inquiry, observation, and test. Inquiry is evidence gained through interviews; observation through, not surprisingly, looking around; and test through means such as running an emergency response drill or through physical sampling. When writing findings in the audit report, the method of discovery should be inherent to the language used. For example, where appropriate, phrases such as "It was observed that..." or

"It was reported that..." should be used. This phrasing gives more meaning to the auditor's findings.

The auditor should be very careful to avoid outright conjecture, or conclusions deduced by guesswork. This is a common pitfall in auditing. There should be at least one solid piece of evidence for each finding and probably two pieces of evidence for each significant finding. Speculation on the part of a maintenance supervisor that there was a fifty-gallon spill of perchloroethylene fifteen years ago is not sufficient evidence. The supervisor may have been reprimanded recently and possibly is looking to damage the reputation of the plant manager. When the auditor does run into this kind of situation, an effort should be made to find corroborative evidence. This could include independent confirmation by another facility staff member, a spill report, or regulatory agency correspondence. Where no confirmation can be made, the auditor should raise this issue at the debriefing. Usually the issue will be resolved one way or the other in this forum.

## Avoid Common Pitfalls

Each of us falls into writing traps that we cannot seem to avoid. Presented below are some of the more typical problems evident in audit reports. Ten sentences or paragraphs are listed and the problems with each are listed directly afterward. There may be other problems as well, but the problems pointed out are probably the most important ones to remember.

### Pitfall Examples

1. The audit team inspected 20 transformers and 10 capacitors, of which 3 showed past signs of leaks.

   *(Indefinite modifier; it is unclear which of the equipment is leaking. "Past signs of leaks" is very imprecise, especially for PCB equipment where a stain along the side can be considered a leak. The sentence is not helpful to facility management; they need to know exactly which of the equipment is leaking.)*

2.  Several of the drums at the hazardous waste accumulation point had no labels, as required by 40 CFR 262.

    *("Several" is imprecise; which drums? It would be helpful to know several of how many drums had no labels. The regulation does not require "no" labels as the sentence suggests.)*

3.  Three PCB transformers at the site did not have the required labels.

    *(Three of how many transformers, and where are they? Same basic problem as No. 1 above.)*

4.  The hazardous waste accumulation point had no secondary containment.

    *(Is this a regulatory requirement or a good management practice? Federal regulations do not require secondary containment for accumulation points, but some states do require it.)*

5.  It is possible that some previous spillage occurred at the bulk loading transfer station.

    *("Some previous spillage" is imprecise. There is no citing of evidence. Was it observed or was the information gained through interviews?)*

6.  A unit operator reported that the wastewater treatment plant is bypassed during severe storm events. This violates the site's NPDES permit.

    *(This is a significant issue and before it is written in an audit report, a second, confirming source is needed. "Violation" is an inappropriate and volatile word.)*

7.  Because of the possibility of solvent releases into the sewers, the wastewater treatment ponds might have to be permitted as a RCRA surface impoundment.

*(A mixture of a finding and a recommendation in the same sentence. This is broad conjecture and is not appropriate for an audit report. Also, evidence is not presented.)*

8. The permitted hazardous waste storage facility at Building 51 was recently inspected by NJDEP and found to be in compliance with 40 CFR 264 standards. Used oil at the site is accumulated at Building 52 and is disposed of monthly by Used Oil Reclamation Co. Used oil is exempt from RCRA regulations.

*(Auditor relies completely on regulatory agency inspection; this is a very dangerous approach, since an inspection next month by a "more seasoned" agency inspector could result in numerous NOVs. The analysis is incomplete; used oil is a hazardous waste in New Jersey.)*

9. The fire training area reportedly has been contaminated with waste solvents, used as recently as five years ago as supplementary fuel for the fires. The audit team recommends that the facility file for a RCRA permit as a treatment/disposal facility and immediately excavate the contaminated soils.

*(Broad conjecture and a completely inappropriate recommendation. Further study should be undertaken before excavation is initiated. Also, facilities don't file, facility managers file.)*

10. As part of the hazardous waste audit we went to the point where the hazardous wastes are accumulated at the site and we reviewed whether the drums had labels and whether there was enough aisle space for a fork truck and whether the hazardous waste accumulation point was inspected weekly and whether all the other RCRA regs were complied with. Their were a plethora of violations ...

*(A wonderful example with a few small problems, including a run-on sentence, use of slang (e.g., regs), misspelled word (their), use of pretentious word (plethora), use of volatile word (violations), and ending a sentence with a preposition (with). Also, the auditor is*

*writing the report as he did the audit; he should separate the process from the product. Finally, the report is written in the first person; most audit reports are written in the third person.)*

## REPORT FOLLOW-UP

With the report now completed, one could assume that the process is complete. It is not. Audit follow-up is a crucial part of the entire process. Without adequate follow-up, the system is likely to fail. One should pay special attention to a few items in particular.

### Assure Legal Review of Reports

Some companies manage their audit programs through their legal departments, due to concerns about protecting the confidentiality of the findings. Potential protections such as attorney-client privilege, work product doctrine, and the principle of self-evaluation privilege were discussed in Chapter 3. Whether or not the audits are done under these mechanisms for protection, it is most prudent to have a lawyer review the draft audit report. The lawyer can help to ensure that legal references in the report are correctly stated and applied. More importantly, the lawyer can make suggestions relative to the wording and tone of the report from a legal view. It is a good rule of thumb to assume that the audit report may be discoverable in a subsequent legal proceeding against the company. Review of the draft report by the lawyer may be very helpful in the event that he or she is required to defend the company in environmental litigation.

### Limit Distribution of the Report

The draft report should be sent to the facility manager and his immediate supervisor for review. This report should be distributed within two weeks of the audit. A time limit should be stipulated for responses so that any disagreements can be resolved quickly. The plant staff should have no longer than a week to respond. Upon review, a final report should be issued, together with an action plan. The final report should be distributed to upper management as well as to the facility manager and his immediate supervisor.

Producing more copies of the audit report than those needed, as stated above, should be discouraged. These extra copies have a way of finding their way to people who don't necessarily have a need to know. Many companies number the copies so that distribution can be controlled and persons held accountable for their copies of the report. Once a final report has been issued, many companies have adopted a policy of destroying the draft copies. Final reports are kept either indefinitely or until the next audit report is issued for that site. The auditors' notes and working papers are normally destroyed after the audit report has been issued. Obviously, it is a good idea to develop a formal records retention policy covering audit reports and associated documents and papers.

## Accept No Mistakes

Not much needs to be said here. Audit reports should be mistake-free. Even the smallest typographical error should not be allowed, not even in a draft report. And, in fact, in the audit business, much like the consulting business, there is really no such thing as a draft report. If it is to be sent to the client or customer, it should be considered final, no matter if words like draft or preliminary are used.

Every report should have a third-party review by someone not on the audit team. Also, do not rely solely on word processor software spell checks to eliminate all typos; they will not. A human must conduct the final check.

And, based on experience, it is better to take the heat for a report being a day or two late than for it being sloppy. It is quite interesting that people will remember that report with all the typos for ten years afterward and will forget the perfect report within a month.

## Remove Barriers to Efficiency

Auditors must be given the tools to be able to develop a quality report. This includes a sample audit report to be used as a guide. Many companies, such as Kodak, have also developed a separate audit report-writing guide, which typically includes a discussion of the writing process, pitfalls to avoid, and a sample report with sample findings. Kodak has gone one step further and has developed a database of over one thousand written findings from past reports, codified by category, which assist auditors in the field. Also, in-field use of laptop computers has done much to improve the efficiency of the process. Draft reports can be developed "on-the-fly."

## Develop Action Plans

Approaches to follow-up actions on audit findings vary among companies. Still, this is viewed as one of the key factors in having a successful program. As discussed elsewhere, each audit report typically contains an action-item table that lists each finding, the associated recommended remedial action, the responsible party, and the expected date of completion. Tracking can be accomplished by follow-up phone calls to facility management or by expedited follow-up audits.

Recently, the action-item tables quite often have been entered onto a computer database, and monthly tracking reports are generated to monitor progress. Responsibility for this monitoring usually rests with the audit program leader, who may delegate it to audit team leaders for particular sites. Whether plant staff have the ability to revise information on the database is an option that is handled in a variety of ways. In some cases, the system is secured from entry by anyone but an auditor.

Regardless of which method is used, the audit process is not complete unless some formal measures are instituted to ensure that each of the audit findings and recommendations are being addressed by the facility staff. Clearly, some recommendations made in an audit report may involve large capital expenditures that cannot be made immediately. In this case, placing the item on a subsequent year's capital expenditure plan is an appropriate step to take, as long as this action does not implicitly condone continuing noncompliance. All corrective actions do not necessarily have to take place immediately. As long as the facility management has a schedule for completion and is tracking the status of corrective measures, it can demonstrate a good faith effort to operate in an environmentally responsible manner.

## Train the Auditors

Train, train and retrain! All auditors should receive formal training before they embark on their first audit. And a good part of this training should be on report writing, or at minimum, on the writing of individual findings. Providing on-the-job training on how to write findings at an inopportune time (*e.g.*, on a Thursday night at 11:00 p.m.) for someone's first audit is no fun. Unfortunately, though, it happens too much.

Preparing a quality audit report is a most difficult task. It takes time, effort, and lucidity. But the effort is well worth it. The report is *the* decision-making document for the environmental audit program.

---

*Note: Any discussion of a company's individual environmental audit program approaches in this chapter is based on information provided to the public at large through technical papers, presentations, and the like.*

# Figure 18.1
## Facility Environmental Compliance Status

**ATTACHMENT**

**FACILITY ENVIRONMENTAL
COMPLIANCE STATUS**

| | Compliance Area | Deficiencies[1,2] | | | Compliance[3] Summary |
|---|---|---|---|---|---|
| | | Sig | Maj | Min | |
| 1. | Air Quality | ☐ | ☐ | ☐ | _____ |
| 2. | Hazardous Wastes | ☐ | ☐ | ☐ | _____ |
| 3. | Wastewater | ☐ | ☐ | ☐ | _____ |
| 4. | Stormwater | ☐ | ☐ | ☐ | _____ |
| 5. | PCB's | ☐ | ☐ | ☐ | _____ |
| 6. | TSCA | ☐ | ☐ | ☐ | _____ |
| 7. | Electromagnetic Radiation | ☐ | ☐ | ☐ | _____ |
| 8. | Low-Level Radio-active Waste | ☐ | ☐ | ☐ | _____ |
| 9. | Pesticides | ☐ | ☐ | ☐ | _____ |
| 10. | Underground Storage Tanks | ☐ | ☐ | ☐ | _____ |
| 11. | Hazardous Materials | ☐ | ☐ | ☐ | _____ |
| 12. | Pathological Waste | ☐ | ☐ | ☐ | _____ |

**Figure 18.1** *(cont'd)*

## FACILITY ENVIRONMENTAL COMPLIANCE STATUS

| Compliance Area | Deficiencies[1,2] Sig Maj Min | Compliance[3] Summary |
|---|---|---|
| 13. Community Right-to-Know | ☐☐ ☐☐ ☐☐ | _____ |
| **Others** | | |
| 14. _____ | ☐☐ ☐☐ ☐☐ | _____ |
| 15. _____ | ☐☐ ☐☐ ☐☐ | _____ |
| 16. _____ | ☐☐ ☐☐ ☐☐ | _____ |
| Total Deficiencies | ☐☐ ☐☐ ☐☐ | _____ |

Explanatory Notes:

1.  | A | B |

   A.  <u>Regulatory Deficiencies</u>. A regulatory deficiency is one involving regulations external to          (Federal, state, or local environmental requirements). A regulatory deficiency may be classified as significant, major or minor.

## Figure 18.1 *(cont'd)*

**FACILITY ENVIRONMENTAL
COMPLIANCE STATUS**

Explanatory Notes: (Continued)

    B.   Procedural Deficiencies. A procedural deficiency is one involving requirements. A procedural deficiency may be classified as significant, major, or minor.

2.   SIG/MAJ/MIN

    A.   Significant Deficiencies. A problem categorized as significant requires immediate action. It poses, or has a high likelihood to pose, a direct and immediate threat to human health, safety, the environment, or the facility's mission. The EPA publishes listings of significant violations by media program. (See Attachment). Some administrative issues can be categorized as significant. For example, failure to insure that hazardous waste is destined for a permitted facility, failure to report when required, and failure to meet a compliance schedule are all significant deficiencies.

    B.   Major Deficiencies. A problem categorized as major requires action, but not necessarily immediately. This category of deficiencies usually result in a notice of violation from regulatory agencies. Major deficiencies may pose a future threat to human health, safety, or the environment. Immediate threats must be categorized as significant.

    C.   Minor Deficiencies. Minor deficiencies are mostly administrative in nature. They may also involve temporary or occasional instances of non-compliance.

## Figure 18.1 *(cont'd)*

3.  Compliance Summary System - This summary will be used in all evaluation reports.

    B - Blue        No deficiencies.

    G - Green       Minor deficiencies.

    Y - Yellow      Major and minor deficiencies only.

    R - Red         At least one significant deficiency.

**Figure 18.1** *(cont'd)*

**EXAMPLE DEFINITIONS
OF
SIGNIFICANT NONCOMPLIANCE**

**U.S. EPA[1]**

| Media Program | Term/Definition |
|---|---|
| Water/<br>NPDES (Continued) | - A pass-through of pollutants which causes or has the potential to cause a water quality problem or health problems. |
| Hazardous Waste/<br>RCRA | ° Significant noncompliance are land disposal facilities with Class I violations of groundwater monitoring, closure, post-closure or financial responsibility requirements. A Class I violation is defined as a violation that results in a release or serious threat of release of hazardous waste to the environment, or involves the failure to assure that groundwater will be protected, that proper closure and post-closure activities will be undertaken, or that hazardous wastes will be destined for and delivered to permitted or interim status facilities. |
| Hazardous Waste/<br>CERCLA | ° The Superfund equivalent of a significant noncomplier is those facilities on the National Priorities List (NPL) that will require remedial and/or removal actions. |
| Toxic Substances/<br>TSCA | ° Significant noncompliance is a violation for which the enforcement action, is at minimum, an administrative complaint, and includes but is not limited to the following: |

## Figure 18.1 *(cont'd)*

### EXAMPLE DEFINITIONS
### OF
### SIGNIFICANT NONCOMPLIANCE

### U.S. EPA[1]

| Media Program | Term/Definition |
|---|---|
| Toxic Substances/ TSCA (Continued) | - Any PCB violation involving improper disposal, manufacturing, processing, distribution, improper use, storage, recordkeeping and/or marking violations. |
| | - Any PCB disposal resulting in contamination of surface or groundwater, food or feeds. |
| | - Any violation of testing requirements under TSCA Section 4. |
| | - Any violation of a premanufacturing notification under TSCA Section 5. |
| | - Any violation of TSCA Section 13 including: failure to either certify that all imported chemical substances are in compliance with TSCA or are not subject to TSCA, and falsification of a certification report. |
| | - Any violation of TSCA Section 8 including: failure to submit required records, falsification or records, and incomplete reporting and/or recordkeeping. |

**Figure 18.1** *(cont'd)*

## EXAMPLE DEFINITIONS
## OF
## SIGNIFICANT NONCOMPLIANCE

## U.S. EPA[1]

| Media Program | Term/Definition |
|---|---|
| Air/<br>CAA | °  <u>Significant violator</u> meets any one of the following requirements:<br><br>- A source in violation of a hazardous air pollutant standard, other than a source violating asbestos demolition and renovation requirements.<br><br>- A source in violation of new source requirements, including NSPS, PSD, and Part D nonattainment permitting requirements.<br><br>- A Class A source in violation of a SIP if the source is located so as to impact a nonattainment area and is in violation for the pollutant for which the area is nonattainable.<br><br>- A source in violation of a Federal consent decree or Administrative Order. |

## Figure 18.1 *(cont'd)*

### TYPICAL FORMAT OF AN ENVIRONMENTAL AUDIT REPORT

---

**I. INTRODUCTION**

1. Who conducted the audit?

2. Where was the audit conducted?

3. When was it conducted?

4. Why was it conducted?

5. What was the scope of the audit - what compliance areas did it include?

   - Wastewater discharges
   - Air emissions
   - Hazardous waste
   - PCB
   - Solid waste
   - Oil spill containment
   - Other

   Note: This is a relatively straightforward section, probably no more than 1-2 pages in length.

**II. EXECUTIVE SUMMARY**

A brief 1 page summary of major findings and recommendations intended for upper management review.

**III. AUDIT FINDINGS**

a. Activity/operations - Brief statement of what occurs at the facility (i.e., "The facility operates one 250 MBtu boiler, one 50-ton solid waste incinerator, and three 30,000 fuel tanks, which are a source of VOC air emissions.")

b. Requirement - Brief statement of citation and what it requires (i.e., "COMAR 10.18.06 requires fuel burning facilities to produce no more than 20 percent opacity for more than 6 minutes during start-up operations.")

c. Findings - Brief statement of what was observed during the audit (i.e. "During start-up of No. 1 boiler, it was observed that opacity reached about 50 percent and remained so for about 20 minutes.")

---

**Figure 18.1** *(cont'd)*

## TYPICAL FORMAT OF AN ENVIRONMENTAL AUDIT REPORT
(continued)

2.  Wastewater discharges

    Same sections a, b, and c as above.

3.  Hazardous waste

    Same sections a, b, and c as above.

4.  PCB management

    Same sections a, b, and c as above.

5.  Any other compliance areas that were audited would similarly be documented.

Note:     These sections can consist of many different findings. For example, air emissions could have as many as 10 different citations listed with associated findings. Wastewater could have only 1 or 2 findings or as many as 10. The length of this section can typically vary anywhere from 5-20 pages.

IV.  RECOMMENDATIONS

1.  Air emissions

    a.   "It is recommended that boiler plant operators be trained in the requirements for limited visible emissions during boiler operations."

    b.   "A written procedure should be established specifying proper start-up procedures to limit visible emissions to no greater than 20 percent for no more than 6 minutes. The procedure should include periodic monitoring of visible emissions by plant environmental staff using the Ringelman chart method."

Note:     The above recommendation is an example of how the findings identified in Section III of the report could be addressed. Similar recommendations would be documented here for other noncompliance.findings listed in Section III, Audit Findings.

V.  SUPPLEMENTARY DATA AND INFORMATION

If appropriate, copies of analytical data, NOV's, schematics, and photographs would be included here.

# 19

# ENVIRONMENTAL AUDITING: THE GOOD AND THE BAD[1]

# A MODERN FABLE

---

*"What You Don't Know Will Hurt You."*[2]

---

## THE COMPANY

The year was 2050 and times were good at PEP, Inc. The firm was indeed living up to its name (Purchase Everything Possible) and had just passed the two largest law firms in the United States to assume the number one position on the Fortune 500 list. It was the first time in 10 years a law firm wasn't number one.

A not insignificant reason for the rapid growth of PEP was the performance of its newest subsidiary—Moon Rock, Inc. The founders of MRI had discovered that if certain rocks on the moon were ground to dust, refined using a special solvent, and combined with several "active" ingredients, the resultant product (sold as Klairol's Hairall) would arrest male-pattern-baldness. Not surprisingly, since the discovery five years ago, MRI's revenues had grown by 1,000 percent each year.

---

[1] This story is a fictional case study. Any terms that may relate to existing organizations or products is purely coincidental.

[2] F. Friedman, Vice President, Occidental Petroleum Corporation, in Speech to Chemical Manufacturers' Association, 1983.

Yet, as with most things in life, Hairall could not be manufactured without some risk. Processing had to take place in an oxygen-free environment, so both mining and manufacturing had to occur on the moon. This wouldn't have been so bad if the special solvent, methyl ethyl death, and one of the active ingredients, pandemonium trichloride, weren't so difficult to transport and handle. As a consequence, MRI was having difficulty booking cargo space on the weekly moon shuttle, and several environmental groups, particularly Citizens for a Clean Universe, were questioning the benefit of hair to men when compared with the potential cost of a spill in transit or processing. Risk assessments and cost/benefit analyses conducted to date were inconclusive in this regard.

These and other concerns resulted in the establishment of a corporate environmental audit program by PEP management. This program had been recently certified by the U.S. EPA under its voluntary registration procedure, which was first proposed just before the turn of the century but only recently implemented. There was after all a twenty-five year public comment period for all new environmental regulations.

## THE AUDIT

As luck would have it, MRI had thus far escaped being audited, mostly due to the somewhat remote location. PEP had first focused on domestic facilities, then international, and now, finally, the corporation was hitting its NEB (non-earth-bound) facilities. It was soon to be MRI's turn.

Two of PEP's more experienced auditors had been assigned to the MRI audit. Although experienced, each typically had a different approach to the work. I.M. Fair (Irene, as she was known) viewed audits as a way for facility managers to help themselves do the right thing. On the other hand C.M. Sweat (affectionately known as Charlie, or by others as "that Cretin") saw auditing as a temporary stop-over on his way to top management; he was determined to demonstrate his ability to implement the "lean and mean" philosophy of the current management.

Irene understood the importance of the assignment and was already in the midst of her preparations. Although she was on temporary duty in

Europe, she called Charlie at corporate headquarters in Washington, D.C. (PEP had relocated to the long-since abandoned Waterside Mall complex.)

"Charlie," she said, "We really should get ready for this MRI audit. I'm pretty excited by the trip to the moon, and I want to do a great job. I'll be getting my checklists, a completed pre-visit questionnaire, and a plant profile sent here to London. And..."

"Whoa," said Charlie, "Aren't we getting ahead of ourselves. After all, the audit isn't for two months."

"I don't think so," responded Irene, "You remember how much more preparation we needed to conduct the international audits, don't you?"

"Oh yeah," said Charlie.

"If it's okay with you," said Irene, "I'll take responsibility for hazardous materials, water and sanitary wastes. That will leave you with solid and hazardous wastes. Oh, by the way, I need you to coordinate with the MRI plant manager. He needs to know our schedule, agenda, and scope. He also needs to help us with logistics. You know, like providing an escort, security clearances, access to a phone, availability of an office, and so on. Charlie, are you still there?..."

Charlie was still there but he was beginning to think that this trip might not be the cake-walk he originally had thought. These guys at MRI are supposed to be the hot-shots of PEP; why should he spend his precious time coordinating the trip with *them*. It should be the other way around. And if Irene thinks she's going to unload all this preparatory work on him, well...

"...Charlie?"

"Yeah Irene, I hear you," responded Charlie. "Don't you worry your pretty little head. I'll take care of those details for us. I'll meet you on the shuttle leaving on the 14th. See you now."

As she heard the click, Irene was now starting to think along the same lines as Charlie—this was not going to be an easy assignment. It would be the last time that Irene and Charlie would have anything close to similar thoughts.

## THE BEGINNING

It was now the 14th, and although Charlie had not gone over his checklist or the pre-visit questionnaire, or anything else for that matter, he had had his secretary pack those materials and had also had her call the MRI plant manager, Dr. Demanding, to coordinate the audit with him. He felt good about his preparation steps and his ability to delegate. He could always go over the paperwork on the trip.

Irene, on the other hand, was edgy. Much like an athlete before the big game, she never quite settled down until the audit had begun. She had read all her materials but felt uncomfortable having left the plant coordination to Charlie. As it turned out, she had very good instincts.

The trip to MRI proved uneventful except for one problem. Although Charlie and Irene spent considerable useful time discussing the forthcoming audit, it occurred to Charlie that he had left his background materials in his luggage, which was not retrievable during the flight. He was able to review Irene's copy of the questionnaire and the plant profile but Irene had only brought the checklists and regulatory citations for the areas she was to cover. Charlie had always prided himself in his ability to "wing it," and this assignment was to once again tax that skill.

## THE FIRST MEETING

It was the 15th, the morning of the first day of the audit, although "morning" on the moon was strictly a man-made convention. Charlie and Irene had had a good breakfast at the visitors' quarters, and they now were waiting in the MRI reception area for the arrival of Dr. Demanding. Irene was reading that day's edition of the *Moon News* to see if there was anything of note that might help her on the audit. Charlie was reviewing his stocks in the *Wall Street Journal*.

Suddenly Dr. Demanding turned the corner, walked up to them, and said brusquely, "Your secretary told me you were to be here yesterday. Where the hell were you?"

Irene looked at Charlie who was a bit flustered, but upon regrouping, said, "Doctor, I think she probably confused the arrival date with the date we were to start the audit. You see, we arrived yesterday and..."

"Well, if you and I had talked," Dr. Demanding interrupted, "This might not have happened. I'll be able to make adjustments today, but we won't be able to get started for another two hours. You can sit in my office until then."

Attempting to calm things down, Charlie said, "Well, we don't want to inconvenience you, so why don't we sit in the office you've allocated to us for our stay."

"Say what!?," responded Dr. Demanding.

"You know, the office we'll work out of while...didn't my secretary mention we'll need an office?"

"No, she didn't! What else is it that she didn't tell me?"

Irene thought it was about time that things returned to normal so she interjected.

"Sorry for the misunderstanding, Dr. Demanding. But it would be easier on all of us if we had a work space to spread-out in and access to a phone, and if you're too busy I'm sure one of your staff could be an effective escort for us. We really do want to minimize any disruptions to your operations."

This seemed to quell Dr. Demanding's mounting anger. Although maybe all it did was transfer it to Charlie who was thinking he would "get" Demanding for being so unreasonable and would certainly make Irene's life miserable for making him look bad.

Two hours later the briefing to plant management seemed to be going well. Everyone came to understand the scope of the audit and Irene was able to ask several very relevant questions because of the research she had done on the facility prior to the audit.

While things were going better, Charlie was still having his problems. In an attempt at conciliation, he began his opening remarks with his favorite cliché—"We're from corporate and we're here to help you." This went over like the proverbial lead balloon.

## THE FIELD WORK

Upon completion of the briefing, Charlie and Irene stated that they would next like to review the environmental records and then begin their physical inspection of the facilities. More problems surfaced due to the lack of preparation, of course. They were told that some of the permits were kept back on earth for security reasons and there were no available spacesuits for either Charlie or Irene so their inspections of the mining operations would have to be canceled. However, Irene was once again ahead of the game. In her research she became aware of both problems and obtained copies of the permits and had been fitted on earth for a spacesuit, which was due the next day. Her phone call one month ago to Charlie on these issues went unanswered.

They both reviewed the paperwork and most everything seemed to be in order. Charlie did note that a new waste disposal cell had been constructed and was about to receive its first load of waste material that week. He suggested that Irene schedule her field work in a way that would allow her to see this operation. Irene agreed wholeheartedly. Charlie also reviewed the hazardous waste disposal permit and a dozen manifests, one of which had two errors (*i.e.*, an incorrect abbreviation for drum and the lack of a signature). He noted these in his log book. Irene noted that the Spill Plan was close to three years old and would be in need of updating and recertification within three months. Although this was not strictly a non-compliance finding, she thought she might mention it in the debriefing and note it in the report. She felt that plant management should begin the update process right away in order not to miss the forthcoming expiration date.

It took pretty much all day for the team to review the paperwork. Both Charlie and Irene were following the prescribed checklists during their reviews. Notes were put into their bound journals as required by corporate procedure.

## THAT NIGHT

Towards the end of the day Irene went searching for Dr. Demanding to see if he would be available for an early dinner that evening. The good Doctor was very receptive to the idea since his family had remained on earth during his six months of moon duty, and he still had three months to go. Irene always attempted to have at least one dinner with the site or environmental manager as she saw this as an opportunity to learn more about the person and the facility she was auditing. She also made sure this dinner was not on the evening before the closing conference, as this was usually a late night involving significant preparation.

Of course, Irene assumed Charlie would join them. She was learning, however, that assuming anything with Charlie was a mistake.

"No, I don't think I can join you tonight," he said. "I really need to go over my checklists and regulatory abstracts before we begin the field work tomorrow."

"Charlie!" shouted Irene, "If you had done that when you should have, you wouldn't have to miss this dinner! It's important."

"Oh, I'm sure you can handle it for the both of us," Charlie responded. "It's just not that critical."

That night Irene learned more about the facility's operations than she had in all her previous work. Dr. Demanding turned out to be quite a character and very sociable outside the work environment. He even took Irene to see the plant on the night shift, which raised some questions in both their minds about the adequacy of the Emergency Response Plan. Meanwhile, Charlie ordered room service, ate, watched television (100-year-old reruns of the Honeymooners), and promptly fell asleep without reviewing his checklists.

## THE NEXT DAY

Charlie and Irene were now ready to undertake their field-work. They were to review the tank farms, the processing areas, the materials storage areas, and the waste disposal areas. In addition, Irene was to take a moon-rover out to the mining operations. Each had an escort as they went their separate ways. Charlie relied principally on observation and spent little time asking questions of his escort or the unit operators. And when he did ask questions, they went something like:

"Do you inspect weekly?"

"Are staff here competent?"

"Have you ever had a spill that you didn't report?"

After a while Charlie even stopped asking these types of questions because, for some reason, the responses were not helpful.

Based on his evaluation against the checklists, Charlie noted several problems:

> Poorly labeled drums
>
> Lack of containment in the tank farms
>
> Insufficient liners in the waste disposal cells.

He was finally starting to enjoy the audit.

Meanwhile, Irene was having more success with her approach. While each auditor's observation skills were roughly equivalent, Irene was having better luck discussing operations with staff. She was using open-ended questions like:

> "Can you explain to me what you do here?"
>
> "What are some of your concerns in achieving successful operational performance?"
>
> "Do the environmental requirements you face always make sense operationally?"

You see, Irene, believed that if she was able to get staff to open up she might understand better what transpires on a typical day as opposed to the day or days when there is an audit or inspection. She felt that the one-day snapshot approach was too restrictive. She also believed that audits were not just inspections of records and physical facilities, but of management systems as well; for she had noticed on other audits that if management and staff were not environmentally sensitive and aware, there would eventually be a problem no matter how sound the physical systems.

In her evaluation of sanitary discharges and hazardous materials she did observe:

The potential for water/wastewater sewer cross-connections

Potentially uncontrolled releases through drains

Poor inventory control practices (last-in, first-out) resulting in the accumulation of expired raw materials in deteriorating drums.

She noted these in her journal and on plant drawings so she would remember exactly where in the facility she observed the problems. It was a big place.

## FINISHING

Upon completion of three days of inspection, our auditor friends, Charlie and Irene, sat down to discuss their findings. They agreed that during the debriefing Irene would take the lead as she seemed to have fostered a stronger relationship with Dr. Demanding, and just about everybody else for that matter.

They each developed a detailed set of notes from which they would speak on the next morning. Irene found the going easier since each evening she spent some time in her room compiling the previous day's findings. As a result, Irene felt that her findings were sound but that Charlie was being less than precise in several areas. She pressed him on these issues but his defensiveness caused her to back down. She would wait until tomorrow and do her best to support Charlie in the meeting. Her patience was wearing thin and she was glad tomorrow was the final day of the audit.

Charlie, on the other hand, couldn't understand what was bothering Irene. He had seen all he needed to. Yes, he had observed several significant issues, and, yes, he was going to fix somebody's wagon tomorrow.

## THE BRIEFING

The next morning came and the 10:00 A.M. ESET (Eastern Standard Earth Time) meeting began promptly. Irene began by thanking the plant people for their hospitality and stated, by way of summary, that the plant is generally in compliance.

Irene then highlighted her findings, taking pains to be thorough and accurate but not overemphasizing unnecessary details. She realized that most of the people in the room wanted to hear the "bottom line" and would be willing to review the details later in the report. Her main points were the following:

"The Spill Plan must be updated and recertified in three months. Much has changed at the facility in the past three years, so the update will not be a trivial exercise. You should begin this task immediately."

"Your 97-6A hazardous materials storage area contains several deteriorating drums with potentially expired materials. These drums should be removed and the materials recovered or disposed before the situation worsens. This is not strictly a compliance problem now but it could be if left unattended."

"Based on a review of drawings and my observation of the facility, you have a potential for a cross-connection of drinking water lines with sanitary lines in area 92-7B. There are also five floor drains in this area not on the drawings. No one on the unit knew where these drains discharged. They could be the cause of the contamination detected by the continuous monitoring soil detection (CMSD) system installed below the foundation of the AY Complex."

And so Irene went on....

Dr. Demanding responded favorably to Irene's findings. She was precise, thorough, anticipatory, and offered new insights into problems that had been a "thorn in his saddle," as he liked to say.

Now it was Charlie's turn. He was bolstered by management's response to Irene and was really feeling good about himself. And he began.

"Your hazardous waste manifests are not filled out correctly, and this could result in fines and citations. Further..."

"Hold on," Dr. Demanding interrupted, "We must fill out a thousand manifests a year. Are you telling me that they're all filled out incorrectly? And what is it about them that's wrong?"

"Well," Charlie began, "Of those I reviewed, several had problems including no signature and the wrong abbreviation for drum."

"Certainly the signature being missing is a problem," Dr. Demanding responded, "But I need to know if this is a chronic problem or not. How many manifests did you review and how many had problems?"

"Uh, I only had time to review a dozen, but it was enough for me to come to a conclusion."

"You mean to tell me," shouted Dr. Demanding, "That of the 5,000 manifests we've generated in the last five years you inspected 12? You must be joking!...Oh, never mind, go on."

"Well, I'm sure you understand we only have a short time up here. And besides my other findings are better substantiated. Irene will back me up in that regard."

Irene was beginning to feel like the heroine in the movie she had recently seen in the Hollywood Archives Museum. What was it called, she asked herself? Oh, *Aliens*, that was it.

And Charlie continued, feeling a nervousness he had not anticipated.

"I also noted that several of your hazardous waste drums were not labeled. This is, as you know, a serious violation that has cost PEP some money at other plants."

"I agree with you Charlie about the seriousness of the violation," said Dr. Demanding. "If you can tell me the details of where and how many, I will send one of my staff out now to remedy the problem."

Charlie immediately began shuffling through his notes but could not find any reference to where his "problem drums" were. His nervousness was increasing, coupled now with a sense of agitation caused by Dr. Demanding's continuous challenges.

"I can't find my notes on the labeling problem right now. I'll have to get back to you on that one. Let me go on if I can," said Charlie.

"Okay," said Dr. Demanding, "But I want these issues resolved before you return home."

"Fine," responded Charlie, "I only have one more finding and then we can wrap up."

About this time, Dr. Demanding was thinking of wrapping *Charlie* up and jettisoning him into deep space.

> "With respect to your hazardous waste storage and disposal, I noticed that your three tanks do not have the appropriate diking and that the disposal cells do not have double electrosynthetic liners. This added protection is required by the 2044 amendments to the Resource Conservation and Recovery Act Scrolls. Actions to remedy the situations should take place immediately."

As Charlie was rather smugly stating his final finding, Dr. Demanding was reaching for his tranquilizers. Demanding could not believe the idiocy portrayed by this corporate hot-shot. He vowed that if he ever had the chance, Charlie Sweat would do just that in the last remaining coal mine on earth.

Dr. Demanding began, having lost all patience by now, "Mr. Sweat, how can you be so...so...stupid!? The tanks you mentioned do not need diking because they are double-walled tanks on a special foundation. We have an exemption from the diking requirement, which if you had asked anybody out at the storage area they would have gladly shown it to you. Didn't you talk to anybody out there?"

"As for the disposal cell," he continued, "The preemptive local regulations allow for single liners on the moon since a leak presents few risks to us or the environment. We have no groundwater to protect, remember!? Didn't you review the local regulations before you began this audit?"

Charlie was shell-shocked. His only thought was of something his brother the poet had once said to him. He had quoted an ancient philosopher named Thoreau who had said at one time—"The mass of men lead lives of quiet desperation." He hadn't understood it then, but he understood it now.

Dr. Demanding wasn't about to wait for Charlie's response. He said as he was leaving the meeting taking his staff with him, "You two had best do some regrouping before you leave this facility. If you have anything else to report, do it to me privately. I don't want to waste my staff's time. Otherwise, I'll wait to see the draft report, which by the way, had best be more accurate than your briefing. Good day!"

Irene was furious! Not only was she caught up with this loser, but she was now going to have to stay an extra day to fix the problems Charlie had caused. This would mean that she would miss her husband's birthday.

Charlie was now regaining this composure. He turned to Irene and said matter of factly, "I don't understand what his problem is. So I missed a few things. Nobody's perfect."

"Charlie, haven't you learned anything from this mess you've created?," asked Irene.

"Well, yeah," Charlie responded, "Plant people sure can be arrogant."

"No, Charlie," shouted Irene, "This is what you should have learned:

> **Attend to Details.** Don't leave important communications and logistics to your secretary. If you have no choice, at least follow-up with a phone call.

> **Be Prepared.** Spend some time up-front understanding the facility and the regulations. Bring what you need to get the job done, even if it means an extra suitcase.

> **Communicate.** Spend the time it takes to understand the problems these people face. For God's sake, talk to them.

> **Commit the Effort.** This is not a 9 to 5 assignment. Meet these people for dinner, visit the plant during the off-shifts, read the local newspaper.

**Be Thorough.** Reviewing a dozen of 5,000 manifests is not sufficient. Pick a much larger sample, particularly if you begin seeing problems.

**Be Precise.** If you see a problem make sure you're able to describe it so management can act on it. "Some drums somewhere" just won't do.

**Be Supportive.** We're all on the same team. Dr. Demanding is clearly trying to do the right thing. We are here to help him follow through on that commitment.

And lastly,

**Be a Little Scared.** Don't be so cavalier about these audits. These assignments are not as simple as you make them out to be. And there's a lot riding on our ability to do a good job, particularly here on the moon, which many people on earth believe should have never been commercially mined in the first place. PEP faces some serious liabilities if MRI is operating outside of standards, whether it's intentional or not."

Charlie did his best to pay attention, but he couldn't help but think Irene was over-reacting. What was the "Be a little scared" nonsense all about, anyway? What could possibly happen to PEP if they had missed a couple of items? However, to placate Irene, he mentioned that he had indeed learned his lesson.

And so it appeared for the remainder of the audit. Charlie and Irene re-audited the hazardous waste area on the next day, successfully debriefed Dr. Demanding, and made plans to return to earth.

## THE REPORT

On the shuttle on the way back, Irene and Charlie talked about the audit report. They agreed that the outline proposed in the corporate guidelines would work well for the MRI audit. They decided on the following responsibilities:

Introduction and Summary

- Audit Background (C)
- Facility Overview (C)
- Regulatory Setting (I)
- Summary of Findings (I)

Findings of the Audit

- Paperwork (C)
- Environmental Management (I)
- Water and Wastewater (I)
- Hazardous Materials (I)
- Hazardous Wastes (C)
- Solid Wastes (C)
- Housekeeping and Nuisances (C)

The audit background would discuss the who, what, when where, and why of the audit. This was to be followed by two brief sections—one presenting an overview of the facility and the second presenting any special regulatory requirements associated with MRI's operation on the moon. These two sections were generally well received because many of the report reviewers typically had either never been to the audited facility or had not been there for quite some time.

The findings section would address both noncompliance and poor management practices, clearly delineating between the two. Also, recommendations on corrective actions would not be included as it was corporate policy to leave this responsibility to facility management. Most people in the audit program felt the company could have gone either way on this particular issue but, in the end, management decided that because of its decentralized organizational approach this responsibility was best placed at the plant level. A plant follow-up action item report was formally instituted as a result of this decision.

Based on her experience to date, Irene agreed to take responsibility for producing the report. A draft was due to corporate and MRI in 30 days and somehow she felt Charlie just might miss that deadline if he had to pull it all together.

Upon landing, they went their separate ways; Charlie believing the assignment was pretty much over, and Irene thinking that, to some extent, the most difficult part was ahead of her. It turned out she was right but not for the reason she expected.

Two weeks after the audit, Irene hadn't heard from Charlie and so she called him to find out how he was progressing.

"Hello?"

"Charlie, this is Irene. How's the MRI report coming? I'm done with all my sections except for the executive summary."

That figures, thought Charlie. "Well, Irene, I'm about half done with my sections, and I should be finished in two weeks. My boss has really been pushing me on some other projects since we got back."

"Charlie!" shouted Irene, who realized she shouted more at Charlie over a two-week period than she did at her husband over a year. She was thinking that says something about Charlie, or my husband, or both. She wasn't sure.

"Charlie," she repeated, "The report is due in two weeks. How am I supposed to turn it in on time if I get your material on the date it's due?"

"Hey, no sweat," responded Charlie, "My sections won't be that long. Just leave room for them on the word processor and have a secretary insert them, and then use electronic mail to distribute the report. You shouldn't be more than a day late."

What is this "*You* shouldn't be more than a day late," nonsense, Irene thought? It was clear Charlie was already disassociating himself from the audit.

They finished their phone conversation but it left Irene unsettled. She had reason to be. Charlie had told a pretty common white lie. He was not really half done. He had only finished one of his six sections, but he always felt that getting started was half the battle, so, therefore, he told her he was half done.

This sounded pretty logical to Charlie especially given his present difficulty. You see, Charlie was having a real problem completing his sections because he realized, too late, that his notes were not detailed enough to write a precise and accurate report. He had less than a photographic memory, and he was having difficulty recalling several important observations he had made. He was in a bind because he was not about to admit to Irene or Dr. Demanding that he needed more information. If only he had done what Irene had done...develop an annotated outline before he had left.

Charlie eventually worked his way through his sections. They turned out to be very superficial, and he was two weeks late in submitting them to Irene. Dr. Demanding, based on his previous experience, accurately suspected the root cause of the delay but corporate management was laying the blame on Irene since she had the responsibility for producing the report. Maintaining silence like a good corporate soldier, she could only hope that all good things come to those who wait. It turned out it was well worth the wait.

## EPILOGUE

It was three months later and Charlie had been summoned by the Chief Executive Officer of PEP, Inc. It was 7:00 P.M. and he had no idea what Mr. Vance Hall Wharton wanted, but presumed it was to congratulate him on the fine work he was doing. Charlie continued to view himself as a "fast-tracker."

Charlie was surprised to find Dr. Demanding and Irene both in the meeting as well. Their facial expressions did not bode well.

Mr. Wharton began, "Charlie, I understand that you conducted an environmental audit of the MRI operation a few months ago. Is that right?"

"Yes, sir, with Irene."

"Was it you who was to assess compliance with hazardous waste regulations?"

"Yes, why?"

"Well, it seems we have a problem on our hands. You see, we recently had a U.S. EPA inspection and we were told that, although our waste disposal cells were in compliance with all standards, we were disposing a waste in the cells that is prohibited. It seems that six months ago EPA issued a land disposal restriction on the newest formulation of methyl ethyl death (MED-1,1,4), our special solvent used in manufacturing Hairall. Fortunately, we've only been using 1,1,4 for about four months; we began about a month before you did your audit. However, the restriction applies especially to the moon because trace amounts of MED-1,1,4 can set up a chain reaction in the moon's geologic formations, causing a breakdown in the structural integrity of the strata. Are you still with me, Charlie?"

"Uh, yes, I think?"

"What this means Charlie is that if our liners fail we could have the entire processing facility swallowed up in a sink hole, people and *all*. We have now been operating at risk for over four months. Didn't you know about this restriction when you did the audit? Didn't you do any research on recently issued regulations that might not have made it into our checklists yet?"

"Well, I did some research, but..."

"Obviously, you overlooked something fairly obvious and critically important, don't you agree?"

"Yessir."

"Fortunately Dr. Demanding and Irene found a way to identify, excavate and treat the MED-1,1,4 before any problems occurred. However, the U.S. EPA has decided to assess the maximum penalty possible for the violation, which because of the risks posed is $500,000 per day, or $60 million." (It turned out that the dollar amount of statutory environmental penalties had pretty much tracked with inflation over the past 50 years.)

"Charlie, because of your incompetence, you have cost us $60 million and put scores of people at risk. Firing is too good for you. You still have three years on your contract with us, is that right?"

"Yessir."

"Well, let me have Dr. Demanding, our new Executive Vice President for Environmental Management, tell you what we have in mind for your next assignment."

"Charlie," began Dr. Demanding, "I must admit I can't take full credit for this idea. Miss Fair, here, my new Director of Compliance Management is really responsible.

Mr. Wharton and I must get to another meeting so I'll leave it to Irene to brief you on the assignment. Oh, and Charlie, don't even think about reneging on your contract because we'll see to it that you will not find another job on the face of the earth!"

As they left, Irene began. "I won't belabor this Charlie. You have been designated the permanent environmental coordinator for the new sulfur mine on Venus...."

"You can't do this to me!," protested Charlie, "That mine is a hell hole. Temperatures on Venus reach 480 degrees centigrade, and I hear that the operation uses mostly convicts to mine the sulfur. And women aren't allowed because some unknown source of radiation affects their reproductive capabilities. And..."

"Oh Charlie," responded Irene gleefully, "I'm sure it can't be all that bad. You're a tough guy, you can put up with anything for three years, can't you?"

"Good night, Charlie."

"Good night, Irene, good night."

# PART III

# SPECIAL AUDITING TOPICS

# 20

# PROPERTY TRANSFER ASSESSMENTS

## INTRODUCTION

In the past few years, there has been a remarkable increase in property transfers and acquisitions and divestitures in the U.S., resulting in new ventures for the players. In the 1980s, Exxon entered the electric motor business (Reliance); Dupont, the oil business (Conoco); U.S. Steel, the petrochemical business (Marathon Oil); Mobil, the retail consumer products business (Montgomery Ward); and R.J. Reynolds, the food business (Nabisco), all through acquisitions. In many years, there are several transactions of at least a billion dollars, and in today's U.S. economic setting of uncertainty, the changing patterns of ownership likely will continue for some time.

Yet, even in this "pro-merger climate, it is significant that it has been estimated that one-third of all corporate acquisitions fail, and seventy percent are disappointing to the acquiring corporation."[1] As evidenced by the three noteworthy examples below, undisclosed environmental problems have had some impact on this lack of success:

In 1968, Occidental Petroleum Corporation purchased Hooker Chemical Company and the subsequent liabilities associated with Love Canal. Less than ten years later, the Canal became the genesis of the federal Superfund program.

---

[1] B. Kennedy, "The Need for Environmental Audits," *Pollution Engineering*, July, 1982.

Also in 1968, Northwest Industries, Inc. purchased Velsicol Chemical Company, and throughout the 1970s experienced incidents "like the inadvertent use of polybrominated biphenyl for cattle-feed supplement in Michigan, the claims of neurological injury to plant works in Bayport, Texas, and charges of contamination from dump sites in several locations." Velsicol was subsequently divested by Northwest.[2]

In 1966, Nalco Chemical Company acquired Industrial Biotest Laboratories (IBT), which has been swamped with lawsuits claiming laboratory analyses throughout the early 1970s had been purposefully and improperly performed and documented. Nalco has taken a $13 million write-off and has legal suits pending.

These past problems with property transfers have not gone unnoticed by regulators, particularly given the ever-expanding problem of abandoned waste sites. A handful of states enacted laws designed to prevent certain property transfers without government approval. For example, the State of New Jersey, in 1984, passed the Environmental Cleanup and Responsibility Act (ECRA), Public Law 1983, Chapter 330, which required "that any place of business where hazardous materials were stored or handled in any way be given a clean bill of health by the state before a transfer of ownership or change of operation take place." ECRA was supplanted with the passage of the Industrial Site Recovery Act (ISRA), Public Law 1993, Chapter 139, in 1993. In essence, ISRA and its predecessor require companies to notify the New Jersey Department of Environmental Protection and Energy, within thirty days of an announcement of a pending sale, that it has completed an evaluation of its property—groundwater as well as surface conditions—including a history of spills, together with a detailed description of its testing protocols, and so on. The state then reviews the submission, and it can

---

[2] "Velsicol Still Striving to Live Down Old Image," *Chemical Week*, May 14, 1980, p. 29.

hold up any sale until it issues an approval that the site is not contaminated.[3] As a result of the New Jersey initiative, other states have enacted similar statutes (*e.g.*, buyer protection laws and superlien provisions) for their jurisdictions.

At the federal and state level, property owners also face tremendous liabilities under environmental cleanup statutes such as the federal Comprehensive Environmental Response, Compensation and Liability Act (CERCLA), otherwise known as Superfund. Superfund, and its state counterparts, allow for recovery of cleanup costs by the enforcing agencies at sites contaminated by past hazardous waste disposal. There are now over 75,000 sites nationwide that are under investigation by the U.S. EPA and state environmental agencies for potential cleanup, and almost 7,000 sites which currently require cleanup under one or more of the environmental cleanup programs. Surveys sponsored by the U.S. Congress suggest that the number of cleanup sites could approach a half million before all is said and done. Even worse, the U.S. EPA estimates that the average cost of cleaning up a Superfund site under CERCLA currently exceeds $35 million; but even "small" cleanups can result in seven-figure costs.

A more perplexing problem for the business community is that cost recovery under these laws is based on joint and several, and strict liability. In other words, liability for cleanup is independent of the volume of waste disposed by a particular company or whether the disposing company was acting in compliance with all rules and regulations in effect at the time. In fact, a property owner is liable under Superfund based solely on ownership of the property and regardless of whether the owner contributed to the contamination or whether the owner knew the property was contaminated at the time of acquisition. Yet, the Superfund Amendments and Reauthorization Act of 1986 (SARA) does provide some protection for property owners. The "innocent landowner" provision of SARA can shield an owner from environmental liabilities if "all appropriate inquiry into the previous ownership and uses of the property"

---

[3] "Environmental Audit Letter," May and June, 1984, p. 1.

is made prior to the purchase. Many of the state cleanup statutes have similar "innocent landowner" provisions.

All this activity in mergers and acquisitions has given senior corporate managers pause for thought. Most companies now have an internal policy requiring that various environmental investigations be conducted prior to completion of an acquisition. In addition, because of several recent court cases[4], banks and lending institutions are requiring similar due diligence investigations where they are providing refinancing or initial financing of a project or leveraged buy out. Developers, as well, are assessing "virgin" properties for signs of indiscriminate past waste disposal, excessive use of pesticides, or the presence of wetlands.

## APPROACH OVERVIEW

Superfund and similar laws have driven most parties involved in commercial real estate transactions to conduct an environmental assessment of the property, usually called a Phase I environmental site assessment, prior to transfer. These assessments are driven by due diligence considerations similar to the more traditional facility compliance audits, but they differ significantly in both the objective and the scope. These differences are outlined below.

## SCOPE AND THE ASTM STANDARD

The scope of a property transfer assessment varied from one region of the country and from one transaction to the next until 1993. In May of 1993, ASTM (formerly the American Society of Testing and Materials) published standard E-1527 for Phase I Environmental Site Assessments in

---

[4] See: *Environmental Law and Real Estate Handbook*, published by Government Institutes, Inc. Precedent cases include *Midatlantic National Bank v. N.J. Department of Environmental Protection (1986), United States v. Mirabile (1985), United States v. Maryland Bank and Trust (1986), and United States v. Fleet Factors (1988)*.

Commercial Real Estate Transactions. The ASTM standard, in essence, codified the current practice for conducting Phase I assessments, with certain limitations. By design, the standard sets out a minimum scope of services for the assessment and, depending on the nature of the property or the transaction, a more thorough scope of work may be necessary. As a minimum scope of work, the standard implicitly assumes that no environmental risks are known or suspected at the start of the project, so that the ASTM procedure can appear extremely elementary on industrial sites with known practices involving the use and disposal of hazardous substances. Also, the ASTM standard only addresses risks associated with hazardous substances and petroleum products. Other common environmental issues associated with property transfers and ownership, such as asbestos, radon, wetlands, and lead in paint, are not covered by the standard.

## ASSESSMENT TEAM

On property transfer, or Phase I, assessments, use of senior staff is essential. Acquisitions, in particular, often require the utmost discretion from staff involved in the process. Disclosure of sensitive information can significantly affect the purchase price or even the likelihood of a successful agreement. Thus, environmental auditors involved in the acquisition must appreciate the distinction between a Phase I assessment for a real estate transfer, and a facility compliance audit. Auditors will often be required to review the facility(ies) on very short notice, in the absence of any pre-assessment information and in the presence of a hostile or unaware host facility. Auditors may also be reviewing facilities or operations with which they have little familiarity because the acquisition may be an attempt at significant diversification (*e.g.*, U.S. Steel and Marathon Oil). Finally, on occasion, external third-party auditors may be necessary where: (1) special expertise (*e.g.*, groundwater monitoring and modeling) is required; (2) internal staff are otherwise occupied; or (3) external financing or insurance coverage might demand it. (The ASTM standard requires that a Phase I be performed by an "environmental professional," which is defined as someone with the training and experience to form conclusions about environmental risk on the property and an opinion as to the impact of the environmental risks to the property.)

## EXTERNAL CONTACTS

During the course of a compliance audit, environmental audit teams might not call regulatory agencies. This is done for one of two reasons: either to "let sleeping dogs lie" or because it is thought that regulatory agency contact is best left in the hands of facility staff. However, under the ASTM Phase I procedure, a review of certain federal and state agency databases is required. Also, calls to certain local agencies may be necessary. In the case of the federal and state databases, the information can be obtained from private commercial companies that specialize in environmental information to thereby avoid "tipping off" the agency through the inquiry. The ASTM standard identifies specific databases that must be reviewed and designates search distances for each database. The private services can generate a report that satisfies these requirements and provides a map of the area that shows the locations of the risk sites in the neighborhood.

There are also ways regulatory agency contacts can be made without revealing the purpose of the call. For example, the assessor can make a blind call stating that a survey of the region's industry is being conducted and that a part of that survey is assessing the reputation of the major firms relative to one another. This way, no one firm is singled out as the target of the call, and the regulator is put in a more comfortable position. Further, in some states, compliance and compliance history records on any firm are commonly made available to the public if a request is made. These are usually readily available, except that some agencies are now requiring that formal Freedom of Information Act (FOIA) requests be made. This can delay the process from four to eight weeks. Once again, private information companies often collect compliance information in addition to the Phase I databases, and they can typically produce reports in 24 to 72 hours.

Finally, for this facet of the assessment, it may be best to use a third-party (*e.g.*, consultant or law firm) so that internal staff are not compromised when asked the name of their employer. Also, each caller should understand the agreed to protocol (*e.g.*, whether the company's name is specifically mentioned and in what context) before the calls are made.

## THE THREE PHASES

Within these guidelines, the approach to property transfer assessments is to divide the program for a site into three sequential phases of effort:

**Phase I.** The ASTM assessment consists of historical property use research, government records research, interviews with key site managers and local agency officials, and a detailed site inspection.

**Phase II.** Selective sampling of areas suspected of being potential environmental liabilities.

**Phase III.** Strategy for site utilization and/or potential environmental remediation or clean up.

It is important to understand, as shown in Figure 20.1, that all three phases will not be required for each property. For example, at a minimum, every property will undergo a Phase I assessment. If the records, available data, interviews, and inspection show no indications of environmental concerns associated with hazardous substances or toxic wastes and materials use and disposal on the property, then there is no need to proceed with Phases II and III. However, if the Phase I activities show that selected areas of the property may now, or in the future, pose environmental questions and/or liabilities, then samples (soil, groundwater, surface water, vegetation, etc.) should be taken and analyzed as part of Phase II efforts.

Based upon the Phase II analytical results, conclusions can be drawn as to the environmental condition of the property. It is possible that in isolated cases, a purchaser may wish to proceed with a project even though the site does, or most likely will, represent environmental (*i.e.*, clean-up) responsibilities. In these instances, as a part of a Phase III level of service, the assessor would undertake a more comprehensive sampling and site assessment program, and would estimate the remedial actions and capital and operating costs for the property as a basis of negotiations for lease or purchase. In many instances, this type of property transaction must be consummated not only with the full cognizance of owner and buyer, but also with approvals from appropriate state, regional, and federal agencies.

# Figure 20.1
# Approach to Conducting Site Environmental Assessments

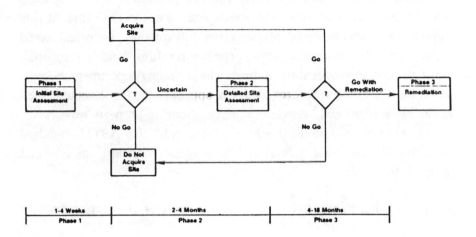

# PHASE I

As previously discussed, a Phase I assessment includes a compilation and review of available historical records pertaining to the specific property uses and surrounding property uses. The ASTM standard lists eight sources for historical property use information: historical aerial photographs, fire insurance maps, property tax files, land title records, historical topo maps, local street directories, building department records, and zoning/land-use records. Other sources may also be consulted, such as site plans/maps and interviews with neighboring property owners.

In addition to the historical use research, the ASTM standard identifies the following federal and state agency databases as essential research sources[5]:

| FEDERAL DATABASES | SEARCH DISTANCES |
|---|---|
| National Priorities List (NPL) | 1 mile |
| CERCLIS List | .5 miles |
| RCRA TSDs | 1 mile |
| RCRA Generators | Prop. & Adj. Prop. |
| Emergency Response Notification System | Prop. only |

**STATE DATABASES**

| State Priorities List (SPL) | 1 mile |
|---|---|
| Registered Underground Storage Tanks | Prop. & Adj. Prop. |
| Leaking Underground Storage Tanks | .5 miles |
| Solid Waste/Landfill Facilities | .5 miles |

---

[5] CERCLIS files are available, with some struggle, from U.S. EPA. However, several consulting firms and law firms also maintain a current CERCLIS database that can be accessed for a fee.

Evaluation of these files can provide a chronology of site operations and conditions for at least as long as the relevant agency has regulated the facility. The search distance refers to a search distance around the subject property, and it is measured from the border of the subject property. "Prop. & Adj. Prop." refers to a search distance that encompasses properties adjoining the subject property, which are those that are contiguous to, or separated only by a street or other right-of-way from, the subject property.

The next step in an ASTM Phase I is interviews with key site managers (typically the same person who is interviewed for a compliance audit at the facility) and local agency officials. The contacts with the key site manager should also include a request for documents such as prior assessment and audit reports; environmental permits, registrations, and notices; MSDSs; community right-to-know plans and safety and preparedness plans; reports of hydrogeological conditions; and notices of violation or other environmental compliance issues.

Local agency officials should be asked for general information regarding the existence or use of hazardous substances or petroleum products on the property or in the neighborhood around the property. (Use the federal and state government database search distances as a guide.) Following the historical and government records research and the interviews, a visual site inspection is conducted to observe the most recent conditions of the site. For small parcels, the visual inspection should be conducted on foot. For large properties, a vehicle search or even a brief helicopter overflight can be a first step and, if necessary, can be followed by a more selective and closer examination of specific areas of concern viewed from the air. If very recent aerial photos are available, it may not be necessary to perform another visual inspection from the air.

As discussed above, the ASTM standard does not address certain issues which may be appropriate for a comprehensive assessment. Nevertheless, the Phase I assessment should be designed to identify the potential presence of the following hazards:

- Asbestos
- Soil or groundwater contamination due to hazardous substances and petroleum products

- Leaking underground storage tanks
- PCBs
- Lead-based paint
- Urea formaldehyde
- Radon
- Contaminated drinking water.

These issues are not solely the problem of industry. Environmental contamination has become every property owner's concern in view of real cases where: (1) groups of homes are sinking because they were built over landfills (Philadelphia), (2) major office buildings are being abandoned because of PCB contamination caused by electrical explosions and fires (San Francisco), and (3) public water supplies have been contaminated (Woburn, Massachusetts).

As a guide to those conducting Phase I assessments, a hierarchical assessment of five property types is presented at the end of this chapter. The five property types are:

- Undeveloped land
- Single-family residences
- Multi-family residences
- Commercial buildings
- Industrial manufacturing facilities.

The table, Figure 20.2, assesses the potential investment consequences of the most commonly observed risk-posing situations, such as the presence of discarded drums on the property. The potential risks to the investors are described and a recommendation is made for each situation as to whether a follow-up, and more detailed, site assessment by an environmental specialist is essential, advisable, or not generally required.

# Figure 20.2
## Potential Risks Posed by Real Estate Transactions

POTENTIAL ENVIRONMENTAL RISKS POSED TO THE PURCHASER/LENDER
BY REAL ESTATE TRANSACTIONS:
A HIERARCHICAL ASSESSMENT OF FIVE PROPERTY TYPES

| Hazard Class | Situation | Potential Consequences | Risks Posed | Need For A Detailed Assessment* |
|---|---|---|---|---|
| **UNDEVELOPED LAND** | | | | |
| Waste Disposal | (1) Property is abandoned waste disposal site or former military/industrial site | Contamination of soils and/or groundwater from past usage | Potential deal killer; especially if neighborhood water supply is on individual wells | Essential |
| | (2) Located within 1 mile of active waste disposal site | Past or future contamination of groundwater | No direct culpability but potential for serious legal entanglements | Advisable |
| | (3) Located within 1 mile of abandoned waste disposal site | Past or future contamination of groundwater | No direct culpability but potential for serious legal entanglements | Advisable |
| | (4) Located adjacent to high-risk neighbor (e.g., gas station) | Past or future contamination of soils or groundwater | No direct culpability but potential for serious legal entanglements | Advisable |
| | (5) Presence of discarded drums, cans and/or evidence of stressed vegetation and stained soils | Contamination of soils and/or groundwater from past usage | Potential deal killer; especially if neighborhood is on individual wells | Essential |

*Evaluates the need for a detailed assessment (to include a review by an
environmental specialist and physical sampling) if the described situation
exists or has a good probability of existing.

**Figure 20.2** *(cont'd)*

| Hazard Class | Situation | Potential Consequences | Risks Posed | Need For A Detailed Assessment |
|---|---|---|---|---|
| Underground Storage Tanks | (1) Existence of abandoned UST's on property | Contamination of soils and/or groundwater from past leaks | Potential deal killer if tanks have leaked in the past | Essential |
| Pesticides | (1) Past significant use of agricultural or landscaping pesticides | Contamination of soils and/or groundwater from past practices | Potential deal killer; especially if neighborhood is on individual wells | Essential |

**SINGLE FAMILY RESIDENCES**

All of the above plus....

| Hazard Class | Situation | Potential Consequences | Risks Posed | Need For A Detailed Assessment |
|---|---|---|---|---|
| Asbestos | (1) Use of asbestos spray-on or bulk insulation or other construction materials (only likely in pre-1975) construction | If asbestos is damaged (i.e., friable), it can become air-borne and inhaled by occupants | Asbestos can be removed or encapsulated but remediation could be costly | Essential |
| Drinking Water | (1) Chemical contamination from on-site plumbing (especially lead) | Drinking water contamination | Lead plumbing can be replaced, but at high cost | Advisable |
| Radon | (1) Possible radon pocket | Buildings could accumulate radon; especially in the basements | Ventilation improvements in basements might be required | Advisable |
| Others | (1) Use of lead based paint | Children have tendency to eat paint | Risks only in cases where children have access | Not Generally Required |
| | (2) Use of area formaldehyde insulation | Possibility of airborne formaldehyde particles | Uncertain; conclusions on the relative risks of formaldehyde unclear | Not Generally Required |

**Figure 20.2** *(cont'd)*

| Hazard Class | Situation | Potential Consequences | Risks Posed | Need For A Detailed Assessment |
|---|---|---|---|---|
| **MULTI-FAMILY RESIDENCES** | All of the above plus.... | | | |
| PCB's | (1) Presence of PCB's in electrical transformers and/or fluorescent light fixture ballasts (only likely in pre-1976 equipment) | Past or future contamination of soils, groundwater or buildings from leaks (transformers may be owned by local electric utility) | Soils must be cleaned to 1 ppm in most cases; could be expensive, even if principal responsible party is utility | Advisable |
| Septic Tanks | (1) Poor operation and/or disposal of chemicals down the drain; especially where commercial establishments (e.g., dry cleaners) occupy building | Contamination of soils and/or groundwater from past practices | Potential deal killer if substantial quantities are involved; especially if neighborhood is on individual wells | Essential |
| **COMMERCIAL BUILDINGS** | All of the above plus.... | | | |
| Hazardous Chemicals | (1) Use, spillage, or indiscriminate disposal of chemicals and/or oils: o schools & laboratories o dry cleaners o auto repair shops o paint shops o gas stations | Contamination of building, soils, and/or groundwater from past practices | Potential deal killer; especially if neighborhood is on individual wells | Essential |
| Trash | (1) Disposal of substantial quantities of hazardous chemicals in trash stream | Hazardous wastes being improperly/illegally sent to local landfill | Potential suit or involvement in landfill cleanup | Advisable |

**Figure 20.2** *(cont'd)*

| Hazard Class | Situation | Potential Consequences | Risks Posed | Need For A Detailed Assessment |
|---|---|---|---|---|
| **INDUSTRIAL MANUFACTURING FACILITIES** | | | | |
| | All of the above plus.... | | | |
| Air Emissions | (1) Presence of regulated emission sources (e.g., stacks, vents) | Exceedance of U.S. EPA emissions standards (community) | Possible need for installation of expensive control equipment | Essential |
| | (2) Presence of workplace odors and/or emissions | Exceedance of U.S. OSHA workplace standards | Same as above plus workmen's compensation suits where health is impacted | Essential |
| Stormwater | (1) Presence of combined sewers or possibility of spills entering storm/floor drains | Discharge of contaminated storm runoff to soils, surface or groundwater | Potential deal killer if substantial quantities are involved; especially if neighborhood is on individual wells | Essential |
| Wastewater | (1) Process wastewater discharge to sewers | Exceedance of U.S. EPA or municipal effluent standards | Possible need for installation of expensive control equipment | Essential |
| | (2) On-site generation of wastewater sludge | Land disposal of contaminated sludge | Possible need to remedy land disposal site; very costly | Essential |
| Waste Disposal | (1) Presence of active or closed on-site land disposal facilities | Contamination of soils and/or groundwater from past practices | Potential deal killer; especially if neighborhood is on individual wells | Essential |
| | (2) Past use of troubled contractor's waste disposal site | Contamination of soils and/or groundwater from past practices | Could be brought into Superfund case as responsible party | Essential |
| Chemicals/Waste Management | (1) Significant use, storage or manufacture of toxic and hazardous chemicals | Noncompliance with CAA, CWA, RCRA, SARA, TSCA, FIFRA, SDWA or other environmental statutes | Possible historic noncompliance and future need for costly new programs and facilities | Essential |

In general, each of the five property types poses the same liabilities as that type preceding it, along with some additional liabilities. For example, in the hierarchy, industrial manufacturing facilities present the most comprehensive set of potential liabilities one might discover with any of the other property types (*e.g.*, asbestos or radon accumulation). Thus, the assessment tables build on one another, beginning with an assessment of undeveloped land.

Lenders and buyers can use the tables to screen investment properties to identify real and potential environmental liabilities. The tables can also be used to assure that, where an environmental specialist is used to assess a site, all potential liabilities are investigated and reported.

At the conclusion of Phase I, a draft report is issued documenting the data and information gathered and the conclusions regarding the environmental status of the site. (In the parlance of the ASTM standard, the report must identify the existence of "recognized environmental conditions," which are defined as a past, present, or threat of a future release of hazardous substances or petroleum products on the property. The assessor also must state his or her opinion as to the potential impact of those conditions on the property. In other words, a "recognized environmental condition" could be found, but its potential impact will vary depending on many other conditions at the property.) It also must be appreciated that Phase I results may be inconclusive due to the limited extent of available data and/or may include the recommendation to proceed to Phase II activities in order to reach more definitive conclusions of the site's environmental condition (or the impact of recognized environmental conditions on the property).

## PHASE II

If, at the conclusion of the Phase I, it appears that more information is required to explain such things as soil discoloration, large earth disturbances, or nearby/surrounding property concerns (*e.g.*, well water odors or contamination, colored seepage, abandoned dump sites), then Phase II soil and/or groundwater sampling should be conducted.

A sampling plan (locations, type, chemical analyses, etc.), schedule, and cost estimate need to be proposed. Analyses should be conducted by a laboratory that is accredited in the state where the site is located.

Normally, any underground storage tanks (USTs) that had not been tested in the year prior to the assessment will either be tested or the subsurface soils analyzed under a Phase II program. Also, any suspected asbestos-containing materials (ACMs) will be sampled and analyzed under a Phase II effort. If the site is known to contain USTs and/or suspected ACMs prior to the Phase I assessment, then the tank testing and ACM sampling can be accomplished during the Phase I site visit.

An ASTM committee has been working for the past two years on a Phase II standard. The draft standard is well along, as of this publication date, but it has not passed the final ASTM balloting process. The draft standard proposes broad guidelines for Phase II activities but, unlike the Phase I standard, it does not include detailed Phase II procedures.

Presently, there is no universal agreement about whether shallow soil or soil gas sampling should be conducted on *all* assessments. Many professionals feel that sampling during the Phase I assessment, without additional information about contaminants, is a wasted effort, since you cannot answer the questions of what to test for and where to test. Nonetheless, some companies require limited soil sampling regardless of the results of a Phase I effort. In extreme cases, some companies also require groundwater monitoring (one up-gradient and two down-gradient wells) for *any* acquisition as a matter of policy. In general, however, the standard property transfer assessment approach is to test all USTs, sample all suspected ACMs, and to sample/monitor soils and groundwater only where the results of a Phase I assessment warrant it.

## PHASE III

In certain cases, the purchaser may decide to proceed with a purchase even though the site represents present or future environmental liabilities as determined by Phase I or II assessments. If so, an engineer would develop a work plan, schedules, and fee estimates to conduct a comprehensive site assessment, including capital and operating cost estimates for future site remedial or cleanup actions. It should be clearly understood that such a detailed site assessment is relatively costly and time-consuming, and the resultant cost estimates are approximations.

## ASSESSMENT ISSUES

The use of environmental assessments in real estate transactions has given rise to a number of thought-provoking issues. And these are not strictly technical issues, as is evidenced by those questions posed below. If you are about to undertake an assessment using a third party, you should consider the following:

### How Do I Know What Rules and Regulations Apply?

This is, unfortunately, a moving target. Superlien, buyer protection, and state transfer statutes are being passed and revised regularly in states throughout the U.S. In addition, the more traditional environmental laws (*e.g.*, the Resource Conservation and Recovery Act) and their related rules are being modified as well. As of early 1989, the U.S. EPA alone had almost 300 separate regulations under development, revision, or review. The best bet is to consult with an attorney who is familiar with environmental laws and laws in the state in which a deal is being consummated. A good alternate source is the environmental assessment consultant.

### How Can I Be Sure That the Consultant I Retain is Qualified?

There are several ways. For example, in some states special certifications are required to do asbestos assessment work. Check these certifications. If underground storage tanks are to be tested, the tester should be familiar with the techniques. Also, consultants will be more than willing to provide statements of qualifications. But beware—you may be inundated with paper. Probably the best way to attain a certain level of comfort is to check references. Any consultant should be able to provide you with three references for whom he or she performed similar work.

## Do I Need a Big Firm or Can an Individual Consultant Suffice?

Within the constraints listed above, individual consultants could be perfectly acceptable. However, they need broad-based capabilities; the skills needed to assess asbestos issues are often very distinct from those of soil contamination. Also, larger firms might have turn-key capabilities that you might need (*e.g.*, the ability to sample *and* analyze soils, the ability to remove friable asbestos or test underground storage tanks). Larger firms will also have multiple offices, so if the investments cover a wide geographic range, you might want to consider this fact. One might also contract only with firms that have errors and omissions insurance coverage, in the event that a particular liability is not uncovered during the assessment. Firms that do not provide this protection can only protect the client with its assets.

## Can I Get a "Clean Bill of Health" Certification?

No. Nor should you look for a certification that the site is clean. Any consultant willing to broadly certify sites, by definition, has suspect qualifications. This is particularly true with respect to soil or groundwater contamination. A site cannot be assumed to be universally clean without turning the landscape into "Swiss cheese," in an effort to test every particle of soil for every conceivable contaminant. There could always be that small pocket of undetected contamination where a ten-gallon pail of solvent was spilled three years ago. What you *can* get from your consultant, however, is a qualified certification, one that typically states that "using generally accepted sampling and testing techniques, the site was found to be essentially free of contamination."

## If There is Some Contamination, How Do I Know if There is a Problem? In Other Words, "How Clean is Clean?"

This is a tough question. The U.S. EPA has yet to resolve it conclusively in determining cleanup standards for Superfund sites. "How clean is clean?" is often a matter of judgment and may have to be negotiated with regulatory agencies. For groundwater, there are federal drinking water standards, but these presently cover only a very few contaminants. Many states today have developed "action levels" or guidelines for acceptable levels of contaminants in soil or groundwater.

## Some Deals Develop Quickly. How Much Advance Warning Do I Have to Give the Consultant?

The more the better, of course. You should be able to obtain a consultant for a Phase I assessment with only a few days' notice, and a Phase I on a small site can be completed in a week, with certain limitations on access to public records. Consultants are prepared to mobilize quickly. There have been cases where several hundred sites have been assessed in four to six weeks. It is best for both parties, however, if the client has a standing contract with the consultant so that contractual terms and conditions don't impede progress.

## What if the Assessment Identifies that Corrective Actions are Necessary?

The need for corrective action should not necessarily be a deal killer. Removal of a damaged UST or ACMs can often be accomplished in a way as to eliminate any residual liability. Groundwater or soil remediation can be another story, however, due to the complexity of the problem and the time needed for remediation. Yet, even these issues can be resolved with the creative use of contractual warranties, covenants, and indemnifications; that is, the seller may have to assume some ongoing cleanup responsibilities or be contractually responsible for part or all of any contamination discovered in subsequent years. Or, in the alternative, the selling price could be discounted based on estimated liabilities. (The classic case is the Exxon corporate headquarters building in New York, which was sold in the 1980s for substantially less than its market value because of the widespread presence of asbestos.) The key is to understand the costs of continuing liabilities and to factor these into the agreement of sale.

## What Situations Might Constitute Deal Killers?

This can be a very judgmental decision. Some lenders, for instance, will not provide financing on any property that contains asbestos, regardless of its condition. Others take a more reasoned approach. In general, the presence of one or more of the following conditions could be considered unacceptable to the buyer or lender:[6]

- The structure is built over a closed sanitary landfill or other solid, hazardous, or municipal waste disposal site.

- There is a presence of friable asbestos-containing materials; or substantial amounts of non-friable asbestos that can't be safely encapsulated and/or removed or won't be routinely inspected and maintained by the owner/operator.

- There is a presence of high-risk neighbors with evidence of spills or soil/groundwater contamination on or around their properties.

- There is documented soils/groundwater contamination on the subject property and/or a documented tank leak greater than 0.05 gal/hr (the National Fire Protection Association standard), and any of the following four situations exist:

  - Physical constraints posed by the site-specific geology, geohydrology, or subsurface structure render corrective actions technically unfeasible

---

[6] This material is abstracted from guidelines developed by the Federal National Mortgage Association for its Multifamily Delegated Underwriting and Servicing Program.

- Constraints render treatment processes or disposal options prohibitively expensive; that is, beyond the financial capabilities of the owner

- Potentially responsible parties are unwilling or financially incapable of instituting corrective actions on neighboring properties

- Soil or groundwater sampling values are above the state or federal guidelines.

■ There is PCB contamination where:

- Physical constraints posed by the site-specific geology, geohydrology, or subsurface structure render corrective actions technically impossible

- Constraints render treatment processes or disposal options prohibitively expensive.

■ High radon levels (*i.e.*, >4 pCi/l) can only be corrected through large capital improvements and/or extensive on-going maintenance programs that are beyond the financial, organizational, or technical capability of the owner.

■ Conditions represent material violations of applicable local, state or federal environmental or public health statutes or laws (although this is not a function of a typical Phase I assessment).

■ Properties are currently the subject of environmental or public health litigation or administrative action from private parties or public agencies.

## What Do I Do With the Report if the Deal is Killed?

If there were no significant environmental issues identified, then this becomes a moot issue. If, however, some significant issues were raised, then you should consult your lawyers. Many environmental laws have specific reporting requirements where real or potential environmental contamination could occur or has occurred. Although you may have no regulatory reporting obligations yourself, the property owner might have had to do so, and now you have knowledge of this fact. This has even more significance in the unlikely event that an imminent health hazard has been identified.

## Who Should Hire the Assessor?

In cases where only a buyer and seller are involved, each party might want to hire its own assessor. More typically, however, the buyer will contract with an assessor (or use in-house staff) and the seller will make sure that its representatives (*i.e.*, trained in-house or consulting staff) are on-site during the site inspection—in essence, assessing the assessors.

More complicated assessments could include a buyer, a seller, attorneys representing each, and a bank providing financial support to the buyer. In these cases, the contractual relationship among the parties can become quite confused. This is particularly true where the assessor is working for both the buyer and its bank. These two parties have similar, but not always identical, objectives in the transaction. More often than not, the buyer will be willing to tolerate more risk and/or unknowns than the bank. In these cases, it should be made clear for whom the assessor is working.

Experience seems to indicate that the most troublesome relationship occurs when the assessor is hired by an owner, at the request of a bank, where the owner is in need of financing. In that event, it is obviously in the owner's best interest to have the assessor report the site to be essentially free of environmental liabilities. On the other hand, the bank that required the assessment and probably recommended the firm being used by the owner prefers to have every single liability identified and priced, regardless of significance. Given the judgmental nature of these assessments, this often puts the assessor in a direct conflict-of-interest situation. There have even been cases where owners have offered bribes to assessors to downplay the significance of the findings. The best solution to this problem is to have the bank choose and hire the consultant and be willing to absorb the costs if the deal falls through.

## What if an Imminent Hazard is Identified During the Assessment?

During the course of an assessment, an assessor might uncover an imminent hazard. A leaking PCB transformer or the presence of significant quantities of friable asbestos in a work area would be two examples. Assessors are not likely to uncover these hazards very often, but they should be prepared to respond appropriately when and if it happens.

Typically, responses will have to be handled on a case-by-case basis, and attorneys should be consulted. In the absence of any better guidance, the policy should be to inform the owner/operator immediately. This notification does not have to be made by the assessor directly, but can be made through an intermediary such as the owner's legal counsel or the lender.

Under most reporting requirements, the assessor has no personal responsibility to report an incident to a regulatory agency. However, some jurisdictions today impose reporting requirements for releases of hazardous substances on *anyone* with knowledge. Also, in cases where assessors are not satisfied that an intermediary will carry the word forward or that the response will be adequate, the situation becomes much more cloudy. In these situations, in particular, legal counsel should be sought.

## What are the General Approaches for Assessing Asbestos?

Asbestos is a mineral fiber that, prior to the mid-1970s, was considered a "miracle" product used in building construction due to its insulation, acoustic, tensile, and decorative properties. In buildings, asbestos has been used as sprayed-on acoustic material; pipe, boiler, tank, and duct insulation; adhesives, fillers, and sealers; and roofing felt and shingles. If material containing asbestos remains intact, the asbestos presents no immediate hazards to those who live or work in its vicinity. However, when the material is damaged or disturbed, asbestos fibers are released, presenting a potential health risk to building occupants. While the use of asbestos-containing materials in construction has been mostly eliminated since roughly 1975, some products are still made with asbestos as a component (e.g., asbestos-cement pipe, vinyl asbestos floor tile, pipe joint compound, automobile brake pads). Material is considered asbestos-containing if asbestos constitutes more than one percent of the total composite.

The presence or absence of asbestos can be determined in several ways, including:

- Review of "as built" engineering drawings and specifications
- Review of a previous asbestos assessment
- Certification that a newly constructed building (e.g., newer than 1975-1979) was completed without use of asbestos-containing products
- A bulk sampling program.

The challenge, then, for assessors on many sites is to be able to rely on the first three approaches listed above during a Phase I assessment. Simple physical observation will not be sufficient to make a negative finding, because much asbestos material can be hidden above dropped-ceiling panels or behind interior walls. The Phase I assessment is generally a "non-intrusive" procedure where, because of the risks of exposure, ceiling panels and the like are not routinely disturbed. Dried, friable, fallen sprayed-on asbestos can sometimes be laying on top of the panels, and any disturbance could release substantial amounts of fibers into the living or work space.

The assessor should recommend a bulk sampling program as part of or after a Phase I assessment when there is no documentation that the site is asbestos-free; observation of visible insulation is inconclusive (recall that materials with more than one percent asbestos are considered asbestos-containing materials); or the assessor is unable to access all areas of the building safely. Should a bulk sampling program be recommended, it should be completed by a certified asbestos inspector. In some locations, such as New York City, this is required.

## How is Radon Assessed?

Radon is a naturally occurring gas produced by the radioactive decay of the element radium. Recently, the occurrence of radon in buildings, especially in residential structures, has received increasing attention. Radon can be found in the air space in buildings and in the water supply. With the construction of "tighter" homes and the decreased ventilation as a result, indoor air pollutants such as radon can accumulate to unsafe levels. U.S. EPA estimates that radon exposure causes as many as 20,000 lung cancer deaths a year in the U.S.

Radon is not a localized issue. It can be found at high levels throughout the country. Radon concentrations are typically highest in basements, although there are some exceptions to this rule. Radon can also be very selective; neighboring homes can have decidedly different concentrations. Average annual radon concentrations below 4 pCi/l are thought to be safe.

In a Phase I assessment, the assessor can conduct a risk screening for radon that could include:

- Contacting the local or county agencies to determine whether the area is believed to be a radon "hot spot"

- Assessing the relative risks of the structure (*e.g.*, determining whether there are basement apartments or offices where radon could accumulate).

In addition to the approaches listed above, there are also specific measurement techniques recommended by the U.S. EPA. In general, the method of choice is the charcoal canister, readily available in the marketplace for $15–$25. This canister should be left in place for two to three days, with winter conditions being simulated (*i.e.*, windows and doors remaining closed). The canisters are then closed and sent to a laboratory for analysis.

In using the canisters, the assessor faces two problems. Because the assessor may not be on-site for two to three days, he or she must rely on occupants to reseal the canister and mail it to the laboratory. Second, the assessor will have no verification that the canister was not moved to a lower-risk location during the two-to-three-day period. This lack of control can jeopardize the integrity of the results. Thus, canisters should be used with some discretion until such time as cost-effective, real-time monitoring equipment is available.

## Are Wetlands Important?

Yes. The wetlands issue is receiving considerably more attention from regulatory agencies such as the U.S. EPA. In the 1980s, EPA adopted a national goal of no-net-loss for wetland areas. There are now many cases where development projects have been stopped dead in their tracks because wetlands had been destroyed during construction and no permit had been obtained. In several instances, the developers have been required to purchase and dedicate, in perpetuity, other wetlands on a two-for-one trade; that is, two dedicated acres for each acre destroyed.

Thus, assessors may need to determine whether greenfield properties and properties where facility expansions are being planned contain wetlands. Determining whether a given property contains a wetland may require the eyes of an expert or consultation with the U.S. Army Corps of Engineers or county agencies. Wetlands are not always easily defined by the casual observer, or even a trained environmental assessor.

## Can Farmland Pose Liabilities?

Assessors may be requested to conduct assessments of farmland, particularly in cases where the land will be converted to residences, commercial buildings, or industrial plants. Farmland can present substantial environmental liabilities. The issues of importance at these sites are underground storage tanks, spills of oils, lubricants and hazardous chemicals, and past application of pesticides. Many common pesticides are classified as acutely hazardous wastes under RCRA, and therefore they present significant health risks to humans, even at low exposures.

With regard to the pesticide issue, the assessor can conduct a Phase I screening that would address past pesticide use (*e.g.*, type—particularly restricted-use or banned pesticides—and application rate and method), crop type, soil type, depth to groundwater, and prevalence of use of groundwater as potable water. This screening should provide some sense of whether past practices involved significant use of fairly toxic pesticides (*e.g.*, chlordane, DDT) and whether this use has a low or high probability of contaminating an important groundwater source. Yet, without actual groundwater testing, one can never be sure. Because of the extreme toxicity of many pesticides, sampling groundwater is usually recommended where significant long-term application of pesticides is found, or where the level of use is unknown.

## What About Urea Formaldehyde?[7]

Urea formaldehyde foam insulation (UFFI) and certain other building products such as some paneling and particle-board flooring have been used historically in building construction. UFFI, in particular, has been used extensively as a retrofit wall insulation. Volatile releases of formaldehyde from these products can cause respiratory problems in humans sensitive to the material. Trailer homes have been of particular concern due to the high use of formaldehyde products and the "tightness" of the homes.

In 1982, the Consumer Product Safety Commission banned the use of UFFI in residential and school buildings. However, due to the lack of definitive public health data, the ban was challenged and overturned by the courts almost immediately. Nonetheless, the consequences of the regulatory attention paid to formaldehyde makes it an issue in property

---

[7] It is interesting to note that, in addition to asbestos and urea formaldehyde foam insulation, U.S. EPA is considering listing fiberglass as a possible human carcinogen and mineral wool as a probable human carcinogen.

transfer assessments. As a consequence, the assessor should identify whether any formaldehyde products, especially UFFI, are present on the property. The assessor should not overstate the impact of the presence of UFFI; it is presently not an issue on a par with asbestos. But the buyer or lender should be aware that the material is present so that susceptible individuals can be alerted.

## What About Lead-Based Paint?

The presence of paints with high lead concentrations can create liabilities in buildings, particularly residences. Some children have an "abnormal craving to eat substances not fit for food, such as clay, soils, and paints." This abnormality is known as pica. Thus, if lead-based paints exist in a building where children can be exposed to them, there is a legal risk to the building owner. Children are especially susceptible to lead poisoning. This issue has received special attention from the U.S. Department of Housing and Urban Development (HUD), and new regulations, effective October 31, 1995, require that lead in paint be tested in multifamily property. HUD housing project managers also must provide notification to tenants where lead-based paint exists.

Fortunately, use of paints with high lead content in residential applications was prohibited over ten years ago. However, older buildings where multiple coats of paints still exist do present a hazard if the paint layers are peeling. This issue should not arise that often; yet, the assessor should be aware of it when assessing an older single-family or multifamily residence. Use of lead-based paints in apartments and hallways should be identified, particularly on wall areas less than five feet off the floor.

## What About Drinking Water?

The drinking water for a facility can be supplied through bottled water, an on-site well, or by a private or municipal water distribution system. The riskiest option is the on-site well, and where this is the case, some lenders will not, as a matter of policy, provide financing. However, even the presence of a municipal system cannot guarantee a safe water supply since some contaminants may not be removed effectively by a treatment system, or contaminants may enter the distribution system on route to the property through line-break infiltration or pipe corrosion and scaling.

Lead is a contaminant of particular concern. Several recent surveys have detected lead levels in public drinking water supplies significantly above the 50ppb standard. In many cases, the contamination is coming from lead solder used to connect copper tubing. Thus, the municipality may be providing a perfectly safe supply, but piping within the building may be the cause of contamination.

The lead issue does not mean that every site must have its drinking water sampled as part of a property transfer assessment. However, testing might have to be done where children are present (the lead poisoning issue), the piping system is old, or there have been complaints about the municipal drinking water or the drinking water in the building.

## CONCLUSIONS

In summary, conducting property transfer or Phase I assessments is essential in these days of increasing corporate and personal liability. This assessment will do much to identify any liabilities that might come hand-in-hand with the assets of the target; yet, assessments are not guarantees. If something is overlooked, there is typically recourse through the agreement of sale, which may contain a statement made by the seller that no material information has been withheld. Of course, this is a legal remedy with the potential for protracted litigation. It is best to understand the potential liabilities up-front through both financial and environmental assessments.

# TOP TEN REASONS WHY PHASE I ENVIRONMENTAL ASSESSMENT REPORTS MISS THE MARK[1]

Recently, there has been considerable professional concern over many of the issues associated with environmental assessments, including report writing.

In this article, the authors show how reports can be vastly improved to help companies evaluate risks and make better business decisions. Their guidance can be applied to a broad range of auditing and assessment activity to improve environmental quality performance.

Environmental assessments of property transactions have become routine in the past few years. Typically, buyers and investors wish to identify any environmental liabilities that could affect the value of the acquisition or investment. Even sellers have a vested interest in having assessments conducted because the assessments establish an environmental baseline that can protect the seller should future regulatory agency investigations identify significant problems at the site.

Almost all environmental assessments result in a written report. In virtually no cases will these reports state that there are absolutely no environmental problems posed by the site. Thus, the document is the principal tool that parties use to determine if the identified environmental issues at the site are significant enough to affect the financial structure of the deal. As such, the report is critical to the buyers and investors in making an intelligent business decision.

---

[1] Originally published in the Summer 1993 Issue of *Total Quality Environmental Management,* Executive Enterprises, Inc., NY, NY.

Unfortunately, there are no nationally recognized published guidelines on what constitutes an acceptable report. And experience dictates that the report quality varies considerably. Buyers and investors are quite often left with the unenviable task of hiring a second environmental consultant to interpret the report of the consultant who conducted the initial assessment. The reports are sometimes so deficient that the second consultant must revisit the site, a costly and unnecessary expense.

There has been considerable activity in the assessment profession to come to grips with many of the issues associated with environmental assessments, including report writing. Both Congress and the U.S. EPA have addressed the issue of defining "due diligence" as related to environmental assessments of acquired property in SARA 1986. SARA allows for an "innocent landowner defense" where the property owner has been diligent in identifying and remedying environmental liabilities. The EPA has attempted to further define diligence (all due inquiry) in published guidelines. More recently, the American Society for Testing and Materials (ASTM) has developed a draft set of guidelines for conducting environmental assessments, which include a recommended report outline (Standard Practices for Environmental Site Assessments: Phase I Environmental Site Assessment Process, Standard E.50.02.2, to be published spring 1993). And finally, many lending institutions have developed their own guidelines on what constitutes an acceptable assessment and report.

These efforts notwithstanding, there remains considerable variability in assessment and report quality. Simply providing an outline for the assessment report, as the ASTM standard does, can only go so far in assuring a product that meets current and future expectations. In an effort to provide guidance on what constitutes a good assessment report, this article discusses some of the most common report writing problems and how to deal with them.

# FAILURE TO MAINTAIN INDEPENDENCE AND OBJECTIVITY

The assessor must be diligent in attaining a high degree of objectivity when writing the report. This is easier said than done once the draft report is submitted to the client. The seller typically would prefer to see a clean bill of health and, in many cases, will put considerable pressure on the assessor to soften or eliminate findings. There are instances where the seller has refused to pay the assessor unless changes are made. This presents an ethical dilemma for the assessor that can only be prevented if full payment is made in advance. After the fact, the choices are to make the changes, write off the client and the costs, or sue for damages. This is why doing assessments for sellers can be risky.

Conversely, the buyer will wish to see every single environmental problem identified and costed, in order to use the information to reduce the price of the property. The risk here for the assessor is that the site will be so fraught with problems that the buyer will walk away with no intention of throwing good money after bad—another bad debt for the assessor.

Assessors should also remember that their roles are different from environmental attorneys representing buyers, sellers, or lenders. Attorneys are advocates for their clients, and as the client's agent, may put considerable pressure on the assessor to make changes or use softer language in the report. Assessors should be independent, third-party analysts providing an accurate accounting of the environmental liabilities posed by the site. There is a distinct difference.

Notwithstanding the risks of maintaining objectivity, the assessor is generally better off, in the long run, holding firm on technically sound findings. There is less chance for a damaged reputation and less likelihood of a lawsuit in which the assessor must defend an unsound report.

Aside from this concern about pressure being applied to modify report findings, there is a more fundamental reason to use independent and objective assessors. Every effort should be made to use environmental professionals who are not familiar with the property. Without even meaning to, people who are too familiar with the property may overlook problems that have become background conditions to them. Using independent assessors helps to assure a fresh set of eyes and ears are not going to miss things from overfamiliarity.

# FAILURE TO DEFINE THE EXACT SCOPE OF WORK

In spite of the ASTM standards of practice and the other Phase I assessment guidelines, there still can be miscommunication regarding the scope and format of the Phase I report. There should be a thorough discussion of the contents of the report before the actual work begins. One way of clearing up any differences is for the consultant to include a table of contents of the Phase I report in the proposal so that the client understands early in the process what to expect when the report is submitted. Yet, as discussed above, a table of contents only helps the client understand the format, not the content. Developing and submitting a "sanitized" report from a previous assessment is also an approach worth considering, although consultants are usually reticent to provide a past report due to standing confidentiality agreements and/or the risk of leaving something in the edited version that would permit identification of the site or client. On assignments where multiple sites are to be assessed, it is a good practice to conduct a pilot assessment of one site and submit the report for that site before beginning any other site reports. This minimizes the need to substantially revise multiple reports.

Some report issues that need to be resolved early in the process include the level of detail of the findings, whether only exceptions or problems found will be presented, and whether all investigated areas will be reported, even if no problems were encountered. Recommendations are often requested and presented in Phase I reports, but sometimes they are not required or desired. These and related issues need to be discussed before the report is submitted to avoid costly and time consuming rewrites.

It should also be noted that simply adopting a scope proposed by a recognized authority, such as ASTM, may not be sufficient in all cases. For example, because the ASTM standard focuses exclusively on CERCLA-defined liabilities, adopting the ASTM approach can lead to omission of key environmental areas of concern. That is, the ASTM standard does not address investigation of asbestos, radon, lead-based paint, lead in drinking water, and wetlands because these are not considered CERCLA issues. This would be considered a serious over sight by most investors and buyers who still may be held liable for these sources of potential environmental liability.

## USE OF CONJECTURE IN REPORT FINDINGS

Hearsay evidence and unsubstantiated information are often incorrectly used in judgments and opinions in Phase I reports. This can lead to findings that may either understate or overstate the environmental problem described. For example, a maintenance person at a site might suggest to the assessor that several drums of hazardous waste were dumped in a nearby culvert some months ago. While this might have been the case, it also might not be true. The interviewee could be mistaken (the drums could have been filled with rain water), or he could be deliberately misrepresenting the facts because his job is in jeopardy if the company is sold. It is the assessor's responsibility to obtain confirming evidence before presenting this spill as an undisputed fact.

It is often difficult to obtain completely reliable and thorough information when conducting property assessments. Partial and incomplete information is often the norm. It is important, however, that the parties conducting the Phase I assessment try to obtain as much corroborating evidence as possible about a potential environmental problem before including it in the Phase I assessment report. All sources of evidence, including personnel interviewed at the site as well as at neighboring properties, records and files reviewed, and physical inspections performed must be used to insure that the findings are defensible. If real uncertainty exists about the extent of the environmental problem, some level of Phase II investigation may be needed. But all efforts to identify and describe the environmental problem using Phase I evidence should be exerted first. Do not jump into Phase II recommendations when more diligence in the Phase I tasks will answer the question.

# USE OF IMPRECISE LANGUAGE

In the rush to complete environmental assessments on schedule, the assessor may not be diligent enough in gathering field data, and this will result in an imprecise report. Assessors must strive to provide quantitative site information where possible. Too often statements like those given below are included in audit reports:

- There were a few drums behind the maintenance shed. (How many drums? Were they protected from rain on a contained pad? Were they full or empty? What shape were they in? What was in them?)

- There was evidence of past spills on the ground. (What exactly was the evidence? How big an area was affected? What materials might have been involved? Is runoff a problem or infiltration into the soils and groundwater or both?)

- The building was constructed in the seventies. (When in the seventies? Asbestos and PCB manufacture were prohibited in the mid-seventies. There could be a big difference between 1971 and 1979.)

- There were a number of stacks and vents protruding from the roof. (How many stacks and vents? What shape were they in? Did any show signs of deterioration? What sources were associated with each stack? Did any need permits?)

When conducting field investigations, assessors must spend the time to quantify what they observe. This quantification should be translated directly into the report.

On the other hand, in an effort to be complete and thorough, the assessor must be careful not to use categorical phrases that are not supported by the evidence. For example, the statement "there are no underground storage tanks on the site" should never be made in an assessment report. One could almost never be completely assured of the truth of this statement. And if, at a later date, a fifty-year-old abandoned tank is discovered twenty feet below the surface in a remote area of the property, one can be assured that litigation will follow shortly thereafter. There are better ways to state this particular finding, such as "Based on interviews with the site supervisor, a review of the site's records, and a walk over, there is no indication of any underground fuel oil tanks at the site."

## FAILURE TO DISTINGUISH "COMPLIANCE" FINDINGS FROM "LIABILITY" FINDINGS

In the mid-1980s, when Phase I environmental assessments were first becoming a part of real estate transactions, the focus was typically on identifying major sources of contamination at the site (spills, leaks, dumps). These sources were often seen as the "big ticket" items that could make or break a deal. Gradually, evaluating the property and the operations and activities at the property for compliance with environmental, health, and safety regulations has become of equal interest to buyers, sellers, and lenders. Buyers and lenders, in particular, have become keenly aware of the potential for $25,000-per-day fines for compliance violations, which can quickly add up to substantial amounts. Moreover, compliance issues can result from what might be considered progressive environmental actions, and therefore can be masked at a site that seems to be doing all the right things. For example, many sites over the past few years have been removing their fuel oil underground storage tanks and replacing them with above ground tanks. Although this practice helps to prevent ground water contamination, many sites now have to comply with Spill Prevention Control and Countermeasure (SPCC) requirements promulgated under the Clean Water Act. This has occurred because the SPCC below ground regulatory threshold ( >42,000 gallons) is so much greater than the above-ground threshold ( >1,320 gallons). Given that SPCC regulations could require secondary containment systems for all tank and drum storage of oils and fuels, it would be a substantial oversight to omit this issue from an assessment report.

Thus, it is important that for Phase I assessments, the report clearly presents the findings in categories that distinguish them according to their type and significance. A useful scheme is to break findings and recommendations into three categories:

- **Required by regulation.** Environmental concerns identified as regulatory deficiencies are primarily compliance items mandated by federal, state, county, or even municipal laws and/or regulations. Regulatory deficiencies may be related to paper violations (*e.g.*, failure to submit an annual hazardous waste generation report), which carry only monetary penalties, while others may pose a significant environmental concern as well (*e.g.*, unpermitted discharges of pollutants to ground water).

- **Environmental liabilities.** Environmental concerns identified as potential liabilities are items or operational practices that are not currently regulated by federal, state, county, or local agencies but may have currently or historically adversely impacted the environment (*e.g.*, an abandoned industrial septic system on-site or the use of a CERCLA's-listed site for disposal of hazardous waste). Usually these issues are candidates for Phase II investigations.

- **Best management practices.** These are items identified during the course of the Phase I assessment that, although not required by law or regulation, would limit the potential liability associated with the operation of the facility (*e.g.*, secondary containment around above-ground storage tanks).

## DOCUMENTED SOURCES

Evidence is gathered on environmental assessments in one of three ways: interviews, records review, physical observation. The assessment report must be clear about the source of evidence for each conclusion. Where evidence is not fully provided in the report itself, the project's files must contain complete documentation. There are now cases where poor document management has jeopardized a client's and consultant's position in subsequent litigation.

Names, titles, and affiliations of people interviewed must be obtained and the dates of the interviews must be documented as well as where the interview took place and whether it was face-to-face or over the phone. Interview logs and forms are excellent for assuring that all appropriate background information is obtained.

Written sources must be documented accurately and thoroughly. Written records are typically obtained from a variety of sources: the site, local municipal authorities (*e.g.*, the county health department, the city water department), state and federal regulatory agencies (U.S. EPA), federal information agencies, and third-party providers of general site and area compliance data. Written records must be referenced thoroughly in the report and copies included where it would be helpful to the reader. For example, where the records show that the site had never received a fine for a hazardous waste violation, the effective date of that finding would be important to include. Some written records are not current and if there had been a recent fine that the dated records did not reflect, the assessor would be rightfully challenged if this information were to be unearthed by someone else through other means.

And finally, any physical observation must be described accurately and should include appropriate parameters (*e.g.*, time of day, weather, status of the site's manufacturing operations) when necessary to understand the issue.

## FAILURE TO STATE ASSUMPTIONS REGARDING COST ESTIMATES IN ASSESSMENT REPORTS

There is usually high interest by the buyer, seller, or lender for the Phase I report to include estimates of costs to conduct additional Phase II investigations or costs to clean up or remediate the environmental problems found. And this is not unjustified. After all, the real reason for conducting the assessment is to determine how environmental liabilities might affect the value of the property.

This issue often leads to conflict. The buyer or seller would prefer to have the assessment conducted as cheaply as possible but often expects any cost estimates for remedies to be quite precise. These two objectives are, for the most part, conflicting. The assessment firm's cost estimates are typically order-of-magnitude or approximate estimates. They often are based on generic unit cost estimating factors or rules of thumb to give the property owner some idea of the potential future costs. The buyer, seller, or lender who has requested the assessment is outraged at the assessment firm when the actual cleanup or remediation costs solicited from construction or remedial firms are significantly higher than the cost estimates in the report.

To prevent this conflict, it is imperative that all cost estimates for future Phase II or cleanup activities be presented with the complete list of assumptions that were used to generate the estimate. Three types of assumption that should be clearly stated for each cost estimate are:

**Quantity assumptions.** This refers to cost estimates that are based on volume of soil to be excavated, area of asbestos to be removed, quantity of liquids in tanks, and similar types of quantity-derived cost estimates. All assumptions (including whether quantities are based on actual measurement, calculated mathematically, or taken from an other source of information) that provide a basis for the reader to understand the reliability of the estimate should be stated in the report. This will also provide a defense for the assessment firm if a subsequent cost estimate by another party is different from the cost estimate in the Phase I report.

**Replacement cost assumptions.** This mistake is often made by assessment report writers because they typically are viewing the cost estimates only as environmental cleanup costs, not construction costs. Replacing tanks, fuel lines, or new ceiling tiles and flooring that were removed to get to a source of contamination are examples of real costs that the owner will see in a construction company's estimate. Care should be taken to list the replacement or field-construction related costs that will be incurred. Or simply make clear in the Phase I report that these costs are not a part of the estimate.

**Unit cost assumptions.** There are several cost estimating guides that provide unit cost estimates for items such as:

- Cubic foot of soil removed
- Hundred gallons of water pumped
- Drum removed
- Square foot of asbestos removed

It is important that the source of any unit costs provided be clearly stated in the Phase I report. If the costs came from a published cost estimating guide, the specific name of the document and the edition used should be listed. If the unit costs came from quotes from construction or remedial firms, they should be noted in the report.

Stating assumptions used in deriving costs based on unit cost factors, whatever their source, will help to prevent conflicts between the owner of the property and the assessment firm about the validity and reliability of cost estimates presented in the Phase I report.

## LACK OF EDITING AND A QUALITY ASSURANCE REVIEW

Every Phase I report should be reviewed by a qualified third party. This must be someone qualified to give an independent and objective reaction to the report in three key areas:

- Technical accuracy
- Readability
- Defensibility

The quality assurance review should normally be conducted by a senior member of the assessment firm. The writer of the report needs to leave his ego at the door and accept the comments of the editor for the purpose intended—to improve the value of the report to the client.

Certainly the report should be technically accurate. As stated earlier, words such as "some," "widespread," and "a few" should be eliminated.

The writer should also demonstrate a clear understanding of how state and local regulations, in particular, might affect the site and any proposed remedies.

Readability means just that. Is the report simple and clear? Are acronyms defined? Are too many polysyllabic words used? Writing in the active voice is not critical, but it tends to be more direct and descriptive for the reader. Remember, the readers of the Phase I report have probably not been to the site. The report needs to help the reader visualize the issues and conditions.

For the report to be defensible, the findings must be backed up by hard evidence. The editor of the report should challenge the writer if findings appear to be unsubstantiated. Defensibility also can be viewed in the legal sense. More and more transactions that have gone bad are being argued in court based on what the Phase I environmental assessment report did or did not say.

## DISPUTES OVER DISCLAIMERS

This issue typically becomes a tug of war between the buyer, seller, or lender—which prefers that no disclaimers be included in the report—and the firm conducting the assessment—which prefers as complete a disclaimer as possible. As with most things in life, an equitable compromise usually can be reached.

There are typically three disclaimer issues that arise in an assessment report. The first is certification. From the assessment firm's perspective, there should be no expectation that they will "certify" or "guarantee" that the site is "clean." This is simply too absolute and no responsible firm will agree to make the certification. There are softer certification statements that can be made that both parties would find acceptable. These usually center on the certification that the Phase I assessment was conducted "in accordance with industry standards."

Second, the assessment firm should not be expected to warrant that documents, records, and reports prepared by others not under the direction of the assessment firm, and which are used by the assessment firm in determining findings, are accurate or up-to-date. These records are necessary for the assessment process, but they may contain errors or may not be complete. This can happen with records obtained from federal, state, and local regulatory agencies as well as materials prepared by private companies or their consultants.

And third, the assessment firm should be able to disclaim knowledge of any environmental problems that were not apparent during the assessment, but came to light some time after the assessment was completed. Judgment is, of course, needed here. Not identifying that a leaking underground tank was on the property during the audit, which eventually creates cleanup liability for the owner months later when it is discovered, should not be covered under a disclaimer. But the presence of twenty rusty, leaking drums behind a warehouse, which may have been placed there months after the assessment, should be viewed as a possible disclaimer issue for the assessment firm.

Other than the three areas of potential disclaimer use mentioned above, the assessment firm should be held accountable for all findings or lack of findings in the assessment report.

## FAILURE TO WRITE THE REPORT AS A BUSINESS-DECISION TOOL

Engineers and scientists generally write Phase I assessment reports. They are primarily concerned about environmental impacts and risks. Consequently, their reports are chock full of details and information about everything they did and observed during the Phase I assessment. Too often this information, although well intended and documented, is extraneous and not useful to the reader, who is generally a business person simply trying to make a business decision.

As an example, a report that forces the reader to struggle through three pages of a description of the geologic subsurface conditions and each underlying rock strata and hydrogeologic characteristic of the site is really not serving the client's needs. The important point that might need to be made here is that "local ground water, which is not a drinking water source, is about 200 feet below the surface with a ten-foot layer of dense clay in between, making potential contamination of ground water virtually impossible."

Another good example comes from a report that stated that the soils underlying an underground storage tank were tested and found to contain one percent total petroleum hydrocarbons (tph). In this case, the buyer, unalarmed, was left to interpret the meaning of that statement. When he discovered later that the 10,000 ppm concentrations were 100 times that state's guidelines for tph in soil, he was rightfully angry.

Another trap that writers of Phase I reports sometimes fall into is to quantify the environmental risk. Unlike health-based risk assessments, for example, where quantification of risk in incremental cancer-caused deaths is a normal practice, quantification of environmental risks in Phase I assessments is much less practiced. Many lenders and buyers would like to see statements like "there is a ninety-five percent chance that the underground tank will not leak in the next five years or there is a ten percent risk that ground water will be contaminated from the chemical spill." But these are difficult to defend unless a clearly defined methodology based on statistics or computer modeling has been used to generate the numbers. Using qualitative definitions of risk potential and severity of risk, based on the assessment firm's professional judgment, is generally a better approach. This still gives the owner a sense of the risk potential in a way that helps him or her make the business decision.

Business people are used to making decisions based on risks. Writing the report in the simplest way possible allows them to make informed judgments about the risks associated with the transaction.

# WASTE CONTRACTOR AUDITS

---

*"All companies that generate waste and send it off-site for handling and disposal should understand how well the waste management facilities that they use are operated and financed. Periodic audits of these facilities are essential for making sound waste management decisions."*[1]

---

It has become the norm, rather than the exception, for generators of solid and hazardous wastes to be concerned about the companies that treat, store, or dispose of their solid and hazardous wastes. Many corporations spend hundreds of thousands of dollars annually to ensure that their waste is being properly handled. Until the late 1970s, before the proliferation of environmental legislation controlling these activities, the transportation and off-site disposal of what was then referred to as "garbage and chemical wastes" represented little more than a minor line item on a corporation's annual operating budget. In many cases, small contractors and vendors who operated local waste disposal businesses handled almost any type of waste offered to them, including trash, drums of waste oils and solvents, treatment sludges, and other by-products of manufacturing and industrial operations.

Many of these chemical wastes generated from manufacturing operations, such as solvents and oils, were mixed with domestic trash and placed in municipal landfills. Worse yet, they were taken to inactive quarries, warehouses, farmland, or other remote areas and simply abandoned or buried. There was little concern on behalf of the waste generator about the ultimate disposition and fate of these wastes, since

---

[1] Membership Information Pamphlet, Commercial Hazardous Waste Management Evaluation Group.

their disposal costs were low, and regulations did not exist at the time to prevent or control these activities. In fact, many companies sought to find the most inexpensive waste disposal option, regardless of the technology employed.

In the late 1970s and early 1980s, three key factors combined to abruptly change corporate America's waste disposal practices forever: (1) the discovery of toxic waste disposal sites such as Love Canal and Kin-Buc landfill, where the illicit disposal of industrial wastes posed a real threat to human health and welfare; (2) the passage of landmark environmental legislation controlling the generation, handling, and treatment/disposal of toxic and hazardous wastes (RCRA); and (3) the establishment of legislation aimed at cleaning up those properties already contaminated with hazardous substances (CERCLA, or Superfund).

The Superfund legislation allows the U.S. EPA, under certain conditions, to impose severe, retroactive joint and several liability upon any party responsible for the release of hazardous substances into the environment. Joint and several liability means that any one responsible party can be held liable for the entire cost of site investigation, cleanup, and other damages. In 1994, the average cost to investigate and remediate a contaminated site on the CERCLA National Priorities List, or NPL, was approximately $25 million, exclusive of legal costs. And there are many former waste disposal sites, such as the Stringfellow site in southern California and the Lowery Landfill in Colorado, where the costs are much greater. Responsible parties can include current or previous owners or operators of a waste treatment, storage or disposal (TSD) facility; transporters or contractors that selected the disposal facilities; and, most importantly, the generators who provided the hazardous wastes. Waste generators are the most susceptible to joint and several liability for several reasons, which are as follows:

- The previous or current site owners may not have the financial resources to meet the cost of the damages.

- Due to the complex nature of many Superfund sites, in some cases it becomes very difficult to accurately allocate responsibilities among owners, operators, and generators, so all of the responsible parties are forced to contribute. Legal costs for prolonged negotiation and discovery can prove insurmountable, forcing companies to settle.

■ Industrial corporations that generated the wastes often represent the "deep pocket" at the site, making them the most likely target for both CERCLA cost-recovery actions and private party "toxic tort" suits.

It should be noted that potential liabilities for generators associated with the disposal of waste are not solely limited to the use of the off-site TSD facilities. CERCLA liability can also be assigned to parties that transported or arranged for the disposal of the wastes. As a result, transporters and waste "brokers" can be held accountable as "potentially responsible parties" (PRPs) under CERCLA. In today's environment, with a decreasing number of waste disposal options available to a generator, many generators are relying more heavily on the use of multiple transporters and waste brokers to remove and dispose of wastes. In many cases, waste brokers are actually selecting the disposal site for the generator's waste, potentially exposing the generator to additional future liability if the site is poorly managed, operated, or financed. Accordingly, generators must be cognizant of the increased risks associated with the use of these multiple vendors.

There are also numerous sites on the NPL that handled wastes considered to be nonhazardous at the time, such as empty drums, waste oil, asbestos, and batteries, where the generators of these "nonhazardous" wastes have been held accountable as PRPs under CERCLA. Waste generators must also consider the risks associated with the treatment or disposal of these nonhazardous wastes (which today can include precious metal bearing wastes, Ni-Cd batteries, and other electronic scrap), which becomes difficult since the vendors handling these non-RCRA wastes may not be as carefully monitored or controlled as TSD facilities that are regulated under RCRA, TSCA, or state equivalents.

Considering the range of potential liabilities associated with waste management, responsible solid and hazardous wastes generators are anxious to avoid using poorly operated, improperly managed, and insufficiently financed TSD facilities in order to limit their future exposure. In addition to the potential for future financial liability under CERCLA, the possibility of both financial impact and negative public relations exposure—due to private party "toxic tort" suits for real or perceived health impacts, or to property devaluation—is forcing corporations to take a closer look at the commercial TSD facilities that handle their wastes.

Waste vendor audit programs are also increasingly driven by simple economics. As options for waste treatment and disposal become more limited, many generators are forced to pay higher prices for waste disposal and/or transport their wastes much greater distances to reach their selected TSD facility. In some cases, generators start waste vendor audit programs simply as a way to identify and evaluate less expensive but equally suitable alternatives for the management of the their wastes.

## OBJECTIVE AND SCOPE OF A TSD FACILITY AUDIT PROGRAM

The basic objective of any TSD facility audit program should be a simple one: to identify and minimize the generator's liabilities that are associated with the off-site treatment, storage, and disposal of their wastes. The vast majority of generators do not have captive treatment or disposal facilities and, therefore, must rely on commercial facilities to handle some, if not all, of these wastes. Accordingly, control over a significant portion of the liabilities associated with the treatment and disposal of hazardous wastes thereby rests in the hands of third-party contractors (transporters, brokers, and the TSD facility), although under RCRA the generator is responsible in perpetuity for assuring their proper fate. The TSD audit program should be designed to identify the potential liabilities associated with the use of these third-party vendors. Once the potential problems have been identified, the program should allow for some type of quantitative analysis by which different TSD facilities can be evaluated, compared, and selected.

Ideally, the selection of TSD facilities to be audited should not be limited to those facilities that are currently handling a generator's RCRA hazardous wastes, because RCRA-regulated facilities tend to be more closely monitored by regulatory agencies than solid waste, oil, or precious metal recovery operations. Many of these non-RCRA facilities present significant environmental, operational, and demographic concerns which could ultimately result in generator liability. Since the financial assurance requirements for non-RCRA facilities are typically less stringent than for RCRA-regulated facilities (or are nonexistent), the management and financial stability of these facilities may be of concern to a large deep-pocket generator in the event of an environmental problem.

The TSD facility audit program should assist waste generators in identifying those transporters, brokers, and/or TSD facilities that are best suited to handle the generator's particular waste streams. By thoroughly assessing the capabilities and operations of these vendors, generators can often reduce the number of contractors that they utilize for waste transportation, treatment, and disposal, resulting in a more focused and cost-effective waste management program. As an additional benefit, the commercial waste vendor audit program allows the generator to identify and eliminate the use of contractors who present unreasonable environmental risks that otherwise would not have been evident without a comprehensive, risk-based review.

## INTERNAL PROGRAMS VS. EXTERNAL PROGRAMS

Waste generators typically conduct commercial waste vendor and audit programs in one of two different ways: (1) through the use of internal staff or (2) by using independent professional environmental consultants.

The use of company employees to conduct a commercial waste vendor audit program usually requires a lengthy start-up period for the training of audit staff, selection of the program elements (*i.e.*, what will be covered in the audit), development of audit checklists and report formats, and establishment of the audit infrastructure. Depending upon the scope of the program, the development of an internal, comprehensive commercial waste vendor audit program may require dedicated staff and a significant amount of resources. Once the program is implemented, large expenditures are infrequent, but the program management can be overwhelming, especially for corporate environmental staff with other responsibilities. For this reason, most internal commercial waste vendor audit programs are very limited in scope.

Although using consultants to develop and implement a commercial waste vendor audit program will minimize the managerial burden on corporate staff, program start-up costs can be large, and the program implementation costs become directly proportional to the number of audits that are assigned. The design and product of company-specific audit protocols, checklists, scoring models and financial analysis models can be expensive if a truly comprehensive audit program is desired. Of course, the amount of information that the generator needs to assess the suitability of a waste vendor will ultimately determine the cost of program start-up. Once the program is in place, the costs to have the consultant conduct the audits can range from several thousand dollars to $20,000 per site, based upon the characteristics of the TSD facility, its location, the complexity of the audit, and the scope of the audit report. The advantage to having an independent, outside consultant conduct the audits is that experienced and trained engineers and scientists can conduct as many audits as the generator is willing to assign without draining the generator's available manpower and resources.

In some cases, generators with common interests in conducting commercial waste vendor audits can arrange to have one contractor develop the program and conduct the audits on behalf of the group. The information can then be distributed on a cost-sharing basis. There are several major industry-oriented programs in place today which employ external consultants to conduct commercial waste vendor audits using pre-established, proprietary and confidential audit protocols, report formats, and risk-ranking models.

For example, the Commercial Hazardous Waste Management Evaluation Group (CHWMEG) is an unincorporated association of Fortune 500 companies formed in 1987 to obtain independent evaluations of TSD facilities. An independent administrator and escrow agent are employed to operate CHWMEG and handle all transactions on a confidential basis. Member companies' decisions to audit or utilize a particular TSD facility are strictly confidential. Each member company selects TSD facilities to be audited, and the audits are then conducted according to a CHWMEG-developed protocol by independent, nationally recognized environmental consulting firms. The cost of each audit is shared by the member companies who have chosen to sponsor the audit of that particular facility. Each sponsor then receives a complete copy of the audit report. The CHWMEG group has sponsored the development of a specially designed database management system (DBMS), which allows computer access to key audit information, data sorts, and extractions. CHWMEG's audit procedure is very comprehensive, and includes a proprietary financial analysis model as well. CHWMEG has also sponsored the development of a quantitative Risk Assessment System (RAS), which considers the environmental, operational and financial risks posed by a TSD facility. CHWMEG routinely evaluates as many as thirty different TSD facilities annually.

The CHWMEG administrator can be reached at the Center for Hazardous Materials Research, University of Pittsburgh Applied Research Center, Pittsburgh, PA, (412) 826-5320.

## CONDUCTING THE TSD FACILITY AUDIT

It is recommended that waste generators design their commercial waste vendor audit programs to incorporate *all* of their waste vendors, including transporters and brokers. However, since the core aspects of most commercial waste vendor programs center around auditing and evaluating TSD facilities, for the purposes of this chapter we will use the example of a TSD facility audit to describe the basic steps in the waste vendor audit process. The same audit tasks and principles will apply to waste transporters and brokers as well, but the specific risk management issues associated with these vendors will focus on waste handling and transportation rather than treatment and disposal.

Waste vendor facility audits are similar to internal plant audits in that they have several different, distinct phases: the pre-audit, on-site, and post-audit activities. However, the approach to conducting waste vendor facility audits differs in some key respects:

- Although regulatory compliance is an important factor in evaluating the relative liability posed by a waste vendor, the audit must focus on risk-related aspects of the waste management operation as well.

- The relative financial strength of the waste vendor (and its parent company, if applicable) is an important area of evaluation.

- Staff cooperation, available auditing time, and access to data may be limited by the fact that you or your consultant are not an employee of the waste management company or a regulatory agency inspector. Therefore, proper pre-audit preparation takes on an even more critical role in conducting waste vendor facility audits. Trying to complete a comprehensive audit of a very large commercial TSD facility without proper pre-audit preparation is akin to attempting to complete your federal income tax return on April 14th without reading the instructions first.

The following sections discuss the elements of conducting a successful waste vendor audit at a commercial TSD facility. The discussion is meant to serve as an addition to the basic audit instructions presented earlier, by highlighting the areas which are appropriate to auditing commercial waste management facilities.

## PRE-AUDIT PREPARATION

Once a TSD facility has been selected as a candidate to be audited, the audit team leader or project manager should implement the following steps:

- Obtain basic location, operation, and regulatory agency information about the TSD facility (the facility pre-audit screening).

■ Obtain basic financial information about the TSD facility (ownership, financial health, insurance information, litigation status).

■ Speak directly with TSD facility representatives to confirm facility operations and to schedule the audit date and activities.

■ Review the appropriate federal, state, and local regulations pertinent to the TSD facility operations.

■ Select the audit team staff most suited for the TSD facility, based upon the items above.

The importance of the pre-audit preparation steps described above cannot be overemphasized. The first three steps may, in some cases, serve to immediately eliminate some sites from further consideration due to major regulatory problems, scheduled phase-out or closure of certain operations, financial distress, or shutdown of the entire site. More often, however, the completion of the pre-audit activities prior to arriving on-site for the audit will ensure that the audit objectives are met.

Several methodologies are usually employed in order to obtain location, operational, and regulatory information about a TSD facility. The use of telephone interviews, electronic databases, and other readily available sources of information such as maps, aerial photographs, and publications is presented in Figure 22.1. The information obtained during the pre-audit activities should be sufficient to:

■ Determine if the facility has been the subject of legal action such as Administrative Consent Orders or private party litigation.
■ Determine the legal status and financial condition of the company owner(s) and operator(s).
■ Identify the basic waste management operations and corresponding environmental compliance issues at the facility.
■ Determine the specific outside contractor or client audit policies that the TSD facility may require you follow before, during, and after the on-site audit.

- Give the TSD facility staff an understanding of the scope, complexity, time constraints, and objectives of your audit activities.

- Give the audit team a good understanding of the facility location and demographics with respect to nearby potential environmental and human receptors.

- Determine if a serious (major groundwater contamination, permit application denial) or recurring (repeated spills and releases, poor inspection history) environmental condition or regulatory compliance problem exists at the facility.

Key regulatory compliance parameters that should be ascertained during the pre-audit phase include permit status, facility-specific permit-mandated compliance requirements, recurring/serious compliance problems, existing or suspected environmental contamination issues, RCRA corrective actions status and/or CERCLA involvement, recent regulatory agency inspection results, and closure/post-closure financial assurance information.

Be sure to leave enough time to obtain the pre-audit information prior to commencing the on-site portion of the audit. In some states, regulatory agency personnel are literally inundated with requests for compliance information on TSD facilities and will not answer phone inquiries regarding TSD facility compliance. These agencies may require you to submit a formal Freedom of Information Act (FOIA) request in order to obtain the information you need. This process can take up to several weeks. Certain agencies may allow you to review the compliance records for a particular facility in person if you make an appointment ahead of time. If at all possible, this is a valuable way to both review the actual inspection records and correspondence as well as possibly question the agency contact in person. Remember, you'll need to get information regarding air, water, and other regulatory compliance issues, as well as the RCRA compliance files, so allow time to coordinate between the different federal, state, and local contacts.

The use of a brief, pre-audit questionnaire to obtain information about the TSD facility is a good idea for several reasons. First, it gives the TSD facility a good indication of the type of information you will need to complete the audit. To that end, the questionnaire should contain a list of the information and documents that you will be requesting to review or copy during the audit. Second, it will provide you with a facility contact (or contacts) who will be able to assist you during the information gathering stage. Third, it will provide a good indication of both how well the facility is prepared for an audit and what level of cooperation you can expect to receive from the TSD facility staff. Because of the increasing number of generators and interested third parties requesting to conduct waste vendor audits, many waste management contractors have prepared "audit information packages" that are designed to provide a prospective auditor with the pertinent facts regarding the facility, such as waste acceptance criteria, operations, permit status, and so on.

Figure 22.2 lists the topics that are most often addressed in a TSD facility audit program.

## SPECIAL PRE-AUDIT CONSIDERATIONS

At this stage of the pre-audit activities, several other important aspects of the auditor-TSD facility relationship must be defined, such as the need for confidentiality agreements, the audit schedule, and the exchange of information following the audit.

It is not uncommon for TSD facilities to request that a detailed confidentiality statement be drafted and signed prior to granting a request for an audit. Many small, privately owned non-RCRA waste management companies (such as precious metal recyclers, drum reconditioners, and brokers) may be reluctant to provide details concerning special waste handling, treatment, or disposal processes, financial data and management-related information, or groundwater monitoring results, unless they are satisfied that the information will remain confidential.

## Figure 22.1
## Sources for Obtaining the Pre-Audit Information

### Location

- U.S. Geological Survey maps
- Local historical and current aerial photographs
- U.S. Census data
- Local historical societies, real estate groups, and chamber of commerce
- County and city municipal government interviews

### Regulatory Status and Compliance

- Personal or telephone interviews with federal, state, regional, and local regulatory agencies
- Freedom of Information Act (FOIA) requests
- U.S. EPA and private subscription databases regarding permit status and compliance
- Independent commercial database services that monitor TSD facilities, such as Environmental Information Limited, ERIIS, EDR-Toxicheck, and VISTA.
- Regulation and compliance-oriented publications such as *Hazardous Waste Intelligence Reporter, Inside EPA, Hazardous Waste News,* and others

### TSD Facility Operations

- Direct telephone contact with TSD facility technical representatives
- Review of corporate promotional literature and pre-audit information packages
- Review of federal, state, regional, and local regulatory agency files
- Use of a pre-audit questionnaire

## TSD Facility Financial Information

- Annual reports and audited balance sheets
- SEC 10-Q and 10-K forms and other corporate filings
- Financial information services such as Dun and Bradstreet, SEC Online, Dow Jones News Retrieval, and Robert Morris and Associates
- Credit, banking, and vendor references.

**Figure 22.2**
**Areas Most Often Addressed in Commercial**
**TSD Facility Audit Programs**

- Compliance with applicable federal, state, and local environmental regulations and permit conditions
- Assessment of the historical operations at the site, especially pre-RCRA or non-RCRA regulated activities
- An evaluation of the facility's environmental setting, such as hydrogeology, topography, distance to surface waters, wind patterns, climate, critical habitats, etc.
- An evaluation of the facility's demographic setting, including distance to nearest human receptors, population density, critical receptors, and other information
- A review of the adequacy and implementation of the facility's key waste management operations and controls, such as the waste analysis plan, emergency and contingency plans, operational plan, training plans, and environmental monitoring systems
- A detailed review of the waste management operations, including processing, treatment and disposal of all incoming waste and residuals
- An evaluation of the company's management practices and attitudes
- A financial analysis of the facility/parent corporation, including a determination of the company's financial status, applicable insurance coverages, and disclosure of contingent environmental liabilities and litigation
- A detailed review of all operational/regulatory compliance documentation, such as Part B and TSCA operating permits, NPDES permits, closure/post-closure plan, inspection records, waste analysis plan and records, manifests, etc.
- A review of any ongoing or anticipated RCRA corrective actions investigations
- A review of any consent agreements, consent orders, or other legal documents affecting the site operations.

You should be prepared for such a request by either preparing your own confidentiality statement in advance, or by asking the TSD facility if they will require you to sign their own agreement before they release any information. It is recommended that this determination be made well in advance of the scheduled date of the on-site visit, since negotiating a confidentiality agreement or other additional legal agreements between waste contractor and a generator can be a lengthy and time-consuming process. An additional confidentiality request may be required if the generator is using a third party, such as an environmental consultant, to conduct the audit. One common waste contractor auditing horror story tends to repeat itself time and time again: the audit team shows up at the TSD facility, is denied access to the site, and is asked to sign a confidentiality agreement before they are allowed to conduct the on-site audit or review sensitive documents. Is there a lawyer in the house?

In order to ensure that the on-site activities flow smoothly, it is necessary to establish a schedule for the audit activities well in advance. This means clearly identifying a mutually agreeable date for the audit as well as defining the particular operations you will want to see, the audit scope and level of detail, and the TSD facility staff with whom you intend to speak. Most commercial TSD facilities are audited by federal, state, and local agencies on a fairly routine basis, as well as by other generators, corporate environmental staff members, civic groups, and concerned parties. It is often difficult to schedule a generator audit because of these other concerns, so be prepared to be flexible in your schedule. You don't want to show up on-site the same day as the EPA's National Enforcement Investigation Center (NEIC) auditors, or on the day that the site technical manager is leaving for vacation. Also, remember that you may be one of a hundred waste generators that want to schedule their audit at a TSD facility in southern California in February. Similarly, it is important to let the TSD facility know in advance whether you intend to conduct a comprehensive facility audit, or if you merely want to tour the site in the van, watch the promotional video, have lunch, and leave.

Prior to conducting the on-site audit, you will need to establish what the TSD facility will require in terms of reviewing copies of the audit report, photographs, or other audit-generated information. Some facilities will require that, at a minimum, they are provided with copies of all photographs taken, whereas others may allow you to take pictures only if the film is developed and screened by the site management to eliminate unfavorable pictures. Many waste management facilities will permit a generator-sponsored audit only if they receive a copy of the audit checklist and/or report. Ask if there are any special limitations on making copies of permits and other compliance documentation, or accessing generator-related information such as manifests. Knowing these limitations in advance will prevent misunderstandings and potential confrontations during the on-site audit.

## SELECTING THE AUDIT TEAM

The selection of the audit team, irrespective of whether the audit is conducted by the generator or an environmental consultant, is an important element in ensuring the success of the audit. Some key factors that should be considered include:

- *Never* send an audit team composed entirely of auditors who have never been to a TSD facility before. This is the cardinal rule of selecting audit teams. The facility staff is not going to make it easy for you in the first place. A TSD facility is quite unlike any operating facility that your company might own; simply put, there is too much to observe and too much liability at stake to risk having an entire audit team overwhelmed and/or undertrained.

- Choose team members who have educational and career backgrounds commensurate with the type of operations to be audited. It doesn't make sense to send two or three hydrogeologists to audit a TSD facility that has a complex incinerator or chemical treatment process. Ideally, the audit team should be composed of no more than three professionals that have compatible skills which are commensurate to the operations at the TSD facility. As an example, a team composed of one hydrogeologist, one chemical engineer, and one regulatory specialist would be able to successfully understand and audit most of the typical TSD facilities in the country.

■ Concerning team size, the same guidelines apply for in-house audits. Usually a team of two or three auditors will be sufficient. Four or more auditors are not only difficult for the facility to accommodate, but they are also likely to overpower an already overworked and limited TSD facility staff. Conversely, in most cases, there is simply too much to see and read for only one auditor to handle. Unless the facility is very small or the scope of the audit is quite minimal, consider two auditors to be the starting point when selecting the size of the team.

■ Although it sounds trite to say it, choose TSD facility auditors that work well together and are most likely not to be rattled or annoyed by scheduling problems, abrasive or uncooperative facility staff, cramped conditions, and other typical audit problems. It is imperative to select audit team members who are not prone to confrontational standoffs. It's no fun to audit a landfill during a heavy rainstorm, but a well-picked audit team with "foul weather fortitude" will at least make the exercise less miserable and ensure that the audit objectives are met. (By the way, auditing a TSD facility in the rain is a *good* idea, since it can give you an indication of surface runoff and stormwater management practices, which are important in evaluating the risks posed by releases that may occur during the waste handling and transfer operations.)

As recently as the mid-1980s, generator audits of waste management facilities were the exception rather than the rule. As such, TSD facility staff were receptive to generator audits in order to attract and maintain new customer relationships. In the past several years, however, TSD facility auditing has increased at a rapid rate, with more and more companies requesting to conduct time-consuming and potentially damaging audits. At the same time, many of the smaller waste treatment and disposal contractors have fallen on hard times as a result of increased regulatory vigilance, inability to meet stringent technical and financial requirements, and stiffened regional and national competition by larger, better financed firms. The waste treatment and disposal market today is dominated by these larger firms, and the entrance of new hazardous waste management firms into the market is extremely difficult, if not impossible, due to the reasons cited above as well as increased public awareness and reluctance on the part of communities to accept these types of facilities (*i.e.*, the "not in my backyard," or NIMBY, syndrome).

The end product of this changed market is increased reluctance on behalf of waste management companies to provide *carte blanche* privileges to generators requesting TSD facility audits. Although most TSD facilities respond favorably to inquiries and understand the generator's desire for requesting the audit, the open door policy has somewhat cooled. This change in perspective manifests itself via overly restrictive confidentiality agreements, limited scheduling windows, a reluctance to share details of proprietary treatment technologies, and a reserved approach to meeting auditor's requests for key audit information. Accordingly, the audit team must be cognizant of these factors and must be highly adaptable to somewhat less than ideal auditing conditions as they arise.

## THE ON-SITE AUDIT

Although the auditing process is similar, conducting a TSD facility audit is fundamentally different in nature from doing an internal environmental compliance audit at one of your own facilities. Because the primary interest of the TSD facility is to either attract new business or keep an existing customer satisfied, it is natural for the facility management to downplay any problems that exist at the facility. As a result, the auditors must rely more heavily on their ability to uncover potential problems through document review, interviewing key employees, and their direct observation of facility practices and operations. Figure 22.3 lists a recommended protocol for conducting the on-site audit.

**Figure 22.3**
**Proposed TSD Facility Audit Agenda**

---

## DAY ONE

- Auditors arrive at TSD facility
- TSD facility representatives summarize facility capabilities and operations to familiarize auditors with the site (½ – 1 hour)
- TSD facility tour guide and audit liaison are assigned to the audit team
- TSD facility tour is conducted, preferably in the following sequence:

    - Waste arrival and acceptance procedures at site
    - Waste sampling procedure
    - Manifest verification
    - Waste analysis procedures, including a tour of the laboratory, description of QA/QC, equipment and capabilities, and treatment/disposal option decisions
    - Treatment, storage, and disposal procedures, including a tour of operations to cover all regulated units and a discussion of design and operation controls
    - Training records and inspection logs.

- Auditors conduct interviews with TSD facility representative(s). The purpose of the interviews is to summarize all regulatory correspondence and permits, as well as to identify any discrepancies between current operations and permit conditions. General interview areas will include:

    - Discussion of company structure
    - Current use of the facility by generator(s)
    - The environmental setting of the facility
    - Site history

- Site operations
- Waste volumes and types
- Regulatory compliance history
- Environmental incidents or releases of waste
- Involvement in third-party litigation
- Financial assurance
- Management qualifications
- Recordkeeping
- Transportation capabilities
- Process specifications
- Air, surface water, and groundwater monitoring results
- Proposed facility modifications.

■ Auditors review RCRA Part B permit application, other permits, and regulatory correspondence to evaluate TSD facility procedures and determine whether risk factors are adequately mitigated.

## DAY TWO

■ Auditors continue document review
■ Particular areas of interest are revisited to observe waste management practices, safety procedures, environmental monitoring, waste sampling, etc.
■ Follow-up interviews are conducted with key TSD facility personnel
■ Closing meeting is done with facility personnel.

The document review portion of the audit should focus on several key topics. First, the auditors must determine if the proper documentation (permits, reports, records, etc.) is in place so as to ensure facility compliance. Secondly, the auditors should observe if the documents are complete and adequate, not only to ensure compliance, but also to protect the generator's interests. For example, auditors should carefully review procedures and records which directly involve the generator, such as manifests, the waste analysis plan, and the ultimate disposal records for wastes and residuals, as well as treatment, solvent recovery, or landfill disposal logs.

In order to assess the facility's waste tracking system, it is helpful to track random waste shipments from initial receipt at the facility through ultimate treatment, disposal, and, most importantly, trans-shipment or treatment at another site. The TSD facility should be able to easily document the path of any incoming waste shipment through the facility. Many generators test the adequacy of the TSD facility's waste tracking and recordkeeping system by bringing several manifests (from both old and recent shipments) to determine whether the facility can rapidly and accurately determine the ultimate fate or current location of each particular manifest item. Records should be carefully reviewed to assess whether employees have been properly trained, whether the contingency and emergency response plans are appropriate for the risks presented by the facility operations, and whether the environmental monitoring systems in place are adequate to determine if contamination is leaving the site via groundwater, surface water, or air pathways. A careful review of the completeness and organization of a facility's recordkeeping system will tell a savvy auditor much about the site's compliance status, safety, and general management attitudes.

Interviewing facility management and operations staff is a critical part of any audit, but it is particularly informative when assessing the relative risks at a TSD facility. Facility management should be intimately familiar with the site operations and should be willing to discuss how waste handling, treatment, and disposal decisions are made. Concerns over confidentiality are often presented when auditors request information concerning waste treatment decisions, so it is important to identify if there will be any such problems during the pre-audit phase. Ask to interview the employees who actually conduct the waste handling activities, such as sampling technicians, waste processing staff, and laboratory personnel, not just the site manager or audit liaison. The purpose of interviewing these employees is to make sure that they know exactly what it is they are responsible for, instead of assuming that they have been trained in accordance with the written training plan in the RCRA Part B permit the auditors have reviewed. Be sure to question employees on key risk areas such as waste receipt and analysis, waste handling, and treatment decision-making (how it is determined what wastes are sent where), training, preventative maintenance, and emergency response procedures. Remember that your overall goal is to ensure that your waste is being handled properly and that your liabilities are being minimized through the use of sound, well conceived operational practices.

The visual observation of the general appearance and housekeeping practices at the TSD facility will provide a good initial overall picture as to the safety and operation of the site. Look for evidence of poor oversight or inattentive management such as sloppy housekeeping, indication of repeated spills and releases, nonexistent or poorly maintained emergency response equipment, or accumulation of precipitation in secondary containment areas. If possible, observe actual waste handling operations such as tanker truck transfers, container decanting, waste stabilization, active landfilling, or waste sampling practices. Obvious signs of poor facility management may include evidence of repeated spillage, damaged or ineffective secondary containment areas, damaged or unsecured groundwater monitoring wells, poor segregation of incompatible materials, and absence of stormwater runoff control practices. Some typical findings that should lead the auditor to ask additional questions include multiple, confusing labels on containers; containers which are staged outside of containment areas; lack of respect for health and safety warning signs; airborne dispersion of vapors or particulates during transfer operations; and cracked or poorly maintained containment areas. Figure 22.4 provides some examples of typical problems encountered during TSD facility audits.

**Figure 22.4**
**Typical Problems Encountered During**
**TSD Facility Audits**

---

## REGULATORY COMPLIANCE ISSUES

- Failure to conduct or document proper inspections as required
- Improper/inaccurate/illegible labeling practices
- Segregation problems
- Inadequate/improperly designed secondary containment areas
- Minimal employee training programs and/or incomplete training records
- Poor implementation of personal protective equipment requirements, such as failure to wear face shields, gloves, respirators, long-sleeved shirts, etc.
- Poorly organized or incomplete recordkeeping system, or an inability to easily track waste shipments to final disposition
- Drums/bungs left open at all times, uncovered roll-off boxes, or lack of vapor controls for volatile liquids
- Containers in contact with accumulated rain water
- Inadequate/incomplete waste characterization, or failure to identify and reject contaminated loads
- Inability to carefully track ultimate disposal for trans-shipped wastes or residuals, such as tank bottoms, empty drums, or wastewater treatment sludges
- Closure/post-closure plans that do not reflect accurate waste volumes or technical requirements, or plans containing insufficient closure cost estimates or financial assurance.

## SAFETY AND ENVIRONMENTAL RISK ISSUES

- Improper electrical grounding during tanker or drum transfers of flammable liquids

- Questionable groundwater monitoring programs, in terms of well construction, sampling, and QA/QC procedures
- Lack of management oversight for emergency response procedures, such as failure to remove rainfall or accumulated spillage from secondary containment areas
- Inadequate supply and/or poor location of emergency response equipment
- Inadequate employee health monitoring programs
- Lack of compliance with facility safety requirements in restricted areas (hard hats, steel toe shoes, respirators, etc.)

---

Upon completion of the on-site audit, some companies prefer to brief the TSD facility staff on the general findings of the audit, in order to provide the TSD facility with the opportunity to either correct misinterpreted information or provide additional verification or evidence of compliance. It is in the generator's best interests to ensure that the information collected during the audit is accurate and complete. Therefore, it's a good idea to discuss, at a minimum, any confusing or unresolved issues prior to leaving the site. The auditors should leave the TSD facility knowing that their information is correct and understanding whether the TSD facility expects to receive a copy of the written report or other conclusions about the generator's findings.

## THE TSD FACILITY AUDIT REPORT

The information obtained during the TSD facility audit should be presented in a fashion that allows a comparison of the positive features of the facility and the existing or potential environmental, operational, and financial risks of the site. The audit report should be concise in nature and should be arranged in the following fashion:

- An overall description of the site history, location, and operational parameters, including an identification of any associated risk factors

- A summary and assessment of the site's demographic and environmental setting, aimed at the development of a source-pathway-receptor analysis

- A description of the site's waste management operations, focusing on the positive and negative issues associated with waste receipt, processing, and ultimate treatment/disposal procedures

- A review of the site's regulatory compliance history and current compliance status

- An evaluation of the site's environmental risk management procedures, operational oversight, and attitudes towards employee safety and community awareness

- An analysis of the adequacy and feasibility of the site closure/post-closure plan and cost estimates

- A financial analysis of the site and its parent company, if applicable

- An overall assessment of the risks posed by the site, using both qualitative and quantitative methods so as to allow direct comparison of different sites.

Remember that the ultimate objective of the audit program is to identify and manage the inherent liabilities associated with waste treatment and disposal, not to prepare a classical treatise of the subject site. The reports should limit the discussion of factors that do not directly affect the decision-making process.

## SUMMARY

All off-site commercial waste treatment, storage, and disposal facilities will present some level of risk to a waste generator. The handling and treatment of hazardous and toxic wastes is by nature fraught with environmental, human health, and regulatory-related risks. It is the proper management of those risks that the commercial TSD audit program is designed to facilitate. Generators should seek to limit their liabilities through both the reduction of the amount of wastes that they generate and through the use of commercial TSD facilities that are located, designed, operated, and managed in a fashion that reduces the generator's risk to an acceptable level.

# 23

# WASTE MINIMIZATION OR POLLUTION PREVENTION AUDITS

---

*"Clearly the answer to our problems of waste has got to involve incentives to reduction of waste right at the source, right at the start, and to recycling. And I will be committed to these goals."*[1]

---

As environmental audit programs have evolved over the past few years, companies have very often changed their initial focus to reflect findings of baseline audits. More often than not, this has meant incorporating a significant waste minimization component into the program. There have been several reasons for this development:

- **Historical Noncompliance.** Many of the noncompliance findings in the early audits have been a result of poor waste management practices.

- **Regulatory Requirements.** Hazardous waste manifests require generators to certify that they "have a program in place to reduce the volume and toxicity of waste generated...." The Pollution Prevention Act and the Emergency Planning and Community Right-to-Know Act also call for active pollution prevention programs.

---

[1] William K. Reilly, U.S. EPA Administrator, 1989.

- **Regulatory Trends.** Federal and state rules are prohibiting the land disposal of a wide variety of hazardous wastes, thereby reducing the disposal options available.

- **Economics.** Waste disposal is becoming considerably more expensive.

Successful waste minimization can mean having to meet less stringent regulatory standards, not having to rely on other companies to ultimately dispose of wastes, and paying substantially less for waste disposal. Accomplishment of these goals can do much to improve compliance at individual plants and minimize the liabilities of the corporation.

Conducting waste minimization audits can typically be accomplished in two ways. They can be included as part of the standard audit process, or the company can create waste minimization "hit squads," using staff with specialties in manufacturing and process control who travel from plant to plant attacking the specific problem of waste minimization. The first approach is less resource-intensive but is also less likely to achieve significant reductions.

Either approach can be successful, but one needs to be aware of the distinctions between the more traditional audit approach and what is required to complete a waste minimization audit. Historically, environmental audits have been designed to verify that adequate compliance assurance systems (organizational and physical) are in place at the operating facilities. Therefore, audits are not meant to substitute for good, effective site environmental management, which is required day-in and day-out, not simply when the audit team is on-site.

Waste minimization audits differ in that more is involved than simply verification. The audit team is usually asked to work with the site staff to review waste generation operations, analyze the potential for waste reductions, and jointly identify solutions to the high-priority problem streams. In this way, the audit team is asked to become more intimately involved in the operations of the facility.

Of course, these differences in waste minimization audits bring with them some advantages and disadvantages. On the plus side, the audit can be seen as more helpful than otherwise might be the case. However, it is also more intrusive. Thus, the auditors must be very diplomatic in their investigations; good technical skills will accomplish only so much.

The objective of a waste minimization audit generally is to reduce the volume and/or toxicity of wastes. However, meeting this commendable objective is fraught with frustrations. Auditors should be especially aware of the following roadblocks:

- **Waste minimization audits are very site-specific.** Therefore, it is often very difficult to apply the lessons of the past to the case at hand.

- **Like many things in life, waste minimization suffers from the 90/10 rule.** Ninety percent of the reductions are often achieved with the first ten percent of the effort. Thus, it becomes exponentially more difficult to attain significant reductions as time goes on.

- **Waste minimization can be as much an organizational issue as a technical one.** Seemingly simple steps recommended by the audit team will result in scores of responses by purchasing, manufacturing, and the sales force as to why the steps are impossible to implement. For example, based on a corporate directive, purchasing staff will typically buy materials at the lowest available unit cost. This can often mean obtaining chemicals in 55-gallon drum quantities, even though 10-gallon pails would be easier to handle and each might last as long as two to three months, based on production requirements. Thus, a built-in conflict can exist with purchasing.

Also, research and development (R&D) staff are notorious for experimenting with chemicals and then accumulating the unused portions in a nearby store room that was designed to be a coat room. These materials can accumulate for years and, in many cases, in containers that are inadequately labeled. R&D staff could be directed to use only suppliers who have a chemicals "take-back" policy, but if their favored supplier has no policy, they may fight the change.

Finally, manufacturing and sales staff have been making and selling successful product lines for years. When it is suggested that substitutions should be made for certain chemicals, they often have a million reasons why their customers would not tolerate the switch.

Notwithstanding these limitations, waste minimization audits can be a worthwhile exercise. In general, the analytical approach is similar to conducting a more routine environmental audit.[2] The audit team reviews the available written records, interviews key site environmental and operational staff, and inspects the plant's facilities. In conducting this exercise, the auditors are generally attempting to identify the following:

- **The "Devil's" Chemicals.** In most manufacturing, there will be a few chemicals that are extremely toxic and/or troublesome (*e.g.*, phosgene). A waste audit's goal should be to identify where these chemicals are used and what by-products or waste products their use generates.

- **High-Volume Waste Streams.** The 90/10 rule often applies here as well. In many cases, ten percent of the waste streams generate ninety percent of the volume. Investigating these waste streams first is likely to result in the most cost-effective reductions.

- **Highly Regulated Wastes.** Wastes such as dioxins, PCBs, chlorinated solvents, and RCRA acutely hazardous wastes should receive considerable attention during the audit. If for no other reason, they should receive this attention because, in many cases, they have become "political" wastes.

- **Land-Disposed Wastes.** Wastes that are land disposed present special problems for generators. Strict, joint, and several liability

---

[2] For a more detailed discussion of approaches see the "Facility Pollution Prevention Guide," U.S. EPA Risk Reduction Engineering Laboratory, 1992, published by Government Institutes.

tenets suggest that the generator maintains some liability, even when indemnified by a disposal company.

Identifying waste streams that fall into the classes presented above will help set priorities for the audit, which, by its nature, is a time- and resource-limited exercise. And, there are ways to use readily available data to indeed identify these streams. Some of the more common are listed below:

- Material safety data sheets
- RCRA annual or biennial reports
- RCRA hazardous waste manifests
- The RCRA list of acutely hazardous substances
- SARA Title III lists, reports, and submissions
- State air toxins lists
- EPA Clean Air Act NESHAPS chemicals and candidate chemicals
- Air emissions and asbestos inventories
- PCB electrical equipment inventories
- EPA Clean Water Act priority pollutants
- Waste disposal companies' records and invoices
- Sludge disposal records
- Records of pollution control equipment and operations and maintenance expenditures.

Once these sources are reviewed, each of the priority waste streams can be investigated for its reduction potential. It should be clear from the nature of the sources that the waste minimization audit must be more encompassing than simply a review of RCRA-generated hazardous wastes. Wastewaters, air emissions, and other sources of toxic releases must be considered as well.

Evaluating the reduction potential involves analyzing each waste stream against a number of options. Although it is difficult to make generic suggestions on possible waste minimization strategies, some of the more common reduction solutions generally fall into the control strategies listed below:

- **Material Conservation**. Until the last few years, there has been little incentive to conserve many raw materials. Yet, better management and/or more careful use of paints, solvents, and maintenance chemicals could significantly reduce chemicals being emitted through vents, washed down the drains, or sent out for disposal. Vehicle and machinery maintenance should be scheduled at proper, but not overly conservative, intervals. Where process vessels are used to manufacture different products, line management should schedule runs such that chemical cleaning of the vessels is minimized.

- **Material Substitution.** Significant inroads have been made by many companies using this strategy. Typical substitutions include fiberglass for asbestos and urea formaldehyde insulation, nonchlorinated for chlorinated solvents, and water-based for solvent-based coatings and paints.

- **Inventory Management.** Even at the risk of displeasing the discount-minded purchasing agent, many companies are buying chemicals in much smaller quantities. Where at all possible, five- and ten-gallon pails are often preferred to the tanker truck or ubiquitous 55-gallon drum. In fact, one company has a policy that 55-gallon drums can be purchased on an exception basis only. Companies are also becoming more fastidious about requiring a first-in, first-out inventory policy for chemicals. This minimizes the potential for old drums to accumulate and deteriorate in the rear of a warehouse.

- **Recycling/Segregation.**These two approaches fit hand-in-glove; often, materials cannot be recycled because they have not been segregated properly. Used oils are occasionally so contaminated with solvents that they are not reclaimable. Where wastes must be generated, it is often prudent to search for a waste exchange—one man's waste might be another man's raw material. In this regard, some companies use their corporate environmental department as a clearinghouse. The departments search for both intra-company and commercial users or buyers of excess inventory, past-shelf-life materials, or recoverable wastes.

- **Process Modifications.** Significant gains can be achieved using this strategy. Some potential approaches were mentioned above (*e.g.*, the use of water-based coatings). Other waste reductions can be attained by looking closely at the manufacturing process with trained eyes. One company was able to develop a sophisticated plastics extrusion process that allowed them to fabricate composite conduit directly, as opposed to fabricating two separate pieces and binding them together with adhesive. The new process was especially cost-effective when the cost of waste-adhesive disposal was factored into the analysis.

- **Waste Treatment and Destruction.** This is the waste minimization strategy of last resort. If all else fails, do your best to detoxify or destroy the wastes. Land disposal, or long-term storage if you will, is an extremely questionable approach in this day of Superfund-mandated cleanups; presently, however, some wastes and treatment residuals have no other alternative. Yet, many companies have developed a policy that land disposal of any wastes will be allowed on an exception basis only. Biological, chemical, and physical treatment technologies are evolving rapidly, so it is likely that fewer and fewer waste streams will become exceptions under this type of policy.

Adopting these strategies can result in defining some very specific opportunities for waste minimization. Some of the more common techniques are:

- Using nonchlorinated solvents for cleaning (*e.g.*, Simple Green)
- Using water-based coatings and paints
- Adopting purchase-by-exception for 55-gallon drums
- Adopting purchase-by-exception for toxic ("the devil's") chemicals

- Placing emphasis on maintenance chemicals as well as process chemicals
- Setting schedules of vessel runs to minimize the need for frequent cleanouts
- Adopting a first-in, first-out inventory control
- Minimizing the cross-contamination of used oil with solvents
- Developing an internal clearinghouse function for excess inventory, old chemicals, and recoverable wastes
- Revising processes (*e.g.*, using composite plastics extrusion manufacturing vs. simple extrusion and adhesion to eliminate the generation of adhesive wastes)
- Improving housekeeping (*e.g.*, keeping lids on degreasers, and minimizing spills during materials transfer and dispensing)
- Using oil-water separators in sewer systems
- Requiring R&D staff to purchase chemicals in small quantities, to not accept salesmen samples indiscriminately, and to work with vendors with return policies
- Training R&D staff to be aware of environmental consequences of design
- Requiring all capital improvements to be signed off by corporate environmental departments
- Requiring secondary containment at all fill and dispensing areas (*e.g.*, tank farm manifolds, underground storage tank fill ports)
- Segregating chemical storage from waste storage; isolating solid wastes from hazardous wastes
- Evaluating the integrity of all floor drains and septic tank systems
- Using the full-service capabilities of reputable waste disposal contractors (*e.g.*, Chemical Waste Management's Waste Minimization Service)
- Above all, making waste minimization a *real* corporate policy by instituting an incentive program for plant managers. They will make it happen!

In summary, waste minimization audits are not unlike any analytical exercise. First, high priority waste streams are identified through a review of documents, discussions with site staff, and an inspection of the site. These streams are typically ranked and analyzed for potential reduction strategies, such as material substitution and process modifications. Where potential strategies are identified, a detailed study is made of the costs and benefits. This study includes the concerns of purchasing, manufacturing, and sales. In other words, recommended changes must incorporate use of equivalent and readily available materials and processes that do not materially affect the quality of the product or the cost of its manufacture, including the cost of waste disposal.

# 24

# EVALUATING MANAGEMENT SYSTEMS ON ENVIRONMENTAL AUDITS[1]

## WHY EVALUATE MANAGEMENT SYSTEMS?

For the past few years, there has been much discussion in the environmental auditing profession about evaluating management systems as part of a facility audit. Some people have gone so far as to say that the management systems evaluation should be the *principal* objective of an environmental audit. And it is probably fair to say that many people, while espousing this approach, are not quite sure what it means.

There have been two recent initiatives in Europe, in particular, which have done a lot to further emphasize the management systems aspect of audits and to better define what that means. The European Community has proposed an Eco-Audit Scheme that encourages companies to conduct environmental audits and to focus heavily on systems during the exercise.[2] Secondly, the British Standards Institute has published an Environmental Management Systems Specification that provides guidance on what effective systems should look like and how to evaluate them.[3]

---

[1] Originally published in the Winter 1992/93 Issue of *Total Quality Environmental Management*, Executive Enterprises, Inc., New York, NY.

[2] "Proposal for a Council Regulation (EEC) allowing voluntary participation by companies in the industrial sector in a Community Eco-Audit scheme," (92/C 76/02, March 6, 1992), *Official Journal of the European Communities*, No. C 76/2, March 27, 1992.

[3] "Specification for Environmental Management Systems," British Standard BS 7750:1992, ISBN 0 580 20644 0, March 16, 1992.

So, even with all this recent activity, one might still ask the question: Does this "management systems" approach add value to an audit? Well, it most assuredly does! Focusing on management systems allows one to identify the underlying causes, as opposed to the symptoms, which are typically at the heart of noncompliance at a site.

For example, if an auditor were to observe something as simple as a label missing from a waste solvent drum, the resultant finding could be described as just that: "a drum had a missing label." And surely what would happen is that the site staff would immediately place a label on the drum, and the matter would be considered closed. However, if that same site were to be audited a year later, it is likely that *at least* one drum would be missing labels. Why?

Well, the auditor has failed to address the underlying cause of the problem and, nine times out of ten, this is likely to be the breakdown of a management system. In the case of the problem drum, the label could be missing for a variety of reasons, including:

- The person responsible for drum storage and management has not been trained properly.
- It is unclear as to who maintains responsibility for the drum while it moves to various locations on the site, from accumulation near the point of generation, to a "ninety-day" accumulation pad, to a permitted storage facility.
- The drum has been sitting around for quite a while and no one is sure of its contents. A sample has been sent out for analysis and the decision has been made to not label the drum until the results are known.
- The purchasing department bought "cheap" labels and they cannot withstand the rigors of outdoor storage—they keep falling off the drums.
- The site inadvertently ran out of labels and the normal purchasing process for re-supply takes two weeks.

Now when we look at the problem, it takes on a different light. Maybe the solution is more complicated than simply placing a label on the drum. Maybe, for example, the site's training programs are not including the right people, or job position responsibilities are unclear, and so on. Thus, the corrective actions can now focus on underlying causes and might, for instance, state that the site needs to train its operators better, or assign drum management responsibilities more clearly, or do something as simple as incorporating minimum quality specifications into the label purchasing process. By identifying and remedying the true problem, it is more likely that in a year's time when the next audit takes place, there will be no repeat occurrence. In this way, focusing on management systems can result in long-term environmental compliance improvements, not quick fixes.

## WHAT IS A MANAGEMENT SYSTEM?

One can define an environmental audit as a verification of the existence and use of appropriate on-site management systems. As such, an audit is not meant to substitute for good site environmental management. And "good" management implies that there are systems in place on-site to assure compliance on an on-going basis. These systems can be defined as "the organizational structure, responsibilities, practices, procedures, processes, and resources for implementing environmental management"[4] at an operating site.

"The environmental management system should be designed so that emphasis is placed on the prevention of adverse environmental effects, rather than on detection and amelioration after occurrence. It should:

- Identify and assess the environmental effects arising from the organization's existing or proposed activities, products or services
- Identify and assess the environmental effects arising from incidents, accidents, and potential emergency situations
- Identify the relevant regulatory requirements
- Enable priorities to be identified and pertinent environmental objectives and targets to be set

---

[4] *Ibid.*, page 4.

- Facilitate planning, control, monitoring, auditing, and review activities to ensure both that the policy is complied with and that it remains relevant
- Be capable of evolution to suit changing circumstances."[5]

Although the above elements should be at the core of a site's environmental management system, they are often difficult to audit against because a site's programs frequently are not structured that way. Typically, the site's environmental management system is the sum of separate programs, which would include most, if not all, of the following:

- Specific programs designed to address corporate environmental policies and procedures
- Employee training and statements of job accountabilities
- Regulatory tracking system
- Environmental review of new activities
- Waste minimization planning
- Release prevention/emergency response planning
- Environmental auditing including noncompliance follow-up and reporting
- Community outreach program/complaint management
- Product stewardship
- On-site contractor reviews and evaluations
- Off-site contractor reviews and evaluations.

These are the *auditable* program elements of a site's environmental management system.

## HOW DO YOU DO IT?

To conduct an audit of a management system, an auditor needs to select those programs that are to be evaluated and to audit the following components of each of the selected programs:

---

[5] *Ibid.*, Annex, Page 9.

- Organization
- Administrative procedures
- Staff assignments
- Documentation, reports, and records
- Implementation.

These components would be evaluated against the standards set by regulations or corporate policy. For example, if a site is required to have an emergency response plan, it could be because of Contingency Plan requirements under the Resource Conservation and Recovery Act (RCRA); Spill Prevention Control and Countermeasure (SPCC) requirements under the Clean Water Act (CWA); Hazardous Materials Response requirements under OSHA's Hazardous Waste Operations (HAZWOPER) rule; or requirements under a mandated corporate procedure. The auditor would first have to determine which of these requirements apply and then evaluate the program organization, procedures, documentation, and implementation against those requirements.

When a facility falls under more than one set of requirements in one program area, the auditor would have to determine both if the site has been responsive to each requirement *and* if the overall program is workable. In the emergency response area, for example, some sites will attempt to develop one site emergency response plan that addresses all applicable regulatory requirements. The advantage of this approach is that the site management does not have to determine what kind of incident has occurred before utilizing the actions recommended by the plan. On the other hand, having an individual plan developed for each set of requirements simplifies the regulatory response and allows for a more direct assurance to agency inspectors and others that the regulatory requirements have been met. Yet this approach, while valid, can create response-time problems when an incident does occur. In either case, the workability criterion becomes paramount.

Much of the organizational and procedural review of management systems can be conducted through interviews and evaluations of the programs' documentation. And if the systems can be identified ahead of time, protocols or checklists can be developed to guide the auditors in their investigations.

One of the most important aspects of the audit, however, is to ensure that written procedures are, in fact, carried out effectively. This can best be explained by way of example. Take again, for instance, emergency response planning. One can review the organization and planning documents designed to respond to an emergency and find that they respond well to regulatory requirements. However, the real test of effectiveness is in the implementation. In practice, this effectiveness can be tested either through evaluations of the response to actual emergencies, or through drills or simulations. Thus, the auditor can only determine that the system is effective if there are assurances that it is tested or evaluated on a routine basis. In other words, any audit of management systems should include an evaluation of actual practices as well as documented procedures.

## WHY IS IT SO HARD?

Many companies have considerable experience in assessing management systems on environmental audits. Yet, observation of numerous audits suggests that even those with experience have difficulties in applying consistent review techniques. This is for a number of reasons, including:

- **Performance Appraisal.** An audit of management systems is truly an indicator of personnel performance, and therefore, it will always have a performance appraisal flavor to it. This means that interviews will have an additional tension that must be dealt with by the auditor.

- **Underlying Causes.** Identifying the root causes of a problem requires extra digging and investigation by the auditor. This will often take even a very experienced auditor additional time. These are two luxuries that may not be available to the team.

- **Lack of Standards.** There will be many requirements placed on an organization (*e.g.*, increased environmental awareness), which will have no standards against which they can be evaluated.

- **Cross-Cutting Programs.** Assessing management systems is difficult because responsibilities typically cut across media, and therefore, necessary review techniques would be counter to the more traditional approaches.

In addition, developing management systems findings on an audit is just plain difficult. Building the case for a management breakdown requires a certain mind-set that has to be learned. For example, an auditor might conclude that the "hazardous waste management system at the site is inadequate." And immediately, staff personnel will justifiably ask the question, "Why do you feel that way?" And the response is all too often, "Well, I'm not sure, but I just wasn't comfortable after looking at the records and talking with a few of the staff." This is an insufficient evaluation.

The above conclusion related to the hazardous waste management system may, in fact, be correct. But the conclusion must be verified and substantiated with evidence. Accordingly, a deficiency statement or conclusion should have the structure of the following example:

"The hazardous waste management program at the site is not completely responsive to Corporative Directive HW-100. Deficiencies include:

(1) Five drums at the accumulation point had no labels.

(2) Accumulation point inspection logs had not been completed for the past month.

(3) Two of the maintenance staff had not received their annual training."

Developing findings in this fashion begins to build a strong case that the system is breaking down and needs rebuilding.

## CLOSURE

Evaluating management systems is a crucial component of any environmental audit. Evolving regulations and directives acknowledge this importance. Further, remedying management deficiencies can result in long-term, lasting improvements of environmental compliance with external requirements and internal policies. And this is indeed what environmental audits are designed to accomplish.

# 25

# INTERNATIONAL ENVIRONMENTAL AUDITS

*"The International Chamber of Commerce believes that effective protection of the environment is best achieved by an appropriate combination of legislation/regulation and of policies and programs established voluntarily by industry.... Environmental auditing is an important component of such voluntary policies."*[1]

International environmental audits are receiving increased attention from multinational companies. Recently, the ISO 14000 Environmental Management Systems Standards and the European Eco-Management and Audit Scheme (EMAS) have been the principal drivers. In the United States, most major companies with an international presence have conducted routine audits of their overseas facilities for many years.

Most multinational industrial companies that have worldwide environmental audit programs developed them with objectives that closely parallel objectives and goals of domestic audit programs. Typical objectives of a cross-section of several Fortune 200 companies' worldwide audit programs include:

- To provide independent verification that the corporation's operations are in compliance with the laws of the host nation

- To insure that the corporation's environmental policies, procedures, and practices are being followed in overseas locations

---

[1] ICC Position Paper on Environmental Auditing, adopted by the ICC Executive Board at its 56th Session, 29 November 1988.

- To identify and evaluate all hazardous conditions, operations, or activities that pose risk to the plant and the public

- To insure that procedures have been established for handling environmental incidents and emergencies.

This last objective, concerning incidents and emergencies, might lead one to think that the Bhopal tragedy of December 1984 was the primary factor leading to the development of international audit programs. In fact, many U.S.-based firms with worldwide facilities had already established audit programs prior to Bhopal. What Bhopal did was to make these companies take a harder look at their program objectives and strengthen them where necessary. As a result, most worldwide audit programs today seem to focus more on "risk assessment" of hazardous operations and activities than on the detailed regulatory compliance audits prevalent in many domestic audit programs, especially in developing nations and Third World countries, where environmental regulations are less stringent than those found in the United States and Western Europe. As many multinational corporations seek to set up or expand their presence in the developing nations in Southeast Asia and Latin America, international environmental audit programs are taking on more of a total risk assessment and management-systems focused approach.

Worldwide environmental audit programs are typically structured totally separate from domestic audit programs within a company's organization. In many cases, an individual or group dedicated specifically to operating the worldwide audit program is established within the company's corporate management. This is because the demands of long-distance travel, overseas communications, keeping up with international laws and regulations, and other similar factors require a full-time effort. In addition, a dedicated international audit team develops an understanding over time of the key environmental risk management issues that similar facilities face throughout the world. As a result, the audit team can assist management in developing corporate procedures and programs that are designed to incorporate a broad-spectrum approach to environmental risk management suitable for implementation across social, economic, regulatory and cultural lines.

In most cases, U.S.-based corporate management issues a written policy statement that clearly puts the burden of environmental compliance and protection on the backs of the managing directors or officers of the worldwide facilities. A few excerpts from different Fortune 100 companies' worldwide policies on environmental protection illustrate the clear message:

> "...*Responsible environmental protection is one of the priority areas of accountability against which management performance is measured....*"

> "...*The managements of all companies' manufacturing plants, research laboratories and all other operating units are expected to ensure their operations and facilities meet the requirements of the corporate policy as well as all applicable laws and governmental regulation....*"

One issue that is unique to international audit programs is the degree to which these corporate policies and procedures are enforceable by the firm. Clearly, for worldwide subsidiaries and operating companies in which the U.S.-based parent has a majority ownership, compliance is simply dictated. But many U.S. companies only hold a minority interest in their overseas plants, making mandatory adherence to corporate environmental policy difficult to achieve. In these cases, the management of the minority owner is usually careful to suggest and recommend that the major owner of the facilities adopt the minority owner's policy or issue one of its own that addresses top management support of sound environmental practices. In the most conscientious multinational companies, compliance with applicable local laws and regulations is merely the baseline for environmental risk management policy. It is the acceptance and implementation of sound, standardized environmental risk management practices (that go above and beyond local standards, where appropriate) at all of a corporation's locations worldwide that define an effective global environmental policy.

# ROLE OF CORPORATE MANAGEMENT

The role of the U.S.-based corporate management in worldwide audit programs does not end with the promulgation of the environmental policies discussed above, but includes involvement in other activities as well, such as:

- Providing environmental technical advice and support to foreign subsidiaries or operating units from domestic divisions or units

- Conducting training programs for facility environmental managers and others with related responsibilities

- Providing information and clearinghouse functions to ensure that access to technological information and databases is readily obtainable

- Participating in international conferences, seminars, and symposiums to demonstrate U.S.-based corporate support to foreign nations' environmental programs

- Assisting in the development of written environmental guidelines and procedures to help overseas facilities operate in accordance with good management practices

- Conducting environmental audits and/or risk assessments, either independently or in conjunction with foreign operating units.

The size of the corporate staff dedicated to international environmental programs in most major companies is typically small and ranges from one to three people. Surprisingly, many of the largest U.S. firms have only one corporate person assigned to international programs. With the formal acknowledgment that we now are truly a global economy, some companies have expanded their corporate international staffs or hired outside consulting firms to assist them. In addition, there is a growing trend in multinational corporations to "regionalize" corporate environmental management staff. This practice effectively reduces the number of staff located at corporate headquarters that are dedicated to environmental matters, while placing senior-level managers in regional management positions and assigning them the responsibility of overseeing environmental issues in a specific region such as Europe or Asia.

The number of international audits/risk assessments conducted varies substantially, based on factors such as locations of the plants, degree of risk and concern, stringency of environmental laws in the host country, and financial or resource limitations. Interestingly, several major U.S. multinational companies are focusing only on conducting audits in countries having weak environmental laws, based on the assumption that their risks are greater there than in countries with sound environmental laws. A survey of a representative sample of major U.S. companies revealed the range of worldwide audits completed on an annual basis to be between ten to thirteen facilities.

# DESIGN OF INTERNATIONAL AUDIT PROGRAMS

It is important to understand both the basic approach of international audit programs and the standards against which compliance is evaluated. The common two-phased approach that is used, along with highlights of some important issues that one faces in the international arena, is discussed below.

## Two-Phased Approach

Most international audit programs appear to be designed in two phases. Phase 1 audits are broad-based, general surveys that focus on understanding the environmental requirements and actual operating practices at the facilities. Typically a general checklist is used, which is designed more as a tool to record relevant information about plant operations, chemicals and practices, and information on the surrounding environment (groundwater, surface water, soils). In addition, detailed regulatory compliance information such as copies of pertinent national and local environmental, health and safety laws and regulations are gathered and evaluated to determine the scope and extent of future, compliance-oriented audits.

From this information, a judgment is made by the auditor about the risk to the public and the environment that the facility presents, and about the regulatory compliance requirements that the facility must adhere to. This ranking is a subjective, relative risk rating that allows decisions to be made about which facilities need additional attention and the frequency and scope of subsequent audits.

In Phase 2, facilities that were determined to have a high risk are subjected to a more detailed audit/risk assessment that is tailored to a verification of the potential hazard, developing solutions to mitigating or eliminating existing or potential risk, and assessing regulatory compliance issues. In practice, Phase 1 general surveys for some facilities may be scheduled at the same time that the more detailed Phase 2 risk assessment is occurring for other facilities.

This approach allows corporations to constantly be focusing on those facilities that present the greatest risk to the environment, the public, and the company. As noted earlier, international audit program designs seem to have a common denominator in that initial auditing efforts are focused more heavily on an evaluation of "environmental risk" rather than on compliance with administrative and procedural requirements in the regulations. It is only after these risk issues are identified and addressed that most programs zero in on more traditional regulatory compliance. In countries where environmental laws and regulations are weak and/or poorly enforced, the identification, assessment, and ongoing management of environmental, health and safety risks at the facility remains the primary function of the auditing program, even after multiple facility visits.

## Audit Standards and Criteria

Notwithstanding the risk assessment approach discussed above, international audit programs should include all appropriate environmental regulations promulgated by cognizant agencies and regulatory bodies. Most countries have their own environmental regulations, which should also be consulted when implementing an international audit program. These regulations cover many of the same areas as do U.S. federal and state environmental regulations (hazardous waste, air emissions, waste water discharges, PCBs), but they differ in scope, procedural requirements, limitations, and enforcement policy. For example, the German federal government has no power to collect emissions data directly from chemical companies; it has to ask state governments to provide it. Even developing nations such as China, the Philippines, and Malaysia have both national and local laws and regulations addressing wastewater discharge, air emissions, and other environmental concerns. In many of these locations, the specific permits and approvals that are required prior to construction and operation of new facilities or expansions have specific stipulations governing environmental issues.

There are many significant differences between the United States' and other countries' approaches to environmental regulation. Companies with operating units or subsidiaries in foreign countries need to understand these regulatory philosophies and attitudes and how they may affect their facilities. For example, the reality of regulatory compliance and enforcement practices in many countries differs significantly from the programs described in the published law. Many U.S. and foreign-owned joint venture companies in China have quickly learned that they now face much stricter performance standards, higher discharge fees, and more prevalent enforcement practices than their formerly state-owned entities did before their acquisition or formation of the foreign joint venture. International auditors must attempt to understand these local differences and ensure that environmental policies and programs are designed to reflect and incorporate such conditions.

Perhaps the most significant difference is the non-adversarial nature of the relationship between industry and national governments in Europe and other foreign countries. Foreign industry has easily adapted to environmental regulation because they have been subject to it longer than U.S.-based companies. Europeans believe that "good law is good order," as Aristotle once taught. Philosophically, European industry faced up to the realities of the need for regulation long before the U.S. did because it had little choice. The population density of many European countries alone necessitated central control of new industrial development.

Germany, for example, which is about the size of Oregon and has a population of 60 million people, has a population density over seven times that of the continental United States. Is it any wonder why there is little opposition to stringent air emissions limitations in Germany?

Because European countries have a long tradition of regulation that stretches back as far as the Industrial Revolution, environmental control is characterized more as "compromise and negotiation," as compared to the litigious nature of environmental regulation in the U.S.

However, legislation and implementation of environmental laws within Third World nations has traditionally been well behind more developed nations. Because pollution control and technology requires capital and continued investment, countries where the average wage base and standard of living is well below European or U.S. standards have failed to set aside money and resources for environmental control, as other more pressing priorities (such as raising the country's domestic output and improving poor socioeconomic conditions) are addressed. In these countries, environmental controls, although sorely needed, have been viewed as luxuries that only more wealthy nations could afford. As a result, companies seeking to impose more advanced environmental control and policies may be stymied by both cultural barriers and apathy driven by a low prioritization of environmental issues. Unfortunately, many of the countries with the greatest need of improved environmental regulations are those with the least ability to achieve these goals in the short term. Multinationals with facilities in these locations must seek to achieve commensurate results from their audit programs, starting with increased employee awareness of environmental matters and basic risk management practices such as improved training.

Another difference between the United States' and other countries' environmental regulation is the power and prestige of the environmental bureaucracy. For example, the "best and the brightest" university graduates in European countries seek long-term careers in public administration. This creates an atmosphere of trust, confidence, and consistency in the development and implementation of environmental laws and regulations. Because the people carrying out environmental policy in Europe are well respected and not in the job for only a few years until a higher paying job in the private sector comes along, there is greater stability and predictability to environmental regulation in the European Community. It is simply expected that industry will comply with environmental laws because they have been brought about by a competent, fair, and powerful bureaucracy.

A third aspect of environmental compliance in the international arena is that there is little tradition of litigation in settling environmental conflicts. Unlike in the U.S., where attorneys pervade environmental business, lawyers are not popular, are fewer in number, and do not control the implementation of laws as much as in the U.S. Environmental regulators normally deal with plant technical staff in the private sector and have developed the skill of working out differences without hanging the threat of legal action over industry's head.

A final element of difference between U.S. and international implementation of environmental law is that there is no tradition of the public's right to access to private sector information, as exists in the U.S. There is no formal Freedom of Information Act (FOIA), and the environmental regulators have wide discretion in choosing to consult and inform interested parties on environmental laws and regulations. This is not to say that there is no public disclosure of environmental issues, but it is not implemented with the formality and inflexibility of the "notice and comment" approach taken in the U.S. for new environmental regulations. With its call for disclosure, the Eco-Management and Audit Scheme (EMAS) may alter this historical pattern in Europe.

Largely because of the above factors, international private sector facilities are not subject to as much inspection and monitoring as exists in the U.S. International environmental regulators expect that because environmental control requirements have been brought about in the spirit of negotiation and compromise, industry is therefore complying or is making best faith efforts to comply. This is an underlying reason why U.S.-based international audit programs tend to focus more on risk assessment than on regulatory compliance.

In summary, U.S. companies with overseas operating units should ensure that the appropriate national environmental regulations have been researched and evaluated for any impact on their facilities and operations. In countries where there is an absence of environmental regulations, it is a practice of many U.S. companies operating overseas to extend U.S.

environmental health and safety standards to facilities and operations in all world-wide locations. This is being pushed by several members of the U.S. Congress to become law, although no legislation has been enacted to date.

Many companies feel that the absence of any foreign national environmental regulations puts them at even greater risk; therefore, they have adopted policies that require conformance to U.S.-based standards and requirements as part of their overall corporate risk management philosophy. Interestingly, as witnessed by the Bhopal incident, a U.S. company found itself in litigation in the U.S. courts even though the event occurred in India, a country not known for its tough environmental health and safety regulations.

# CONDUCTING INTERNATIONAL AUDITS

Many of the same steps discussed previously in this book, relative to conducting audits, apply to both domestic and international audits. Therefore, this section will focus only on the differences or unique problems encountered when conducting worldwide audits. Using the same three phases of an audit—(1) pre-audit, (2) on-site activities, and (3) post-audit—the key issues that multinational companies have run into in their audit programs are discussed below.

## Pre-Audit Activities

As noted previously, additional effort is needed to research and evaluate the environmental laws and regulations in the country where the plants and facilities are located. There are several additional sources that can be consulted in the U.S. that can provide this information. The foreign embassies in Washington, DC are helpful in this area, as are some private-sector databases such as Enflex for Windows and the International Environment Reporter of the Bureau of National Affairs.

Notice and confirmation of the audit dates and scope need to occur much earlier in the process than is required for a domestic audit. This is particularly true when pre-audit questionnaires are used (since translation and interpretation issues are likely to arise) and when multiple facility audits are scheduled for one trip (which, as noted below, is a common practice employed by companies to reduce the overall per-site audit cost). Most companies send a letter outlining the requirements of the audit to the overseas facility at least a month in advance, and lead times of up to three months prior to the on-site visit are not unusual. One week prior to the audit, a second confirmation is typically made via the telephone or telex.

As described above, distribution of the audit questionnaire or protocols is usually made well in advance of the on-site visit to acquaint the plant management with the scope of the audit. In many cases, the plant is asked to complete the informational parts of the questionnaire (*i.e.*, permits, discharge points, number and type of air emissions points, etc.) prior to the auditor arriving on-site. This practice establishes a good starting point for the audit, as the facility representatives can be asked to describe the information presented in the pre-visit questionnaire as a kick-off exercise during the on-site portion of the audit (following the opening conference).

Effort should be made to understand, ahead of time, any language barrier problems that may be encountered during the audit. Interestingly, very few multinational companies seem to have a problem with non-English-speaking plant management. However, often second- and third-level supervisors do not understand English well enough for an auditor to conduct a thorough interview. It is apparently not a common practice that the U.S.-based corporate staff conducting international audits are multilingual, so it is important to inquire, prior to being on-site, about the degree to which English is understood by key plant staff. Every effort to ease the translation and interpretation problems should be considered, such as using local employees or consultants to assist the audit team and the facility staff as necessary. The audits will be difficult enough without the uncertainty of knowing whether someone really understands a question or a response.

Being aware of any unique cultural customs or traditions in the country prior to traveling abroad is important so that inadvertent indiscretions that can cause a strain between the auditors and plant staff do not occur. Attitudes and traditions on such things as religious observances, holidays, courtesies, and other common protocols should be known ahead of time. Auditors of a certain religious or ethnic background may feel uncomfortable in certain countries if they are not aware of these sensitivities, and sometimes even the gender of the auditor can be an issue in some worldwide locations. As unfortunate as it is in today's society, in many places throughout the world both age and gender can play an important role in establishing an effective relationship so essential to an effective audit. Also, certain practices and comments, although seemingly innocuous and not meant to be defensive or accusatory, can cause employees to "lose face" with local management, so auditors should be cognizant of these cultural concerns.

## On-Site Activities

Once on-site, there are very few differences in conducting the audit in a domestic facility or in an international plant. They generally seem to follow the same approach.

An initial orientation meeting is held with plant management and key staff to discuss the objectives of the audit and go over a detailed agenda. Typically, international audits seem to be limited to one or two auditors who spend anywhere from one to three days on-site, meaning that the effort expended to conduct the audit ranges from one to six person days. There appears to be a trend among multinational companies to spend less effort per facility than is typically expended on a domestic audit. This is because there is a sensitivity to overpowering the foreign plant with a large group of corporate staff, and secondly, the time and cost to implement an international program can become excessive if audits are not conducted efficiently. Most U.S.-based auditors will plan their trip to visit several plants in several countries when conducting overseas audits. That is another reason why spending more time than is necessary at any one plant is normally not done. The schedule is sometimes very demanding, and travel between several countries and multiple time zones can be much more complicated than is typically experienced between states in the U.S.

The on-site activities of an international audit follow the same general plan as in a domestic audit. Interviews are conducted with key plant staff, physical inspections of operations and facilities are made, and environmental records and files are reviewed. Often this written documentation is in the native language, giving rise to the need for translation to English. The audit questionnaire referred to previously is completed during the course of the on-site review, and copies of any local or municipal laws, ordinances, or regulations are usually obtained so that the auditor can maintain a file of these requirements back in the U.S.

A closing conference is typically held after the auditors are finished, so that plant management is aware of the findings and recommendations. The post-audit conference can take on a huge importance in the international audit process. It is not uncommon for numerous key managers and other personnel deemed to be important by local management to attend the closing conference. Many of these people may perceive that the findings of the auditors may be used as a measuring stick by their managers and will reflect on their standing and level of respect within the local company organization. For this reason, auditors must once again be respectful of local customs and practices and be certain to present the findings in a factual and low-key manner that will both inform and educate the facility representatives present at the closing conference.

## Post-Audit Activities

As in domestic audit programs, an audit report is normally prepared, which presents the findings and recommendations of the auditors. The report is normally submitted to the audited facility for comment at the same time the corporate law department receives a copy. After comments and necessary revisions are made, a final audit report is submitted to the operating unit's senior management and the corporate director of environmental affairs. The final audit report is often transmitted with a letter that requires the facility management to develop an action plan to address the findings and recommendations in the audit report.

Typically, the corporate environmental auditor who conducted the audit receives the action plan, only to confirm that the final report has been understood and that the response is consistent with the findings of the report. Action plans are normally prepared within one to two months of the issuance of the final report. At this point the auditor's role has been completed, and follow-up on the action plan is the responsibility of plant management and its senior management. The subsequent audit of the facility would normally include an evaluation of the progress made in the correction of the problems in the action plan since the previous audit.

Few multinational companies appear to manage their international audit programs through their legal departments or classify their audit reports as "Privileged and Confidential" to protect them from discovery. This may be due to the fact that these U.S.-based legal doctrines are not recognized in international settings as binding anyway, and the U.S. judicial system has no jurisdiction in foreign countries.

## SUMMARY OF KEY ELEMENTS IN INTERNATIONAL AUDITS

Based on several auditors' experience overseas, there are a number of special factors to consider prior to conducting international audits. These are as follows:

- Preparation time for overseas audits needs to be much greater than for domestic audits. Obtaining copies of foreign regulations can take weeks or months and sometimes require translation. Additionally, audit protocols will have to be revised to reflect these often very different regulations. It is important to understand these regulations since they may be completely different in terms of concept and implementation than U.S. regulations.

- As in the United States, many foreign countries enable regional and local jurisdictions to develop additional regulations and/or requirements and implement stricter limitations than those found in the national laws and regulations. In addition, permits to operate may impose specific standards. It is important for auditors to determine whether these conditions exist and to incorporate these relevant local compliance concerns into the audit scope.

- There are great differences in how European countries implement rational regulations in order to meet the European Economic

Community's (EEC's) environmental directives. European Community (EC) members are free to design and implement regulations as long as they meet the intent of the directives. Consequently, there can be problems with important areas such as transfrontier shipments of hazardous waste since the definitions of "hazardous" or "toxic" wastes may vary from country to country, leaving potential loopholes in the "cradle to grave" concept.

■ Although regulations may exist in some countries, there may simply be no easy way for facilities to comply with the stated requirements. As an example, the Philippines has an environmental law patterned after RCRA which requires the manifesting of specific hazardous wastes to a "licensed" disposal facility. At present, no manifest system or licensed hazardous waste disposal facilities are in place in the Philippines, making compliance with an existing law a difficult endeavor.

■ When conducting environmental audits in lesser-developed countries, auditors must be cognizant of the fact that environmental controls will be secondary to socioeconomic concerns. Control and regulation of environmental issues is perceived as being much less critical than raising the country's standard of living. It is difficult to audit facilities in lesser-developed countries using U.S. standards. The overall goal of doing these types of audits should be to minimize the environmental impacts and to educate the employees to make them aware of environmental and health issues.

■ The auditor should take the time to become familiar with the local customs and practices *prior* to arriving on-site. In many countries, and particularly in the Middle East and Asian countries, traditional customs and practices are revered above all other regulations or protocols.

- Local timetables and working hours vary greatly throughout the world, so the auditor should be thoughtful of the employees' schedules. In Spain, for instance, expect to break for several hours in the afternoon and not eat the evening meal until late in the day. Asian countries typically have very strict work habits, so be prepared for a very structured, efficient schedule (in developed Asian countries only).

- The use of a pre-audit questionnaire is critical, particularly in those countries where English is not widely spoken.

- Everything takes longer than you would expect. Plan ahead to spend at least twice as long to conduct the audit as you would expect for a similar U.S. operation. Because there is little enforcement action in many countries, environmental recordkeeping may be scattered and/or disorganized. Interviews may be difficult due to language problems, so plan ahead.

- Although the auditor should be willing to adhere to the local customs and schedules, do not let the scope or time frame of the audit get away from you. It is easy to become distracted and disoriented in unfamiliar facilities where no English language signs are present. Keep a facility plot plan handy at all times as a reference. Stick to the pre-arranged audit schedule and topics as best as possible.

- Although it may sound trite, learning a few simple phrases in the native language will endear you to your hosts and serve as a good ice-breaker. This is particularly helpful when interviewing or debriefing employees and management.

- Do not assume that management and employees are comprehending everything you say, regardless of the nods of approval they may give you. Be prepared to repeat yourself several times, or change the statement to a question to ensure the correct answer is known. Do not hesitate to ask plant management to translate for you. This is often a very effective way to get the attention of your audience.

- Above all, keep things in their proper perspective. Remember that it is difficult to be concerned over secondary containment or storage practices when the majority of the country's population is starving.

- The auditor should be sensitive to conflicts between compliance with national regulations versus corporate policies and/or U.S. standards. It is best to have a corporate policy statement in place regarding these issues prior to conducting the audit program. Be careful not to make too many "good management practice" suggestions when the existing conditions are in compliance with local regulations.

- Get a copy of everything you think you may need, and even the material you don't need. When it comes time for writing the report, it's too late to confirm any questions you may have, as the facility may be over 10,000 miles away. Also, it is a good idea to get photographs, if possible.

Even when the auditor follows these guidelines, an international audit is a difficult assignment at best. Those that have completed them know that the challenges of doing a thorough and accurate job often take the edge off the "glamour of international travel."

# Appendix A

# REFERENCES

Barton, Hugh and Bruder, Noel, *Guide to Local Environmental Auditing*, Island Press, 1995.

Blakeslee, H.W. and Grabowski, T.M., *A Practical Guide to Plant Environmental Audits*, Van Nostrand Reinhold, New York, NY, 1985.

Blumenfeld, Mark, "Conducting an Environmental Audit," Environmental Audit Handbook Series: Volume 1, Executive Enterprises Publications, 1991.

Buckley, R., *Perspectives in Environmental Management*, Springer-Verlag, 1991.

Cheremisinoff, Paul N. and Cheremisinoff, Nicholas P., *Professional Environmental Auditors' Guidebook*, Noyes, 1993.

"Corporate Counsel's Guide to Environmental Compliance & Audits," No. 105, Business Laws Inc., 1992.

Crist, Joseph G., "Reporting, Recordkeeping & Disclosure Requirements for an Environmental Audit," Environmental Audit Handbook Series: Volume 2, Executive Enterprises Publications, 1989.

Grayson, Lesley, *Environmental Auditing: A Guide to Best Practice in the U.K. & Europe*, Stanley Thornes Publishers Ltd., U.K., 1994.

Greeno, J.L., et. al., *Environmental Auditing: Fundamentals and Techniques*, John Wiley & Sons, New York, NY, 1985.

Gunnerson, Charles G., editor, *Post-Audits of Environmental Programs & Projects*, American Society of Civil Engineers, 1989.

Harrison, Henry S., *National Association of Environmental Risk Auditors Environmental Manual*, H Squared Company, New Haven, CT, 1990.

Harrison, L., editor, *Environmental Health and Safety Auditing Handbook*, 2nd edition, McGraw -Hill, New York, NY, 1995.

Hillary, Ruth, "The Eco-Management & Audit Scheme: A Practical Guide," Business & The Environment Practitioner Series (C), Stanley Thornes Publishers Ltd., UK, 1994.

Hoffman, Stephen A., *Planning, Staffing and Contracting for an Environmental Audit*, John Wiley & Sons, 1994.

Hoffman, Stephen D., "Planning, Staffing & Contracting for an Environmental Audit," Environmental Audit Handbook Series: Volume 4, Executive Enterprises Publications, 1989.

Jones, David, et. al., editors, *EC Auditing & Environmental Management*, John Wiley & Sons, 1993.

Ledgerwood, Grant, *Implementing an Environmental Audit: How to Gain a Competitive Advantage Using Quality & Environmental Responsibility*, Irwin Professional Publishing, 1994.

Lotter, Donald W., *Earthscore: Your Personal Environmental Audit & Guide*, Morning Sun Press, 1993.

McGaw, David, *Environmental Auditing & Compliance Manual*, Van Nostrand Reinhold, 1993.

Moskowitz, Joel S., *Environmental Liability & Real Estate Property Transactions: Law & Practice*, 2nd edition, John Wiley & Sons, 1995.

Ruiz, M.A. and Cheremisinoff, Paul N., *Pocket Guidebook on Environmental Auditing*, 2nd Edition, SciTech Publications, 1989.

Shields, J., ed., *Air Emissions, Baselines, & Environmental Auditing*, Van Nostrand Reinhold, 1993.

Smith, Ann C. and Yodis, William A., "Environmental Auditing Quality Management," Environmental Auditing Handbook Series: Volume 3, Executive Enterprises Publications, 1989.

Sullivan, Thomas F.P., et. al., *Environmental Law Handbook*, 13th Edition, Government Institutes, Inc., Rockville, MD, 1995.

Truitt, T.H., et. al., *Environmental Auditing Handbook: Basic Principles of Environmental Compliance Auditing*, Second Edition, Wald, Harkrader & Ross, 1983.

U.S. Environmental Protection Agency Staff, *Multi-Media Investigation Manual*, Government Institutes, Inc., Rockville, MD, 1992.

Willig, John T., ed., *Auditing for Environmental Quality Leadership: Beyond Compliance to Environmental Excellence*, John Wiley & Sons, 1995.

Young, Steven S., *Environmental Auditing*, Gulf Publishing Company, 1994.

# Appendix B

# U.S. ENVIRONMENTAL PROTECTION AGENCY'S AUDIT POLICIES

## July 9, 1986 and December 22, 1995

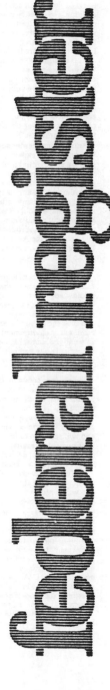

**Wednesday
July 9, 1986**

## Part IV

# Environmental
# Protection Agency

**Environmental Auditing Policy Statement;
Notice**

**25004**          **Federal Register** / Vol. 51, No. 131 / Wednesday, July 9, 1986 / Notices

## ENVIRONMENTAL PROTECTION AGENCY

[OPPE-FRL-3046-6]

**Environmental Auditing Policy Statement**

**AGENCY:** Environmental Protection Agency (EPA).

**ACTION:** Final policy statement.

**SUMMARY:** It is EPA policy to encourage the use of environmental auditing by regulated entities to help achieve and maintain compliance with environmental laws and regulations, as well as to help identify and correct unregulated environmental hazards. EPA first published this policy as interim guidance on November 8, 1985 (50 FR 46504). Based on comments received regarding the interim guidance, the Agency is issuing today's final policy statement with only minor changes.

This final policy statement specifically:

• Encourages regulated entities to develop, implement and upgrade environmental auditing programs;
• Discusses when the Agency may or may not request audit reports;
• Explains how EPA's inspection and enforcement activities may respond to regulated entities' efforts to assure compliance through auditing;
• Endorses environmental auditing at federal facilities;
• Encourages state and local environmental auditing initiatives; and
• Outlines elements of effective audit programs.

Environmental auditing includes a variety of compliance assessment techniques which go beyond those legally required and are used to identify actual and potential environmental problems. Effective environmental auditing can lead to higher levels of overall compliance and reduced risk to human health and the environment. EPA endorses the practice of environmental auditing and supports its accelerated use by regulated entities to help meet the goals of federal, state and local environmental requirements. However, the existence of an auditing program does not create any defense to, or otherwise limit, the responsibility of any regulated entity to comply with applicable regulatory requirements.

States are encouraged to adopt these or similar and equally effective policies in order to advance the use of environmental auditing on a consistent, nationwide basis.

**DATES:** This final policy statement is effective July 9, 1986.

**FOR FURTHER INFORMATION CONTACT:**
Leonard Fleckenstein, Office of Policy, Planning and Evaluation, (202) 382-2728;
or
Cheryl Wasserman, Office of Enforcement and Compliance Monitoring, (202) 382-7550.

**SUPPLEMENTARY INFORMATION:**

## ENVIRONMENTAL AUDITING POLICY STATEMENT

### I. Preamble

On November 8, 1985 EPA published an Environmental Auditing Policy Statement, effective as interim guidance, and solicited written comments until January 7, 1986.

Thirteen commenters submitted written comments. Eight were from private industry. Two commenters represented industry trade associations. One federal agency, one consulting firm and one law firm also submitted comments.

Twelve commenters addressed EPA requests for audit reports. Three comments per subject were received regarding inspections, enforcement response and elements of effective environmental auditing. One commenter addressed audit provisions as remedies in enforcement actions, one addressed environmental auditing at federal facilities, and one addressed the relationship of the policy statement to state or local regulatory agencies. Comments generally supported both the concept of a policy statement and the interim guidance, but raised specific concerns with respect to particular language and policy issues in sections of the guidance.

*General Comments*

Three commenters found the interim guidance to be constructive, balanced and effective at encouraging more and better environmental auditing.

Another commenter, while considering the policy on the whole to be constructive, felt that new and identifiable auditing "incentives" should be offered by EPA. Based on earlier comments received from industry, EPA believes most companies would not support or participate in an "incentives-based" environmental auditing program with EPA. Moreover, general promises to forgo inspections or reduce enforcement responses in exchange for companies' adoption of environmental auditing programs—the "incentives" most frequently mentioned in this context—are fraught with legal and policy obstacles.

Several commenters expressed concern that states or localities might

use the interim guidance to *require* auditing. The Agency disagrees that the policy statement opens the way for states and localities to require auditing. No EPA policy can grant states or localities any more (or less) authority than they already possess. EPA believes that the interim guidance effectively encourages *voluntary* auditing. In fact, Section II.B. of the policy states: "because audit quality depends to a large degree on genuine management commitment to the program and its objectives, auditing should remain a voluntary program."

Another commenter suggested that EPA should not expect an audit to identify all potential problem areas or conclude that a problem identified in an audit reflects normal operations and procedures. EPA agrees that an audit report should clearly reflect these realities and should be written to point out the audit's limitations. However, since EPA will not routinely request audit reports, the Agency does not believe these concerns raise issues which need to be addressed in the policy statement.

A second concern expressed by the same commenter was that EPA should acknowledge that environmental audits are only part of a successful environmental management program and thus should not be expected to cover every environmental issue or solve all problems. EPA agrees and accordingly has amended the statement of purpose which appears at the end of this preamble.

Yet another commenter thought EPA should focus on environmental performance results (compliance or non-compliance), not on the processes or vehicles used to achieve those results. In general, EPA agrees with this statement and will continue to focus on environmental results. However, EPA also believes that such results can be improved through Agency efforts to identify and encourage effective environmental management practices, and will continue to encourage such practices in non-regulatory ways.

A final general comment recommended that EPA should sponsor seminars for small businesses on how to start auditing programs. EPA agrees that such seminars would be useful. However, since audit seminars already are available from several private sector organizations, EPA does not believe it should intervene in that market, with the possible exception of seminars for government agencies, especially federal agencies, for which EPA has a broad mandate under Executive Order 12088 to

Federal Register / Vol. 51, No. 131 / Wednesday, July 9, 1986 / Notices    25805

provide technical assistance for environmental compliance.

## Requests for Reports

EPA received 12 comments regarding Agency requests for environmental audit reports, far more than on any other topic in the policy statement. One commenter felt that EPA struck an appropriate balance between respecting the need for self-evaluation with some measure of privacy, and allowing the Agency enough flexibility of inquiry to accomplish future statutory missions. However, most commenters expressed concern that the interim guidance did not go far enough to assuage corporate fears that EPA will use audit reports for environmental compliance "witch hunts." Several commenters suggested additional specific assurances regarding the circumstances under which EPA will request such reports.

One commenter recommended that EPA request audit reports only "when the Agency can show the information it needs to perform its statutory mission cannot be obtained from the monitoring, compliance or other data that is otherwise reportable and/or accessible to EPA, or where the Government deems an audit report material to a criminal investigation." EPA accepts this recommendation in part. The Agency believes it would not be in the best interest of human health and the environment to commit to making a "showing" of a compelling information need before ever requesting an audit report. While EPA may normally be willing to do so, the Agency cannot rule out in advance all circumstances in which such a showing may not be possible. However, it would be helpful to further clarify that a request for an audit report or a portion of a report normally will be made when needed information is not available by alternative means. Therefore, EPA has revised Section III.A., paragraph two and added the phrase: "and usually made where the information needed cannot be obtained from monitoring, reporting or other data otherwise available to the Agency."

Another commenter suggested that (except in the case of criminal investigations) EPA should limit requests for audit documents to specific questions. By including the phrase "or relevant portions of a report" in Section III.A., EPA meant to emphasize it would not request an entire audit document when only a relevant portion would suffice. Likewise, EPA fully intends not to request even a portion of a report if needed information or data can be otherwise obtained. To further clarify this point EPA has added the phrase,

"most likely focused on particular information needs rather than the entire report," to the second sentence of paragraph two, Section III.A. Incorporating the two comments above, the first two sentences in paragraph two of final Section III.A. now read: "EPA's authority to request an audit report, or relevant portions thereof, will be exercised on a case-by-case basis where the Agency determines it is needed to accomplish a statutory mission or the Government deems it to be material to a criminal investigation. EPA expects such requests to be limited, most likely focused on particular information needs rather than the entire report, and usually made where the information needed cannot be obtained from monitoring, reporting or other data otherwise available to the Agency."

Other commenters recommended that EPA not request audit reports under any circumstances, that requests be "restricted to only those legally required," that requests be limited to criminal investigations, or that requests be made only when EPA has reason to believe "that the audit programs or reports are being used to conceal evidence of environmental non-compliance or otherwise being used in bad faith." EPA appreciates concerns underlying all of these comments and has considered each carefully. However, the Agency believes that these recommendations do not strike the appropriate balance between retaining the flexibility to accomplish EPA's statutory missions in future, unforeseen circumstances, and acknowledging regulated entities' need to self-evaluate environmental performance with some measure of privacy. Indeed, based on prime informal comments, the small number of formal comments received, and the even smaller number of adverse comments, EPA believes the final policy statement should remain largely unchanged from the interim version.

## Elements of Effective Environmental Auditing

Three commenters expressed concerns regarding the seven general elements EPA outlined in the Appendix to the interim guidance.

One commenter noted that were EPA to further expand or more fully detail such elements, programs not specifically fulfilling each element would then be judged inadequate. EPA agrees that presenting highly specific and prescriptive auditing elements could be counter-productive by not taking into account numerous factors which vary extensively from one organization to another, but which may still result in effective auditing programs.

Accordingly, EPA does not plan to expand or more fully detail these auditing elements.

Another commenter asserted that states and localities should be cautioned not to consider EPA's auditing elements as mandatory steps. The Agency is fully aware of this concern and in the interim guidance noted its strong opinion that "regulatory agencies should not attempt to prescribe the precise form and structure of regulated entities' environmental management or auditing programs." While EPA cannot require state or local regulators to adopt this or similar policies, the Agency does strongly encourage them to do so, both in the interim and final policies.

A final commenter thought the Appendix too specifically prescribed what should and what should not be included in an auditing program. Other commenters, on the other hand, viewed the elements described as very general in nature. EPA agrees with these other commenters. The elements are in no way binding. Moreover, EPA believes that most mature, effective environmental auditing programs do incorporate each of these general elements in some form, and considers them useful yardsticks for those considering adopting or upgrading audit programs. For these reasons EPA has not revised the Appendix in today's final policy statement.

## Other Comments

Other significant comments addressed EPA inspection priorities for, and enforcement responses to, organizations with environmental auditing programs.

One commenter, stressing that audit programs are *internal* management tools, took exception to the phrase in the second paragraph of section III.B.1. of the interim guidance which states that environmental audits can 'complement' regulatory oversight. By using the word 'complement' in this context, EPA does not intend to imply that audit reports must be obtained by the Agency in order to supplement regulatory inspections. 'Complement' is used in a broad sense of being in addition to inspections and providing something (i.e., self-assessment) which otherwise would be lacking. To clarify this point EPA has added the phrase "by providing self-assessment to assure compliance" after "environmental audits may complement inspections" in this paragraph.

The same commenter also expressed concern that, as EPA sets inspection priorities, a company having an audit program could appear to be a 'poor performer' due to complete and accurate reporting when measured against a company which reports something less than required by law. EPA agrees that it is important to communicate this fact to Agency and state personnel, and will do so. However, the Agency does not believe a change in the policy statement is necessary.

A further comment suggested EPA should commit to take auditing programs into account when assessing all enforcement actions. However, in order to maintain enforcement flexibility under varied circumstances, the Agency cannot promise reduced enforcement responses to violations at all audited facilities when other factors may be overriding. Therefore the policy statement continues to state that EPA may exercise its decretion to consider auditing programs as evidence of honest and genuine efforts to assure compliance, which would then be taken into account in fashioning enforcement responses to violations.

A final commenter suggested the phrase "expeditiously correct environmental problems" not be used in the enforcement context since it implied EPA would use an entity's record of correcting nonregulated matters when evaluating regulatory violations. EPA did not intend for such an inference to be made. EPA intended the term "environmental problems" to refer to the underlying circumstances which eventually lead up to the violations. To clarify this point, EPA is revising the first two sentences of the paragraph to which this comment refers by changing "environmental problems" to "violations and underlying environmental problems" in the first sentence and to "underlying environmental problems" in the second sentence.

In a separate development EPA is preparing an update of its January 1984 *Federal Facilities Compliance Strategy*, which is referenced in section III. C. of the auditing policy. The Strategy should be completed and available on request from EPA's Office of Federal Activities later this year.

EPA thanks all commenters for responding to the November 8, 1985 publication. Today's notice is being issued to inform regulated entities and the public of EPA's final policy toward environmental auditing: This policy was developed to help (a) encourage regulated entities to institutionalize effective audit practices as one means of improving compliance and sound environmental management, and (b) guide internal EPA actions directly related to regulated entities' environmental auditing programs.

EPA will evaluate implementation of this final policy to ensure it meets the above goals and continues to encourage

better environmental management, while strengthening the Agency's own efforts to monitor and enforce compliance with environmental requirements.

## II. General EPA Policy on Environmental Auditing

### A. Introduction

Environmental auditing is a systematic, documented, periodic and objective review by regulated entities [1] of facility operations and practices related to meeting environmental requirements. Audits can be designed to accomplish any or all of the following: verify compliance with environmental requirements; evaluate the effectiveness of environmental management systems already in place; or assess risks from regulated and unregulated materials and practices.

Auditing serves as a quality assurance check to help improve the effectiveness of basic environmental management by verifying that management practices are in place, functioning and adequate. Environmental audits evaluate, and are not a substitute for, direct compliance activities such as obtaining permits, installing controls, monitoring compliance, reporting violations, and keeping records. Environmental auditing may verify but does not include activities required by law, regulation or permit (e.g., continuous emissions monitoring, composite correction plans at wastewater treatment plants, etc.). Audits do not in any way replace regulatory agency inspections. However, environmental audits can improve compliance by complementing conventional federal, state and local oversight.

The appendix to this policy statement outlines some basic elements of environmental auditing (e.g., auditor independence and top management support) for use by those considering implementation of effective auditing programs to help achieve and maintain compliance. Additional information on environmental auditing practices can be found in various published materials. [2]

[1] "Regulated entities" include private firms and public agencies with facilities subject to environmental regulation. Public agencies can include federal, state or local agencies as well as special-purpose organizations such as regional sewage commissions.

[2] See, e.g., "Current Practices in Environmental Auditing," EPA Report No. EPA–230–09–83–006, February 1984; "Annotated Bibliography on Environmental Auditing," Fifth Edition, September 1985, both available from: Regulatory Reform Staff, PM–223, EPA, 401 M Street SW, Washington, DC 20460.

Environmental auditing has developed for sound business reasons, particularly as a means of helping regulated entities manage pollution control affirmatively over time instead of reacting to crises. Auditing can result in improved facility environmental performance, help communicate effective solutions to common environmental problems, focus facility managers' attention on current and upcoming regulatory requirements, and generate protocols and checklists which help facilities better manage themselves. Auditing also can result in better-integrated management of environmental hazards, since auditors frequently identify environmental liabilities which go beyond regulatory compliance. Companies, public entities and federal facilities have employed a variety of environmental auditing practices in recent years. Several hundred major firms in diverse industries now have environmental auditing programs, although they often are known by other names such as assessment, survey, surveillance, review or appraisal.

While auditing has demonstrated its usefulness to those with audit programs, many others still do not audit. Clarification of EPA's position regarding auditing may help encourage regulated entities to establish audit programs or upgrade systems already in place.

### B. EPA Encourages the Use of Environmental Auditing

EPA encourages regulated entities to adopt sound environmental management practices to improve environmental performance. In particular, EPA encourages regulated entities subject to environmental regulations to institute environmental auditing programs to help ensure the adequacy of internal systems to achieve, maintain and monitor compliance. Implementation of environmental auditing programs can result in better identification, resolution and avoidance of environmental problems, as well as improvements to management practices. Audits can be conducted effectively by independent internal or third party auditors. Larger organizations generally have greater resources to devote to an internal audit team, while smaller entities might be more likely to use outside auditors.

Regulated entities are responsible for taking all necessary steps to ensure compliance with environmental requirements, whether or not they adopt audit programs. Although environmental laws do not require a regulated facility to have an auditing program, ultimate responsibility for the environmental

Federal Register / Vol. 51, No. 131 / Wednesday, July 9, 1986 / Notices **25007**

performance of the facility lies with top management, which therefore has a strong incentive to use reasonable means, such as environmental auditing, to secure reliable information of facility compliance status.

EPA does not intend to dictate or interfere with the environmental management practices of private or public organizations. Nor does EPA intend to mandate auditing (though in certain instances EPA may seek to include provisions for environmental auditing as part of settlement agreements, as noted below). Because environmental auditing systems have been widely adopted on a voluntary basis in the past, and because audit quality depends to a large degree upon genuine management commitment to the program and its objectives, auditing should remain a voluntary activity.

### III. EPA Policy on Specific Environmental Auditing Issues

#### A. *Agency Requests for Audit Reports*

EPA has broad statutory authority to request relevant information on the environmental compliance status of regulated entities. However, EPA believes routine Agency requests for audit reports [3] could inhibit auditing in the long run, decreasing both the quantity and quality of audits conducted. Therefore, as a matter of policy, EPA will *not* routinely request environmental audit reports.

EPA's authority to request an audit report, or relevant portions thereof, will be exercised on a case-by-case basis where the Agency determines it is needed to accomplish a statutory mission, or where the Government deems it to be material to a criminal investigation. EPA expects such requests to be limited, most likely focused on particular information needs rather than the entire report, and usually made where the information needed cannot be obtained from monitoring, reporting or other data otherwise available to the Agency. Examples would likely include situations where: audits are conducted under consent decrees or other settlement agreements; a company has placed its management practices at issue by raising them as a defense; or state of mind or intent are a relevant element of inquiry, such as during a criminal investigation. This list

is illustrative rather than exhaustive, since there doubtless will be other situations, not subject to prediction, in which audit reports rather than information may be required.

EPA acknowledges regulated entities' need to self-evaluate environmental performance with some measure of privacy and encourages such activity. However, audit reports may not shield monitoring, compliance, or other information that would otherwise be reportable and/or accessible to EPA, even if there is no explicit 'requirement' to generate that data.[4] Thus, this policy does not alter regulated entities' existing or future obligations to monitor, record or report information required under environmental statutes, regulations or permits, or to allow EPA access to that information. Nor does this policy alter EPA's authority to request and receive any relevant information—including that contained in audit reports—under various environmental statutes (e.g., Clean Water Act section 308, Clean Air Act sections 114 and 208) or in other administrative or judicial proceedings.

Regulated entities also should be aware that certain audit findings may by law have to be reported to government agencies. However, in addition to any such requirements, EPA encourages regulated entities to notify appropriate State or Federal officials of findings which suggest significant environmental or public health risks, even when not specifically required to do so.

#### B. *EPA Response to Environmental Auditing*

##### 1. General Policy

EPA will not promise to forgo inspections, reduce enforcement responses, or offer other such incentives in exchange for implementation of environmental auditing or other sound environmental management practices. Indeed, a credible enforcement program provides a strong incentive for regulated entities to audit.

Regulatory agencies have an obligation to assess source compliance status independently and cannot eliminate inspections for particular firms or classes of firms. Although environmental audits may complement inspections by providing self-assessment to assure compliance, they are in no way a substitute for regulatory oversight. Moreover, certain statutes (e.g. RCRA) and Agency policies

establish minimum facility inspection frequencies to which EPA will adhere.

However, EPA will continue to address environmental problems on a priority basis and will consequently inspect facilities with poor environmental records and practices more frequently. Since effective environmental auditing helps management identify and promptly correct actual or potential problems, audited facilities' environmental performance should improve. Thus, while EPA inspections of self-audited facilities will continue, to the extent that compliance performance is considered in setting inspection priorities, facilities with a good compliance history may be subject to fewer inspections.

In fashioning enforcement responses to violations, EPA policy is to take into account, on a case-by-case basis, the honest and genuine efforts of regulated entities to avoid and promptly correct violations and underlying environmental problems. When regulated entities take reasonable precautions to avoid noncompliance, expeditiously correct underlying environmental problems discovered through audits or other means, and implement measures to prevent their recurrence, EPA may exercise its discretion to consider such actions as honest and genuine efforts to assure compliance. Such consideration applies particularly when a regulated entity promptly reports violations or compliance data which otherwise were not required to be recorded or reported to EPA.

##### 2. Audit Provisions as Remedies in Enforcement Actions

EPA may propose environmental auditing provisions in consent decrees and in other settlement negotiations where auditing could provide a remedy for identified problems and reduce the likelihood of similar problems recurring in the future.[5] Environmental auditing provisions are most likely to be proposed in settlement negotiations where:

• A pattern of violations can be attributed, at least in part, to the absence or poor functioning of an environmental management system; or

• The type or nature of violations indicates a likelihood that similar noncompliance problems may exist or occur elsewhere in the facility or at other facilities operated by the regulated entity.

---

[3] An "environmental audit report" is a written report which candidly and thoroughly presents findings from a review, conducted as part of an environmental audit as described in section II.A. of facility environmental performance and practices. An audit report is not a substitute for compliance monitoring reports or other reports or records which may be required by EPA or other regulatory agencies.

[4] See, for example, "Duties to Report or Disclose Information on the Environmental Aspects of Business Activities," Environmental Law Institute report to EPA, final report, September 1985.

[5] EPA is developing guidance for use by Agency negotiators in structuring appropriate environmental audit provisions for consent decrees and other settlement negotiations.

Through this consent decree approach and other means, EPA may consider how to encourage effective auditing by publicly owned sewage treatment works (POTWs). POTWs often have compliance problems related to operation and maintenance procedures which can be addressed effectively through the use of environmental auditing. Under its National Municipal Policy EPA already is requiring many POTWs to develop composite correction plans to identify and correct compliance problems.

*C. Environmental Auditing at Federal Facilities*

EPA encourages all federal agencies subject to environmental laws and regulations to institute environmental auditing systems to help ensure the adequacy of internal systems to achieve, maintain and monitor compliance. Environmental auditing at federal facilities can be an effective supplement to EPA and state inspections. Such federal facility environmental audit programs should be structured to promptly identify environmental problems and expenditiously develop schedules for remedial action.

To the extent feasible, EPA will provide technical assistance to help federal agencies design and initiate audit programs. Where appropriate, EPA will enter into agreements with other agencies to clarify the respective roles, responsibilities and commitments of each agency in conducting and responding to federal facility environmental audits.

With respect to inspections of self-audited facilities (see section III.B.1 above) and requests for audit reports (see section III.A above), EPA generally will respond to environmental audits by federal facilities in the same manner as it does for other regulated entities, in keeping with the spirit and intent of Executive Order 12088 and the EPA *Federal Facilities Compliance Strategy* (January 1984, update forthcoming in late 1986). Federal agencies should, however, be aware that the Freedom of Information Act will govern any disclosure of audit reports or audit-generated information requested from federal agencies by the public.

When federal agencies discover significant violations through an environmental audit, EPA encourages them to submit the related audit findings and remedial action plans expeditiously to the applicable EPA regional office (and responsible state agencies, where appropriate) even when not specifically required to do so. EPA will review the audit findings and action plans and either provide written approval or

negotiate a Federal Facilities Compliance Agreement. EPA will utilize the escalation procedures provided in Executive Order 12088 and the EPA *Federal Facilities Compliance Strategy* only when agreement between agencies cannot be reached. In any event, federal agencies are expected to report pollution abatement projects involving costs (necessary to correct problems discovered through the audit) to EPA in accordance with OMB Circular A-106. Upon request, and in appropriate circumstances, EPA will assist affected federal agencies through coordination of any public release of audit findings with approved action plans once agreement has been reached.

**IV. Relationship to State or Local Regulatory Agencies**

State and local regulatory agencies have independent jurisdiction over regulated entities. EPA encourages them to adopt these or similar policies, in order to advance the use of effective environmental auditing in a consistent manner.

EPA recognizes that some states have already undertaken environmental auditing initiatives which differ somewhat from this policy. Other states also may want to develop auditing policies which accommodate their particular needs or circumstances. Nothing in this policy statement is intended to preempt or preclude states from developing other approaches to environmental auditing. EPA encourages state and local authorities to consider the basic principles which guided the Agency in developing this policy:

• Regulated entities must continue to report or record compliance information required under existing statutes or regulations, regardless of whether such information is generated by an environmental audit or contained in an audit report. Required information cannot be withheld merely because it is generated by an audit rather than by some other means.

• Regulatory agencies cannot make promises to forgo or limit enforcement action against a particular facility or class of facilities in exchange for the use of environmental auditing systems. However, such agencies may use their discretion to adjust enforcement actions on a case-by-case basis in response to honest and genuine efforts by regulated entities to assure environmental compliance.

• When setting inspection priorities regulatory agencies should focus to the extent possible on compliance performance and environmental results.

• Regulatory agencies must continue to meet minimum program requirements

(e.g., minimum inspection requirements, etc.).

• Regulatory agencies should not attempt to prescribe the precise form and structure of regulated entities' environmental management or auditing programs.

An effective state/federal partnership is needed to accomplish the mutual goal of achieving and maintaining high levels of compliance with environmental laws and regulations. The greater the consistency between state or local policies and this federal response to environmental auditing, the greater the degree to which sound auditing practices might be adopted and compliance levels improve.

Dated: June 28, 1986.

**Lee M. Thomas,**

*Administrator.*

**Appendix—Elements of Effective Environmental Auditing Programs**

*Introduction:* Environmental auditing is a systematic, documented, periodic and objective review by a regulated entity of facility operations and practices related to meeting environmental requirements.

Private sector environmental audits of facilities have been conducted for several years and have taken a variety of forms, in part to accommodate unique organizational structures and circumstances. Nevertheless, effective environmental audits appear to have certain discernible elements in common with other kinds of audits. Standards for internal audits have been documented extensively. The elements outlined below draw heavily on two of these documents: "Compendium of Audit Standards" (°1983, Walter Willborn, American Society for Quality Control) and "Standards for the Professional Practice of Internal Auditing" (°1981, The Institute of Internal Auditors, Inc.). They also reflect Agency analyses conducted over the last several years.

Performance-oriented auditing elements are outlined here to help accomplish several objectives. A general description of features of effective, mature audit programs can help those starting audit programs, especially federal agencies and smaller businesses. These elements also indicate the attributes of auditing EPA generally considers important to ensure program effectiveness. Regulatory agencies may use these elements in negotiating environmental auditing provisions for consent decrees. Finally, these elements can help guide states and localities considering auditing initiatives.

Federal Register / Vol. 51, No. 131 / Wednesday, July 9, 1986 / Notices    25009

An effective environmental auditing system will likely include the following general elements:

I. *Explicit top management support for environmental auditing and commitment to follow-up on audit findings.* Management support may be demonstrated by a written policy articulating upper management support for the auditing program, and for compliance with all pertinent requirements, including corporate policies and permit requirements as well as federal, state and local statutes and regulations.

Management support for the auditing program also should be demonstrated by an explicit written commitment to follow-up on audit findings to correct identified problems and prevent their recurrence.

II. *An environmental auditing function independent of audited activities.* The status or organizational locus of environmental auditors should be sufficient to ensure objective and unobstructed inquiry, observation and testing. Auditor objectivity should not be impaired by personal relationships, financial or other conflicts of interest, interference with free inquiry or judgment, or fear of potential retribution.

III. *Adequate team staffing and auditor training.* Environmental auditors should possess or have ready access to the knowledge, skills, and disciplines needed to accomplish audit objectives. Each individual auditor should comply with the company's professional, standards of conduct. Auditors, whether full-time or part-time, should maintain their technical and analytical competence through continuing education and training.

IV. *Explicit audit program objectives, scope, resources and frequency.* At a minimum, audit objectives should include assessing compliance with applicable environmental laws and evaluating the adequacy of internal compliance policies, procedures and personnel training programs to ensure continued compliance.

Audits should be based on a process which provides auditors: all corporate policies, permits, and federal, state, and local regulations pertinent to the facility; and checklists or protocols addressing specific features that should be evaluated by auditors.

Explicit written audit procedures generally should be used for planning audits, establishing audit scope, examining and evaluating audit findings, communicating audit results, and following-up.

V. *A process which collects, analyzes, interprets and documents information sufficient to achieve audit objectives.* Information should be collected before and during an onsite visit regarding environmental compliance(*1*), environmental management effectiveness(*2*), and other matters (*3*) related to audit objectives and scope. This information should be sufficient, reliable, relevant and useful to provide a sound basis for audit findings and recommendations.

a. *Sufficient* information is factual, adequate and convincing so that a prudent, informed person would be likely to reach the same conclusions as the auditor.

b. *Reliable* information is the best attainable through use of appropriate audit techniques.

c. *Relevant* information supports audit findings and recommendations and is consistent with the objectives for the audit.

d. *Useful* information helps the organization meet its goals.

The audit process should include a periodic review of the reliability and integrity of this information and the means used to identify, measure, classify and report it. Audit procedures, including the testing and sampling techniques employed, should be selected in advance, to the extent practical, and expanded or altered if circumstances warrant. The process of collecting, analyzing, interpreting, and documenting information should provide reasonable assurance that audit objectivity is maintained and audit goals are met.

VI. *A process which includes specific procedures to promptly prepare candid, clear and appropriate written reports on audit findings, corrective actions, and schedules for implementation.* Procedures should be in place to ensure that such information is communicated to managers, including facility and corporate management, who can evaluate the information and ensure correction of identified problems. Procedures also should be in place for determining what internal findings are reportable to state or federal agencies.

VII. *A process which includes quality assurance procedures to assure the accuracy and thoroughness of environmental audits.* Quality assurance may be accomplished through supervision, independent internal reviews, external reviews, or a combination of these approaches.

**Footnotes to Appendix**

(*1*) A comprehensive assessment of compliance with federal environmental regulations requires an analysis of facility performance against numerous environmental statutes and implementing regulations. These statutes include:
Resource Conservation and Recovery Act
Federal Water Pollution Control Act
Clean Air Act
Hazardous Materials Transportation Act
Toxic Substances Control Act
Comprehensive Environmental Response, Compensation and Liability Act
Safe Drinking Water Act
Federal Insecticide, Fungicide and Rodenticide Act
Marine Protection, Research and Sanctuaries Act
Uranium Mill Tailings Radiation Control Act

In addition, state and local government are likely to have their own environmental laws. Many states have been delegated authority to administer federal programs. Many local governments' building, fire, safety and health codes also have environmental requirements relevant to an audit evaluation.

(*2*) An environmental audit could go well beyond the type of compliance assessment normally conducted during regulatory inspections, for example, by evaluating policies and practices, regardless of whether they are part of the environmental system or the operating and maintenance procedures. Specifically, audits can evaluate the extent to which systems or procedures:

1. Develop organizational environmental policies which: a. implement regulatory requirements; b. provide management guidance for environmental hazards not specifically addressed in regulations;

2. Train and motivate facility personnel to work in an environmentally-acceptable manner and to understand and comply with government regulations and the entity's environmental policy;

3. Communicate relevant environmental developments expeditiously to facility and other personnel;

4. Communicate effectively with government and the public regarding serious environmental incidents;

5. Require third parties working for, with or on behalf of the organization to follow its environmental procedures;

6. Make proficient personnel available at all times to carry out environmental (especially emergency) procedures;

7. Incorporate environmental protection into written operating procedures;

8. Apply best management practices and operating procedures, including "good housekeeping" techniques;

9. Institute preventive and corrective maintenance systems to minimize actual and potential environmental harm;

10. Utilize best available process and control technologies;

11. Use most-effective sampling and monitoring techniques, test methods, recordkeeping systems or reporting protocols (beyond minimum legal requirements);

12. Evaluate causes behind any serious environmental incidents and establish procedures to avoid recurrence;

13. Exploit source reduction, recycle and reuse potential wherever practical; and

14. Substitute materials or processes to allow use of the least-hazardous substances feasible.

(3) Auditors could also assess environmental risks and uncertainties.

[FR Doc. 86-15423 Filed 7-8-86 8:45 am]
BILLING CODE 6560-50-M

Friday
December 22, 1995

## Part III

# Environmental Protection Agency

Incentives for Self-Policing: Discovery, Disclosure, Correction and Prevention of Violations; Notice

**66706** **Federal Register** / Vol. 60, No. 246 / Friday, December 22, 1995 / Notices

**ENVIRONMENTAL PROTECTION AGENCY**

[FRL–5400–1]

**Incentives for Self-Policing: Discovery, Disclosure, Correction and Prevention of Violations**

**AGENCY:** Environmental Protection Agency (EPA).

**ACTION:** Final Policy Statement.

**SUMMARY:** The Environmental Protection Agency (EPA) today issues its final policy to enhance protection of human health and the environment by encouraging regulated entities to voluntarily discover, and disclose and correct violations of environmental requirements. Incentives include eliminating or substantially reducing the gravity component of civil penalties and not recommending cases for criminal prosecution where specified conditions are met, to those who voluntarily self-disclose and promptly correct violations. The policy also restates EPA's long-standing practice of not requesting voluntary audit reports to trigger enforcement investigations. This policy was developed in close consultation with the U.S. Department of Justice, states, public interest groups and the regulated community, and will be applied uniformly by the Agency's enforcement programs.

**DATES:** This policy is effective January 22, 1996.

**FOR FURTHER INFORMATION CONTACT:** Additional documentation relating to the development of this policy is contained in the environmental auditing public docket. Documents from the docket may be obtained by calling (202) 260–7548, requesting an Index to docket #C–94–01, and faxing document requests to (202) 260–4400. Hours of operation are 8 a.m. to 5:30 p.m., Monday through Friday, except legal holidays. Additional contacts are Robert Fentress or Brian Riedel, at (202) 564–4187.

**SUPPLEMENTARY INFORMATION:**

**I. Explanation of Policy**

*A. Introduction*

The Environmental Protection Agency today issues its final policy to enhance protection of human health and the environment by encouraging regulated entities to discover voluntarily, disclose, correct and prevent violations of federal environmental law. Effective 30 days from today, where violations are found through voluntary environmental audits or efforts that reflect a regulated entity's due diligence, and are promptly

disclosed and expeditiously corrected, EPA will not seek gravity-based (i.e., non-economic benefit) penalties and will generally not recommend criminal prosecution against the regulated entity. EPA will reduce gravity-based penalties by 75% for violations that are voluntarily discovered, and are promptly disclosed and corrected, even if not found through a formal audit or due diligence. Finally, the policy restates EPA's long-held policy and practice to refrain from routine requests for environmental audit reports.

The policy includes important safeguards to deter irresponsible behavior and protect the public and environment. For example, in addition to prompt disclosure and expeditious correction, the policy requires companies to act to prevent recurrence of the violation and to remedy any environmental harm which may have occurred. Repeated violations or those which result in actual harm or may present imminent and substantial endangerment are not eligible for relief under this policy, and companies will not be allowed to gain an economic advantage over their competitors by delaying their investment in compliance. Corporations remain criminally liable for violations that result from conscious disregard of their obligations under the law, and individuals are liable for criminal misconduct.

The issuance of this policy concludes EPA's eighteen-month public evaluation of the optimum way to encourage voluntary self-policing while preserving fair and effective enforcement. The incentives, conditions and exceptions announced today reflect thoughtful suggestions from the Department of Justice, state attorneys general and local prosecutors, state environmental agencies, the regulated community, and public interest organizations. EPA believes that it has found a balanced and responsible approach, and will conduct a study within three years to determine the effectiveness of this policy.

*B. Public Process*

One of the Environmental Protection Agency's most important responsibilities is ensuring compliance with federal laws that protect public health and safeguard the environment. Effective deterrence requires inspecting, bringing penalty actions and securing compliance and remediation of harm. But EPA realizes that achieving compliance also requires the cooperation of thousands of businesses and other regulated entities subject to these requirements. Accordingly, in

May of 1994, the Administrator asked the Office of Enforcement and Compliance Assurance (OECA) to determine whether additional incentives were needed to encourage voluntary disclosure and correction of violations uncovered during environmental audits.

EPA began its evaluation with a two-day public meeting in July of 1994, in Washington, D.C.. followed by a two-day meeting in San Francisco on January 19, 1995 with stakeholders from industry, trade groups, state environmental commissioners and attorneys general, district attorneys, public interest organizations and professional environmental auditors. The Agency also established and maintained a public docket of testimony presented at these meetings and all comment and correspondence submitted to EPA by outside parties on this issue.

In addition to considering opinion and information from stakeholders, the Agency examined other federal and state policies related to self-policing, self-disclosure and correction. The Agency also considered relevant surveys on auditing practices in the private sector. EPA completed the first stage of this effort with the announcement of an interim policy on April 3 of this year, which defined conditions under which EPA would reduce civil penalties and not recommend criminal prosecution for companies that audited, disclosed, and corrected violations.

Interested parties were asked to submit comment on the interim policy by June 30 of this year (60 FR 16875), and EPA received over 300 responses from a wide variety of private and public organizations. (Comments on the interim audit policy are contained in the Auditing Policy Docket, hereinafter, "Docket".) Further, the American Bar Association SONREEL Subcommittee hosted five days of dialogue with representatives from the regulated industry, states and public interest organizations in June and September of this year, which identified options for strengthening the interim policy. The changes to the interim policy announced today reflect insight gained through comments submitted to EPA, the ABA dialogue, and the Agency's practical experience implementing the interim policy.

*C. Purpose*

This policy is designed to encourage greater compliance with laws and regulations that protect human health and the environment. It promotes a higher standard of self-policing by waiving gravity-based penalties for

Federal Register / Vol. 60, No. 246 / Friday, December 22, 1995 / Notices          **66707**

violations that are promptly disclosed and corrected, and which were discovered through voluntary audits or compliance management systems that demonstrate due diligence. To further promote compliance, the policy reduces gravity-based penalties by 75% for any violation voluntarily discovered and promptly disclosed and corrected, even if not found through an audit or compliance management system.

EPA's enforcement program provides a strong incentive for responsible behavior by imposing stiff sanctions for noncompliance. Enforcement has contributed to the dramatic expansion of environmental auditing measured in numerous recent surveys. For example, more than 90% of the corporate respondents to a 1995 Price-Waterhouse survey who conduct audits said that one of the reasons they did so was to find and correct violations before they were found by government inspectors. (A copy of the Price-Waterhouse survey is contained in the Docket as document VIII–A–76.)

At the same time, because government resources are limited, maximum compliance cannot be achieved without active efforts by the regulated community to police themselves. More than half of the respondents to the same 1995 Price-Waterhouse survey said that they would expand environmental auditing in exchange for reduced penalties for violations discovered and corrected. While many companies already audit or have compliance management programs, EPA believes that the incentives offered in this policy will improve the frequency and quality of these self-monitoring efforts.

### D. Incentives for Self-Policing

Section C of EPA's policy identifies the major incentives that EPA will provide to encourage self-policing, self-disclosure, and prompt self-correction. These include not seeking gravity-based civil penalties or reducing them by 75%, declining to recommend criminal prosecution for regulated entities that self-police, and refraining from routine requests for audits. (As noted in Section C of the policy, EPA has refrained from making routine requests for audit reports since issuance of its 1986 policy on environmental auditing.)

### 1. Eliminating Gravity-Based Penalties

Under Section C(1) of the policy, EPA will not seek gravity-based penalties for violations found through auditing that are promptly disclosed and corrected. Gravity-based penalties will also be waived for violations found through any documented procedure for self-policing, where the company can show that it has

a compliance management program that meets the criteria for due diligence in Section B of the policy.

Gravity-based penalties (defined in Section B of the policy) generally reflect the seriousness of the violator's behavior. EPA has elected to waive such penalties for violations discovered through due diligence or environmental audits, recognizing that these voluntary efforts play a critical role in protecting human health and the environment by identifying, correcting and ultimately preventing violations. All of the conditions set forth in Section D, which include prompt disclosure and expeditious correction, must be satisfied for gravity-based penalties to be waived.

As in the interim policy, EPA reserves the right to collect any economic benefit that may have been realized as a result of noncompliance, even where companies meet all other conditions of the policy. Economic benefit may be waived, however, where the Agency determines that it is insignificant.

After considering public comment, EPA has decided to retain the discretion to recover economic benefit for two reasons. First, it provides an incentive to comply on time. Taxpayers expect to pay interest or a penalty fee if their tax payments are late; the same principle should apply to corporations that have delayed their investment in compliance. Second, it is fair because it protects responsible companies from being undercut by their noncomplying competitors, thereby preserving a level playing field. The concept of recovering economic benefit was supported by public comments by many stakeholders, including industry representatives (see, e.g., Docket, II–F–39, II–F–28, and II–F–18).

### 2. 75% Reduction of Gravity

The policy appropriately limits the complete waiver of gravity-based civil penalties to companies that meet the higher standard of environmental auditing or systematic compliance management. However, to provide additional encouragement for the kind of self-policing that benefits the public, gravity-based penalties will be reduced by 75% for a violation that is voluntarily discovered, promptly disclosed and expeditiously corrected, even if it was not found through an environmental audit and the company cannot document due diligence. EPA expects that this will encourage companies to come forward and work with the Agency to resolve environmental problems and begin to develop an effective compliance management program.

Gravity-based penalties will be reduced 75% only where the company meets all conditions in Sections D(2) through D(9). EPA has eliminated language from the interim policy indicating that penalties may be reduced "up to" 75% where "most" conditions are met, because the Agency believes that all of the conditions in D(2) through D(9) are reasonable and essential to achieving compliance. This change also responds to requests for greater clarity and predictability.

### 3. No Recommendations for Criminal Prosecution

EPA has never recommended criminal prosecution of a regulated entity based on voluntary disclosure of violations discovered through audits and disclosed to the government before an investigation was already under way. Thus, EPA will not recommend criminal prosecution for a regulated entity that uncovers violations through environmental audits or due diligence, promptly discloses and expeditiously corrects those violations, and meets all other conditions of Section D of the policy.

This policy is limited to good actors, and therefore has important limitations. It will not apply, for example, where corporate officials are consciously involved in or willfully blind to violations, or conceal or condone noncompliance. Since the regulated entity must satisfy all of the conditions of Section D of the policy, violations that caused serious harm or which may pose imminent and substantial endangerment to human health or the environment are not covered by this policy. Finally, EPA reserves the right to recommend prosecution for the criminal conduct of any culpable individual.

Even where all of the conditions of this policy are not met, however, it is important to remember that EPA may decline to recommend prosecution of a company or individual for many other reasons under other Agency enforcement policies. For example, the Agency may decline to recommend prosecution where there is no significant harm or culpability and the individual or corporate defendant has cooperated fully.

Where a company has met the conditions for avoiding a recommendation for criminal prosecution under this policy, it will not face any civil liability for gravity-based penalties. That is because the same conditions for discovery, disclosure, and correction apply in both cases. This represents a clarification of the interim policy, not a substantive change.

**4. No Routine Requests for Audits**

EPA is reaffirming its policy, in effect since 1986, to refrain from routine requests for audits. Eighteen months of public testimony and debate have produced no evidence that the Agency has deviated, or should deviate, from this policy.

If the Agency has independent evidence of a violation, it may seek information needed to establish the extent and nature of the problem and the degree of culpability. In general, however, an audit which results in prompt correction clearly will reduce liability, not expand it. Furthermore, a review of the criminal docket did not reveal a single criminal prosecution for violations discovered as a result of an audit self-disclosed to the government.

**E. Conditions**

Section D describes the nine conditions that a regulated entity must meet in order for the Agency not to seek (or to reduce) gravity-based penalties under the policy. As explained in the Summary above, regulated entities that meet all nine conditions will not face gravity-based civil penalties, and will generally not have to fear criminal prosecution. Where the regulated entity meets all of the conditions except the first (D(1)), EPA will reduce gravity-based penalties by 75%.

**1. Discovery of the Violation Through an Environmental Audit or Due Diligence**

Under Section D(1), the violation must have been discovered through either (a) an environmental audit that is systematic, objective, and periodic as defined in the 1986 audit policy, or (b) a documented, systematic procedure or practice which reflects the regulated entity's due diligence in preventing, detecting, and correcting violations. The interim policy provided full credit for any violation found through "voluntary self-evaluation," even if the evaluation did not constitute an audit. In order to receive full credit under the final policy, any self-evaluation that is not an audit must be part of a "due diligence" program. Both "environmental audit" and "due diligence" are defined in Section B of the policy.

Where the violation is discovered through a "systematic procedure or practice" which is not an audit, the regulated entity will be asked to document how its program reflects the criteria for due diligence as defined in Section B of the policy. These criteria, which are adapted from existing codes of practice such as the 1991 Criminal Sentencing Guidelines, were fully

discussed during the ABA dialogue. The criteria are flexible enough to accommodate different types and sizes of businesses. The Agency recognizes that a variety of compliance management programs may develop under the due diligence criteria, and will use its review under this policy to determine whether basic criteria have been met.

Compliance management programs which train and motivate production staff to prevent, detect and correct violations on a daily basis are a valuable complement to periodic auditing. The policy is responsive to recommendations received during public comment and from the ABA dialogue to give compliance management efforts which meet the criteria for due diligence the same penalty reduction offered for environmental audits. (See, e.g., II–F–39, II–E–18, and II–G–18 in the Docket.)

EPA may require as a condition of penalty mitigation that a description of the regulated entity's due diligence efforts be made publicly available. The Agency added this provision in response to suggestions from environmental groups, and believes that the availability of such information will allow the public to judge the adequacy of compliance management systems, lead to enhanced compliance, and foster greater public trust in the integrity of compliance management systems.

**2. Voluntary Discovery and Prompt Disclosure**

Under Section D(2) of the final policy, the violation must have been identified voluntarily, and not through a monitoring, sampling, or auditing procedure that is required by statute, regulation, permit, judicial or administrative order, or consent agreement. Section D(4) requires that disclosure of the violation be prompt and in writing. To avoid confusion and respond to state requests for greater clarity, disclosures under this policy should be made to EPA. The Agency will work closely with states in implementing the policy.

The requirement that discovery of the violation be voluntary is consistent with proposed federal and state bills which would reward those discoveries that the regulated entity can legitimately attribute to its own voluntary efforts.

The policy gives three specific examples of discovery that would not be voluntary, and therefore would not be eligible for penalty mitigation: emissions violations detected through a required continuous emissions monitor, violations of NPDES discharge limits found through prescribed monitoring,

and violations discovered through a compliance audit required to be performed by the terms of a consent order or settlement agreement.

The final policy generally applies to any violation that is voluntarily discovered, regardless of whether the violation is required to be reported. This definition responds to comments pointing out that reporting requirements are extensive, and that excluding them from the policy's scope would severely limit the incentive for self-policing (see, e.g., II–C–48 in the Docket).

The Agency wishes to emphasize that the integrity of federal environmental law depends upon timely and accurate reporting. The public relies on timely and accurate reports from the regulated community, not only to measure compliance but to evaluate health or environmental risk and gauge progress in reducing pollutant loadings. EPA expects the policy to encourage the kind of vigorous self-policing that will serve these objectives, and not to provide an excuse for delayed reporting. Where violations of reporting requirements are voluntarily discovered, they must be promptly reported (as discussed below). Where a failure to report results in imminent and substantial endangerment or serious harm, that violation is not covered under this policy (see Condition D(8)). The policy also requires the regulated entity to prevent recurrence of the violation, to ensure that noncompliance with reporting requirements is not repeated. EPA will closely scrutinize the effect of the policy in furthering the public interest in timely and accurate reports from the regulated community.

Under Section D(4), disclosure of the violation should be made within 10 days of its discovery, and in writing to EPA. Where a statute or regulation requires reporting be made in less than 10 days, disclosure should be made within the time limit established by law. Where reporting within ten days is not practical because the violation is complex and compliance cannot be determined within that period, the Agency may accept later disclosures if the circumstances do not present a serious threat and the regulated entity meets its burden of showing that the additional time was needed to determine compliance status.

This condition recognizes that it is critical for EPA to get timely reporting of violations in order that it might have clear notice of the violations and the opportunity to respond if necessary, as well as an accurate picture of a given facility's compliance record. Prompt disclosure is also evidence of the regulated entity's good faith in wanting o achieve or return to compliance as oon as possible.

In the final policy, the Agency has added the words, "or may have occurred," to the sentence, "The regulated entity fully discloses that a specific violation has occurred, or may have occurred * * *." This change, which was made in response to comments received, clarifies that where an entity has some doubt about the existence of a violation, the recommended course is for it to disclose and allow the regulatory authorities to make a definitive determination.

In general, the Freedom of Information Act will govern the Agency's release of disclosures made pursuant to this policy. EPA will, independently of FOIA, make publicly available any compliance agreements reached under the policy (see Section H of the policy), as well as descriptions of due diligence programs submitted under Section D.1 of the Policy. Any material claimed to be Confidential Business Information will be treated in accordance with EPA regulations at 40 C.F.R. Part 2.

### 3. Discovery and Disclosure Independent of Government or Third Party Plaintiff

Under Section D(3), in order to be "voluntary", the violation must be identified and disclosed by the regulated entity prior to: the commencement of a federal state or local agency inspection, investigation, or information request; notice of a citizen suit; legal complaint by a third party; the reporting of the violation to EPA by a "whistleblower" employee; and imminent discovery of the violation by a regulatory agency.

This condition means that regulated entities must have taken the initiative to find violations and promptly report them, rather than reacting to knowledge of a pending enforcement action or third-party complaint. This concept was reflected in the interim policy and in federal and state penalty immunity laws and did not prove controversial in the public comment process.

### 4. Correction and Remediation

Section D(5) ensures that, in order to receive the penalty mitigation benefits available under the policy, the regulated entity not only voluntarily discovers and promptly discloses a violation, but expeditiously corrects it, remedies any harm caused by that violation (including responding to any spill and carrying out any removal or remedial action required by law), and expeditiously certifies in writing to appropriate state, local and EPA

authorities that violations have been corrected. It also enables EPA to ensure that the regulated entity will be publicly accountable for its commitments through binding written agreements, orders or consent decrees where necessary.

The final policy requires the violation to be corrected within 60 days, or that the regulated entity provide written notice where violations may take longer to correct. EPA recognizes that some violations can and should be corrected immediately, while others (e.g., where capital expenditures are involved), may take longer than 60 days to correct. In all cases, the regulated entity will be expected to do its utmost to achieve or return to compliance as expeditiously as possible.

Where correction of the violation depends upon issuance of a permit which has been applied for but not issued by federal or state authorities, the Agency will, where appropriate, make reasonable efforts to secure timely review of the permit.

### 5. Prevent Recurrence

Under Section D(6), the regulated entity must agree to take steps to prevent a recurrence of the violation, including but not limited to improvements to its environmental auditing or due diligence efforts. The final policy makes clear that the preventive steps may include improvements to a regulated entity's environmental auditing or due diligence efforts to prevent recurrence of the violation.

In the interim policy, the Agency required that the entity implement appropriate measures to prevent a recurrence of the violation, a requirement that operates prospectively. However, a separate condition in the interim policy also required that the violation not indicate "a failure to take appropriate steps to avoid repeat or recurring violations"—a requirement that operates retrospectively. In the interest of both clarity and fairness, the Agency has decided for purposes of this condition to keep the focus prospective and thus to require only that steps be taken to prevent recurrence of the violation after it has been disclosed.

### 6. No Repeat Violations

In response to requests from commenters (see, e.g., II–F–39 and II–G–18 in the Docket), EPA has established "bright lines" to determine when previous violations will bar a regulated entity from obtaining relief under this policy. These will help protect the public and responsible companies by ensuring that penalties are not waived

for repeat offenders. Under condition D(7), the same or closely-related violation must not have occurred previously within the past three years at the same facility, or be part of a pattern of violations on the regulated entity's part over the past five years. This provides companies with a continuing incentive to prevent violations, without being unfair to regulated entities responsible for managing hundreds of facilities. It would be unreasonable to provide unlimited amnesty for repeated violations of the same requirement.

The term "violation" includes any violation subject to a federal or state civil judicial or administrative order, consent agreement, conviction or plea agreement. Recognizing that minor violations are sometimes settled without a formal action in court, the term also covers any act or omission for which the regulated entity has received a penalty reduction in the past. Together, these conditions identify situations in which the regulated community has had clear notice of its noncompliance and an opportunity to correct.

### 7. Other Violations Excluded

Section D(8) makes clear that penalty reductions are not available under this policy for violations that resulted in serious actual harm or which may have presented an imminent and substantial endangerment to public health or the environment. Such events indicate a serious failure (or absence) of a self-policing program, which should be designed to prevent such risks, and it would seriously undermine deterrence to waive penalties for such violations. These exceptions are responsive to suggestions from public interest organizations, as well as other commenters. (See. e.g., II–F–39 and II–G–18 in the Docket.)

The final policy also excludes penalty reductions for violations of the specific terms of any order, consent agreement. or plea agreement. (See. II–E–60 in the Docket.) Once a consent agreement has been negotiated, there is little incentive to comply if there are no sanctions for violating its specific requirements. The exclusion in this section applies to violations of the terms of any response. removal or remedial action covered by a written agreement.

### 8. Cooperation

Under Section D(9), the regulated entity must cooperate as required by EPA and provide information necessary to determine the applicability of the policy. This condition is largely unchanged from the interim policy. In the final policy, however, the Agency has added that "cooperation" includes

**66710**     **Federal Register** / Vol. 60, No. 246 / Friday, December 22, 1995 / Notices

assistance in determining the facts of any related violations suggested by the disclosure, as well as of the disclosed violation itself. This was added to allow the agency to obtain information about any violations indicated by the disclosure, even where the violation is not initially identified by the regulated entity.

### F. Opposition to Privilege

The Agency remains firmly opposed to the establishment of a statutory evidentiary privilege for environmental audits for the following reasons:

1. Privilege, by definition, invites secrecy, instead of the openness needed to build public trust in industry's ability to self-police. American law reflects the high value that the public places on fair access to the facts. The Supreme Court, for example, has said of privileges that, "[w]hatever their origins, these exceptions to the demand for every man's evidence are not lightly created nor expansively construed, for they are in derogation of the search for truth." *United States* v. *Nixon*, 418 U.S. 683 (1974). Federal courts have unanimously refused to recognize a privilege for environmental audits in the context of government investigations. See, *e.g.*, *United States* v. *Dexter*, 132 F.R.D. 8, 9–10 (D.Conn. 1990) (application of a privilege "would effectively impede [EPA's] ability to enforce the Clean Water Act, and would be contrary to stated public policy.")

2. Eighteen months have failed to produce any evidence that a privilege is needed. Public testimony on the interim policy confirmed that EPA rarely uses audit reports as evidence. Furthermore, surveys demonstrate that environmental auditing has expanded rapidly over the past decade without the stimulus of a privilege. Most recently, the 1995 Price Waterhouse survey found that those few large or mid-sized companies that do not audit generally do not perceive any need to; concern about confidentiality ranked as one of the least important factors in their decisions.

3. A privilege would invite defendants to claim as "audit" material almost any evidence the government needed to establish a violation or determine who was responsible. For example, most audit privilege bills under consideration in federal and state legislatures would arguably protect factual information—such as health studies or contaminated sediment data—and not just the conclusions of the auditors. While the government might have access to required monitoring data under the law, as some industry commenters have suggested, a privilege of that nature would cloak underlying facts needed to determine whether such data were accurate.

4. An audit privilege would breed litigation, as both parties struggled to determine what material fell within its scope. The problem is compounded by the lack of any clear national standard for audits. The *"in camera"* (i.e., non-public) proceedings used to resolve these disputes under some statutory schemes would result in a series of time-consuming, expensive mini-trials.

5. The Agency's policy eliminates the need for any privilege as against the government, by reducing civil penalties and criminal liability for those companies that audit, disclose and correct violations. The 1995 Price Waterhouse survey indicated that companies would expand their auditing programs in exchange for the kind of incentives that EPA provides in its policy.

6. Finally, audit privileges are strongly opposed by the law enforcement community, including the National District Attorneys Association, as well as by public interest groups. (See, *e.g.*, Docket, II–C–21, II–C–28, II–C–52, IV–G–10, II–C–25, II–C–33, II–C–52, II–C–48, and II–G–13 through II–G–24.)

### G. Effect on States

The final policy reflects EPA's desire to develop fair and effective incentives for self-policing that will have practical value to states that share responsibility for enforcing federal environmental laws. To that end, the Agency has consulted closely with state officials in developing this policy, through a series of special meetings and conference calls in addition to the extensive opportunity for public comment. As a result, EPA believes its final policy is grounded in common-sense principles that should prove useful in the development of state programs and policies.

As always, states are encouraged to experiment with different approaches that do not jeopardize the fundamental national interest in assuring that violations of federal law do not threaten the public health or the environment, or make it profitable not to comply. The Agency remains opposed to state legislation that does not include these basic protections, and reserves its right to bring independent action against regulated entities for violations of federal law that threaten human health or the environment, reflect criminal conduct or repeated noncompliance, or allow one company to make a substantial profit at the expense of its law-abiding competitors. Where a state has obtained appropriate sanctions needed to deter such misconduct, there is no need for EPA action.

### H. Scope of Policy

EPA has developed this document as a policy to guide settlement actions. EPA employees will be expected to follow this policy, and the Agency will take steps to assure national consistency in application. For example, the Agency will make public any compliance agreements reached under this policy, in order to provide the regulated community with fair notice of decisions and greater accountability to affected communities. Many in the regulated community recommended that the Agency convert the policy into a regulation because they felt it might ensure greater consistency and predictability. While EPA is taking steps to ensure consistency and predictability and believes that it will be successful, the Agency will consider this issue and will provide notice if it determines that a rulemaking is appropriate.

### II. Statement of Policy: Incentives for Self-Policing

*Discovery, Disclosure, Correction and Prevention*

#### A. Purpose

This policy is designed to enhance protection of human health and the environment by encouraging regulated entities to voluntarily discover, disclose, correct and prevent violations of federal environmental requirements.

#### B. Definitions

For purposes of this policy, the following definitions apply:

"Environmental Audit" has the definition given to it in EPA's 1986 audit policy on environmental auditing, i.e., "a systematic, documented, periodic and objective review by regulated entities of facility operations and practices related to meeting environmental requirements."

"Due Diligence" encompasses the regulated entity's systematic efforts, appropriate to the size and nature of its business, to prevent, detect and correct violations through all of the following:

(a) Compliance policies, standards and procedures that identify how employees and agents are to meet the requirements of laws, regulations, permits and other sources of authority for environmental requirements;

(b) Assignment of overall responsibility for overseeing compliance with policies, standards, and procedures, and assignment of specific responsibility for assuring compliance at each facility or operation;

Federal Register / Vol. 60, No. 246 / Friday, December 22, 1995 / Notices        **66711**

(c) Mechanisms for systematically assuring that compliance policies, standards and procedures are being carried out, including monitoring and auditing systems reasonably designed to detect and correct violations, periodic evaluation of the overall performance of the compliance management system, and a means for employees or agents to report violations of environmental requirements without fear of retaliation;

(d) Efforts to communicate effectively the regulated entity's standards and procedures to all employees and other agents;

(e) Appropriate incentives to managers and employees to perform in accordance with the compliance policies, standards and procedures, including consistent enforcement through appropriate disciplinary mechanisms; and

(f) Procedures for the prompt and appropriate correction of any violations, and any necessary modifications to the regulated entity's program to prevent future violations.

"Environmental audit report" means the analysis, conclusions, and recommendations resulting from an environmental audit, but does not include data obtained in, or testimonial evidence concerning, the environmental audit.

"Gravity-based penalties" are that portion of a penalty over and above the economic benefit., i.e., the punitive portion of the penalty, rather than that portion representing a defendant's economic gain from non-compliance. (For further discussion of this concept, see "A Framework for Statute-Specific Approaches to Penalty Assessments", #GM–22, 1980, U.S. EPA General Enforcement Policy Compendium).

"Regulated entity" means any entity, including a federal, state or municipal agency or facility, regulated under federal environmental laws.

C. Incentives for Self-Policing

1. No Gravity-Based Penalties

Where the regulated entity establishes that it satisfies all of the conditions of Section D of the policy, EPA will not seek gravity-based penalties for violations of federal environmental requirements.

2. Reduction of Gravity-Based Penalties by 75%

EPA will reduce gravity-based penalties for violations of federal environmental requirements by 75% so long as the regulated entity satisfies all of the conditions of Section D(2) through D(9) below.

3. No Criminal Recommendations

(a) EPA will not recommend to the Department of Justice or other prosecuting authority that criminal charges be brought against a regulated entity where EPA determines that all of the conditions in Section D are satisfied, so long as the violation does not demonstrate or involve:

(i) a prevalent management philosophy or practice that concealed or condoned environmental violations; or

(ii) high-level corporate officials' or managers' conscious involvement in, or willful blindness to, the violations.

(b) Whether or not EPA refers the regulated entity for criminal prosecution under this section, the Agency reserves the right to recommend prosecution for the criminal acts of individual managers or employees under existing policies guiding the exercise of enforcement discretion.

4. No Routine Request for Audits

EPA will not request or use an environmental audit report to initiate a civil or criminal investigation of the entity. For example, EPA will not request an environmental audit report in routine inspections. If the Agency has independent reason to believe that a violation has occurred, however, EPA may seek any information relevant to identifying violations or determining liability or extent of harm.

D. Conditions

1. Systematic Discovery

The violation was discovered through:
(a) an environmental audit; or
(b) an objective, documented, systematic procedure or practice reflecting the regulated entity's due diligence in preventing, detecting, and correcting violations. The regulated entity must provide accurate and complete documentation to the Agency as to how it exercises due diligence to prevent, detect and correct violations according to the criteria for due diligence outlined in Section B. EPA may require as a condition of penalty mitigation that a description of the regulated entity's due diligence efforts be made publicly available.

2. Voluntary Discovery

The violation was identified voluntarily, and not through a legally mandated monitoring or sampling requirement prescribed by statute, regulation, permit, judicial or administrative order, or consent agreement. For example, the policy does not apply to:
(a) emissions violations detected through a continuous emissions monitor (or alternative monitor established in a permit) where any such monitoring is required;
(b) violations of National Pollutant Discharge Elimination System (NPDES) discharge limits detected through required sampling or monitoring;
(c) violations discovered through a compliance audit required to be performed by the terms of a consent order or settlement agreement.

3. Prompt Disclosure

The regulated entity fully discloses a specific violation within 10 days (or such shorter period provided by law) after it has discovered that the violation has occurred, or may have occurred, in writing to EPA;

4. Discovery and Disclosure Independent of Government or Third Party Plaintiff

The violation must also be identified and disclosed by the regulated entity prior to:
(a) the commencement of a federal, state or local agency inspection or investigation, or the issuance by such agency of an information request to the regulated entity;
(b) notice of a citizen suit;
(c) the filing of a complaint by a third party;
(d) the reporting of the violation to EPA (or other government agency) by a "whistleblower" employee, rather than by one authorized to speak on behalf of the regulated entity; or
(e) imminent discovery of the violation by a regulatory agency;

5. Correction and Remediation

The regulated entity corrects the violation within 60 days, certifies in writing that violations have been corrected, and takes appropriate measures as determined by EPA to remedy any environmental or human harm due to the violation. If more than 60 days will be needed to correct the violation(s), the regulated entity must so notify EPA in writing before the 60-day period has passed. Where appropriate, EPA may require that to satisfy conditions 5 and 6, a regulated entity enter into a publicly available written agreement, administrative consent order or judicial consent decree, particularly where compliance or remedial measures are complex or a lengthy schedule for attaining and maintaining compliance or remediating harm is required;

6. Prevent Recurrence

The regulated entity agrees in writing to take steps to prevent a recurrence of the violation, which may include improvements to its environmental auditing or due diligence efforts;

## 7. No Repeat Violations

The specific violation (or closely related violation) has not occurred previously within the past three years at the same facility, or is not part of a pattern of federal, state or local violations by the facility's parent organization (if any), which have occurred within the past five years. For the purposes of this section, a violation is:

(a) any violation of federal, state or local environmental law identified in a judicial or administrative order, consent agreement or order, complaint, or notice of violation, conviction or plea agreement; or

(b) any act or omission for which the regulated entity has previously received penalty mitigation from EPA or a state or local agency.

## 8. Other Violations Excluded

The violation is not one which (i) resulted in serious actual harm, or may have presented an imminent and substantial endangerment to, human health or the environment, or (ii) violates the specific terms of any judicial or administrative order, or consent agreement.

## 9. Cooperation

The regulated entity cooperates as requested by EPA and provides such information as is necessary and requested by EPA to determine applicability of this policy. Cooperation includes, at a minimum, providing all requested documents and access to employees and assistance in investigating the violation, any noncompliance problems related to the disclosure, and any environmental consequences related to the violations.

## E. Economic Benefit

EPA will retain its full discretion to recover any economic benefit gained as a result of noncompliance to preserve a "level playing field" in which violators do not gain a competitive advantage over regulated entities that do comply. EPA may forgive the entire penalty for violations which meet conditions 1 through 9 in section D and, in the Agency's opinion, do not merit any penalty due to the insignificant amount of any economic benefit.

## F. Effect on State Law, Regulation or Policy

EPA will work closely with states to encourage their adoption of policies that reflect the incentives and conditions outlined in this policy. EPA remains firmly opposed to statutory environmental audit privileges that shield evidence of environmental violations and undermine the public's right to know, as well as to blanket immunities for violations that reflect criminal conduct, present serious threats or actual harm to health and the environment, allow noncomplying companies to gain an economic advantage over their competitors, or reflect a repeated failure to comply with federal law. EPA will work with states to address any provisions of state audit privilege or immunity laws that are inconsistent with this policy, and which may prevent a timely and appropriate response to significant environmental violations. The Agency reserves its right to take necessary actions to protect public health or the environment by enforcing against any violations of federal law.

## G. Applicability

(1) This policy applies to the assessment of penalties for any violations under all of the federal environmental statutes that EPA administers, and supersedes any inconsistent provisions in media-specific penalty or enforcement policies and EPA's 1986 Environmental Auditing Policy Statement.

(2) To the extent that existing EPA enforcement policies are not inconsistent, they will continue to apply in conjunction with this policy. However, a regulated entity that has received penalty mitigation for satisfying specific conditions under this policy may not receive additional penalty mitigation for satisfying the same or similar conditions under other policies for the same violation(s), nor will this policy apply to violations which have received penalty mitigation under other policies.

(3) This policy sets forth factors for consideration that will guide the Agency in the exercise of its prosecutorial discretion. It states the Agency's views as to the proper allocation of its enforcement resources. The policy is not final agency action, and is intended as guidance. It does not create any rights, duties, obligations, or defenses, implied or otherwise, in any third parties.

(4) This policy should be used whenever applicable in settlement negotiations for both administrative and civil judicial enforcement actions. It is not intended for use in pleading, at hearing or at trial. The policy may be applied at EPA's discretion to the settlement of administrative and judicial enforcement actions instituted prior to, but not yet resolved, as of the effective date of this policy.

## H. Public Accountability

(1) Within 3 years of the effective date of this policy, EPA will complete a study of the effectiveness of the policy in encouraging:

(a) changes in compliance behavior within the regulated community, including improved compliance rates;

(b) prompt disclosure and correction of violations, including timely and accurate compliance with reporting requirements;

(c) corporate compliance programs that are successful in preventing violations, improving environmental performance, and promoting public disclosure;

(d) consistency among state programs that provide incentives for voluntary compliance.

EPA will make the study available to the public.

(2) EPA will make publicly available the terms and conditions of any compliance agreement reached under this policy, including the nature of the violation, the remedy, and the schedule for returning to compliance.

## I. Effective Date

This policy is effective January 22, 1996.

Dated: December 18, 1995.

**Steven A. Herman,**

*Assistant Administrator for Enforcement and Compliance Assurance.*

[FR Doc. 95-31146 Filed 12-21-95; 8:45 am]

BILLING CODE 6560-50-P

# Appendix C

# SAMPLE HAZARDOUS WASTE AUDIT PROTOCOL

**(Courtesy of the ERM Group)**

# HAZARDOUS WASTE

## TABLE OF CONTENTS

**Part 1**  Introduction...............................................HW/Part 1 - 1

Applicability of this Module................................................HW/Part 1 - 1

Federal, State, and Local Regulations .....................................HW/Part 1 - 1

Key Compliance Definitions ................................................HW/Part 1 - 4

Table 1: Maximum concentration of contaminants for the
"toxicity" characteristic ("D" list)............................................ HW/Part 1 - 10

Table 2: Hazardous waste generated by generic
processes ("F" list)................................................................ HW/Part 1 - 11

Table 3: Hazardous waste from specific sources ("K" list)............ HW/Part 1 - 15

Table 4: Hazardous waste "P" list............................................ HW/Part 1 - 19

Table 5: Hazardous waste "U" list........................................... HW/Part 1 - 24

**Part 2**  Previsit Preparation .......................................HW/Part 2 - 1

**Part 3**  Hazardous Waste Rulebook..............................HW/Part 3 - 1

1.   General ...........................................................HW/Part 3 - 1

2.   Generators: General................................................HW/Part 3 - 1

3.   Generators: Hazardous Waste Laboratories .....................HW/Part 3 - 3

4.   Generators: Transportation and Manifesting ....................HW/Part 3 - 4

5.   Generators: Storage of Wastes....................................HW/Part 3 - 7

6.   Generators: Inspections............................................. HW/Part 3 - 14

7.   Generators: Emergency Preparedness and Prevention ....... HW/Part 3 - 15

8.   Generators: Training ............................................... HW/Part 3 - 17

9.   Generators: Land Disposal ....................................... HW/Part 3 - 18

10.  TSD Facilities: General ........................................... HW/Part 3 - 20

11.  TSD Facilities: Waste Identification ............................ HW/Part 3 - 21

12.  TSD Facilities: Transportation and Manifesting............... HW/Part 3 - 22

13.  TSD Facilities: Storage of Wastes ............................... HW/Part 3 - 24

14.  TSD Facilities:  Training.......................................... HW/Part 3 - 31

15.  TSD Facilities: Emergency Preparedness and Prevention ... HW/Part 3 - 32

16.  TSD Facilities: Recordkeeping.................................... HW/Part 3 - 36

17.  TSD Facilities: Groundwater Monitoring...................... HW/Part 3 - 37

18.  TSD Facilities: Land Disposal for Treatment Facilities....... HW/Part 3 - 39

19.  TSD Facilities: Land Disposal for Land Disposal
Facilities .............................................................. HW/Part 3 - 40

20.  TSD Facilities: Closure/Post-Closure .......................... HW/Part 3 - 40

21. TSD Facilities: Financial Assurance and Liability Insurance............................................................HW/Part 3 - 41
22. TSD Facilities: Location and Construction......................HW/Part 3 - 42
23. TSD Facilities: Surface Impoundments.........................HW/Part 3 - 42
24. TSD Facilities: Waste Piles ......................................HW/Part 3 - 44
25. TSD Facilities: Landfills...........................................HW/Part 3 - 45
26. TSD Facilities: Incinerators......................................HW/Part 3 - 47
27. TSD Facilities: Thermal Treatment.............................HW/Part 3 - 49
28. TSD Facilities: Boilers and Industrial Furnaces ...............HW/Part 3 - 51
29. TSD Facilities: Chemical, Physical, and Biological Treatments .............................................................HW/Part 3 - 52
30. TSD Facilities: Containment Buildings.........................HW/Part 3 - 54
31. TSD Facilities: Land Treatment.................................HW/Part 3 - 55
32. TSD Facilities: Management of Remediation Wastes .........HW/Part 3 - 56
33. TSD Facilities: Drip Pads .......................................HW/Part 3 - 57
34. TSD Facilities: Miscellaneous Units ...........................HW/Part 3 - 58
35. TSD Facilities: Air Emissions Standards.......................HW/Part 3 - 58
36. Universal Wastes .................................................HW/Part 3 - 59

# HAZARDOUS WASTE

# PART 1: INTRODUCTION

## Applicability of this Module

This module applies to facilities that generate, store, treat, or dispose of any type of hazardous waste. This module is necessarily more complex than others as all evaluation items will not be applicable to all facilities; guidance provided in the worksheets directs the evaluator to the rulebook sections related to the type of hazardous waste activities/facilities at the site. This module does not cover issues specific to radioactive or mixed hazardous/radioactive waste. Medical waste and used oil are not covered here, but in the solid waste module. The underground injection of hazardous waste under the Clean Water Act (CWA) is dealt with in the wastewater module, while treatment of polychlorinated biphenyls (PCBs) under the Toxic Substances Control Act (TSCA) is covered in the module on special pollutants.

## Federal, State, and Local Regulations

The Resource Conservation and Recovery Act (RCRA) and the Hazardous and Solid Waste Amendments of 1984 (HSWA) are the enabling legislation that authorize federal hazardous waste regulations. These federal laws have been codified in 49 USC 6901 - 6965. The federal regulations that implement these laws are organized into the following major sections:

Σ 40 CFR 260   Hazardous Waste Management System: General;
Σ 40 CFR 261   Identification and Listing of Hazardous Waste;
Σ 40 CFR 262   Standards Applicable to Generators of Hazardous Waste;
Σ 40 CFR 263   Standards Applicable to Transporters of Hazardous Waste;
Σ 40 CFR 264   Standards for Owners and Operators of Hazardous Waste Treatment, Storage and Disposal Facilities;
Σ 40 CFR 265   Interim Status Standards for Owners and Operators of Hazardous Waste Treatment Storage and Disposal Facilities.;
Σ 40 CFR 266   Standards for the Management of Specific Hazardous Wastes and Specific Types of Hazardous Waste Management Facilities;
Σ 40 CFR 267   Interim Standards for Owners and Operators of New Hazardous Waste Land Disposal Facilities;
Σ 40 CFR 268   Land Disposal Restrictions;
Σ 40 CFR 270   EPA Administered Permit Programs: The Hazardous Waste Permit Program
Σ 40 CFR 271   Requirements for Authorization of State Hazardous Waste Programs;
Σ 40 CFR 280   Technical Standards and Corrective Action Requirements for Owners of Underground Storage Tanks.

## *Intro to Hazardous Waste*

All hazardous waste generators and facilities must comply with these regulations unless a state has an authorized hazardous waste management program and associated regulations. Many states have met U.S. Environmental Protection Agency (EPA) requirements in 40 CFR 271 and have been authorized to manage their own state programs. RCRA encourages states to develop their own hazardous waste statutes and to operate regulatory programs in lieu of the federal EPA-managed program. Many of the states have adopted the EPA regulations by reference or have promulgated regulations that are identical to EPA regulations. Several other states have developed hazardous waste regulatory programs that are substantially equivalent to the federal program, and a few states have implemented programs that are significantly more stringent than the EPA program. In addition, EPA may directly enforce new regulations authorized by HSWA that do not have equivalent state requirements.

This difference between individual state regulations and the federal program requires that auditors check the status of the state's authorization and then determine which regulations apply. Because the worksheets are based exclusively on the requirements of the federal RCRA/EPA program, it is necessary to determine in what ways the applicable state program differs from the RCRA/EPA program.

### Generator Requirements

Any facility that generates at least 100 kilograms per month (100 kg/month; equivalent to 220 pounds or half of a 45-gallon drum) of hazardous waste or 1 kg/month of acutely hazardous waste is required to analyze the waste, properly store or otherwise manage it onsite, prepare transportation manifests, and properly dispose of it offsite. Other requirements include preparedness and prevention, the preparation of a contingency plan, and the provision of worker training.

Hazardous wastes may either be wastes that appear in 40 CFR 261 as "listed" wastes (F, K, U, or P codes) or demonstrate characteristics of ignitability (having a flashpoint less than 140°F), corrosivity (having a pH level less than 2.0 or greater that 12.5), reactivity, or toxicity (exceeding Extraction Procedure (EP)/Toxic Characteristic Leaching Procedure (TCLP) toxicity limits).

A generator of hazardous waste must have an EPA identification (ID) number but does not need a permit if waste is stored for less than 90 days. Small quantity generators (SQGs) and conditionally exempt small quantity generators (CESQGs) may qualify for exceptions to the full scope of the regulations.

### Accumulation Point Management

An accumulation point is an area in or near the workplace where hazardous waste is accumulated prior to disposal. Storage in these areas is temporary and must not exceed 90 days from the time the first waste begins to accumulate in a container. Storage is also limited in quantity to 55 gallons of hazardous waste or 1 quart of acutely hazardous waste in any 1 location. Permits are not required for accumulation points, but certain controls relative to spill containment, labeling, inspections, and training are required.

**Transport Requirements**

Containers of hazardous waste shipped offsite must be properly labeled. The labels should identify the waste and its hazard class. Shipments must also be accompanied by manifests and are subject to the full transportation requirements as stipulated in Department of Transportation (DOT) hazardous materials transportation regulations (49 CFR). DOT packaging requirements can be found in the HM-181 standards of the DOT Regulations.

**Permitted Treatment, Storage, and Disposal (TSD) Facility Requirements**

The operation of a treatment, storage, and/or disposal system is subject to regulation and permitting under federal or state regulations. These regulations are both administrative and technical in nature. The administrative standards require that various plans be developed to ensure that emergencies can be dealt with, that waste received is properly identified and tracked, and that operating personnel are adequately trained to operate the facility and respond to emergencies. These administrative standards also include requirements that the facility be inspected routinely, that records of operations are compiled and maintained, and that reports of both routine and contingency operations are made to the applicable regulatory agency. The administrative standards also require that a plan for ceasing operations and closing the facility be developed, kept on-hand, and updated at least annually.

The technical standards that are applicable to TSD facilities fall into 2 classes: general standards that apply to all TSD facilities, and specific standards that apply to various types of facilities, i.e., container storage areas, tanks, surface impoundments, waste piles, land treatment facilities, incinerators, landfills, thermal treatment facilities, and chemical, physical. or biological treatment facilities.

Administrative and technical standards are applied to a particular facility through a RCRA permit issued to a facility. Existing facilities that were not issued a RCRA permit when the regulations were promulgated in 1980 were considered to have interim status and allowed to continue to operations if they complied with RCRA-mandated Interim Status Standards (ISS). These ISS (which are contained in 40 CFR 265) are similar in scope to the permit standards contained in 40 CFR 264 but are generally less stringent and required fewer facility modifications or improvements. TSD facilities with land disposal or incineration units were required to submit Part B permit applications or withdraw their interim status by specified dates. Surface impoundments became subject to new facility standards on a specified date (HSWA 3005). Most interim status facilities were required to be permitted or commence closure by November 9, 1992. However, some facilities still operate under interim status.

For compliance planning purposes, environmental managers should note that EPA has recently proposed changes to closure and post-closure requirements (59 *Federal Register* 55778, November 4, 1994). Changes in the definitions of hazardous and solid waste are under development. Consult the *Federal Register* for further information.

*ntro to Hazardous Waste*

| Key Compliance Definitions |
|---|

Most of these definitions were obtained from the federal regulations cited previously. Those erms not specifically defined in the regulations are defined here as they are used in this nodule.

*Aboveground Tank or Aboveground Storage Tank (AST)* - A device that meets the definition of "tank" in 40 CFR 260.10 and is situated in such a way that the entire surface area of the tank s completely above the plane of the adjacent surrounding surface, and the entire surface area of he tank (including the tank bottom) can be visually inspected (40 CFR 260.10).

*Accumulation Point* - A designated area where hazardous waste is temporarily stored in tanks or containers for 90 days or less at a facility (40 CFR 262.34). See also "Satellite Accumulation Area".

*Acute Hazardous Waste* - Any waste listed under 40 CFR 261.31 - 33 with a hazard code of "H". These include EPA hazardous waste numbers F020, F021, F022, F023, F026, and F027, and all P-listed wastes. These waste lists are provided in this introduction (40 CFR 261.30(b) and 261.33(e)).

*Ancillary Equipment* - Any device, including but not limited to such devices as piping, fittings, flanges, valves, and pumps, that is used to distribute, meter, or control the flow of hazardous waste from its point of generation to a storage or treatment tank(s), or between hazardous waste storage and treatment tanks to a point of disposal onsite, or to a point of shipment for disposal offsite (40 CFR 260.10).

*Component* - Either the tank or ancillary equipment of the tank system (40 CFR 260.10).

*Conditionally Exempt Small Quantity Generator (CESQG)* - A generator that produces less than 100 kg/month of hazardous waste and less than 1 kg/month of acutely hazardous waste (40 CFR 261.5).

*Container* - Any portable device in which a material is stored, transported, treated, disposed of, or otherwise handled (40 CFR 260.10).

*Containment Building* - A hazardous waste management unit that is used to store or treat hazardous waste under the provisions of 40 CFR 264 or 265 Subpart DD (40 CFR 260.10).

*Corrective Action Management Unit (CAMU)* - An area within a facility that is designated by the EPA Regional Administrator under 40 CFR 264 Subpart S for the purpose of implementing corrective action requirements under 40 CFR 264.101 and RCRA 3008(h). A CAMU shall only be used for the management of remediation wastes pursuant to implementing such corrective action requirements at the facility (40 CFR 260.10).

*Corrosion Expert* - A person who, by reason of his/her knowledge of the physical sciences and the principles of engineering and mathematics acquired by a professional education and related practical experience, is qualified to engage in the practice of corrosion control on buried or submerged metal piping systems and metal tanks. Such a person must be certified as being qualified by the National Association of Corrosion Engineers (NACE) or be a registered professional engineer who has certification and licensing that includes education and experience in corrosion control and/or buried or submerged metal piping systems or tanks [40 CFR 260.10].

*Disposal* - The discharge, deposit, injection, dumping, spilling, leaking, or placing of any solid waste or hazardous waste into or on any land or water so that such solid waste or hazardous waste or any constituent thereof may enter the environment or be emitted into the air or discharged into any waters, including groundwaters (40 CFR 260.10).

*Disposal Facility* - A facility or part of a facility where hazardous waste is intentionally placed into or on any land or water and will remain after closure. The term disposal facility does not include a CAMU into which remediation wastes are placed (40 CFR 260.10).

*Drip Pad* - An engineered structure consisting of a curbed, free-draining base, constructed of nonearthen materials and designed to convey preservative kick-back or drippage from treated wood, precipitation, and surface water run-on to an associated collection system at wood preserving plants (40 CFR 260.10).

*Elementary Neutralization Unit* - A device used for neutralizing only those hazardous wastes that exhibit corrosivity (as defined in 40 CFR 261.22) or are listed in Subpart D of 40 CFR 261 only because of corrosivity. An elementary neutralization unit must meet the definition in 40 CFR 261.10 of tank, tank system, container, transport vehicle, or vessel.

*Existing Tank System or Existing Component* - A tank system or component, used for the storage or treatment of hazardous waste, that is in operation or for which installation commenced on or prior to July 14, 1986. Installations are considered to have commenced if the owner or operator obtained all federal, state, and local approvals or permits necessary to begin physical construction of the site or installation of the tank system, and if either (40 CFR 260.10):

Σ a continuous onsite physical construction of the site or installation program began; or
Σ the owner or operator entered into contractual obligations for physical construction of the site or installation of the tank system to be completed within a reasonable time, and those obligations could not be canceled or modified without substantial loss.

*Facility* - All contiguous land, structures, other appurtenances, and improvements on the land used for treating, storing, or disposing of hazardous waste. A facility may consist of several treatment, storage, or disposal operational units (e.g., 1 or more landfills, surface impoundments, or combinations of them). For the purpose of implementing corrective action under 40 CFR 264.101, "facility" means all contiguous property under the control of the owner or operator seeking a permit under Subtitle C of RCRA. This definition also applies to facilities implementing corrective action under RCRA 308(h) (40 CFR 260.10).

# Intro to Hazardous Waste

*Generator* - Any person, by site, whose acts or processes produce hazardous waste identified or listed in 40 CFR 261 or whose act first causes a hazardous waste to become subject to regulation (40 CFR 260.10).

*Good Management Practice (GMP)* - Practices that, although not mandated by law, are encouraged to promote safe operating procedures.

*Individual Generation Site* - The contiguous site on which 1 or more hazardous wastes are generated. An individual generation site, such as a large manufacturing plant, may have 1 or more sources of hazardous waste, but is considered a single or individual generation site if the site or property is contiguous (40 CFR 260.10).

*Land Disposal* - Includes, but is not limited to, any placement of hazardous waste in a landfill, surface impoundment, waste pile, injection well, land treatment facility, salt dome formation, salt bed formation, underground mine or cave, or placement in a concrete vault or bunker intended for disposal purposes (40 CFR 268.2).

*Land Treatment Facility* - A facility or part of a facility at which hazardous waste is applied onto or incorporated into the soil surface. Such facilities are disposal facilities if the waste will remain after closure (40 CFR 260.10).

*Landfill* - A disposal facility or part of a facility where hazardous waste is placed in or on land that is not a land treatment facility, a surface impoundment, an injection well, a salt dome or salt bed formation, an underground mine, a cave, or a CAMU (40 CFR 260.10).

*Large Quantity Hanlder of Universal Waste (LQHUW )*– A univeral waste handler who accumulates 5,000 kilograms or more total of universal waste at any time. This designation as an LQHUW is retained through the end of the calendar year in which 5,000 kilograms or more total of universal waste is accumulated (40 CFR 273.6).

*Leak Detection System* - A system capable of detecting the failure of either the primary or secondary containment structure or the presence of a release of hazardous waste or accumulated liquid in the secondary containment structure. Such a system must employ operational controls (e.g., daily visual inspections for releases into the secondary containment system of ASTs) or consist of an interstitial monitoring device designed to detect continuously and automatically the failure of the primary or secondary containment structure or the presence of a release of hazardous waste into the secondary containment structure (40 CFR 260.10).

*Manifest* - The shipping document designated as EPA Form No. 8700-22 and, if necessary, EPA Form No. 8700-22A, or the state equivalents, originated and signed by the generator in accordance with the instructions included in the Appendix to 40 CFR 262 (40 CFR 260.10).

*Manifest Document Number* - The 12-digit EPA ID number assigned to the manifest plus a unique 5-digit document number assigned to the manifest by the generator for recording and reporting purposes (40 CFR 260.10).

*Miscellaneous Unit* - A hazardous waste management unit where hazardous waste is treated, stored, or disposed of and that is not a container, tank, surface impoundment, pile, land treatment unit, landfill, incinerator, boiler, industrial furnace, underground injection well with appropriate technical standards under 40 CFR 146, containment building, CAMU, or unit eligible for a research, development, and demonstration permit under 40 CFR 270.65 (40 CFR 260.10).

*Movement* - Hazardous waste transported to a facility in an individual vehicle (40 CFR 260.10).

*New Tank System or New Tank Component* - A tank system or component that will be used for the storage or treatment of hazardous waste and for which installation commenced after July 14, 1986; except, however, for purposes of 40 CFR 264.193(g)(2) and 265.193(g)(2), a new tank system is 1 for which construction commences after July 14, 1986 (see also "Existing Tank System") (40 CFR 260.10).

*Onground Tank* - A device meeting the definition of "tank" in 40 CFR 260.10 that is situated in such a way that the bottom of the tank is on the same level as the adjacent surrounding surface so that the external tank bottom cannot be visually inspected (40 CFR 260.10).

*Onsite* - The same or geographically contiguous property. This property may be divided by a public or private right-of-way, provided that access between the separate portions is provided by crossing directly as opposed to going along the right-of-way. Noncontiguous properties owned by the same person but connected by a right-of-way that he/she controls and to which the public does not have access are also considered to comprise onsite property (40 CFR 260.10).

*Pile* - Any noncontainerized accumulation of solid, nonflowing hazardous waste that is used for treatment or storage and that is in a containment building (40 CFR 260.10).

*Remediation Wastes* - All solid and hazardous wastes and all media (including groundwater, surface water, soils, and sediments) and debris that contain listed hazardous wastes or themselves exhibit a hazardous waste characteristic, that are managed for the purpose of implementing corrective action requirements under 40 CFR 264.101 and RCRA 3008(h). For a given facility, remediation wastes may originate only from within the facility boundary, but they may include waste managed in implementing RCRA 3004(v) or 3008(h) for releases beyond the facility boundary (40 CFR 260.10).

*Restricted Wastes* - Those categories of hazardous wastes that are prohibited from land disposal either by regulation or by statute (RCRA 3004 and 40 CFR 268).

*Satellite Accumulation Area* - An area where up to 55 gallons of hazardous waste or 1 quart of acutely hazardous waste are initially collected, said area being at or near the area where the waste is generated. Once these limits are reached, the waste must be transferred to the hazardous waste accumulation point, and the 90-day clock begins (see "Accumulation Point") (40 CFR 262.34(c)).

## Intro to Hazardous Waste

*Small Quantity Generator (SQG)* - A generator that produces hazardous waste in quantities of more than 100 kg/month but less than 1,000 kg/month (40 CFR 262.34(d) and 262.44).

*Small Quantity Handler of Universal Waste (SQHUW)* - A universal waste hander who does not accumulate more than 5,000 kilograms total of universal waste at any time (40 CFR 273.6).

*Soil* - Materials that are primarily geologic in origin, such as silt, loam, or clay, and that are indigenous to the natural geologic environment. Soils DO NOT include wastes withdrawn from active hazardous waste management units (40 CFR 268 proposed).

*Storage* - The holding of hazardous waste for a temporary period, at the end of which the hazardous waste is treated, disposed of, or stored elsewhere (40 CFR 260.10).

*Sump* - Any pit or reservoir that meets the definition of tank (including those troughs and trenches connected to it) and serves to collect hazardous waste for transport to hazardous waste TSD facilities. When used in relation to landfill, surface impoundment, and waste pile rules, "sump" means any lined pit or reservoir that serves to collect liquids drained from a leachate collection and removal system or leak detection system for subsequent removal from the system (40 CFR 260.10).

*Surface Impoundment or Impoundments* - A facility or part of a facility that is a natural topographic depression, man-made excavation, or diked area formed primarily of earthen materials (although it may be lined with man-made materials) that is designed to hold an accumulation of liquid wastes or wastes containing free liquids and is not an injection well. Examples of surface impoundments are holding, storage, settling, and aeration pits, ponds, and lagoons (40 CFR 260.10).

*Tank* - A stationary device designed to contain an accumulation of hazardous waste that is constructed primarily of nonearthen materials (e.g., wood, concrete, steel, plastic) that provide structural support (40 CFR 260.10).

*Tank System* - A hazardous waste storage or treatment tank and its associated ancillary equipment and containment system (40 CFR 260.10).

*Thermal Treatment* - The treatment of hazardous waste in a device that uses elevated temperature as the primary means to change the chemical, physical, or biological character or composition of the hazardous waste (40 CFR 260.10).

*Transporter* - A person engaged in the offsite transportation of hazardous wastes by air, rail, highway, or water (40 CFR 260.10).

*Treatability Study* - A study in which a hazardous waste is subjected to a treatment process to determine:

Σ whether the waste is amenable to the treatment process;
Σ what pretreatment, if any, is required;
Σ the optimal process conditions needed to achieve the desired treatment;
Σ the efficiency of a treatment process for a specific waste or wastes; or
Σ the characteristics and volumes of residuals from a particular treatment process.

*Treatment* - Any method, technique, or process, including neutralization, designed to change the physical, chemical, or biological character or composition of any hazardous waste so as to neutralize such waste, or so as to recover energy or material resources from the waste, or so as to render such waste nonhazardous or less hazardous; safer to transport, store, or dispose of; or amenable for recovery or storage, or to reduce it in volume (40 CFR 260.10).

*Treatment Zone* - A soil area of the unsaturated zone of a land treatment unit within which hazardous constituents are degraded, transformed, or immobilized (40 CFR 260.10).

*Underground Injection* - The subsurface emplacement of fluids through a bored, drilled, or driven well, or through a dug well where the depth of the dug well is greater than the largest surface dimension (see also "Injection Well") (40 CFR 260.10).

*Underground Tank or Underground Storage Tank (UST)* - A device meeting the definition of "tank" in 40 CFR 260.10 whose entire surface area is totally below the surface of and covered by the ground (40 CFR 260.10).

*Universal Waste* - Any hazardous wastes that are subject to universal waste requirements, including batteries, pesticides, and thermostats (40 CFR 273.6).

*Unfit-for-Use Tank System* - A tank system that has been determined through an integrity assessment or other inspection to be no longer capable of storing or treating hazardous waste without posing a threat of release of hazardous waste to the environment (40 CFR 260.10).

*Wastewater Treatment Unit* - A device that:

Σ is part of a wastewater treatment facility that is subject to regulation under either Section 402 or 307(b) of CAA;

Σ meets the definition of tank or tank system in 40 CFR 260.10; and

Σ does 1 of the following:

    1) receives and treats or stores an influent wastewater that is a hazardous waste as defined in 40 CFR 261.3;

    2) generates and accumulates a wastewater treatment sludge that is a hazardous waste as defined in 40 CFR 261.3; or

    3) treats or stores a wastewater treatment sludge that is a hazardous waste as defined in 40 CFR 261.3.

*Zone of Engineering Control* - An area under the control of a facility owner/operator that, upon detection of a hazardous waste release, can be readily cleaned up prior to the release of hazardous waste or hazardous constituents to groundwater or surface water (40 CFR 260.10).

## *Intro to Hazardous Waste*

| Table 1: Maximum concentration of contaminants for the "toxicity" characteristic ("D" list) | | | |
|---|---|---|---|
| HW No. | Contaminant | CAS No. | Regulatory Level (mg/L) |
| D004 | Arsenic | 7440-38-2 | 5.0 |
| D005 | Barium | 7440-39-3 | 100.0 |
| D0018 | Benzene | 71-43-2 | 0.5 |
| D006 | Cadmium | 7440-43-9 | 1.0 |
| D019 | Carbon tetrachloride | 56-23-5 | 0.5 |
| D020 | Chlordane | 57-74-9 | 0.03 |
| D021 | Chlorobenzene | 108-90-7 | 100.0 |
| D022 | Chloroform | 67-66-3 | 6.0 |
| D007 | Chromium | 7440-47-3 | 5.0 |
| D023 | o-Cresol | 95-48-7 | 200.0** |
| D024 | m-Cresol | 108-39-4 | 200.0** |
| D025 | p-Cresol | 106-44-5 | 200.0** |
| D026 | Cresol | ---------- | 200.0** |
| D016 | 2,4-D | 94-75-7 | 10.0 |
| D027 | 1,4-Dichlorobenzene | 106-46-7 | 7.5 |
| D028 | 1,2-Dichloroethane | 107-06-2 | 0.5 |
| D029 | 1,1-Dichloroethylene | 75-35-4 | 0.7 |
| D030 | 2,4-Dinitrotoluene | 121-14-2 | 0.13* |
| D012 | Endrin | 72-20-8 | 0.02 |
| D031 | Heptachlor | 76-44-8 | 0.008 |
| D032 | Hexachlorobenzene | 118-74-1 | 0.13* |
| D033 | Hexachlorobutadiene | 87-68-3 | 0.5 |
| D034 | Hexachloroethane | 67-72-1 | 3.0 |
| D008 | Lead | 7439-92-1 | 5.0 |
| D013 | Lindane | 58-89-9 | 0.4 |
| D009 | Mercury | 7439-97-6 | 0.2 |
| D014 | Methoxychlor | 72-43-5 | 10.0 |
| D035 | Methyl ethyl ketone | 78-93-3 | 200.0 |
| D036 | Nitrobenzene | 98-95-3 | 2.0 |
| D037 | Pentachlorophenol | 87-86-5 | 100.0 |
| D038 | Pyridine | 110-86-1 | 5.0* |
| D010 | Selenium | 7782-49-2 | 1.0 |
| D011 | Silver | 7740-22-4 | 5.0 |
| D039 | Tetrachloroethylene | 127-18-4 | 0.7 |
| D015 | Toxaphene | 8001-35-2 | 0.5 |
| D040 | Trichloroethylene | 79-01-6 | 0.5 |
| D041 | 2,4,5-Trichlorophenol | 95-95-4 | 400.0 |
| D042 | 2,4,6-Trichlorophenol | 88-06-2 | 2.0 |
| D017 | 2,4,5-TP (Silvex) | 93-72-1 | 1.0 |
| D043 | Vinyl Chloride | 74-01-4 | 0.2 |

\* Quantitation limit is greater than the calculated regulatory level. The quantitation limit therefore becomes the regulatory level.

\*\* If o-, m-, and p-Cresol concentrations cannot be differentiated, the total cresol (D026) concentration is used. The regulatory level of total cresol is 200 milligrams per liter (mg/L).

| | Table 2: Hazardous waste generated by generic processes ("F" list) | |
|---|---|---|
| **EPA Waste No.** | **Hazardous waste** | **Hazard code** |
| F001 | The following spent halogenated solvents used in degreasing: tetrachloroethylene, trichloroethylene, methylene chloride, 1,1,1-trichloroethane, carbon tetrachloride, and chlorinated fluorocarbons; all spent solvent mixtures/blends used in degreasing that contain, before use, a total of 10% or more (by volume) of 1 or more of the above halogenated solvents or those solvents listed in F002, F004, and F005; and still bottoms from the recovery of these spent solvents and spent solvent mixtures. | (T) |
| F002 | The following spent halogenated solvents: tetrachloroethylene, methylene chloride, trichloroethylene, 1,1,1-trichloroethane, chlorobenzene, 1,1,2-trichloro-1,2,2 trifluoroethane, ortho-dichlorobenzene, trichlorofluoromethane, and 1,1,2- trichloro-ethane; all spent solvent mixtures/blends containing, before use, a total of 10% or more (by volume) of 1 or more of the above halogenated solvents or those listed in F001, F004, and F005; and still bottoms from the recovery of these spent solvents and spent solvent mixtures. | (T) |
| F003 | The following spent non-halogenated* solvents: xylene, acetone, ethyl acetate, ethyl benzene, ethyl ether, methyl isobutyl ketone, n-butyl alcohol, cyclohexanone, and methanol; all spent solvent mixtures/blends containing, before use, only the above spent non-halogenated solvents; and all spent solvent mixtures/blends containing, before use, 1 or more of the above non-halogenated solvents, and a total of 10% or more (by volume) of 1 or more of those solvents listed in F001, F002, F004, and F005; and still bottoms from the recovery of these spent solvents and spent solvent mixtures. | (I) |
| F004 | The following spent non-halogenated solvents: cresols, cresylic acid, and nitrobenzene; all spent solvent mixtures/blends containing, before use, a total of 10% or more (by volume) of 1 or more of the above non-halogenated solvents or those solvents listed in F001, F002, and F005; and still bottoms from the recovery of those spent solvents and spent solvent mixtures. | (T) |
| F005 | The following spent non-halogenatedsolvents: toluene, methyl ethyl ketone, carbon disulfide, isobutanol, pyridine, benzene, 2-ethoxyethanol, and 2-nitropropane; all spent solvent mixtures/blends containing, before use, a total of 10% more (by volume) of 1 or more of the above non-halogenated solvents or those solvents listed in F001, F002, and F004; and still bottoms from the recovery of these spent solvents and spent solvent mixtures. | (I,T)* |
| F006 | Wastewater treatment sludges from electroplating operations except from the following processes: 1) sulfuric acid anodizing of aluminum; 2) tin plating on carbon steel; 3) zinc plating (segregated basis) on carbon steel; 4) aluminum or zinc-aluminum plating on carbon steel; 5) cleaning/stripping associated with tin, zinc and aluminum plating on carbon steel; and 6) chemical etching and milling of aluminum. | (T) |
| F007 | Spent cyanide plating bath solutions from electroplating operations. | (R,T) |
| F008 | Plating bath residues from the bottom of plating baths from electroplating operations where cyanides are used in the process. | (R,T) |
| F009 | Spent stripping and cleaning bath solutions from electroplating operations where cyanides are used in the process. | (R,T) |
| F010 | Quenching bath residues from oil baths from metal heat treating operations where cyanides are used in the process. | (R,T) |
| F011 | Spent cyanide solutions from salt bath pot cleaning from metal heat treating operations. | (R,T) |
| F012 | Quenching waste water treatment sludges from metal heat treating operations where cyanides are used in the process. *(continued)* | (T) |

## *ntro to Hazardous Waste*

| Table 2: | Hazardous waste generated by generic processes ("F" list) *(continued)* | |
|---|---|---|
| )19 | Wastewater treatment sludges from the chemical conversion coating of aluminum except from zirconium phosphating in aluminum can washing when such phosphating is an exclusive conversion coating process. | (T) |
| )20 | Wastes (except wastewater and spent carbon from hydrogen chloride purification) from the production or manufacturing use (as a reactant, chemical intermediate, or component in a formulating process) of tri- or tetrachlorophenol, or of intermediates used to produce their pesticide derivatives. (This listing does not include wastes from the production of Hexachlorophene from highly purified 2,4,5-trichlorophenol.) | (H) |
| )21 | Wastes (except wastewater and spent carbon from hydrogen chloride purification) from the production or manufacturing use (as a reactant, chemical intermediate, or component in a formulating process) of pentachlorophenol, or of intermediates used to produce its derivatives. | (H) |
| )22 | Wastes (except wastewater and spent carbon from hydrogen chloride purification) from the manufacturing use (as a reactant, chemical intermediate, or component in a formulating process) of tetra-, penta-, or hexachlorobenzenes under alkaline conditions. | (H) |
| )23 | Wastes (except wastewater and spent carbon from hydrogen chloride purification) from the production of materials on equipment previously used for the production or manufacturing use (as a reactant, chemical intermediate, or component in a formulating process) of tri- and tetrachlorophenols. (This listing does not include wastes from equipment used only for the production or use of Hexachlorophene from highly purified 2,4,5-trichlorophenol.) | (H) |
| )24 | Process wastes, including but not limited to, distillation residues, heavy ends, tars, and reactor clean-out wastes, from the production of certain chlorinated aliphatic hydrocarbons by free radical catalyzed processes. These chlorinated aliphatic hydrocarbons are those having carbon chain lengths ranging from 1 to and including 5, with varying amounts and positions of chlorine substitution. (This listing does not include wastewaters, waste-water treatment sludges, spent catalysts, and wastes listed in RCRA Sections 261.31 or 261.32). | (T) |
| )25 | Condensed light ends, spent filters and filter aids, and spent desiccant wastes from the production of certain chlorinated aliphatic hydrocarbons, by free radical catalyzed processes. These chlorinated aliphatic hydrocarbons are those having carbon chain lengths ranging from 1 to and including 5, with varying amounts and positions of chlorine substitution. | (T) |
| 26 | Wastes (except wastewater and spent carbon from hydrogen chloride purification) from the production of materials on equipment previously used for the manufacturing use (as a reactant, chemical intermediate, or component in a formulating process) of tetra-, penta-, or hexachlorobenzene under alkaline conditions. | (H) |
| 27 | Discarded unused formulations containing tri-, tetra-, or pentachlorophenol discarded unused formulations containing compounds derived from these chlorophenols. (This listing does not include formulations containing Hexachlorophene synthesized from prepurified 2,4,5-trichlorophenol as the sole component.) | (H) |
| 28 | Residues resulting from the incineration or thermal treatment of soil contaminated with EPA hazardous waste numbers F020, F021, F022, F023, F026, and F027. *(continued)* | (T) |

| Table 2: | Hazardous waste generated by generic processes ("F" list) *(continued)* | |
|---|---|---|
| F032 | Wastewaters (except those that have not come into contact with process contaminants), process residuals, preservative drippage, and spent formulations from wood preserving processes generated at plants that currently use or have previously used chlorophenolic formulations (except potentially cross-contaminated wastes that have had the F032 waste code deleted in accordance with RCRA Section 261.35 or potentially cross-contaminated wastes that are otherwise currently regulated as hazardous wastes (i.e., F034 or F035), and where the generator does not resume or initiate use of chlorophenolic formulations). This listing does not include K001 bottom sediment sludge from the treatment of wastewater from wood preserving processes that use creosote and/or pentachlorophenol. | (T) |
| F034 | Wastewaters (except those that have not come into contact with process contaminants), process residuals, preservative drippage, and spent formulations from wood preserving processes generated at plants that use creosote formulations. This listing does not include K001 bottom sediment sludge from the treatment of wastewater from wood preserving processes that use creosote and/or pentachlorophenol. | (T) |
| F035 | Wastewaters (except those that have not come into contact with process contaminants), process residuals, preservative drippage, and spent formulations from wood preserving processes generated at plants that use inorganic preservatives containing arsenic or chromium. This listing does not include K001 bottom sediment sludge from the treatment of wastewater from wood preserving processes that use creosote and/or pentachlorophenol. | (T) |
| F037 | Petroleum refinery primary oil/water/ solids separation sludge - Any sludge generated from the gravitational separation of oil/water/solids during the storage or treatment of process wastewaters and oily cooling wastewaters from petroleum refineries. Such sludges include, but are not limited to, those generated in: oil/water solids separators; tanks and impoundments; ditches and other conveyances; sumps; and stormwater units receiving dry weather flow. Sludge generated in stormwater units that do not receive dry weather flow, sludges generated from non-contact once-through cooling waters segregated for treatment from other process or oily cooling waters, sludges generated in aggressive biological treatment units as defined in RCRA Section 261.31(b)(2) (including sludges generated in 1 or more additional units after wastewaters have been treated in aggressive biological treatment units) and K051 wastes are not included in this listing. | (T) |
| F038 | Petroleum refinery secondary (emulsified) oil/water/solids separation sludge - Any sludge and/or float generated from the physical and/or chemical separation of oil/water/solids in process wastewaters and oily cooling wastewaters from petroleum refineries. Such wastes include, but are not limited to, all sludges and floats generated in: induced air flotation (IAF) units, tanks and impoundments, and all sludges generated in dissolved air flotation (DAF) units. Sludges generated in storm water units that do not receive dry weather flow, sludges generated from non-contact once-through cooling waters segregated for treatment from other process or oily cooling waters, sludges and floats generated in aggressive biological treatment units as defined in RCRA Section 261.31(b)(2) (including sludges and floats generated in 1 or more additional units after wastewaters have been treated in aggressive biological treatment units) and F037, K048, and K051 wastes are not included in this listing. *(continued)* | |

*ntro to Hazardous Waste*

| Table 2: Hazardous waste generated by generic processes ("F" list) *(continued)* | |
|---|---|
| ·039 | Leachate (liquids that have percolated through land disposed wastes) resulting from the disposal of more than 1 restricted waste classified as hazardous under Subpart D (Leachate resulting from the disposal of 1 or more of the following EPA hazardous wastes and no other hazardous wastes retains its EPA hazardous waste number(s): F020, F021, F022, F026, F027, and/or F028). | (T) |

Notes:

(I,T) should be used to specify mixtures containing ignitable and toxic constituents.

₂ For the purposes of the F037 and F038 listings, oil/water/solids is defined as oil and/or water and/or solids.

₂ For the purposes of the F037 and F038 listings, aggressive biological treatment units are defined as units that employ 1 of the following 4 treatment methods: activated sludge; trickling filter; rotating biological contactor for the continuous accelerated biological oxidation of wastewaters; or high-rate aeration. High-rate aeration is a system of surface impoundments or tanks in which intense mechanical aeration is used to completely mix the wastes and enhance biological activity. High-rate aeration units employ a minimum of 6 horsepower per million gallons of treatment volume, and either the hydraulic retention time of the unit is no longer than 5 days, or the hydraulic retention time is no longer than 30 days and the unit does not generate a sludge that is a hazardous waste by the toxicity characteristic. Generators and TSD facilities have the burden of proving that their sludges are exempt from listing as F037 and F038 wastes under this definition. Generators and TSD facilities must maintain, in their operating or other onsite records, documents and data sufficient to prove that the unit is an aggressive biological treatment unit and the sludges sought to be exempted were actually generated in the aggressive biological treatment unit.

**Table 3: Hazardous waste from specific sources ("K" list)**

**Wood Preservation**

K001(T)    Bottom sediment sludge from the treatment of wastewaters from wood-preserving processes that use creosote and/or pentachlorophenol.

**Inorganic Pigments**

K002(T)    Wastewater treatment sludge from the production of chrome yellow and orange pigments.

K003(T)    Wastewater treatment sludge from the production of molybdate orange pigments.

K004(T)    Wastewater treatment sludge from the production of zinc yellow pigments.

K005(T)    Wastewater treatment sludge from the production of chrome green pigments.

K006(T)    Wastewater treatment sludge from the production of chrome oxide green pigments (anhydrous and hydrated).

K007(T)    Wastewater treatment sludge from the production of iron blue pigments.

K008(T)    Oven residues from the production of chrome oxide green pigments.

**Organic chemicals**

K009(T)    Distillation bottoms from the production of acetaldehyde from ethylene.

K010(T)    Distillation side cuts from the production of acetaldehyde from ethylene.

K011(R,T)    Bottom stream from the wastewater stripper in the production of acrylonitrile.

K013(R,T)    Bottom stream from the acrylonitrile column in the production of acrylonitrile.

K014(T)    Bottoms from the acetonitrile purification column in the production of acrylonitrile.

K015(T)    Still bottoms from the distillation of benzyl chloride.

K016(T)    Heavy ends or distillation residues from the production of carbon tetrachloride.

K017(T)    Heavy ends (still bottoms) from the purification column in the production of epichlorohydrin.

K018(T)    Heavy ends from the fractionation column in ethyl chloride production.

K019(T)    Heavy ends from the distillation of ethylene dichloride production.

K020(T)    Heavy ends from the distillation of vinyl chloride in vinyl chloride monomer production.

K021(T)    Aqueous spent antimony catalyst waste from flouromethanes production

K022(T)    Distillation bottom tars from the production of phenol/acetone from cumene.

K023(T)    Distillation light ends from the production of phthalic anhydride from naphthalene.

K024(T)    Distillation bottoms from the production of phthalic anhydride from naphthalene.

K025(T)    Distillation bottoms from the production of nitrobenzene by the nitration of benzene.

K026(T)    Stripping still tails from the production of methyl ethyl pyridines.

K027(R,T)    Centrifuge and distillation residues from toluene diisocyanate production.

K028(T)    Spent catalyst from the hydrochlorinator reactor in the production of 1,1,1-trichloroethane.

K029(T)    Waste from the product steam stripper in the production of 1,1,1-trichloroethane.

K030(T)    Column bottoms or heavy ends from the combined production of trichloroethylene and perchloroethylene.

K083(T)    Distillation bottoms from aniline production.

K085(T)    Distillation or fractionation column bottoms from the production of chlorobenzenes.

K093(T)    Distillation light ends from the production of phthalic anhydride from ortho-xylene.

K094(T)    Distillation bottoms from the production of phthalic anhydride from ortho-xylene.

K095(T)    Distillation bottoms from the produciton of 1,1,1-trichloroethane.

K103(T)    Process residues from aniline extraction from the production of aniline.

K104(T)    Combined wastewater streams from generated from nitrobenzene/aniline production.

K105(T)    Separated aqueous stream from the reactor product washing step in the production of chlorobenzenes.

K107(C,T)    Column bottoms from product separation from the production of 1,1-dimethyl-hydrazine (UDMH) from carboxylic acid hydrazines.

K108(I,T)    Condensed column overheads from product separation and condensed reactor vent gases for the production of UDMH from carboxylic acid hydrazides.

*(continued on next page)*

## Intro to Hazardous Waste

| | |
|---|---|
| K109(T) | Spent filter cartridges from the product purification from the production of UDMH from carboxylic acid hydrazides. |
| K110(T) | Condensed column overheads from intermediate separation from the production of UDMH from carboxylic acid hydrazides. |
| K111(C,T) | Product washwaters from the production of dinitrotoluene via nitration of toluene. |
| K112(T) | Reaction by-product water from the drying column in the production of toluenediamine via hydrogenation of dinitrotoluene. |
| K113(T) | Condensed liquid light ends from the purification of toluenediamine via hydrogenation of dinitrotoluene. |
| K114(T) | Vicinals from the purification of toluenediamine in the production of toluenediamine via hydrogenation of dinitrotoluene. |
| K115(T) | Heavy ends from the purification of toluenediamine in the production of toluenediamine via hydrogenation of dinitrotoluene. |
| K116(T) | Organic condensate from the solvent recovery in the production of toluene diisocyanate via phosgenation of toluenediamine. |
| K117(T) | Wastewater from reactor vent gas scrubber in the production of ethylene dibromide via bromination of ethylene. |
| K118(T) | Spent adsorbent solids from the purification of ethylene dibromide in the production of ethylene dibromide via bromination of ethylene. |
| K136(T) | Still bottoms from the purification of ethylene dibromide in the production of ethylene dibromide via bromination of ethylene. |
| K149(T) | Distillation bottoms from the production of alpha- (or methyl-) chlorinated toluenes, ring-chlorinated toluenes, benzyl chlorides, and compounds with mixtures of these functional groups. (This waste does not include still bottoms from the distillation of benzyl chloride.) |
| K150(T) | Organic residuals, excluding spent carbon adsorbent, from the spent chlorine gas and hydrochloric acid recovery processes associated with the production of alpha- (or methyl-) chlorinated toluenes, ring-chlorinated toluenes, benzyl chlorides, and compounds with mixtures of these functional groups. |
| K151(T) | Wastewater treatment sludges, excluding neutralization and biological sludges, generated during the treatment of wastewaters from the production of alpha- (or methyl-) chlorinated toluenes, ring-chlorinated toluenes, benzyl chlorides, and compounds with mixtures of these functional groups. |

### Inorganic Chemicals

| | |
|---|---|
| K071(T) | Brine purification muds from the mercury cell process in chlorine production, where separately prepurified brine is not used. |
| K073(T) | Chlorinated hydrocarbon waste from the purification step of the diaphragm cell process using graphite anodes in chlorine production. |
| K106(T) | Wastewater treatment sludge from the mercury cell process in chlorine production. |

### Pesticides

| | |
|---|---|
| K031(T) | By-product salts generated in the production of Methanearsonic Acid (MSMA, in the production of chlordane. |
| K034(T) | Filter solids from the filtration of hexachlorocyclapentadiene in the production of chlordane. |
| K035(T) | Wastewater treatment sludges generated in the production of creosote. |
| K036(T) | Still bottoms from toluene reclamation distillation in the production of disulfoton. |
| K037(T) | Wastewater treatment sludges from the production of disulfoton |
| K038(T) | Wastewater from the washing and stripping of phorate production. |
| K039(T) | Filter cake from the filtration of diethylphosphorodithioic acid in the production of phorate. |
| K040(T) | Wastewater treatment sludge from the production of phorate. |
| K041(T) | Wastewater treatment sludge from the production of toxaphene. |
| K042(T) | Heavy ends of distillation residues from the distillation of tetrachlorobenzene in the production of 2,4,5-T. |
| K043(T) | 2,6-Dichlorophenol waste from the production of 2,4-D. |
| K097(T) | Vacuum stripper discharge from the chlordane chlorinator in the production of chlordane. |
| K098(T) | Untreated process wastewater from the production of toxaphene. |
| K099(T) | Untreated wastewater from the production of 2,4-D. |
| K123(T) | Process wastewater (including supermates, filtrates, and washwaters) from the production of ethylebisdithiocarbamic acid and it salt. |
| K124(C,T) | Reactor vent scrubber water from the production of ethylebisdithiocarbamic acid and it salts. |
| K125(T) | Filtration, evaporation, and centrifugation solids from the production of ethylebisdithiocarbamic acid and it salts. |
| K126(T) | Baghouse dust and floor sweepings in milling and packaging operations from the production or |

formulation of ethylebisdithiocarbamic acid and it salts.

K131(C,T)    Wastewater from the reactor and spent sulfuric acid from the acid dryer from the production of methyl bromide.

K132(T)    Spent absorbent and wastewater separator solids from the production of methyl bromide.

## Explosives

K044(R)    Wastewater treatment sludges from the manufacturing and processing of explosives.

K045(R)    Spent carbon from the treatment of wastewater containing explosives.

K046(T)    Wastewater treatment sludges from the manufacturing, formulation, and loading of lead-based initiating compounds.

K047(R)    Pink/red water from trinitrotoluene (TNT) operations.

## Petroleum refining

K048(T)    DAF float from the petroleum industry.

K049(T)    Slop oil emulsion solids from the petroleum refining industry.

K050(T)    Heat exchanger bundle cleaning sludge from the petroleum refining industry.

K051(T)    American Petroleum Institute (API) separator sludge from the petroleum refining industry.

K052(T)    Tank bottoms (leaded) from the petroleum refining industry.

## Iron and Steel

K061(T)    Emission control dust/sludge from the primary production of steel in electric furnaces.

K062(C,T)    Spent pickle liquor generated by steel finishing operations from facilities within the iron and steel industry (Standard Industrial Classification (SIC) codes 331 and 332).

## Primary Copper

K064(T)    Acid plant blowdown slurry/sludge resulting from the thickening of blowdown slurry from primary copper production.

## Primary Lead

K065(T)    Surface impoundment solids contained in and dredged from surface impoundments at primary lead smelting facilities.

## Primary Zinc

K066(T)    Sludge from treatment of process wastewater and/or acid plant blowdown from primary zinc production.

## Primary Aluminum

K088(T)    Spent potliners from primary aluminum reduction.

## Ferroalloys

K090(T)    Emission control dust or sludge from ferrochromiumsilicon production.

K091(T)    Emission control dust or sludge from ferrochromium production.

*(continued on next page)*

## *Intro to Hazardous Waste*

### Secondary Lead
K069(T)    Emission control dust/sludge from secondary lead smelting. (NOTE: This listing is stayed administratively for sludge generated from secondary acid scrubber systems. The stay will remain in effect until further administrative action is taken. If EPA takes further action affecting this stay, EPA will publish a notice of the action in the *Federal Register*.)

K100(T)    Waste leaching solution from acid leaching of emission control dust sludge from secondary lead smelting.

### Veterinary Pharmaceuticals
K084(T)    Wastewater treatment sludges generated during the production of veterinary pharmaceuticals from arsenic or organo-arsenic compounds.

K101(T)    Distillation tar residues from the distillation of aniline-based compounds in the production of veterinary pharmaceuticals from arsenic or organo-arsenic compounds.

K102(T)    Residue from the use of activated carbon for decolorization in the production of veterinary pharmaceuticals from arsenic or organo-arsenic compounds.

### Ink Formulation
K086(T)    Solvent washes and sludges, caustic washes and sludges, or water washes and sludges from cleaning tubs and equipment used in the formulation of ink from pigments, driers, soaps, and stabilizers containing chromium and lead.

### Coking
K060(T)    Ammonia still lime sludge from coking operations.

K087(T)    Decanter tank tar sludge from coking operations.

K141(T)    Process residues from the recovery of coal tar, including, but not limited to, collecting sump residues from the production of coke from coal or the recovery of coke byproducts produced from coal. This listing does not include K087 (decanter tank tar sludges from coking operations).

K142(T)    Tar storage tank residues from the production of coke from coal or from the recovery of coke by-products produced from coal.

K143(T)    Process residues from the recovery of light oil, including, but not limited to, those generated in stills, decanters, and wash oil recovery units from the recovery of coke by-products produced from coal.

K144(T)    Wastewater sump residues from light oil refining, including but not limited to, intercepting or contamination sump sludges from the recovery of coke by-products produced from coal.

K145(T)    Residues from napthalene collection and recovery operations from the recovery of coke by-products produced from coal.

K147(T)    Tar storage tank residues from coal tar refining.

K148(T)    Residues from coal tar distillation, including but not limited to, still bottoms.

Key:  (T) = toxic; (C) = corrosive; (R) = reactive.

Table 4: Hazardous waste "P" list

| HW No. | CAS No. | Substance |
|---|---|---|
| P023 | 107-20-0 | Acetaldehyde, chloro |
| P002 | 591-08-2 | Acetamide, N-(aminothioxomethyl)- |
| P057 | 640-19-7 | Acetamide, 2-fluoro |
| P058 | 62-74-8 | Acetic acid, fluoro, sodium salt |
| P002 | 591-08-2 | 1-Acetyl-2-thiourea |
| P003 | 107-02-8 | Acrolein |
| P070 | 116-06-2 | Aldicarb |
| P004 | 309-00-2 | Aldrin |
| P005 | 107-18-6 | Allyl alcohol |
| P006 | 20859-73-8 | Aluminum phosphide (R,T) |
| P007 | 2763-96-4 | 5-(Aminomethyl)-3-isoxazolol |
| P008 | 504-24-5 | 4-Aminopyridine |
| P009 | 131-74-8 | Ammonium picrate (R) |
| P119 | 7803-55-6 | Ammonium vanadate |
| P099 | 506-61-6 | Argentate(1-), bis(cyano-C)-potassium |
| P010 | 7778-39-4 | Arsenic Acid H(3)AsO(4) |
| P012 | 1327-53-3 | Arsenic oxide As(2)O(3) |
| P011 | 1303-28-2 | Arsenic pentoxide |
| P012 | 1327-53-3 | Arsenic trioxide |
| P038 | 692-42-2 | Arsine, diethyl- |
| P036 | 696-28-6 | Arsonous dichloride, phenyl- |
| P054 | 151-56-4 | Aziridine |
| P067 | 75-55-8 | Aziridine, 2-methyl- |
| P013 | 542-62-1 | Barium cyanide |
| P024 | 106-47-8 | Benzenamine, 4-chloro- |
| P077 | 100-01-6 | Benzenamine, 4-nitro- |
| P028 | 100-44-7 | Benzene, (chloromethyl)- |
| P042 | 51-43-4 | 1,2-Benzenediol, 4-[1-hydroxy-2-(methylamino)ethyl]-, (R)- |
| P046 | 122-09-8 | Benzeneethanamine, alpha, alpha-dimethyl- |
| P014 | 108-98-5 | Benzenethiol |
| P001 | [1]81-81-2 | 2H-1-Benzopyran-2-one, 4-hydroxy-3-(3-oxo-1-phenylbutyl)- and salts when present at concentrations greater than 0.3% |
| P028 | 100-44-7 | Benzyl chloride |
| P015 | 7440-41-7 | Beryllium Powder |
| P017 | 598-31-2 | Bromoacetone |
| P018 | 357-57-3 | Brucine |
| P045 | 39196-18-4 | 2-Butanone, 3,3-dimethyl-1 -(methylthio)-, O-[methylamino) carbonyl] oxime |
| P021 | 592-01-8 | Calcium cyanide |
| P022 | 75-15-0 | Carbon disulfide |
| P095 | 75-44-5 | Carbonic dichloride |
| P023 | 107-20-0 | Chloroacetaldehyde |
| P024 | 106-47-8 | p-Chloroaniline |
| P026 | 5344-82-1 | 1-(o-Chlorophenyl)thiourea *(continued)* |

## *Intro to Hazardous Waste*

| Table 4: Hazardous waste "P" list *(continued)* | | |
|---|---|---|
| HW No. | CAS No. | Substance |
| P027 | 542-76-7 | 3-Chloropropionitrile |
| P029 | 544-92-3 | Copper cyanide |
| P030 | ——— | Cyanides (soluble cyanide salts), not otherwise specified |
| P031 | 460-19-5 | Cyanogen |
| P033 | 506-77-4 | Cyanogen chloride |
| P034 | 131-89-5 | 2-Cyclohexyl-4,6-dinitrophenol |
| P016 | 542-88-1 | Dichloromethyl ether |
| P036 | 696-28-6 | Dichlorophenylarsine |
| P037 | 60-57-1 | Dieldrin |
| P038 | 692-42-2 | Diethylarsine |
| P041 | 311-45-5 | Diethyl-p-nitrophenyl phosphate |
| P040 | 297-97-2 | O,O-Diethyl O-pyrazinyl phosphorothioate |
| P043 | 55-91-4 | Diisopropylfluorophosphate (DFP) |
| P004 | 309-00-2 | 1,4,5,8-Dimethanonaphthalene, 1,2,3,4,10,10-hexachloro-1,4,4a,5,8,8a-hexahydro-, (1alpha,4alpha, 4abeta, 5alpha, 8alpha,8abeta)- |
| P060 | 465-73-6 | 1,4,5,8-Dimethanonaphthalene, 1,2,3,4,10,10-hexachloro-1,4,4a,5,8,8a-hexahydro-,(1alpha,4alpha,4abeta,5beta, 8beta,8abeta)- |
| P037 | 60-57-1 | 2,7:3,6-Dimethanonaphth[2,3b]oxirane,3,4,5,6,9,9-hexachloro-1a,2,2a,3,6,6a,7,7a-octahydro-,(1aalpha,2beta,2aalpha,3beta,6beta,6aalpha,7beta,7aalpha)- |
| P051 | 72-20-5 | 2,7,3,6-Dimethanonaphth[2,3b] oxirine,3,4,5,6,9,9-hexachloro-1a,2,2a,3,6,6a,7,7a-octahydro (1aalpha,2beta,2abeta,3alpha,6alpha,6abeta,7beta,7aalpha)-, & metabolites |
| P044 | 60-51-5 | Dimethoate |
| P046 | 122-09-8 | alpha, alpha-Dimethylphenethylamine |
| P047 | [1]534-52-1 | 4,6-Dinitro-o-cresol, and salts |
| P048 | 51-28-5 | 2,4-Dinitrophenol |
| P020 | 88-85-7 | Dinoseb |
| P085 | 152-16-9 | Diphosphoramide, octamethyl- |
| P111 | 107-49-3 | Diphosphoric acid, tetraethyl ester |
| P039 | 298-04-4 | Disulfoton |
| P049 | 541-53-7 | Dithiobiuret |
| P050 | 115-29-7 | Endosulfan |
| P088 | 145-73-3 | Endothall |
| P051 | 72-20-8 | Endrin, & metabolites |
| P042 | 51-43-4 | Epinephrine |
| P031 | 460-19-5 | Ethanedinitrile |
| P066 | 16752-77-5 | Ethanimidothioic acid, N[[(methylamino) carbonyl]oxy]-, methyl ester |
| P101 | 107-12-0 | Ethyl cyanide |
| P054 | 151-56-4 | Ethyleneimine |
| P097 | 52-85-7 | Famphur |
| P056 | 7782-41-4 | Fluorine |
| P057 | 640-19-7 | Fluoroacetamide |
| P058 | 62-74-8 | Fluoroacetic acid, sodium salt *(continued)* |

| HW No. | CAS No. | Substance |
|---|---|---|
| | | **Table 4: Hazardous waste "P" list** *(continued)* |
| P065 | 628-86-4 | Fluminic acid, mercury(2+) salt (R,T) |
| P059 | 76-44-8 | Heptachlor |
| P062 | 757-58-4 | Hexaethyl tetraphosphate |
| P116 | 79-19-6 | Hydrazinecarbothioamide |
| P068 | 80-34-4 | Hydrazine, methyl- |
| P063 | 74-90-8 | Hydrocyanic acid (Hydrogen cyanide) |
| P096 | 7803-51-2 | Hydrogen phosphide |
| P060 | 465-73-6 | Isodrin |
| P007 | 2763-96-4 | 3(2H)-Isoxazolone, 5-(aminomethyl)- |
| P092 | 62-38-4 | Mercury, (acetato-O)phenyl |
| P065 | 628-86-4 | Mercury fulminate (R,T) |
| P082 | 62-75-9 | Methanamine, N-methyl-N-nitroso- |
| P064 | 624-83-9 | Methane, isocyanato- |
| P016 | 542-88-1 | Methane, oxybis[chloro- |
| P112 | 509-14-8 | Methane, tetranitro- (R) |
| P118 | 75-70-7 | Methanethiol, trichloro- |
| P050 | 115-29-7 | 6,9-Methano-2,4,3-benzodioxathiepin,6,7,8,9,10,10-hexachloro-1,5,5a,6,9,9a-hexahydro-, 3-oxide |
| P059 | 76-44-8 | 4,7-Methano-1H-indene, 1,4,5,6,7,8,8-heptachloro-3a,4,7,7a-tetrahydro |
| P066 | 16752-77-5 | Methomyl |
| P068 | 60-34-4 | Methyl hydrazine |
| P064 | 624-83-9 | Methyl isocyanate |
| P069 | 75-86-5 | 2-Methyllactonitrile |
| P071 | 298-00-0 | Methyl parathion |
| P072 | 86-88-4 | alpha-Naphthylthiourea |
| P073 | 13463-39-3 | Nickel carbonyl |
| P074 | 557-19-7 | Nickel cyanide |
| P075 | [1]54-11-5 | Nicotine and salts |
| P076 | 10102-43-9 | Nitric oxide |
| P077 | 100-01-6 | p-Nitroaniline |
| P078 | 10102-44-0 | Nitrogen dioxide |
| P076 | 10102-43-9 | Nitrogen oxide NO |
| P081 | 55-63-0 | Nitroglycerine (R) |
| P082 | 62-75-9 | N-Nitrosomethylamine |
| P084 | 4549-40-0 | N-Nitrosomethylvinylamine |
| P085 | 152-16-9 | Octamethylpyrophosphoramide |
| P087 | 20816-12-0 | Osmium tetroxide) |
| P088 | 145-73-3 | 7-Oxabicyclo[2.2.1]heptane-2,3-dicarboxylic acid |
| P089 | 56-38-2 | Parathion |
| P034 | 131-89-5 | Phenol, 2-cyclohexyl-4,6-dinitro- |
| P048 | 51-28-5 | Phenol, 2,4-dinitro- |
| P047 | (1) 534-52-1 | Phenol, 2-methyl-4,6-dinitro- and salts |
| P020 | 88-85-7 | Phenol, 2-(1-methylpropyl)-4,6-dinitro- |
| P009 | 131-74-8 | Phenol, 2,4,6-trinitro-, ammonium salt (R)  *(continued)* |

## *Intro to Hazardous Waste*

| HW No. | CAS No. | Substance |
|--------|---------|-----------|
| Table 4: Hazardous waste "P" list *(continued)* | | |
| P092 | 62-38-4 | Phenylmercury acetate |
| P093 | 103-85-5 | Phenylthiourea |
| P094 | 298-02-2 | Phorate |
| P095 | 75-44-5 | Phosgene |
| P096 | 7803-51-2 | Phosphine |
| P041 | 311-45-5 | Phosphoric acid, diethyl 4-nitrophenyl ester |
| P039 | 298-04-4 | Phosphorodithioic acid, O,O-diethyl S-[2-(ethylthio)ethyl]ester |
| P094 | 296-04-2 | Phosphorodithioic acid, O,O-diethyl S-[(ethylthio)methyl] ester |
| P044 | 60-51-5 | Phosphorodithioic acid, O,O-dimethyl S-[2-(methylamino)-2-oxoethyl] ester |
| P043 | 55-91-4 | Phosphorofluoridic acid, bis-(1-methylethyl) ester |
| P089 | 56-38-2 | Phosphorothioic acid, O,O-diethyl O-(4-nitrophenyl) ester |
| P040 | 297-92-2 | Phosphorodithioic acid, O,O-diethyl O-pyrazinyl ester |
| P097 | 52-85-7 | Phosphorodithioic acid, O-O,4 [(diimethylamino)sulfonyl])phenyl]O,O-dimethyl ester |
| P071 | 296-00-0 | Phosphorodithioic acid, O,O-dimethyl O-(4-nitrophenyl)ester |
| P110 | 78-00-2 | Plumbane, tetraethyl- |
| P098 | 151-50-8 | Potassium cyanide |
| P099 | 506-61-6 | Potassium silver cyanide |
| P070 | 116-06-3 | Propanal, 2-methyl-2-(methylthio)-,O-[(methylamino)carbonyl] oxime |
| P101 | 107-12-0 | Propanenitrile |
| P027 | 542-76-7 | Propanenitrile,3-chloro- |
| P069 | 75-86-5 | Propanenitrile, 2-hydroxy-2-methyl- |
| P081 | 55-63-0 | 1,2,3-Propanetriol, trinitrate (R) |
| P017 | 598-31-2 | 2-Propanone, 1-bromo- |
| P102 | 107-19-7 | Propargyl alcohol |
| P003 | 107-02-8 | 2-Propenal |
| P005 | 107-18-6 | 2-Propen-1-ol |
| P067 | 75-55-8 | 1,2-Propylenimine |
| P102 | 107-19-7 | 2-Propyn-1-ol |
| P008 | 504-24-5 | 4-Pyridinamine |
| P075 | [1]54-11-5 | Pyridine, 3-(1-methyl-2-pyrrolidinyl)-, (S)-, and salts |
| P114 | 12039-52-0 | Selenious acid, dithallium(1+) salt |
| P103 | 630-10-4 | Selenourea |
| P104 | 506-64-9 | Silver cyanide |
| P105 | 26628-22-8 | Sodium azide |
| P106 | 143-33-9 | Sodium cyanide |
| P108 | [1]57-24-9 | Strychnidin-10-one (Strychnine), and salts |
| P018 | 357-57-3 | Strychnidin-10-one, 2,3-dimethoxy- |
| P115 | 7446-18-6 | Sulfuric acid, dithallium(1+) salt |
| P109 | 3689-24-5 | Tetraethyldithiopyrophosphate |
| P110 | 78-00-2 | Tetraethyl lead |
| P111 | 107-49-3 | Tetraethyl pyrophosphate |
| P112 | 509-14-8 | Tetranitromethane (R) *(continued)* |

| HW No. | CAS No. | Substance |
|--------|---------|-----------|
| | | Table 4: Hazardous waste "P" list *(continued)* |
| P062 | 757-58-4 | Tetraphosphoric acid, hexaethyl ester |
| P113 | 1314-32-5 | Thallic oxide˙ |
| P114 | 12039-52-0 | Thallium(I) selenite |
| P115 | 7446-18-6 | Thallium(I) sulfate |
| P109 | 3689-24-5 | Thiodiphosphoric acid, tetraethyl ester |
| P045 | 39196-18-4 | Thiofanox |
| P049 | 541-53-7 | Thiomidodicarbonic diamide [(H(2)N)C(S)](2)NH |
| P014 | 108-98-5 | Thiophenol |
| P116 | 79-19-6 | Thiosemicarbazide |
| P026 | 5344-82-1 | Thiourea, (2-chlorophenyl)- |
| P072 | 86-88-4 | Thiourea, 1-naphthalenyl- |
| P093 | 103-85-5 | Thiourea, phenyl- |
| P123 | 8001-35-2 | Toxaphene |
| P118 | 75-70-7 | Trichloromethanethiol |
| P119 | 7803-55-6 | Vanadic acid, ammonium salt |
| P120 | 1314-62-1 | Vanadium pentoxide |
| P084 | 4549-40-0 | Vinylamine, N-methyl-N-nitroso |
| P001 | [1]81-81-2 | Warfarin, & salts, when present at concentrations greater than 0.3% |
| P121 | 557-21-1 | Zinc cyanide |
| P122 | 1314-84-7 | Zinc phosphide Zn(3)P(2), when present at concentrations greater than 10% (R,T) |

Notes:

Key:  T = toxicity; R = reactivity; I = ignitability; C = corrosivity.  Absence of a letter indicates that the compound is only listed for toxicity.

[1]  CAS No. given for parent compound only.

(f)  The commercial chemical products, manufacturing chemical intermediates, or off-specification commercial chemical products referred to above are identified as toxic wastes (T), unless otherwise designated, and are subject to the SQG exclusion defined in 40 CFR 261.5(a) and (g).

## *Intro to Hazardous Waste*

| Hazardous Waste # | CAS No. | Substance |
|---|---|---|
| | | **Table 5. Hazardous waste "U" list** |
| U001 | 75-07-0 | Acetaldehyde (I) |
| U034 | 75-87-6 | Acetaldehyde,trichloro- |
| U187 | 62-44-2 | Acetamide,N-(4-ethoxyphenyl)- |
| U005 | 53-96-3 | Acetamide,N-9H-fluoren-2-yl- |
| U240 | (1)94-75-7 | Acetic acid,(2-4-dichlorophenoxy)-salts & esters |
| U112 | 141-78-6 | Acetic acid,ethylester (I) |
| U144 | 301-04-2 | Acetic acid,lead(2+)salt |
| U214 | 563-68-8 | Acetic acid,thallium(1+) salt See 93-76-5 Acetic acid, (2,4,5-trichlorophenoxy)-F027 |
| U002 | 67-64-1 | Acetone (I) |
| U003 | 75-05-8 | Acetonitrile (I,T) |
| U004 | 98-86-2 | Acetophenone |
| U005 | 53-96-3 | 2-Acetylaminofluorene |
| U006 | 75-36-5 | Acetylchloride (C,R,T) |
| U007 | 79-06-1 | Acrylamide |
| U008 | 79-10-7 | Acrylic acid (I) |
| U009 | 107-13-1 | Acrylonitrile |
| U011 | 61-82-5 | Amitrole |
| U012 | 62-53-3 | Aniline (I,T) |
| U136 | 75-60-5 | Arsinic acid,dimethyl |
| U014 | 492-80-8 | Auramine |
| U015 | 115-02-6 | Azaserine |
| U010 | 50-07-7 | Azirino[2',3':3,4]pyrrolo[1,2-a]indole-4,7-dione,6-amino-8-[[(aminocarbonyl)oxy]methyl]-1,1a,2,8,8a,8b-hexahydro-8a-methoxy-5-methyl-,[1aS-(1aalpha,8beta,8aalpha,8balpha)]- |
| U157 | 50-49-5 | Benz[j]aceanthrylene,1,2-dihydro-3-methyl- |
| U016 | 225-51-4 | Benz(c)acridine |
| U017 | 98-87-3 | Benzalchloride |
| U192 | 23950-58-5 | Benzamide,3,5-dichloro-N-(1,1-diethyl-2-propynyl)- |
| U018 | 56-55-3 | Benz[a]anthracene |
| U094 | 57-97-6 | Benz[a]anthracene,7,12-dimethyl- |
| U012 | 62-53-3 | Benzenamine (I,T) |
| U014 | 492-80-8 | Benzenamine,4,4-carbonimidoylbis (N,N-dimethyl- |
| U049 | 3165-93-3 | Benzenamine,4-chloro-2-methyl-hydrochloride |
| U093 | 60-11-7 | Benzenamine,N,N-dimethyl-4-(phenylazo)- |
| U328 | 95-53-4 | Benzenamine,2-methyl- |
| U353 | 106-49-0 | Benzenamine,4-methyl- |
| U158 | 101-14-4 | Benzenamine,4,4'-methylenebis[2-chloro- |
| U222 | 636-21-5 | Benzenamine,2-methyl-,hydrochloride |
| U181 | 99-55-8 | Benzenamine,2-methyl-5-nitro |
| U019 | 71-43-2 | Benzene (I,T) |
| U038 | 510-15-6 | Benzeneacetic acid,4-chloro-alpha-(4-chlorophenyl)-alpha-hydroxy-ethylester |
| U030 | 101-55-3 | Benzene,1-bromo-4-phenoxy- |
| U035 | 305-03-3 | Benzenebutanoic acid,4-[bis(2-chloroethyl)amino]- |
| U037 | 108-90-7 | Benzene,chloro |
| U221 | 25376-45-8 | Benzenediamine,ar-methyl-  *(continued)* |

| Hazardous Waste # | CAS No. | Substance |
|---|---|---|
| \multicolumn{3}{|c|}{Table 5. Hazardous waste "U" list *(continued)*} |
| U028 | 117-81-7 | 1,2-Benzenedicarboxylic acid,bis(2-ethylhexyl)ester |
| U069 | 84-74-2 | 1,2-Benzenedicarboxylic acid,dibutylester |
| U088 | 84-66-2 | 1,2-Benzenedicarboxylic acid,diethylester |
| U102 | 131-11-3 | 1,2-Benzenedicarboxylic acid,dimethylester |
| U107 | 117-84-0 | 1,2-Benzenedicarboxylic acid,dioctyl |
| U070 | 95-50-1 | Benzene,1,2-dichloro- |
| U071 | 541-73-1 | Benzene,1,3-dichloro- |
| U072 | 106-46-7 | Benzene,1,4-dichloro- |
| U060 | 72-54-8 | Benzene,1,1'-(2,2-dichloroethylidene)bis[4-chloro- |
| U017 | 98-87-3 | Benzene,(dichloromethyl)- |
| U223 | 26471-62-5 | Benzene,1,3-diisocyanatomethyl-(R,T) |
| U239 | 1330-20-7 | Benzene,dimethyl-(I,T) |
| U201 | 108-46-3 | 1,3-Benzenediol |
| U127 | 118-74-1 | Benzene,hexachloro- |
| U056 | 110-82-7 | Benzene,hexahydro-(I) |
| U220 | 108-88-3 | Benzene,methyl- |
| U105 | 121-14-2 | Benzene,1-methyl-2,4-dinitro- |
| U106 | 606-20-2 | Benzene,2-methyl-1,3-dinitro- |
| U055 | 98-82-8 | Benzene,(1-methylethyl)-(I) |
| U169 | 98-95-3 | Benzene, nitro- |
| U183 | 608-93-5 | Benzene, pentachloro- |
| U185 | 82-68-8 | Benzene, pentachloronitro- |
| U020 | 98-09-9 | Benzene sulfonic acid chloride (C,R) |
| U020 | 98-09-9 | Benzene sulfonyl chloride (C,R) |
| U207 | 95-94-3 | Benzene,1,2,4,5-tetrachloro- |
| U061 | 50-29-3 | Benzene,1,1'-(2,2,2-trichloroethylidene)bis[4-chloro- |
| U247 | 72-43-5 | Benzene,1,1'-(2,2,2-trichloroethylidene)bis[4-methoxy- |
| U023 | 98-07-7 | Benzene,(trichloromethyl)- |
| U234 | 99-35-4 | Benzene,1,3,5-trinitro- |
| U021 | 92-87-5 | Benzidine |
| U202 | (1)81-07-2 | 1,2-Benzisothiazol-3(2H)-one,1,1-dioxide, and salts |
| U203 | 94-59-7 | 1,3-Benzodioxole,5-(2-propenyl)- |
| U141 | 120-58-1 | 1,3-Benzodioxole,5-(1-propenyl)- |
| U090 | 94-58-6 | 1,3-Benzodioxole,5-propyl- |
| U064 | 189-55-9 | Benzo[rst]pentaphene |
| U248 | (1)81-81-2 | 2H-1-Benzopyran-2-one,4-hydroxy-3-(3-oxo-1-phenyl-butyl)-,&salts,when present at concentrations of 0.3%orless |
| U022 | 50-32-8 | Benzo[a]pyrene |
| U197 | 106-51-4 | p-Benzoquinone |
| U023 | 96-07-7 | Benzotrichloride(C,R,T) |
| U085 | 1464-53-5 | 2,2'-Bioxirane |
| U021 | 92-87-5 | [1,1'-Biphenyl]-4,4'-diamine |
| U073 | 91-94-1 | [1,1'-Biphenyl]-4,4'-diamine, 3,3'-dichloro- |
| U091 | 119-90-4 | [1,1'-Biphenyl]-4,4'-diamine, 3,3'-dimethoxy- |
| U095 | 119-93-7 | [1,1'-Biphenyl]-4,4'-diamine, 3,3'-dimethyl- |
| U225 | 75-25-2 | Bromoform |
| U030 | 101-55-3 | 4-Bromophenyl phenyl ether *(continued)* |

## *Intro to Hazardous Waste*

| Hazardous Waste # | CAS No. | Substance |
|---|---|---|
| U128 | 87-68-3 | 1,3-Butadiene,1,1,2,3,4,4-hexachloro- |
| U172 | 924-16-3 | 1-Butanamine,N-butyl-N-nitroso- |
| U031 | 71-36-3 | 1-Butanol (I) |
| U159 | 78-93-3 | 2-Butanone (I,T) |
| U160 | 1338-23-4 | 2-Butanone peroxide (R,T) |
| U053 | 4170-30-3 | 2-Butenal |
| U074 | 764-41-0 | 2-Butene,1,4-dichloro-(I,T) |
| U143 | 303-34-4 | 2-Butenoic acid,2-methyl-,7-[[2,3-dihydroxy-2-(1-methoxyethyl)-3-methyl-1-oxobutoxy]methyl]-2,3,5,7a-tetrahydro-1H-pyrrolizin-1-ylester,[1S-[1alpha(Z),7(2S*,3R*),7aalpha]]- |
| U031 | 71-36-3 | n-Butylalcohol (I) |
| U136 | 75-60-5 | Cacodylic acid |
| U032 | 13765-19-0 | Calcium chromate |
| U238 | 51-79-6 | Carbamic acid,ethylester |
| U178 | 615-53-2 | Carbamic acid,methylnitroso-,ethylester |
| U097 | 79-44-7 | Carbamic chloride,dimethyl |
| U114 | (1)111-54-6 | Carbamodithioic acid,1,2-ethanediylbis-,saltsandesters |
| U062 | 2303-16-4 | Carbamothioic acid,bis(1-methylethyl)-,S-(2,3-dichloro-2-propenyl)ester |
| U215 | 6533-73-9 | Carbonic acid,dithallium(1+)salt |
| U033 | 353-50-4 | Carbonic difluoride |
| U156 | 79-22-1 | Carbonochloridic acid,methylester (I,T) |
| U033 | 353-50-4 | Carbonoxyfluoride (R,T) |
| U211 | 56-23-5 | Carbon tetrachloride |
| U034 | 75-87-6 | Chloral |
| U035 | 305-03-3 | Chlorambucil |
| U036 | 57-74-9 | Chlordane, alpha and gamma isomers |
| U026 | 494-03-1 | Chlornaphazin |
| U037 | 108-90-7 | Chlorobenzene |
| U038 | 510-15-6 | Chlorobenzilate |
| U039 | 59-50-7 | p-Chloro-m-cresol |
| U042 | 110-75-8 | 2-Chloroethyl vinyl ether |
| U044 | 67-66-3 | Chloroform |
| U046 | 107-30-2 | Chloromethyl methyl ether |
| U047 | 91-58-7 | beta-Chloronaphthalene |
| U048 | 95-57-8 | o-Chlorophenol |
| U049 | 3165-93-3 | 4-Chloro-o-toluidine, hydrochloride |
| U032 | 13765-19-0 | Chromic acid,H(2)CrO(4)calcium salt |
| U050 | 218-01-9 | Chrysene |
| U051 | – | Creosote |
| U052 | 1319-77-3 | Cresol(Cresylicacid) |
| U053 | 4170-30-3 | Crotonaldehyde |
| U055 | 98-82-8 | Cumene(I) |
| U246 | 506-68-3 | Cyanogenbromide (CN)Br |
| U197 | 106-51-4 | 2,5-Cyclohexadiene-1,4-dione |
| U056 | 110-82-7 | Cyclohexane (I) |
| U129 | 58-89-9 | Cyclohexane,1,2,3,4,5,6-hexachloro-,(1alpha,2alpha,3beta,4alpha,5alpha,6beta)- *(continued)* |

Table 5. Hazardous waste "U" list *(continued)*

| Hazardous Waste # | CAS No. | Substance |
|---|---|---|
| | Table 5. Hazardous waste "U" list *(continued)* | |
| U057 | 108-94-1 | Cyclohexanone(I) |
| U130 | 77-47-4 | 1,3-Cyclopentadiene,1,2,3,4,5,5-hexa-chloro- |
| U058 | 50-18-0 | Cyclophosphamide |
| U240 | (1) 94-75-7 | 2,4-D,salts and esters |
| U059 | 20830-81-3 | Daunomycin |
| U060 | 72-54-8 | DDD |
| U061 | 50-29-3 | DDT |
| U062 | 2303-16-4 | Diallate |
| U063 | 53-70-3 | Dibenz[a,h]anthracene |
| U064 | 189-55-9 | Dibenzo[a,i]pyrene |
| U066 | 96-12-8 | 1,2-Dibromo-3-chloropropane |
| U069 | 84-74-2 | Dibutylphthalate |
| U070 | 95-50-1 | o-Dichlorobenzene |
| U071 | 541-73-1 | m-Dichlorobenzene |
| U072 | 106-46-7 | p-Dichlorobenzene |
| U073 | 91-94-1 | 3,3'-Dichlorobenzidine |
| U074 | 764-41-0 | 1,4-Dichloro-2-butene (I,T) |
| U075 | 75-71-8 | Dichlorodifluoromethane |
| U078 | 75-35-4 | 1,1-Dichloroethylene |
| U079 | 156-60-5 | 1,2-Dichloroethylene |
| U025 | 111-44-4 | Dichloroethyl ether |
| U027 | 108-60-1 | Dichloroisopropyl ether |
| U024 | 111-91-1 | Dichloromethoxyethane |
| U081 | 120-83-2 | 2,4-Dichlorophenol |
| U082 | 87-65-0 | 2,6-Dichlorophenol |
| U084 | 542-75-6 | 1,3-Dichloropropene |
| U085 | 1464-53-5 | 1,2:3,4-Diepoxybutane (I,T) |
| U108 | 123-91-1 | 1,4-Diethyleneoxide |
| U028 | 117-81-7 | Diethylhexylphthalate |
| U086 | 1615-80-1 | N,N'-Diethylhydrazine |
| U087 | 3288-58-2 | O,O-DiethylS-methyldithiophosphate |
| U088 | 84-66-2 | Diethylphthalate |
| U089 | 56-53-1 | Diethylstilbestrol |
| U090 | 94-58-6 | Dihydrosafrole |
| U091 | 119-90-4 | 3,3'-Dimethoxybenzidine |
| U092 | 124-40-3 | Dimethylamine (I) |
| U093 | 60-11-7 | p-Dimethylaminoazo benzene |
| U094 | 57-97-6 | 7,12-Dimethylbenz[a]anthracene |
| U095 | 119-93-7 | 3,3'-Dimethylbenzidine |
| U096 | 80-15-9 | alpha,alpha-Dimethylbenzylhydroperoxide (R) |
| U097 | 79-44-7 | Dimethylcarbamoylchloride |
| U098 | 57-14-7 | 1,1-Dimethylhydrazine |
| U099 | 540-73-8 | 1,2-Dimethylhydrazine |
| U101 | 105-67-9 | 2,4-Dimethy lphenol |
| U102 | 131-11-3 | Dimethyl phthalate |
| U103 | 77-78-1 | Dimethyl sulfate |
| U105 | 121-14-2 | 2,4-Dinitrotoluene *(continued)* |

## Intro to Hazardous Waste

| Hazardous Waste # | CAS No. | Substance |
|---|---|---|
| Table 5.  Hazardous waste "U" list  (continued) | | |
| U106 | 606-20-2 | 2,6-Dinitrotoluene |
| U107 | 117-84-0 | Di-n-octylphthalate |
| U108 | 123-91-1 | 1,4-Dioxane |
| U109 | 122-66-7 | 1,2-Diphenylhydrazine |
| U110 | 142-84-7 | Dipropylamine (I) |
| U111 | 621-64-7 | Di-n-propylnitrosamine |
| U041 | 106-89-8 | Epichlorohydrin |
| U001 | 75-07-0 | Ethanal (I) |
| U174 | 55-18-5 | Ethanamine,N-ethyl-N-nitroso- |
| U155 | 91-80-5 | 1,2-Ethanediamine,N,N-dimethyl-N'-2-pyridinyl-N'-(2thienylmethyl)- |
| U067 | 106-93-4 | Ethane,1,2-dibromo- |
| U076 | 75-34-3 | Ethane,1,1-dichloro- |
| U077 | 107-06-2 | Ethane,1,2-dichloro- |
| U131 | 67-72-1 | Ethane,hexachloro |
| U024 | 111-91-1 | Ethane,1,1'-[methylenebis-(oxy)]bis[2-chloro- |
| U117 | 60-29-7 | Ethane,1,1'-oxybis-(I) |
| U025 | 111-44-4 | Ethane,1,1'-oxybis[2-chloro- |
| U184 | 76-01-7 | Ethane,pentachloro- |
| U208 | 630-20-6 | Ethane,1,1,1,2-tetrachloro- |
| U209 | 79-34-5 | Ethane,1,1,2,2-tetrachloro- |
| U218 | 62-55-5 | Ethanethioamide |
| U226 | 71-55-6 | Ethane,1,1,1-trichloro- |
| U227 | 79-00-5 | Ethane,1,1,2-trichloro- |
| U359 | 110-80-5 | Ethanol,2-ethoxy- |
| U173 | 1116-54-7 | Ethanol,2,2'-(nitrosoimino)bis- |
| U004 | 98-86-2 | Ethanone,1-phenyl- |
| U043 | 75-01-4 | Ethene,chloro- |
| U042 | 110-75-8 | Ethene,(2-chloroethoxy)- |
| U078 | 75-35-4 | Ethene,1,1-dichloro- |
| U079 | 156-60-5 | Ethene,1,2-dichloro-, (E)- |
| U210 | 127-18-4 | Ethene,tetrachloro- |
| U228 | 79-01-6 | Ethene,trichloro- |
| U112 | 141-78-6 | Ethylacetate (I) |
| U113 | 140-88-5 | Ethylacrylate (I) |
| U238 | 51-79-6 | Ethylcarbamate (urethane) |
| U117 | 60-29-7 | Ethylether (I) |
| U114 | (1)111-54-6 | Ethylenebisdithiocarbamic acid,salts & esters |
| U067 | 106-93-4 | Ethylene dibromide |
| U077 | 107-06-2 | Ethylene dichloride |
| U359 | 110-80-5 | Ethylene glycol monoethyl ether |
| U115 | 75-21-8 | Ethylene oxide (I,T) |
| U116 | 96-45-7 | Ethylene thiourea |
| U076 | 75-34-3 | Ethylidene dichloride |
| U118 | 97-63-2 | Ethylmethacrylate |
| U119 | 62-50-0 | Ethylmethane sulfonate |
| U120 | 206-44-0 | Fluoranthene |
| U122 | 50-00-0 | Formaldehyde  (continued) |

| Hazardous Waste # | CAS No. | Substance |
|---|---|---|
| U213 | 109-99-9 | Tetrahydrofuran (I) |
| U214 | 563-68-8 | Thallium (I) acetate |
| U215 | 6533-73-9 | Thallium (I) carbonate |
| U216 | 7791-12-0 | Thallium (I) chloride |
| U216 | 7791-12-0 | Thallium chlorideTlcl |
| U217 | 10102-45-1 | Thallium (I) nitrate |
| U218 | 62-55-5 | Thioacetamide |
| U153 | 74-93-1 | Thiomethanol (I,T) |
| U244 | 137-26-8 | Thioperoxydicarbonicdiamide [(H(2)N)C(S)](2)S(2)tetramethyl- |
| U219 | 62-56-6 | Thiourea |
| U244 | 137-26-8 | Thiram |
| U220 | 108-88-3 | Toluene |
| U221 | 25376-45-8 | Toluene diamine |
| U223 | 26471-62-5 | Toluene diisocyanate (R,T) |
| U328 | 95-53-4 | o-Toluidine |
| U353 | 106-49-0 | p-Toluidine |
| U222 | 636-21-5 | o-Toluidine hydrochloride |
| U011 | 61-82-5 | 1H-1,2,4-Triazol-3-amine |
| U227 | 79-00-5 | 1,1,2-Trichloroethane |
| U228 | 79-01-6 | Trichloroethylene |
| U121 | 75-69-4 | Trichloromonofluoromethane See 95-95-42,4,5-Trichlorophenol F027 See 88-06-22,4,6-Trichlorophenol F027 |
| U234 | 99-35-4 | 1,3,5-Trinitrobenzene (R,T) |
| U182 | 123-63-7 | 1,3,5-Trioxane,2,4,6-trimethyl- |
| U235 | 126-72-7 | Tris(2,3-dibromopropyl) phosphate |
| U236 | 72-57-1 | Trypan blue |
| U237 | 66-75-1 | Uracil mustard |
| U176 | 759-73-9 | Urea,N-ethyl-N-nitroso- |
| U177 | 684-93-5 | Urea,N-methyl-N-nitroso |
| U043 | 75-01-4 | Vinyl chloride |
| U248 | (1)81-81-2 | Warfarin,when present at concentrations of 0.3% or less |
| U239 | 1330-20-7 | Xylene (1) |
| U200 | 50-55-5 | Yohimban-16-carboxylicacid,11,17-dimethoxy-18-[(3,4,5-tri-methoxybenzoyl)oxy]-,methy lester,(3 beta,16 beta,17 alpha,18 beta,20 alpha)- |
| U249 | 1314-84-7 | Zinc phosphide,Zn(3)P(2),when present at concentrations of 10% or less. |

**Table 5. Hazardous waste "U" list** *(continued)*

Notes:
(1) CAS Number given for parent compound only.

## Table 5. Hazardous waste "U" list *(continued)*

| Hazardous Waste # | CAS No. | Substance |
|---|---|---|
| U123 | 64-18-6 | Formic acid (C,T) |
| U124 | 110-00-9 | Furan (I) |
| U125 | 98-01-1 | 2-Furancarboxaldehyde (I) |
| U147 | 108-31-6 | 2,5-Furandione |
| U213 | 109-99-9 | Furan,tetrahydro- (I) |
| U125 | 98-01-1 | Furfural (I) |
| U124 | 110-00-9 | Furfuran (I) |
| U206 | 18883-66-4 | Glucopyranose,2-deoxy-2-(3-methyl-3-nitrosoureido)-D |
| U206 | 18883-66-4 | D-Glucose,2-deoxy-2-[[(methylnitrosoamino)carbonyl]amino]- |
| U126 | 765-34-4 | Glycidylaldehyde |
| U163 | 70-25-7 | Guanidine,N-methyl-N-nitro-N-nitroso- |
| U127 | 118-74-1 | Hexachlorobenzene |
| U128 | 87-68-3 | Hexachlorobutadiene |
| U130 | 77-47-4 | Hexachlorocyclopentadiene |
| U131 | 67-72-1 | Hexachloroethane |
| U132 | 70-30-4 | Hexachlorophene |
| U243 | 1888-71-7 | Hexachloropropene |
| U133 | 302-01-2 | Hydrazine (R,T) |
| U086 | 1615-80-1 | Hydrazine,1,2-diethyl- |
| U098 | 57-14-7 | Hydrazine,1,1-dimethyl- |
| U099 | 540-73-8 | Hydrazine,1,2-dimethyl- |
| U109 | 122-66-7 | Hydrazine,1,2-diphenyl- |
| U134 | 7664-39-3 | Hydrofluoricacid (C,T) |
| U135 | 7783-06-4 | Hydrogen sulfide |
| U096 | 80-15-9 | Hydroperoxide,1-methyl-1-phenylethyl-(R) |
| U116 | 96-45-7 | 2-Imidazolidinethione |
| U137 | 193-39-5 | Indeno[1,2,3-cd]pyrene |
| U190 | 85-44-9 | 1,3-Isobenzofurandione |
| U140 | 78-83-1 | Isobutyl alcohol (I,T) |
| U141 | 120-58-1 | Isosafrole |
| U142 | 143-50-0 | Kepone |
| U143 | 303-34-4 | Lasiocarpine |
| U144 | 301-04-2 | Lead acetate |
| U146 | 1335-32-6 | Lead,bis(acetato-O)tetrahydroxytri |
| U145 | 7446-27-7 | Lead phosphate |
| U146 | 1335-32-6 | Lead subacetate |
| U129 | 58-89-9 | Lindane |
| U163 | 70-25-7 | MNNG |
| U147 | 108-31-6 | Maleicanhydride |
| U148 | 123-33-1 | Maleichydrazide |
| U149 | 109-77-3 | Malononitrile |
| U150 | 148-82-3 | Melphalan |
| U151 | 7439-97-6 | Mercury |
| U152 | 126-98-7 | Methacrylonitrile (I,T) |
| U092 | 124-40-3 | Methanamine,N-methyl-(I) |
| U029 | 74-83-9 | Methane,bromo- |
| U045 | 74-87-3 | Methane,chloro-(I,T) *(continued)* |

## *Intro to Hazardous Waste*

| Hazardous Waste # | CAS No. | Substance |
|---|---|---|
| | | **Table 5. Hazardous waste "U" list** *(continued)* |
| U046 | 107-30-2 | Methane,chloromethoxy- |
| U068 | 74-95-3 | Methane,dibromo- |
| U080 | 75-09-2 | Methane,dichloro- |
| U075 | 75-71-8 | Methane,dichlorodifluoro- |
| U138 | 74-88-4 | Methane,iodo- |
| U119 | 62-50-0 | Methane sulfonic acid, ethyl ester |
| U211 | 56-23-5 | Methane,tetrachloro- |
| U153 | 74-93-1 | Methanethiol (I,T) |
| U225 | 75-25-2 | Methane,tribromo- |
| U044 | 67-66-3 | Methane,trichloro- |
| U121 | 75-69-4 | Methane,trichlorofluoro- |
| U036 | 57-74-9 | 4,7-Methano-1H-indene,1,2,4,5,6,7,8,8-octachloro-2,3,3a,4,7,7a hexahydro- |
| U154 | 67-56-1 | Methanol (I) |
| U155 | 91-80-5 | Methapyrilene |
| U142 | 143-50-0 | 1,3,4-Metheno-2H-cyclobuta[cd]pentalen-2-one,1,1a,3,3a,4,5,5a,5b,6-decachlorooctahydro- |
| U247 | 72-43-5 | Methoxychlor |
| U154 | 67-56-1 | Methyl alcohol (I) |
| U029 | 74-83-9 | Methyl bromide |
| U186 | 504-60-9 | 1-Methyl butadiene (I) |
| U045 | 74-87-3 | Methyl chloride (I,T) |
| U156 | 79-22-1 | Methylchlorocarbonate (I,T) |
| U226 | 71-55-6 | Methylchloroform |
| U157 | 56-49-5 | 3-Methyl cholanthrene |
| U158 | 101-14-4 | 4,4-Methylene bis(2-chloroaniline) |
| U068 | 74-95-3 | Methylene bromide |
| U080 | 75-09-2 | Methylene chloride |
| U159 | 78-93-3 | Methyl ethyl ketone (MEK) (I,T) |
| U160 | 1338-23-4 | Methyl ethyl ketone peroxide (R,T) |
| U138 | 74-88-4 | Methyliodide |
| U161 | 108-10-1 | Methyl isobutyl ketone (I) |
| U162 | 80-62-6 | Methyl methacrylate (I,T) |
| U161 | 108-10-1 | 4-Methyl-2-pentanone (I) |
| U164 | 56-04-2 | Methyl thiouracil |
| U010 | 50-07-7 | MitomycinC |
| U059 | 20830-81-3 | 5,12-Naphthacenedione,8-acetyl-10-[(3-amino-2,3,6-trideoxy)-alpha-L-lyxo hexopyranosyl)oxyl]-7,8,9,10-tetrahydro-6,8,11-trihydroxy-1-methoxy-,(8S-cis)- |
| U167 | 134-32-7 | 1-Naphthalenamine |
| U168 | 91-59-8 | 2-Naphthalenamine |
| U026 | 494-03-1 | Naphthalenamine,N,N'-bis(2-chloroethyl)- |
| U165 | 91-20-3 | Naphthalene |
| U047 | 91-58-7 | Naphthalene,2-chloro- |
| U166 | 130-15-4 | 1,4-Naphthalenedione |
| U236 | 72-57-1 | 2,7-Naphthalenedisulfonicacid,3,3'-[(3,3'-dimethyl[1,1'-biphenyl]-4,4'-diyl)bis (azo)bis[5-amino-4-hydroxy]-, tetrasodium salt *(continued)* |

*Intro to Hazardous Waste*

| Hazardous Waste # | CAS No. | Substance |
|---|---|---|
| Table 5.   Hazardous waste "U" list   (*continued*) | | |
| U166 | 130-15-4 | 1,4,Naphthoquinone |
| U167 | 134-32-7 | alpha-Naphthylamine |
| U168 | 91-59-8 | beta-Naphthylamine |
| U217 | 10102-45-1 | Nitric acid,thallium(1+)salt |
| U169 | 98-95-3 | Nitrobenzene (I,T) |
| U170 | 100-02-7 | p-Nitrophenol |
| U171 | 79-46-9 | 2-Nitropropane (I,T) |
| U172 | 924-16-3 | N-Nitrosodi-n-butylamine |
| U173 | 1116-54-7 | N-Nitrosodiethanolamine |
| U174 | 55-18-5 | N-Nitrosodiethylamine |
| U176 | 759-73-9 | N-Nitroso-N-ethylurea |
| U177 | 684-93-5 | N-Nitroso-N-methylurea |
| U178 | 615-53-2 | N-Nitroso-N-methylurethane |
| U179 | 100-75-4 | N-Nitrosopiperidine |
| U180 | 930-55-2 | N-Nitrosopyrrolidine |
| U181 | 99-55-8 | 5-Nitro-o-toluidine |
| U193 | 1120-71-4 | 1,2-Oxathiolane,2,2-dioxide |
| U058 | 50-18-0 | 2H-1,3,2-Oxazaphosphorin-2-amine,N,N-bis(2-chloroethyl)tetrahydro-,2-oxide |
| U115 | 75-21-8 | Oxirane (I,T) |
| U126 | 765-34-4 | Oxirane carboxyaldehyde |
| U041 | 106-89-8 | Oxirane,(chloromethyl)- |
| U182 | 123-63-7 | Paraldehyde |
| U183 | 608-93-5 | Pentachlorobenzene |
| U184 | 76-01-7 | Pentachloroethane |
| U185 | 82-68-8 | Pentachloronitrobenzene(PCNB)See 87-86-5PentachlorophenolF027 |
| U161 | 108-10-1 | Pentanol,4-methyl- |
| U186 | 504-60-9 | 1,3-Pentadiene (I) |
| U187 | 62-44-2 | Phenacetin |
| U188 | 108-95-2 | Phenol |
| U048 | 95-57-8 | Phenol,2-chloro- |
| U039 | 59-50-7 | Phenol,4-chloro-3-methyl- |
| U081 | 120-83-2 | Phenol,2,4-dichloro- |
| U082 | 87-65-0 | Phenol,2,6-dichloro- |
| U089 | 56-53-1 | Phenol,4,4'-(1,2-diethyl-1,2-ethenediyl) bis-,(E)- |
| U101 | 105-67-9 | Phenol,2,4-dimethyl- |
| U052 | 1319-77-3 | Phenol, methyl- |
| U132 | 70-30-4 | Phenol,2,2'-methylenebis[3,4,6-trichloro- |
| U170 | 100-02-7 | Phenol,4-nitro-See 87-86-5Phenol,pentachloro-F027 See58-90-2Phenol,2,3,4,6-tetrachloro-F027 See 88-06-2Phenol,2,4,6-trichloro-F027 |
| U150 | 148-82-3 | L-Phenylalanine,4-[bis(2-chloroethyl)amino]- |
| U145 | 7446-27-7 | Phosphoric acid, lead (2+) salt(2:3) |
| U087 | 3288-58-2 | Phosphorodithioic acid,0,0-diethylS-methylester |
| U189 | 1314-80-3 | Phosphorus sulfide (R) |
| U190 | 85-44-9 | Phthalic anhydride |
| U191 | 109-06-8 | 2-Picoline  (*continued*) |

## Intro to Hazardous Waste

**Table 5. Hazardous waste "U" list** *(continued)*

| Hazardous Waste # | CAS No. | Substance |
|---|---|---|
| U179 | 100-75-4 | Piperidine,1-nitroso- |
| U192 | 23950-58-5 | Pronamide |
| U194 | 107-10-8 | 1-Propanamine (I,T) |
| U111 | 621-64-7 | 1-Propanamine,N-nitroso-N-propyl- |
| U110 | 142-84-7 | 1-Propanamine,N-propyl-(I) |
| U066 | 96-12-8 | Propane,1,2-dibromo-3-chloro- |
| U083 | 78-87-5 | Propane,1,2-dichloro- |
| U149 | 109-77-3 | Propane dinitrile |
| U171 | 79-46-9 | Propane,2-nitro-(I,T) |
| U027 | 39638-32-9 | Propane,2,2'oxybis[2-chloro- |
| U193 | 1120-71-4 | 1,3-Propanesultone See93-72-1Propanoicacid,2-(2,4,5-F027 trichlorophenoxy)- |
| U235 | 126-72-7 | 1-Propanol,2,3-dibromo-,phosphate(3:1) |
| U140 | 78-83-1 | 1-Propanol,2-methyl-(I,T) |
| U002 | 67-64-1 | 2-Propanone (I) |
| U007 | 79-06-1 | 2-Propenamide |
| U084 | 542-75-6 | 1-Propene,1,3-dichloro- |
| U243 | 1888-71-7 | 1-Propene,1,1,2,3,3,3-hexachloro- |
| U009 | 107-13-1 | 2-Propene nitrile |
| U152 | 126-98-7 | 2-Propene nitrile,2-methyl-(I,T) |
| U008 | 79-10-7 | 2-Propenoic acid (I) |
| U113 | 140-88-5 | 2-Propenoic acid,ethylester (I) |
| U118 | 97-63-2 | 2-Propenoic acid,2-methyl-,ethylester |
| U162 | 80-66-2 | 2-Propenoic acid,2-methyl-,methylester (I,T) |
| U194 | 107-10-8 | n-Propylamine (I,T) |
| U083 | 78-87-5 | Propylene dichloride |
| U148 | 123-33-1 | 3,6-Pyridazinedione,1,2-dihydro- |
| U196 | 110-86-1 | Pyridine |
| U191 | 109-06-8 | Pyridine,2-methyl- |
| U237 | 66-75-1 | 2,4-(1H,3H)-Pyrimidinedione,5-[bis(2-chloroethyl)amino]- |
| U164 | 56-04-2 | 4(1H)-Pyrimidinone, 2,3-dihydro-6-methyl-2-thioxo- |
| U180 | 930-55-2 | Pyrrolidine,1-nitroso- |
| U200 | 50-55-5 | Reserpine |
| U201 | 108-46-3 | Resorcinol |
| U202 | (1)81-07-2 | Saccharin, and salts |
| U203 | 94-59-7 | Safrole |
| U204 | 7783-00-8 | Selenious acid |
| U204 | 7783-00-8 | Selenium dioxide |
| U205 | 7488-56-4 | Selenium sulfide |
| U015 | 115-02-6 | L-Serine,diazoacetate (ester) See93-72-1Silvex(2,4,5-TP)F027 |
| U206 | 18883-66-4 | Streptozotocin |
| U103 | 77-78-1 | Sulfuric acid, dimethyl ester |
| U189 | 1314-80-3 | Sulfurphosphide (R) See93-76-52,4,5-TF027 |
| U207 | 95-94-3 | 1,2,4,5-Tetrachlorobenzene |
| U208 | 630-20-6 | 1,1,1,2-Tetrachloroethane |
| U209 | 79-34-5 | 1,1,2,2-Tetrachloroethane |
| U210 | 127-18-4 | Tetrachloroethylene (see58-90-22,3,4,6-Tetrachlorophenol F027) *(continued)* |

# HAZARDOUS WASTE

## PART 2: PREVISIT PREPARATION

**Items to consider getting in advance:**

Σ  The generator's classification category (large, small, or conditionally exempt).

Σ  The facility's contingency plan (showing layout and emergency response procedures).

Σ  The Biennial Report of Hazardous Waste Activity or other listing of hazardous wastes.

**Items to have facility personnel prepare or gather in advance:**

Σ  For all sites:

1) Manifests and land disposal forms.
2) Exception/discrepancy reports.
3) Inspection records.
4) Job descriptions, employee list, and training records.
5) Waste classification records.
6) The EPA ID number form.
7) The Biennial Report of Hazardous Waste Activity.
8) Spill records and reports.
9) Records of arrangements with police, fire department, and emergency response teams.
10) Tests of fire water.
11) The facility evacuation plan.
12) Tank testing records.

Σ  For hazardous waste laboratories:

1) Waste receiving and storage records.
2) Waste treatment records.

Σ  For TSD facilities:

1) Part A and Part B permit applications.
2) The Part B permit.
3) The waste analysis plan.
4) The preparedness and prevention plan.
5) Treatment and disposal records.
6) Monitoring records.
7) Operating records.

## *Hazardous Waste*

| ACRONYMS AND ABBREVIATIONS USED IN THIS MODULE |
| :--- |

| | |
| :--- | :--- |
| < | less than |
| API | American Petroleum Institute |
| AST | aboveground storage tank |
| BIFs | boilers and industrial furnaces |
| CAMU | corrective action management unit |
| CAS | Chemical Abstracts Service |
| CESQG | conditionally exempt small quantity generator |
| CFR | Code of Federal Regulations |
| cm | centimeter |
| CWA | Clean Water Act |
| DAF | dissolved air flotation |
| DOT | U.S. Department of Transportation |
| EP | Extraction Procedure |
| EPA | U.S. Environmental Protection Agency |
| F | Fahrenheit |
| GMP | good management practice |
| HSWA | Hazardous and Solid Waste Amendments of 1984 |
| HW | hazardous waste |
| IAF | induced air flotation |
| ID | identification |
| ISS | Interim Status Standards |
| kg | kilogram |
| kPa | kilopascals |
| LDR | Land Disposal Restriction (standards) |
| LQG | large quantity generator |
| m$^3$ | cubic meter |
| mg/L | milligrams per liter |
| mm | millimeter |
| NACE | National Association of Corrosion Engineers |
| No. | number |
| PCB | polychlorinated biphenyl |
| pH | *pouvoir hydrogene* (French, hydrogen power) |
| ppm | parts per million |
| RCRA | Resource Conservation and Recovery Act |
| SIC | Standard Industrial Classification |
| SPCC Plan | Spill Prevention, Control, and Countermeasure Plan |
| SQG | small quantity generator |
| TCLP | Toxic Characteristic Leaching Procedure |
| TNT | trinitrotoluene |
| tpy | tons per year |
| TSCA | Toxic Substances Control Act |
| TSD | treatment, storage, and disposal (facility) |
| U.S. | United States |
| UDMH | 1,1-dimethyl-hydrazine |
| USC | United States Code |
| USPS | United States Postal Service |
| UST | underground storage tank |
| VOC | volatile organic compound |

# HAZARDOUS WASTE

## PART 3: RULEBOOK

## 1. General

**1.1** Determine actions taken or changes since the previous review of hazardous waste management.

**Guide Note**

Σ Obtain a copy of the previous hazardous waste review or audit and determine if noncompliance issues have been resolved.

Σ If the site has been subject to compliance orders or other legal actions relating to hazardous waste management practices, determine compliance status with these orders.

**1.2** Copies of all relevant federal, state, and local regulations and guidance documents on hazardous waste should be maintained at the facility. (GMP)

**Guide Note**

Σ Determine from interviews if copies are maintained and kept current at the facility of federal (40 CFR 260 - 271; 49 CFR 172 and DOT regulations at 172, 173, 178, and 179), HM-181, and state hazardous waste management regulations.

Σ Determine if facility environmental staff are familiar with and knowledgeable of regulatory requirements.

Σ Determine how the site stays up to date with changing environmental requirements.

## 2. Generators: General

**2.1** Sites that generate solid wastes must determine if the wastes are hazardous wastes. (40 CFR 262.11)

NOTE: CESQGs are not subject to regulation under 40 CFR 262 - 266, 268, or 270, or to notification under requisite requirements of Section 3010 of RCRA. Provided CESQGs meet certain requirements (see 40 CFR 261.5), they need only comply with requirements for identification of hazardous waste as indicated in this paragraph. (40 CFR 262.11)

**Guide Note**

Σ Sites should have a hazardous waste management program (GMP).

Σ Determine if there is a master list of the types and quantities of hazardous wastes generated, treated, and disposed of at the site (GMP).

Σ Verify that the list is complete and updated regularly (GMP).

Σ Discuss with staff how wastes generated on the site were identified and classified.

Σ Determine if the site followed federal or state criteria for identifying the specific listing or characteristic of hazardous waste, whichever is more stringent (e.g., waste oil is sometimes considered a hazardous waste by state regulations but is exempt under most sections of RCRA).

Σ Record any inconsistencies.

## *Hazardous Waste Rulebook*

**2.2** Sites need to notify the appropriate agency of their hazardous waste activities. (40 CFR 262.12 Subpart A; state regulations)

**Guide Note**

Σ Review the Notification of Hazardous Waste Activity (EPA Form 8700-12) that was sent to EPA and verify the EPA ID number (40 CFR 262.12 and the Appendix to 40 CFR 262). NOTE: CESQGs do not require an EPA ID number under federal law, but may be required to obtain one under state law.

**2.3** Generators are required to submit reports on hazardous waste activities. (40 CFR 262.41; state regulations)

**Guide Note**

Σ Determine if biennial reports (EPA Form 8700-13 A) have been submitted by March 1 of each even-numbered year in accordance with federal regulations (40 CFR 262.41 and the Appendix to 40 CFR 262). NOTE: Federal requirements for biennial reports do not apply to SQGs or CESQGs.
Σ Verify that biennial report were properly compiled.
Σ If state agencies require submittals more often, or require reporting from SQGs, confirm compliance.
Σ Determine if separate reports were prepared for the export of hazardous waste.

**2.4** Sites that generate hazardous wastes should actively seek to minimize waste production. (40 CFR 262.41(a)(6) - (8))

**Guide Note**

Σ Determine if large quantity generators (LQGs) have a written plan for minimizing the toxicity and volume of hazardous waste generation (GMP for SQG; see EPA Guidance at 58 *Federal Register* 31114).
Σ Review biennial reports for descriptions of waste minimization efforts and successes (EPA Form 8700-13A).
Σ Interview staff regarding waste minimization awareness and practices. A waste minimization program should include (GMP; see 58 *Federal Register* 31114):

- top management support;
- characteristics of waste generation and waste management costs;
- periodic waste minimization assessments;
- a cost allocation system; and
- program implementation.

**2.5** Sites must maintain copies of records and reports with regard to hazardous waste disposal for at least 3 years. (40 CFR 262.40)

**Guide Note**

Σ Examine records covering the 3 years prior to the date of review. Check for records of waste disposal. These must include:

- manifests and/or signed copies from designated TSD facilities that received the waste. Signed copies from the designated facility that received the waste must be retained for at least 3 years from the date the waste was received by the initial transporter (40 CFR 262.40(a));
- receipts for hazardous waste from appropriate facilities;
- biennial reports submitted to EPA each even-numbered year (40 CFR 262.40(b));
- land ban restriction forms, which must be kept for 5 years (40 CFR 268.7);
- documentation of methods used to determine the hazardous status of wastes, including records of test results, waste analyses, or other determinations (40 CFR 262.40(c)); and
- exception reports (filed when the manifest copy is not received from the designated TSD facility within 45 days, or 60 days for an SQG) (40 CFR 262.40(b) and 262.42).

**2.6** Samples collected for the purpose of testing to determine waste characteristics may be stored temporarily by generators without meeting the requirements of 40 CFR 261, 262, or 263, or RCRA hazardous waste generator notification requirements, or quantity determinations for SQGs or waste accumulation areas, if proper procedures are followed. (40 CFR 261.4)

**Guide Note**

Σ Review quantities of materials used in testing. Verify that the following quantities are not exceeded:

- 10,000 kg of media contaminated with nonacute hazardous waste;
- 1,000 kg of nonacute hazardous waste;
- 1 kg of acute hazardous waste; or
- 2,500 kg of media contaminated with acute hazardous waste for each process being evaluated for each generated waste stream (40 CFR 261.4(e)(2)(i)).

Σ Verify that laboratories that receive the waste for treatability testing comply with the provisions of paragraph 3.1 (40 CFR 261.4(f)).

---

## 3. Generators: Hazardous Waste Laboratories

**3.1** Laboratories and testing facilities may temporarily store or ship hazardous waste samples for treatability studies without meeting hazardous waste generator or transporter requirements, if they operate in accordance with 40 CFR 261.4(f).

**Guide Note**

Σ Determine if written notification was provided to the administering agency at least 45 days before conducting treatability studies (40 CFR 264.4(f)(1)).

Σ Determine if the laboratory or testing facility has an EPA ID number (40 CFR 261.4(f)(2)).

Σ Review records to verify that no more than 10,000 kg of media contaminated with nonacute hazardous waste, 2,500 kg of media contaminated with acute hazardous waste, or 250 kg of other hazardous waste is subject to initiation of treatability studies in any single day (40 CFR 264.4(f)(3)).

Σ Verify that the quantity of waste stored at the facility for evaluation does not exceed 10,000 kg, which can include 10,000 kg of media contaminated with nonacute hazardous waste, 2,500 kg of media contaminated with acute hazardous waste, 1,000 kg of nonacute hazardous waste, and 1 kg of acute hazardous waste (40 CFR 264.4(f)(4)).

Σ Determine whether samples of untreated waste are stored for less than 90 days following completion of the treatability study (40 CFR 264.4(f)(5)).

Σ Determine if samples of untreated waste are stored less than 1 year following shipment of the waste to the laboratory (2 years for bioremediation studies) (40 CFR 264.4(f)(5)).

Σ Determine if samples of treated material are archived in quantities of 500 kg or less, for periods of 5 years or less (40 CFR 264.4(f)(5)).

Σ Verify that the treatability study does not involve placement of hazardous waste on the land or open burning (40 CFR 264.4(s)).

Σ Determine if the facility maintains records for 3 years that show compliance with treatment rate limits, storage time, and quantity limits (40 CFR 264.4(f)(7)).

Σ Determine if the facility maintains a copy of the treatability study contract and shipping papers for 3 years from the completion date of each study (40 CFR 264.4(f)(8)).

Σ Review the annual report that is sent to the administering agency by March 15 of each year (40 CFR 264.4(f)(9)).

Σ Determine if the facility establishes whether any unused samples or residues are hazardous waste, and returns the hazardous samples to the originator (40 CFR 264.4(f)(10)).

*Hazardous Waste Rulebook*

| 4. | Generators: Transportation and Manifesting |
|----|---------------------------------------------|

**4.1** Plants must not offer hazardous wastes to transporters or to TSD facilities that are not properly licensed. (40 CFR 262.12 (c))

**Guide Note**

Σ Examine records pertaining to disposal contract awards; confirm that all transporters of hazardous wastes have appropriate licensing.

Σ Verify that sites receiving wastes have verified in writing that they are approved as interim status or permitted TSD facilities.

**4.2** Before transporting hazardous waste, or offering hazardous wastes for transportation offsite, the site must package the waste in accordance with requirements. (40 CFR 262.30 - 33; 49 CFR 172 (including Subpart F and DOT regulations), 173, 178, and 179; HM-181)

**Guide Note**

Σ Interview staff relative to pretransport procedures for hazardous waste (GMP).

Σ Inspect samples of containers awaiting transport.

Σ Verify that containers are properly constructed and contain no leaks, corrosion, or bulges (40 CFR 262.30; 49 CFR 173, 178, and 179).

Σ Ensure that the waste is compatible with the container (40 CFR 262.30; 49 CFR 173, 178, and 179).

Σ Verify that the labeling and marking on each container are compatible with the manifests and meet DOT regulations (40 CFR 262.31 and 262.32; 49 CFR 172).

Σ Inspect a random sample of containers of 110 gallons or less to ensure the proper label information is displayed (40 CFR 262.31; 49 CFR 172).

Σ Verify that proper placards are available (40 CFR 262.33; 49 CFR 172 Subpart F).

**4.3** Generators of hazardous waste, including SQGs, who ship their wastes for offsite treatment, storage, or disposal must prepare a manifest prior to transport. (40 CFR 262.20)

**Guide Note**

Σ Interview staff responsible for arranging transportation of hazardous wastes.

Σ Check a random sample of manifests to verify they are filled out completely. Check to see all required blanks are filled (often quantities are omitted), the manifest is signed (40 CFR 262.33(a)), and the waste type is properly defined (40 CFR 261; 49 CFR172).

Σ Examine documentation from EPA for the plant's EPA ID number and ensure that it is correct on the manifest.

Σ Examine manifests and verify that they include (Appendix to 40 CFR 262):

- the name, address, telephone number, and EPA ID number of the installation, transporter, TSD facility, and alternate (if any);
- waste information;
- proper DOT name and classification;
- emergency information;
- certification statements;
- signatures and date of acceptance of manifest; and
- percentage of hazardous waste components, if needed.

Σ Verify that manifests are also filled out for recyclable wastes, or federally exempt wastes if regulated by state law (see 40 CFR 262.20(e) for generators of 100 - 1000 kg/month).

Σ Verify that copies of the manifest are sent to the disposal state or generator state, as required.

Σ NOTE: Although not required by federal regulation, it is considered GMP for CESQGs to complete a manifest for each shipment.

**4.4**    A generator must file an exception report if a signed manifest from the offsite receiving facility is not received from the receiving facility in a specified timeframe. (40 CFR 262.42)

**Guide Note**

Σ  Determine that the site has a system to alert personnel if a manifest has not been returned in the specified time (GMP).

Σ  Determine if the facility contacts the transporter and/or the TSD facility if the manifest is not received within 35 days (40 CFR 262.42(a)(1)).

Σ  Interview staff and review exception reports or manifest records. Determine that exception reports are being filed under the following circumstances, as required (40 CFR 262.42(a) and (b)):

- if a manifest from an offsite receiving facility is not received within 45 days from the date of shipment, or 60 days if the generator is an SQG;
- if a manifest received does not have the handwritten signature of the receiving facility operator.

Σ  Verify exception reports include a copy of the manifest and cover letter explaining efforts taken to locate the waste and results of these efforts (40 CFR 262.42(a)(2)).

Σ  If the generator is an SQG, exception reports only need to indicate in some way that confirmation of delivery was not received. Verify that this was indicated and a manifest was submitted (40 CFR 262.42(b)).

**4.5**    Sites using their own vehicles to transport hazardous waste offsite must meet certain standards. (40 CFR 263; 49 CFR 171 - 179; HM-181)

**Guide Note**

Σ  Determine if site vehicles are used to transport hazardous waste off the property. If no, go to the next paragraph.

Σ  If site vehicles are used, determine if they are properly identified with placards and authorized for transport of hazardous waste with an EPA ID number (40 CFR 263.11) and, usually, a state ID number.

Σ  If practical, the reviewer should observe the placarding of vehicles used to transport hazardous wastes (40 CFR 262.33; 49 CFR 172 Subpart F).

**4.6**    The site should ensure that transportation of hazardous wastes between buildings is accomplished in accordance with GMPs to help ensure against spills, releases, and accidents. (GMP)

**Guide Note**

Σ  Determine if procedures exist to manage movement of hazardous wastes throughout the site.

Σ  Determine if drivers are trained in spill control procedures.

Σ  Determine if provisions have been made for securing wastes in vehicles when transporting.

Σ  Determine if the site contingency plan covers accidents during this transport.

Σ  Determine if transportation between buildings entails crossing a public road (this may trigger the need to obtain an EPA ID number; see 40 CFR 263) .

## *Hazardous Waste Rulebook*

**4.7**   Installations that export hazardous wastes for treatment, storage, or disposal must meet certain standards.   (40 CFR 262.50 - 57)

**Guide Note**

Σ   Determine if the site exports hazardous waste to locations outside the U.S.

Σ   Determine if a notification of intended export was submitted to EPA at least 60 days before initial shipment offsite to a foreign destination (40 CFR 262.53(a)).

Σ   Review the notification of intended export.  Determine if the notification covers export activities for a period of not longer than 12 months (40 CFR 262.53(a)).

Σ   Determine if the notice contains the name, address, phone number, and EPA ID number of the exporter (40 CFR 262.53(a)(1)).

Σ   Determine if the notice contains the following information, by consignee, for each hazardous waste type (40 CFR 262.53(a)):

-   waste description and EPA hazardous waste code;
-   DOT shipping name, hazard class, and ID number;
-   frequency, rate, and period of time for shipments;
-   estimated total quantity of hazardous waste;
-   points of entry and departure for each country; and
-   modes of transport and packaging.

Σ   Verify that EPA was notified of any changes in consignee (40 CFR 262.53(c)).

Σ   Verify that copies of EPA's acknowledgment of consent of the receiving country were received by the exporter (40 CFR 262.53(f)).

Σ   Review copies of waste manifests for export.  Confirm that the manifests include the name and address of the consignee, point of departure, and certification of conformance to the terms of the EPA acknowledgment (40 CFR 262.54).

Σ   Verify that manifests of receiving countries were used if required (40 CFR 262.54(e)).

Σ   Verify that a signed copy of the manifest was received within 45 days from the transporter, or within 90 days from the consignee, or an exception report was submitted.  Verify that an exception report was filed if the waste was returned to the U.S. (40 CFR 262.55).

Σ   Determine if a notification was sent to EPA and an acknowledgment of reassignment of the waste was received if the waste could not be delivered (40 CFR 262.54(g)).

Σ   Determine that an annual report is submitted to EPA by March 1 of each year.  Determine if the report contains a description of hazardous wastes exported, identification of transporters, amounts shipped, number of shipments pursuant to each notification, waste minimization efforts and results (in even-numbered years), and certification (40 CFR 262.56).

Σ   Review records and determine if the following are kept for at least 3 years:

-   notifications of intent to export;
-   acknowledgments of consent;
-   confirmations of delivery; and
-   annual reports (40 CFR 262.57).

**4.8**   Installations that import hazardous wastes for treatment, storage, or disposal must complete a hazardous waste manifest.   (40 CFR 262.60)

**Guide Note**

Σ   Review hazardous waste manifests for imported wastes.  Verify that manifests include the name of the foreign generator and the importer's name, address, EPA ID number, and signature.

**4.9** Samples collected for the purpose of testing to determine waste characteristics may be transported without hazardous waste manifests if proper procedures are followed. (40 CFR 261.4)

Guide Note

Σ Determine if shipments of treatability samples are in compliance with DOT and United States Postal Service (USPS) shipping requirements, or are accompanied by the following required information (40 CFR 264.4(e)(2)(iii)):

- the sample collector's name, mailing address, and telephone number;
- the laboratory's name, mailing address, and telephone number;
- quantity of the sample;
- the date of shipment; and
- a description of the sample.

Σ Determine if samples are packaged so that the material does not leak, spill, or vaporize from the packaging (40 CFR 264.4(e)(2)(iii)).

Σ Determine if the mass of each shipment is less than 10,000 kg (40 CFR 264.4(c)(2)(ii)).

Σ Determine if the generator retains records for 3 years, including shipping documents, contracts, waste quantities, laboratory identification, and information on the disposition of samples (40 CFR 264.4(e)(2)(iv)).

**4.10** Transporters who discharge hazardous waste must clean up spills and notify authorities (40 CFR 263.30).

Guide Note

Σ Examine records of hazardous waste discharges that occurred during transport. Determine if the National Response Center received proper notification (49 CFR 171.15).

Σ Determine if written reports of spills were filed with DOT (49 CFR 171.16).

Σ Determine if reports have been submitted for water (bulk shipment) discharges (33 CFR 153.203).

Σ Determine if hazardous waste discharges have been cleaned up by the transporter so that they no longer present a hazard to human health or the environment (40 CFR 263.31).

---

## 5. Generators: Storage of Wastes

**5.1** Hazardous waste may be temporarily stored at locations within the site provided that certain standards are met. (40 CFR 262.34)

Guide Note

Σ Determine from interviews and/or a site tour the locations of all hazardous waste accumulation points at the site, including satellite accumulation points (40 CFR 265.73(a)).

Σ Determine if the site has assigned a person in each area to oversee waste collection (40 CFR 262.34(c)(1)).

Σ Inspect waste storage areas for compliance with labeling, time and quantity limitations, and container and tank storage requirements as indicated in paragraphs 5.2 - 5.8 below.

## *Hazardous Waste Rulebook*

**5.2** Hazardous waste may be temporarily stored at accumulation points for 90 days or less without a permit, provided that certain standards are met. (40 CFR 262.34 Subpart C)

NOTE: Separate storage time limitations pertain to SQGs (see paragraph 5.4).

**Guide Note**

Σ Inspect each accumulation area. Determine that:
- The start date is clearly labeled on each container/tank (40 CFR 262.34(a)(2)).
- Each container/tank is labeled or marked clearly with the words "HAZARDOUS WASTE" (40 CFR 262.34(a)(3) and 262.34(c)(1)(ii)).
- No waste has been stored longer than 90 days (40 CFR 262.34(b)).

Σ Containers are closed except when adding or removing waste (40 CFR 265.173(a)).

Σ Determine if there are procedures to ensure that wastes are not stored for longer than 90 days in accumulation areas (GMP).

**5.3** Sites may store as much as 55 gallons of hazardous waste or 1 quart of acutely hazardous waste in satellite accumulation areas at or near any point of generation without a permit and without complying with storage standards, provided certain standards are met. (40 CFR 262.34 (c))

**Guide Note**

Σ Determine from interviews or visual inspection where satellite accumulation areas are located.

Σ Inspect each point of generation/satellite accumulation area and verify that containers are:
- in good condition (40 CFR 265.171);
- compatible with the wastes stored (40 CFR 265.172);
- kept closed (40 CFR 265.173(a)); and
- labeled as hazardous waste or with an identification of contents (40 CFR 262.34(c)(1)(ii)).

Σ Verify through staff interviews or review of the areas that satellite accumulation areas are maintained by and under the control of the operator of the process generating the waste (40 CFR 262.34(c)(1)).

Σ Determine that the site has procedures to ensure that, within 3 days of exceeding quantity limits, wastes are transferred to hazardous waste storage areas and containers are marked with the date that the limit was reached.

Σ NOTE: Certain states may limit the use of satellite areas or time of occupation.

**5.4** SQGs may store hazardous waste for 180 days or less without a permit (or no more than 270 days without a permit if waste must be transported more than 200 miles) provided that the quantity of waste stored onsite never exceeds 6,000 kg and that certain standards are met. (40 CFR 262.34(d))

**Guide Note**

Σ Inspect hazardous waste storage. Review inventory and interview staff to confirm that no more than 6,000 kg of hazardous waste has been stored at one time. Inspect manifest records to confirm this limit has been complied with (40 CFR 262.34(d)(1)).

Σ If wastes are stored more than 180 days and not over 270 days, verify that the TSD facility receiving the wastes is more than 200 miles away from the SQG (40 CFR 262.34(e)).

Σ Inspect each storage area and interview the manager. Determine that:
- The start date is clearly labeled on each container/tank (40 CFR 262.34(d)(4)).
- Each container/tank is labeled or marked clearly with the words "HAZARDOUS WASTE" (40 CFR 262.34(d)(4)). (Note that states may require more information on container labels.)
- No waste has been stored longer than 180 days, or 270 days if applicable (40 CFR 262.34(d)and(e)).

Σ Determine if there are procedures to ensure that wastes are not stored for longer than 180 days, or 270 days if applicable, in accumulation areas (GMP).

Σ Ensure that the facility consistently generates less than 1,000 kg/month, maintaining SQG status.

*Hazardous Waste Rulebook*

**5.5**  Hazardous waste in containers at storage areas should meet certain standards. (40 CFR 265 Subpart I)

**Guide Note**

Σ  Inspect the containers and look for the following management practices:

- Containers are tightly sealed (40 CFR 265.173) and not leaking, bulging, rusting, or badly dented (40 CFR 265.171).
- Containers are compatible with waste. In particular, check the condition of containers that hold strong caustics or acids and ensure solvents are not stored in plastic drums (40 CFR 265.172).
- Containers are closed (check bungs on drums; look for funnels) (40 CFR 265.173).
- Containers are handled properly (a drum dolly and bung wrench are available) (GMP).
- Containers stored on top of each other have pallets between them and are not stored more than 2 high (GMP).
- Containers of highly flammable wastes are electrically grounded, especially during material transfer (check for grounding clips and wires; make sure wires actually lead to a ground rod or system) (GMP).
- Wastes are not placed in the same container or unwashed containers that previously held incompatible wastes (check for hydrocarbons in acid drums and other incompatible waste mixing) (40 CFR 265.177).
- Containers holding hazardous wastes incompatible with wastes stored nearby in other containers, open tanks, piles, or surface impoundments are separated or protected from each other by a dike, berm, wall, or other divide (40 CFR 265.177(c)).
- At least 3 feet of aisle space is provided between rows of containers to allow access for inspections and emergency response (GMP; also see hazardous material standards for combustible/flammable materials).
- Containers holding ignitable or reactive wastes are at least 50 feet from the property line (SQGs are exempt) (40 CFR 265.176).

Σ  Ensure that the area:

- is marked with a "No Smoking" sign (40 CFR 265.17(a));
- is contained (GMP);
- has an impervious base (GMP); and
- is protected from runoff and weather (GMP).

**5.6**  Containers that previously held hazardous wastes may be reused for other purposes or discarded as a solid waste, providing certain standards are met. (40 CFR 261.7)

**Guide Note**

Σ  Inspect "empty containers" found that are not labeled as hazardous waste to ensure:

- All wastes have been removed by pouring, pumping, aspirating, or by other practice commonly employed to empty that type of container (40 CFR 261.7(b)(1)(i))
- No more than 3% by weight of total capacity and no more than 1 inch of residue remain in the container or inner liner (containers under 110 gallons) (40 CFR 261.7(b)(1)(ii) and (iii)(A)).
- No more than 0.3% by weight of the total capacity of the container remains in the container or inner liner (containers greater than 110 gallons) (40 CFR 261.7(b)(1)(iii)(B)).
- Containers or inner liners holding acutely hazardous waste have been triple-rinsed using a solvent capable of removing the material, or have been cleaned by another method that achieves equivalent removal.

*Hazardous Waste Rulebook*

**5.7** Waste in tanks at less-than-90-day storage sites should comply with certain storage regulations. (40 CFR 265 Subpart J)

**Guide Note**

Σ Inspect tank storage.

Σ Inspect each tank for ruptures, leaks, corrosion, or other signs of failure (e.g., dead vegetation, wet spots, etc.) (40 CFR 265.195(a)(4) and 265.201(c)(5)).

Σ Inspect each uncovered tank to ensure:

- it is operated with at least 2 feet of freeboard; or
- it is equipped with a dike, trench, or other containment structure and:

1) a drainage control system; or
2) a diversion structure (e.g., standby tank) with a capacity that equals or exceeds the volume of the top 2 feet of the tank (40 CFR 265.201 is applicable to SQGs that accumulate wastes for fewer than 180/270 days; see also 40 CFR 265.194(b)].

Σ Inspect each continuously fed tank to ensure it is equipped with a waste feed cut-off system or a bypass system to a standby tank (40 CFR 265.195(a)(1) and 265.201(b)(4)).

Σ Inspect each covered tank used to store ignitable or reactive wastes to ensure it complies with buffer zone requirements specified in the National Fire Protection Association's (NFPA)"Flammable and Combustible Liquids Code" (see hazardous material rulebook) (40 CFR 265.198(b) and 265.201(e)(2)).

Σ Inspect tanks to ensure they do not contain reactive waste unless (40 CFR 265.198(a) and 265.201(e)(1)):

- The waste has been treated, rendered, or mixed so it is no longer ignitable or reactive.
- The waste is stored or treated to protect it from any materials or conditions that may cause it to ignite or react.
- The tank is used solely for emergencies.

**5.8** Hazardous waste tanks that are no longer used must meet certain requirements for tank closure and disposal of contaminated materials. (40 CFR 265.111 and 265.114)

**Guide Note**

Σ Verify tanks have been or are being closed in accordance with an approved closure plan (40 CFR 265.112; see also 40 CFR 265.201(d)).

Σ Determine if the site has a letter or other communication from the regulatory agency stating that the closure has been approved (GMP).

**5.9.** Facilities must identify wastes that are considered to meet the definitions of volatile organics. (40 CFR 265.1084)

NOTE: Paragraphs 5.9 through 5.15 become effective December 5, 1995.

**Guide Note**

Σ Verify that the average volatile organic compound (VOC) concentration of each waste is determined using direct measurement or knowledge of the waste, both initially and every 12 months or whenever a change is made.

Σ Determine that if wastes are occasionally more than 100 parts per million (ppm) by weight, the average is less. Verify that the maximum and minimum levels are recorded, operating conditions are noted when wastes are more than 100 ppm, and averaging calculations are recorded (40 CFR 265.1084(a)(4)).

Σ If the waste averages more than 100 ppm VOC by weight at the point of origination, then paragraphs 5.10 to 5.15 apply.

**5.10**  Containers greater than 0.1 cubic meter ($m^3$) (26 gallons) holding hazardous wastes must be equipped to control the release of VOCs. (40 CFR 265.1087)

**Guide Note**

Σ  Ensure that containers are equipped with covers that allow no detectable organic emissions and are closed. Ensure testing has been performed per Method 21 of 40 CFR 60 Appendix A for each waste, or the container is equipped with a cover that meets DOT requirements if the container is less than 0.46 $m^3$ (121 gallons).

Σ  Determine if containers that are part of trucks, trailers, or railcars have demonstrated vapor tightness using Method 27 of 40 CFR 60.

Σ  Determine if any waste is being treated in a container by waste stabilization, an exothermic process, or 1 that requires heat. If yes, ensure that the container is only opened in an enclosed area with proper ventilation.

Σ  Ensure that containers meet the following requirements:

- Covers with vents must be designed, installed, maintained, and operated so that they give off no detectable emissions.
- All openings must be in a closed, sealed position except when inspecting or adding waste.
- Vents must meet standards (40 CFR 265.1088).
- If containers are equipped with venting safety devices, they must not be used for routine venting, remaining closed except during unplanned events.

**5.11**  Transfers of wastes into containers greater than 0.46 $m^3$ (121 gallons) must meet certain requirements. (40 CFR 265.1087(b)(3))

**Guide Note**

Σ  Ensure that pumping into the container is done using a tube and that all openings are sealed except the 1 containing the tube.

Σ  Determine if the tube is continuously submerged or is no more than 2 inside diameters of the tube from the bottom, or 15.25 centimeters (cm), whichever is greater, or the tube is connected to a permanent port that is 15.25 cm or less from the bottom.

Σ  For methods other than pumping, ensure that all openings are closed except the 1 through which waste is being added.

Σ  If the tube is leaking, ensure waste is removed and the container repaired.

**5.12**  Tanks containing hazardous wastes must be equipped to control the release of VOCs. (40 CFR 265.1085)

**Guide Note**

Σ  Verify that the tank is equipped with 1 of the following:

- a cover and a closed-vent system. Waste may be placed in a covered tank only if no mixing, stirring, agitation, or circulation is present; the tank is not heated; waste is not treated; and the maximum organic vapor pressure of the waste is less than 5.2 kilopascals (kPa) if the tank has a capacity of 151 $m^3$ or more or less than 27.6 kPa if the tank has a capacity between 75 $m^3$ and 151 $m^3$;
- a fixed roof and an internal floating roof, with the floating roof resting on the waste at all times except during filling or when the tank is emptied. The floating roof must be foam- or liquid-sealed, with 2 seals mounted 1 above the other, or have a mechanical shoe seal (40 CFR 265.1091(a)(1));
- an external floating roof with 2 seals of either the mechanical-shoe type or the liquid-mounted type, mounted 1 above the other (40 CFR 265.1091(a)(2)); or
- a pressure tank designed to operate as a closed system.

*(continued on next page)*

Σ

### *Hazardous Waste Rulebook*

Ensure that tanks meet the following requirements:

- Covers with vents must be designed, installed, maintained, and operated so that they give off no detectable emissions.
- All openings must be in a closed, sealed position except when inspecting or adding waste.
- Vents must meet standards (40 CFR 265.1088).
- Piping systems must exist for transfer to and from the units, and piping must be maintained.
- If tanks are equipped with venting safety devices, they are not to be used for routine venting, remaining closed except during unplanned events.

Σ Verify that tanks with a fixed roof and an internal floating roof are inspected and monitored (40 CFR 265.1091(b)(1)) and the information is recorded. Inspections must be carried out on primary and secondary seals at least every 12 months. Repairs must be made within 45 days or the tank removed from service. An exception exists for double-seal systems, which require inspection every 5 years (40 CFR 265.1091(b)(1)). Inspection records must include the design of equipment and certification of specifications, dates inspected, observed conditions, the nature of any defects, and the dates repairs were made or tanks were removed from service (40 CFR 265.1091(c)(1)).

Σ Verify that tanks with an external floating roof are inspected and monitored (40 CFR 265.1091(b)(2)) and the information recorded. Also determine the following:

- gap distances between the primary seal and the wall of tank, and the secondary seal and the wall of the tank. Measurements must be conducted within 60 days of bringing the tank into service. Subsequently, measurements must be conducted every 5 years for the primary seal and every year for the secondary seal (40 CFR 265.1091(b)(2));
- that records are maintained on the gap distances, and that tanks are repaired or taken out of service within 45 days should the gaps become too extensive (40 CFR 265.1091(b)(2));
- that inspection records include the design of equipment and certification of specifications, gap measurements, dates inspected, observed conditions, the nature of any defects, and the dates repairs were made or tanks were removed from service (40 CFR 265.1091(c)(2)).

**5.13** All closed-vent systems must be installed and operated according to standards. (40 CFR 265.1088)

**Guide Note**

Σ Verify that the system routes all gases to the control device.

Σ Determine if leak detection monitoring was performed at installation and annually thereafter.

Σ If bypass devices are present, ensure a flow indicator is present and a valve is installed at the inlet to the device.

Σ Verify that the control device is:

- designed to reduce the total organic content by at least 95% by weight;
- an enclosed combustion device;
- a flare;
- a carbon absorption system whose carbon is changed regularly;
- operated at all times when gases are vented;
- tested to demonstrate compliance (40 CFR 265.1088(c)(5)).

**5.14** Inspections and monitoring must be performed on volatile organic control equipment. (40 CFR 265.1089)

**Guide Note**

Σ Verify that all units are visually inspected and monitored for detectable organic emissions. Exemptions exist for tank sections that are buried or underground, and for specified containers.

Σ Verify that inspections are performed before placing the unit into service and every 6 months thereafter.

Σ Determine if inspections are conducted per 40 CFR 60 Appendix A Method 21. Determine that, if emissions were detected, leak repairs were attempted within 5 calendar days and completed within 15 days, unless the unit is emptied. If repairs could not be made, verify that the unit was prevented from accepting additional hazardous waste.

Σ Determine if a written plan and schedule to perform inspections has been prepared and implemented (40 CFR 265.1089(e)).

**5.15** Records must be maintained for a minimum of 3 years to demonstrate compliance with volatile organic waste regulations. (40 CFR 265.1090)

**Guide Note**

Σ Review records for the following:

- tank covers (design description and certification of compliance with specifications);
- floating membranes (design description and certification of compliance with specifications);
- enclosures (design description and certification of compliance with specifications);
- closed-vent systems (signed and dated certification stating performance levels, including a design description and certification of compliance with specifications, or performance testing including the test plan and results);
- Method 27 tests;
- visual inspections;
- monitoring for detectable organic emissions;
- the date of each repair attempt, details of repair methods, and the date of successful repair;
- continuous monitoring;
- the management of carbon removed from carbon absorption units;
- inspections of tank covers.

Σ For tanks with fixed roofs only, verify that the following have been recorded:

- the date and time each waste sample was collected for measurement of maximum organic vapor pressure;
- the results of tests;
- tank dimensions and design capacity.

Σ Persons electing not to use controls must record each waste determination including the time, date, and location of sampling.

Σ Operators of boilers or industrial furnaces must record unit ID numbers.

Σ For equipment that is hard to inspect and monitor, maintain the following:

- ID numbers for covers;
- an explanation of why inspection and monitoring have not been carried out; and
- plans and schedules to resolve the issue.

*Hazardous Waste Rulebook*

## 6. Generators: Inspections

6.1 Regular inspections of hazardous waste storage areas are required in accordance with written schedules and checklists. (40 CFR 265.174 and 265.195)

**Guide Note**

Σ Determine if inspections of hazardous waste storage areas are conducted at least as often as the frequencies specified below:

- containers and container storage areas (to look for leakage and signs of deterioration) - weekly (40 CFR 265.174);
- tank waste feed cut-off systems and bypass systems - daily (40 CFR 265.195(a)(1) and 265.201(c)(1));
- tank monitoring equipment (e.g., pressure and temperature gauges) - daily (40 CFR 265.95(a)(3) and 265.201(e)(2));
- waste levels in uncovered tanks - daily (weekly for SQGs) (40 CFR 265.201(e)(3));
- aboveground tank integrity (for signs of corrosion, erosion, leakage of fixtures or seams) - daily (40 CFR 265.195 (a)(2)) or weekly for SQGs (40 CFR 265.201(c)(4)); and
- the area surrounding a tank for signs of leakage (wet spots, dead vegetation) - daily (40 CFR 265.195(a)(4)) or weekly for SQGs (40 CFR 265.201(e)(5)).

Σ The inspection record must include the date and time of the inspection, name of the inspector, observations made, and date and nature of any remedial action. Review inspection records to determine the recording of these items (40 CFR 265.195(c) and 265.15(d)).

Σ Interview inspector(s) to determine if they understand what they are inspecting (GMP).

Σ Inspect a sample of the areas to determine if all defects are identified (compare to the latest inspection) (GMP).

6.2 Inspections and testing of emergency equipment to ensure proper operation in the event of an emergency should be conducted as part of a testing and maintenance program. (GMP; 40 CFR 265.33)

**Guide Note**

Σ Interview the designated person to verify that routine inspections are conducted as part of a testing and maintenance program for the following (40 CFR 265.33):

- communications or alarm systems;
- fire extinguishers and other fire protection equipment;
- spill control equipment; and
- decontamination equipment.

Σ Review testing and inspection records (40 CFR 265.15(d)).

Σ Inspect selected equipment as part of the site tour. Verify inspection tags, if used.

6.3 The site safety manager or other designated person should be responsible for conducting safety evaluations and inspections of the handling and storage of hazardous wastes. (GMP)

**Guide Note**

Σ Interview safety officer and determine inspection frequency, forms, and reporting procedures.

Σ Obtain a list of locations and materials inspected by the safety officer.

Σ Review safety reports and evaluate the safety record.

Σ Verify that any corrective actions recommended in the safety reports have been implemented.

## 7. Generators: Emergency Preparedness and Prevention

**7.1** Sites that generate hazardous wastes must be maintained and operated to minimize the possibility of a fire, explosion, or any unplanned release of hazardous waste, and to meet certain requirements for preparedness and prevention. (40 CFR 262.34(a)(4) and 265.30 - 37)

**Guide Note**

Σ Inspect the site to determine if the following required equipment is easily accessible and in operating condition:

- internal communications or alarm system, such as a horn, klaxon, or public address system (40 CFR 265.32(a));
- a telephone or hand-held 2-way radio (40 CFR 265.32(b));
- portable fire extinguishers and special extinguishing equipment (foam, inert gas, or dry chemicals) (40 CFR 265.32(c));
- spill control equipment (40 CFR 265.32(c));
- decontamination equipment (40 CFR 265.32(c));
- fire hydrants or other source of water (reservoir, storage tank, etc.) with adequate volume and pressure within 500 feet of the facility, foam-producing equipment, sprinkler, or water spray (40 CFR 265.32(d)).

Σ Make a spot check to assess the condition of equipment (e.g., boots without holes, respirators with unused cartridges, etc.) (40 CFR 265.33).

Σ Determine if adequate aisle space (3 feet is GMP) is provided between rows of drums to allow checks for leakage, corrosion, proper labeling, etc.

Σ Review letters of correspondence with local authorities or other records, the emergency response plan, or interview staff of the facility and agencies to determine whether arrangements have been made to familiarize police and fire departments, local hospitals, and emergency response teams with the facility, and whether agreements have been made to coordinate emergency response efforts (40 CFR 265.37).

Σ Determine if emergency response drills or exercises are held and if a follow-up critique is performed (GMP).

**7.2** Sites that generate hazardous waste must have a written contingency plan. (40 CFR 262.34(a) and 265.50 - 56)

NOTE: If the site has already prepared a Spill Prevention, Control, and Countermeasure (SPCC) Plan, amendment of the SPCC Plan to incorporate hazardous waste management is an alternate approach (does not apply to SQGs; see paragraph 7.4)

**Guide Note**

Σ Review the contingency plan to determine if it contains:

- descriptions of actions to be taken in response to fires, explosions, or any unplanned release of hazardous waste (40 CFR 265.52(a));
- descriptions of arrangements agreed to by local police and fire departments, hospitals, contractors, and emergency response teams to coordinate emergency response services (40 CFR 265.52(c));
- an up-to-date list of names, addresses, and telephone numbers (office and home) of primary and alternate emergency coordinators (40 CFR 265.52(d)).
- a list of emergency and decontamination equipment, and location, description, and outline of capabilities (40 CFR 265.52(e)); and
- an evacuation plan for facility personnel, including evacuation procedures and routes (40 CFR 65.32(f).

*(continued on next page)*

Σ

## *Hazardous Waste Rulebook*

Verify that emergency equipment listed in the contingency plan is consistent with what physically is found at the site, including (40 CFR 265.54).:

- fire extinguishers;
- spill control equipment;
- alarm systems; and
- decontamination equipment.

Σ Verify that copies of the contingency plan are maintained at the site and also have been submitted to the local police and fire departments, hospitals, and state and local emergency response teams (documentation is required) (40 CFR 265.53).

Σ Determine if the contingency plan is routinely reviewed and updated when appropriate (e.g., when the permit is revised, the plan fails, there is a change in facility or personnel, etc.) (40 CFR 265.54).

Σ Determine if records containing the time, date, and details of any incident that requires implementation of the contingency plan are kept (GMP).

**7.3**  Site emergency coordinators must follow certain emergency procedures whenever there is an imminent or actual emergency situation.  (40 CFR 265.56)

### Guide Note

Σ Review the contingency plan for the site.  Verify that the emergency coordinator is required to conduct the following emergency procedures:

- identifying the character, exact source, amount, and real extent of any released materials (40 CFR 265.56(b));
- assessing possible hazards to human health or the environment, including direct and indirect effects (e.g., release of gases, surface runoff from water or chemicals used to control fire or explosions, etc.) (40 CFR 265.56(c));
- activating site alarms and evacuating personnel, as necessary (40 CFR 265.56(a)(1)).

Σ Take every reasonable measure to avoid recurrence or spreading of hazardous waste, including the following actions:

- Stop processes and operations at the site.
- Collect and contain the released waste.
- Remove or isolate containers (40 CFR 265.56(e)).
- Monitor for leaks, pressure build-up, gas generation, or ruptures in valves, pipes, or other equipment whenever appropriate (40 CFR 265.56(f)).
- Provide for treatment, storage, or disposal of recovered waste, contaminated soil or surface water, or other material (40 CFR 265.56(g)).
- Ensure that no waste that may be incompatible with the released material is treated, stored, or disposed of until cleanup is completed (40 CFR 265.56(h)(1)).
- Ensure that all emergency response equipment is cleaned or restocked before operations resume (40 CFR 265.56(b)(2)).

Σ Verbally and in writing, notify appropriate federal, state, and local authorities of imminent or actual emergency situations (40 CFR 265.56(a)(2), (d), (i) and (j)].

Σ Provide a written follow-up within 15 days.

**7.4**  SQGs must meet certain requirements for emergency planning.  (40 CFR 262.34(d)(5) and 265 Subpart C)

### Guide Note

Σ Interview staff to determine that at least 1 person responsible for coordinating emergency measures is either onsite or available at all times (40 CFR 262.34(d)(5)(i)).

Σ Verify through interviews or a review of written procedures (documentation is not required by federal regulations) that appropriate procedures are taken in the event of an emergency, including:

- in the case of fire, calling the fire department or extinguishing the fire;
- in the case of a spill, containing the flow of hazardous waste to the extent possible, and cleaning up the waste and contaminated materials or soil;
- in the case of any release that could threaten human health outside the site or that has reached surface water, notifying the National Response Center (40 CFR 262.34(d)(5)(iv)).

Σ Inspect the site to verify that posted next to the telephone is the following information:

- the name and telephone number of the emergency coordinator;
- the location of fire extinguishers, spill control material, and, if present, fire alarms; and
- the telephone number of the fire department, unless the site has a direct alarm (40 CFR 262.34(d)(5)(ii)).

Σ Determine if any event requiring emergency response has occurred at the site and verify that:

- The emergency coordinator or the emergency coordinator's designee responded promptly.
- Appropriate emergency response procedures were employed (40 CFR 262.34).
- The National Response Center was notified if human health outside the site was threatened or a discharge reached surface waters.

## 8. Generators: Training

**8.1** All site personnel involved in handling, storage, or transport of hazardous wastes, or who may be involved in emergency response, should receive certain training. (40 CFR 262.34, 265.16: see also paragraphs in the hazardous material section related to 29 CFR 1910.120)

NOTE: SQGs have different requirements; see paragraph 8.2.

**Guide Note**

Σ Review the written training plan and training records. Interview staff responsible for training.
Σ Determine if the training program includes, at a minimum (40 CFR 265.16(a)(3)):

- equipment operations;
- contingency plan implementation;
- emergency response procedures;
- emergency/monitoring equipment use, inspection, and repair;
- parameters for automatic feed cut-off systems;
- communications and alarms;
- response to fire, explosion, groundwater contamination incidents; and
- site shutdown procedures.

Σ Examine training records to determine if they include:

- job titles, descriptions, and employees' names;
- descriptions of the type and amount of training required for each position;
- documentation of training completed and annual training reviews (40 CFR 265.16(d)); and
- a copy of the training plan.

Σ Determine if all employees who handle hazardous waste are trained.
Σ Determine if employees are trained within 6 months of hire or the assumption of a of new position (40 CFR 265.16(b)).
Σ Determine if annual training is provided to everyone who handles hazardous waste (40 CFR 265.16(c)).
Σ Determine if training records are current for present employees and are maintained for at least 3 years after employees leave the site (40 CFR 265.16(e)).

*Hazardous Waste Rulebook*

**8.2** SQGs must provide training to ensure that all employees are familiar with proper waste-handling procedures and emergency response procedures, relevant to their responsibilities during normal operations and emergencies. (40 CFR 262.34(d)(5)(iii))

**Guide Note**

Σ Interview the emergency coordinator or site manager, or review written documentation (not required), to determine if personnel receive training in the following (GMP):

- equipment operations;
- contingency plan implementation;
- emergency response procedures;
- emergency/monitoring equipment use, inspection, and repair;
- parameters for automatic feed cut-off systems;
- communications and alarms;
- response to fire, explosion, and groundwater contamination incidents; and
- site shutdown procedures.

## 9. Generators: Land Disposal

**9.1** Generators must either test their waste or use knowledge of their waste to determine if the waste is restricted from land disposal. (40 CFR 268.7(a))

**Guide Note**

Σ Interview employees and review records. Has the facility determined which of its waste streams are restricted from land disposal?

Σ If not, obtain a list of waste streams sent offsite and their characteristics. Refer to 40 CFR 268 to determine if the wastes are restricted from land disposal.

Σ Has the facility performed a TCLP on the waste being sent to land disposal to determine if the waste would be restricted from land disposal? If not, determine why the procedure was not performed (40 CFR 261 Appendix II).

Σ If a waste is restricted from land disposal, refer to 40 CFR 268 to determine the appropriate treatment standard. This treatment standard must be met before the waste is land disposed unless the waste is subject to a case-by-case extension under 40 CFR 268.5, a no-migration exemption under 40 CFR 268.6, or a nationwide extension on the effective date of the treatment standard.

**9.2** If a generator is managing a restricted waste that does not meet the treatment standards, the generator must notify the treatment or storage facility of the treatment standards with each shipment. (40 CFR 268.7(a)(1))

**Guide Note**

Σ Review records and determine if notices have been sent with each shipment.

Σ The notice must include the following information:

- the EPA hazardous waste number (40 CFR 268.7(a)(1)(i));
- the corresponding treatment standards and all applicable prohibitions set forth in 40 CFR 268.32 or RCRA section 3004(d) (40 CFR 268.7(a)(1)(ii) and (iv));
- the manifest number associated with the shipment of waste (40 CFR 268.7(a)(1)(iii)); and
- waste analysis data, where available.

**9.3** If a generator is managing a restricted waste that meets the treatment standards, the generator must submit with each shipment a notice and a certification that the waste meets the treatment standards. (40 CFR 268.7(a)(2))

**Guide Note**

Σ Review records and determine if notices and certifications have been sent.

Σ The certification must be signed by an authorized representative and must contain the wording found in 40 CFR 268.7(a)(2)(ii).

Σ The notifications must contain (40 CFR 268.7(a)(2)(i)):

- the EPA hazardous waste number;
- a description of the waste;
- the manifest number; and
- waste analysis data (when available).

**9.4** If the generator's waste is subject to a case-by-case extension, an exemption, or a nationwide variance, with each shipment of waste the generator must submit a notice to the receiving facility that the waste is not prohibited from land disposal. (40 CFR 268.7(a)(3))

**Guide Note**

Σ Review records and determine if notices have been sent.

**9.5** If a generator is managing a restricted waste in tanks or containers and is treating it to meet standards, the generator must develop a written waste analysis plan. (40 CFR 268.7(a)(4))

**Guide Note**

Σ Review records to determine if the plan has been developed (see also 40 CFR 268.45 Table 1).

Σ The plan must be filed with EPA or the state 30 days in advance of treatment (40 CFR 268.7(a)(4)(ii)).

Σ A copy of the plan must be kept onsite (40 CFR 268.7(a)(4)).

Σ Wastes shipped offsite must have the notice specified in 40 CFR 268.7(a)(2) (40 CFR 268.7(a)(4)(iii)).

**9.6** If a generator determines that its waste is restricted based on knowledge or testing, copies of all records related to that determination must be kept on file. (40 CFR 268.7(a)(5))

**Guide Note**

Σ Review records to determine if they are complete (40 CFR 268.7(a)(5)).

**9.7** If a generator determines that its restricted waste is exempt from regulation, the generator must file a notice stating such determination in its files, including notes on the disposition of the waste. (40 CFR 268.7(a)(6))

**Guide Note**

Σ Review records to determine if they are complete (40 CFR 268.7(a)(6); see also 40 CFR 261 to verify the exemption).

**9.8** Generators must retain onsite a copy of all notices, certifications, demonstrations, waste analysis data, and other documentation for at least 5 years. (40 CFR 268.7(a)(7))

**Guide Note**

Σ All supporting data used to determine if a waste is restricted must be retained onsite in the generator's files (40 CFR 268.7(a)(7)).

Σ Review records.

*Hazardous Waste Rulebook*

## 10. TSD Facilities: General

*The remaining sections in this module apply to permitted hazardous waste operations. (If the facility does not store for longer than 90 days, or treat or dispose of hazardous wastes onsite, then skip the remainder of this module.) Note that interim standards are referenced at 40 CFR 265 while the final status standards are at 40 CFR 264.*

**10.1** Copies of Notification of Waste Activity, Part A and Part B permit applications, and Interim Status and Final Part B Permits must be maintained at the facility. (40 CFR 270)

**Guide Note**

Σ Obtain and review available documents.
Σ Note dates of issuance and expiration.
Σ Compare Parts A and B to existing physical facilities.
Σ If the site has a Part B permit, is it in compliance with all the permit requirements?
Σ Do the Part B application and permit documents accurately reflect existing TSD activities?
Σ Is the Part B permit current? What is the date of expiration?
Σ Discuss with environmental staff any problems or notice of deficiencies with permit applications, or negotiations with the state/EPA.

**10.2** A facility must have an EPA ID number for its treatment, storage, or disposal activities. (40 CFR 264.11 and 265.11)

**Guide Note**

Σ Review the facility's permit and obtain an ID number.

**10.3** If the facility receives hazardous waste from an outside source, it must file a notice with the EPA Regional Administrator at least 4 weeks before the waste arrives at the TSD facility. (40 CFR 264.12(a) and 265.12(a))

**Guide Note**

Σ Verify that the written notice is maintained in the facility's operating record.
Σ Verify that the facility was approved to handle the waste type.

## 11. TSD Facilities: Waste Identification

**11.1** TSD facilities should have a waste analysis plan in place and properly implemented. (40 CFR 264.13 and 265.13)

**Guide Note**

Σ Review the written waste analysis plan. Does it include all wastes generated and received at the facility?

Σ Verify that the plan includes:

- parameters for which each waste will be analyzed (40 CFR 264.13(b)(1) and 265.13(b)(1));
- test methods used to test for these parameters (40 CFR 264.13(b)(2) and 265.13(b)(2));
- the sampling method used to obtain representative samples (40 CFR 264.13(b)(3) and 265.13(b)(3));
- the frequency with which the initial analysis will be reviewed or repeated (40 CFR 264.13(b)(4) and 65.13(b)(4));
- provisions for retesting waste when the process or operation generating the waste changes (40 CFR 264.13(a)(3));
- procedures used to inspect and analyze each hazardous waste shipment received (40 CFR 264.13(a)(4)).

Σ Verify that records show confirmation that waste received matches analyses (40 CFR 264.13(b)(3) and 265.13(b)(3)).

Σ Verify that records of waste movement are kept within the facility (40 CFR 264.13(b)(2) and 265.13(b)(2)).

Σ Verify that analyses related to restricted waste disposal are maintained.

**11.2** The waste analysis plan must be updated to reflect requirements applicable to restricted wastes. (40 CFR 265.13(b)(6) and 268.7(c) and (d))

**Guide Note**

Σ Check the waste analysis plan to determine if it contains procedures for the following types of facilities:

- Treatment facilities:

  1) For wastes with treatment standards expressed as concentrations in the waste extract, the TSD facility must use procedures contained in Appendix I of 40 CFR 268 (40 CFR 268.8(b)(1)).
  2) For wastes that are prohibited under 40 CFR 268.32 or RCRA 3004(d) but not subject to any treatment standards under 40 CFR 268 Subpart D, the TSD facility must test the treatment residues according to 40 CFR 268.32 (40 CFR 268.8(b)(2)).
  3) For wastes with treatment standards expressed as concentrations in the waste (40 CFR 268.43), the TSD facility must test the treatment residues, not an extract of such residues (40 CFR 268.8(b)(3)).

- Land disposal facilities must test restricted waste or an extract of the waste or treatment residue using the test method described in 40 CFR 268 Appendix I or using any methods required by generators under 40 CFR 268.32.

## 12. TSD Facilities: Transportation and Manifesting

**12.1** TSD facilities must review information provided on all manifests and shipping papers that they receive. (40 CFR 264.71 and 265.71)

**Guide Note**

Σ Verify that the owner/operator or his/her agent does the following:

- Signs and dates each copy of the manifest to certify that the hazardous waste described was received (40 CFR 264.71(a)(1) and 265.71(a)(1)).
- Notes any significant discrepancies between the manifest and the wastes actually delivered on each copy of the manifest (40 CFR 264.71(a)(2) and 265.71(a)(2)).
- Immediately gives the transporter at least 1 copy of the signed manifest (40 CFR 264.71(a)(3) and 265.71(a)(3)).
- Returns a copy of the signed manifest to the generator within 30 days after delivery (40 CFR 264.71(a)(4) and 265.71(a)(4)).
- Retains a copy of the signed shipping paper onsite for 3 years (40 CFR 264.71(a)(5) and 265.71(a)(5)).

Σ Verify that if the facility receives hazardous waste from a rail or waste (bulk shipment) transporter, it is accompanied by a shipping paper (40 CFR 264.71(b) and 265.71(b)). Verify that the shipping paper contains all the information required on a manifest (except EPA ID numbers, generator's certification, and signatures).

**12.2** TSD facilities must report significant discrepancies on all shipments they receive. (40 CFR 264.72(a) and 265.72(a))

**Guide Note**

Σ Verify that significant discrepancies such as the following are reported:

- quantity variations greater than 10% for bulk waste;
- any variation in piece count for batch waste (e.g., 1 drum missing in a truckload);
- obvious differences of waste type discovered by inspection or waste analysis (i.e., not reporting toxic constituents on the manifest) (40 CFR 264.72 and 265.72).

Σ Verify that significant discrepancies are reconciled with the generator or the transporter within 15 days:

- If discrepancies are not reconciled within 15 days, verify that the owner or operator sent a letter to the EPA Regional Administrator (40 CFR 264.72(b) and 265.72(b)).
- If letters are being sent as necessary, verify that they include a description of the discrepancy and attempts to reconcile it, and a copy of the manifest.

**12.3** Plants must not offer hazardous wastes to transporters or to TSD facilities that are not properly licensed. (40 CFR 262.12 (c))

**Guide Note**

Σ Examine records pertaining to disposal contract awards. Confirm that all transporters of hazardous wastes have appropriate licensing.

Σ Verify that sites receiving wastes have confirmed in writing that they are approved as interim status or permitted TSD facilities.

**12.4** Before transporting hazardous waste or offering hazardous wastes for transportation offsite, the site must package the waste in accordance with requirements. (40 CFR 262.30 - 33; 49 CFR 172 and DOT regulations at 172, 173, 178, and 179; HM-181)

**Guide Note**

Σ Interview staff relative to pretransport procedures for hazardous waste.

Σ Inspect samples of containers awaiting transport.

Σ Verify that containers are properly constructed and contain no leaks, corrosion, or bulges.

Σ Ensure that the waste is compatible with the container.

Σ Verify that labeling and marking on each container is compatible with the manifests and meets DOT regulations.

Σ Inspect a random sample of containers of 110 gallons or less to ensure the proper label information is displayed.

Σ Verify that proper placards are available.

**12.5** A generator must file an exception report if a signed manifest from the offsite receiving facility is not received from the receiving facility in a specified timeframe. This requirement applies to all TSD facilities when they generate and dispose of hazardous wastes. (40 CFR 262.42)

**Guide Note**

Σ Determine that the site has a system to alert personnel if a manifest has not been returned in the specified time (GMP).

Σ Determine if the facility contacts the TSD facility if the manifest is not received within 35 days (40 CFR 262.42(a)(1)).

Σ Interview staff and review exception reports or manifest records to determine that exception reports are being filed under the following circumstances, as required (40 CFR 262.42 and 262.44):

- if the manifest from the offsite receiving facility is not received within 45 days from the date of shipment, or 60 days if the generator is an SQG; and
- if the manifest received does not have the handwritten signature of the receiving facility operator.

Σ Verify that exception reports include a copy of the manifest and cover letter explaining efforts taken to locate the waste and results of these efforts (40 CFR 262.42(a)(2)(ii)).

Σ If the generator is an SQG, exception reports only need to indicate in some way that confirmation of delivery was not received. Verify that this is being indicated and the manifest is being submitted (40 CFR 262.42(b)).

**12.6** Sites using their own vehicles to transport hazardous waste offsite must meet certain standards. (40 CFR 263; 49 CFR 171 - 179; HM-181)

**Guide Note**

Σ Determine if site vehicles are used to transport hazardous waste off the property. If not, go to paragraph 12.7.

Σ If site vehicles are used, determine if the facility to which the vehicles are assigned are properly identified with placards and authorized for transport of hazardous waste (i.e., the facility has an EPA ID number and, usually, a state ID number).

Σ If practical, observe the placarding of vehicles used to transport hazardous wastes (40 CFR 262.33 and 49 CFR Subpart F).

*Hazardous Waste Rulebook*

**12.7** The site should ensure that transportation of hazardous wastes between buildings is accomplished in accordance with GMPs to help ensure against spills, releases, and accidents. (GMP)

**Guide Note**

Σ Determine if procedures exist to manage movement of hazardous wastes throughout the site.
Σ Determine if drivers are trained in spill control procedures.
Σ Determine if provisions have been made for securing wastes in vehicles when transporting.
Σ Determine if the site contingency plan covers accidents during this transport.
Σ Determine if transportation between buildings entails crossing a public road. This may trigger the need for licensing (see 40 CFR 263).

---

## 13. TSD Facilities: Storage of Wastes

*This section applies to storage within a permitted facility. Storage at waste generation sites in plants is generally subject to the standards of section 5 of this module (Generators: Storage of Wastes).*

**13.1** Facilities with hazardous waste facilities that store containers of hazardous waste must comply with certain storage regulations. (40 CFR 264.170 - 178 and 265.170 - 178)

**Guide Note**

Σ Determine if the facility stores hazardous waste in containers. If yes, inspect the containers and look for the following management practices:

- Containers are tightly sealed (40 CFR 264.173 and 265.173) and not leaking, bulging. rusting, or badly dented (40 CFR 264.171 and 265.171).
- Containers are compatible with waste. In particular, check the condition of containers that hold strong caustics or acids and ensure solvents are not stored in plastic drums (40 CFR 265.172).
- Containers are closed. Check bungs on drums and look for funnels (40 CFR 264.173 and 265.173).
- Containers are handled properly (i.e., a drum dolly and bung wrench are available) (GMP).
- Containers stored on top of each other have pallets between them and are not stored more than 2 high (GMP).
- Containers of highly flammable wastes are electrically grounded, especially during material transfer. Check for grounding clips and wires and make sure wires actually lead to a ground rod or system (GMP).
- Wastes are not placed in the same container or unwashed containers that previously held incompatible wastes. Check for hydrocarbons in acid drums and other incompatible waste mixing (40 CFR 264.177 and 265.177).
- Containers holding hazardous wastes incompatible with wastes stored nearby in other containers, open tanks, piles, or surface impoundments are separated or protected from each other by a dike, berm, wall, or other divide (40 CFR 264.177(c) and 265.177(c)).
- At least 3 feet of aisle space is provided between rows of containers to allow access for inspections and emergency response (GMP; also see hazardous material standards for combustible/flammable materials).
- Containers holding ignitable or reactive waste are at least 50 feet from the property line (40 CFR 264.177 and 265.176).

Σ Ensure that the area is:

- marked with a "No Smoking" sign (40 CFR 264.17(a) and 265.17(a));
- contained (GMP; for interim status see paragraph 13.3 for final status facilities);
- has an impervious base (GMP; for interim status see 40 CFR 264.175(b)(1)); and
- protected from runoff and weather (GMP; for interim status see 40 CFR 264.175(a)(4 ·.

**13.2** Containers that previously held hazardous wastes and hazardous materials may be reused for other purposes or discarded as solid waste, providing certain standards are met. (40 CFR 261.7)

**Guide Note**

Σ  Inspect "empty containers" that are not labeled as hazardous waste to ensure:

- All wastes have been removed by pouring, pumping, aspirating, or by other practice commonly employed to empty that type of container (40 CFR 261.7(b)(1)(i)).
- No more than 3% by weight of total capacity and no more than 1 inch of residue remain in the container or inner liner (for containers under 110 gallons) (40 CFR 261.7(b)(1)(ii) and (iii)(A)).
- No more than 0.3% by weight of the total capacity of the container remains in the container or inner liner (for containers greater than 110 gallons) (40 CFR 261.7(b)(1)(iii)(B)).
- Containers or inner liners holding acutely hazardous waste have been triple rinsed using a solvent capable of removing the material, or have been cleaned by another method that achieves equivalent removal.

**13.3** Final-status TSD facilities with container storage areas must have a containment system. (40 CFR 264.170 - 178)

**Guide Note**

Σ  Inspect the container storage area. Determine if the following criteria are met:

- Containers are stored on a base that is free from cracks or gaps and is impervious enough to contain leaks, spills, and precipitation (40 CFR 264.175(b)(1)).
- The base is sloped (or otherwise designed) to drain and remove liquids resulting from leaks, spills, or precipitation (40 CFR 264.175(b)(2)).
- Containers are elevated or protected from contact with accumulated liquids (40 CFR 264.175(b)(2)).
- The containment system has adequate capacity to contain 10% of the volume of the containers or the volume of the largest container, whichever is greatest (40 CFR 264.175(b)(3)).
- Runon into the containment system is prevented, or the system has sufficient capacity to contain any runon that might enter the system (40 CFR 264.175(b)(4)).

**13.4** Installations must conduct weekly inspections of container storage areas and the containment system. (40 CFR 264.174 and 265.174)

**Guide Note**

Σ  Determine if inspections are conducted at least weekly to look for leaking containers and signs of deterioration of containers or the containment system (40 CFR 264.174 and 265.174).

Σ  Make sure inspection forms are signed and dated and indicate any problems discovered. They should also make reference to any repairs completed to resolve problems (40 CFR 264.15(d) and 265.15(d)).

**13.5** Installations with hazardous waste facilities that use tank systems to treat or store hazardous waste should prevent any migration of wastes or accumulated liquid to the soil, groundwater, or surface water. (40 CFR 264.190 - 199 and 265.190 - 199)

**Guide Note**

Σ  Determine if the installation treats or stores hazardous wastes in tanks. If not, go to paragraph 14.1.

Σ  Verify there is a written procedure in place to prevent the addition of wastes that are incompatible with the tank or liner (40 CFR 264.194(a) and 265.194(a)).

Σ  Conduct an inspection of each tank for ruptures, leaks, corrosion, or other signs of failure (e.g., dead vegetation, wet spots, etc.) (40 CFR 264.195(a)(3) and 265.195(a)(4)).

*(continued on next page)*

Σ

## *Hazardous Waste Rulebook*

Inspect each uncovered tank to ensure that it is operated with sufficient freeboard, as follows:

- Uncovered tanks must be protected by maintenance of sufficient freeboard to prevent overtopping by wave or wind action or precipitation (40 CFR 264.194(b)(3) and 265.194(b)(3)).
- TSDs must operate uncovered tanks to ensure at least 2 feet of freeboard, unless the tank is equipped with a containment structure, drainage control system, or division structure having a minimum capacity equal to the volume of the top 2 feet of the tank (40 CFR 265.20(b)(3)).

Σ  Check each tank that does not have secondary containment to ensure:

- Its integrity assessment has been reviewed and certified by an engineer (40 CFR 265.191(a) and 264.191(a)).
- The age of the tank has been documented or estimated (40 CFR 264.19(b)(4) and 265.191(b)(4)).
- The tank's design and strength are adequate and compatible with wastes stored (40 CFR 265.191(b) and 264.191(b)).
- Corrosion protection measures (e.g., coating, cathodic protection) are in place (40 CFR 264.191(b)(3) and 265.191(b)(3)).
- Leak test results are recorded annually (40 CFR 264.193(i) and 265.193(i)).

Σ  Inspect tanks to ensure they do not contain ignitable or reactive waste unless:

- The waste has been treated, rendered, or mixed so it no longer is ignitable or reactive.
- The waste is stored or treated to protect it from any materials or conditions that may cause it to ignite or react.
- The tank is used solely for emergencies (40 CFR 264.198 and 265.198).

Σ  Verify a schedule is in place to ensure that existing tanks systems are equipped with secondary containment systems according to the following schedule:

- for existing tank systems that store or treat hazardous wastes containing dioxin - January 12, 1989 (40 CFR 264.193(a)(2) and 265.193(a)(2));
- for existing tank systems of known and documented age - January 12, 1989 or when the system is 15 years old, whichever occurs later (40 CFR 264.193(a)(3) and 265.193(a)(3));
- for existing tank systems of undocumented age - January 12, 1995 if the tank system is more than 7 years old with secondary containment installed by January 12, 1989 (40 CFR 264.193(a)(4) and 265.193(a)(4)).

Σ  Verify the facility inspects the following each day and documents the results of the inspections in the operating record:

- overfill/spill control equipment (40 CFR 265.195(a)(1));
- aboveground portions of tank systems for corrosion or releases (40 CFR 264.195(b)(1) and 265.195(a)(2));
- data gathered from monitoring and leak detection equipment (40 CFR 264.195(b)(2) and 265.195(a)(3)); and
- construction materials and areas surrounding the externally accessible portion of tank systems for erosion or signs of releases (40 CFR 264.195(b)(3) and 265.195(a)(4)).

Σ  Verify that written procedures are in place for responding to leaks and repairing or replacing tanks that are leaking or otherwise unfit for use (40 CFR 264.196 and 265.196).

Σ  Verify the removal of accumulated liquid from the secondary containment system within 24 hours (40 CFR 264.193(c)(4) and 265.193(c)(4)).

Σ  Verify the owner/operator has ensured closed tank systems have been properly decontaminated and closed in accordance with the closure plan (40 CFR 264.197 and 265.197).

**13.6** New tank systems (installed after July 14, 1986) must meet certain construction criteria. (40 CFR 264.192 and 193 (final status), or 265.192 and 193 (interim status))

**Guide Note**

Σ  Verify the tank system includes 1 or more of the following devices:
- a liner external to the tank (40 CFR 264.193(d)(1) and 265.193(d)(1));
- a vault (40 CFR 264.193(d)(2) and 265.193(d)(2));
- a doublewalled tank (40 CFR 264.193(d)(3) and 265.193(d)(3));
- an equivalent device approved by the EPA Regional Administrator (written documentation must be kept onsite of this approval) designed to prevent the release of hazardous waste or hazardous constituents to the environment (40 CFR 264.193(d)(4) and 265.193(d)(4)).

Σ  Verify that all secondary containment systems that have been installed meet the following requirements:
- They are constructed of or lined with materials compatible with the waste(s) placed in the tank system, and are of sufficient strength and thickness to prevent failure due to pressure gradients, physical contact with the waste(s), climatic conditions, and daily operating stress (40 CFR 264.193(c)(1) and 265.193(c)(1)).
- They are placed on a foundation or base capable of supporting the secondary containment system, providing resistance to pressure gradients above and below the system, and preventing failure due to settlement, compression, or uplift (40 CFR 264.193(c)(2) and 265.193(c)(2)).
- They are provided with a leak detection system to detect within 24 hours the failure of the primary or secondary containment or the presence of released hazardous waste or accumulated liquid (40 CFR 264.193(c)(3) and 265.193(c)(3)).
- They are sloped or otherwise designed to drain and remove within 24 hours liquids from leaks, spills, or precipitation (40 CFR 264.193(c)(4) and 265.173(c)(4)).

**13.7** Facilities must identify wastes that are considered to meet the definitions of volatile organics. (40 CFR 264.1083 and 265.1084)

Note: Paragraphs 13.7 to 13.16 are effective December 5, 1995.

**Guide Note**

Σ  Verify that the average VOC concentration of each waste is determined using direct measurement or knowledge of the waste, both initially and every 12 months or whenever a change is made.
Σ  Determine that if wastes are occasionally more than 100 ppm by weight, the average is less. Verify that the maximum and minimum levels are recorded, operating conditions are noted when wastes are more than 100 ppm, and averaging calculations are recorded (40 CFR 265.1084(a)(4)).
Σ  Determine if alternate methods were used to determine applicability (40 CFR 265.1084).
Σ  If the waste averages more than 100 ppm VOC by weight at the point of origination, then paragraphs 13.8 to 13.16 apply.

**13.8** Containers greater than 0.1 m³ (26 gallons) holding hazardous wastes must be equipped to control the release of VOCs. (40 CFR 264.1086 and 265.1087)

**Guide Note**

Σ  Ensure that containers are equipped with covers that allow no detectable organic emissions and are closed. Ensure testing has been performed per Method 21 of 40 CFR 60 Appendix A for each waste, or the container is equipped with a cover that meets DOT requirements if the container is less than 0.46 m³ (121 gallons).
Σ  Determine if containers that are part of trucks, trailers, or railcars have demonstrated vapor tightness using Method 27 of 40 CFR 60.

*(continued on next page)*

Σ

## *Hazardous Waste Rulebook*

Determine if any waste is being treated in a container by waste stabilization, an exothermic process, or 1 that requires heat. If yes, ensure that the container is only opened in an enclosed area with proper ventilation.
Σ  Ensure that containers meet the following requirements:

- Covers with vents must be designed, installed, maintained, and operated so that they give off no detectable emissions.
- All openings must be in a closed, sealed position except when inspecting or adding waste.
- Vents must meet standards (40 CFR 264.1087 and 265.1088).
- If containers are equipped with venting safety devices, they must not be used for routine venting, remaining closed except during unplanned events.

**13.9**  Transfers of wastes into containers greater than 0.46 m$^3$ (121 gallons) must meet certain requirements.  (40 CFR 264.1086(b)(3) and 265.1087(b)(3))

**Guide Note**

Σ  Ensure that pumping into the container is done using a tube and that all openings are sealed except the 1 containing the tube.

Σ  Determine if the tube is continuously submerged or is no more than 2 inside diameters of the tube from the bottom, or 15.25 centimeters (cm), whichever is greater, or the tube is connected to a permanent port that is 15.25 cm or less from the bottom.

Σ  For methods other than pumping, ensure that all openings are closed except the 1 through which waste is being added.

Σ  If the tube is leaking, ensure waste is removed and the container repaired:

**13.10**  Tanks containing hazardous wastes must be equipped to control the release of VOCs.  (40 CFR 264.1086 and 265.1085)

**Guide Note**

Σ  Verify that the tank is equipped with 1 of the following:

- a cover and a closed-vent system.  Waste may be placed in a covered tank only if no mixing, stirring, agitation, or circulation is present; the tank is not heated; waste is not treated; and the maximum organic vapor pressure of the waste is less than 5.2 kilopascals (kPa) if the tank has a capacity of 151 m$^3$ or more or less than 27.6 kPa if the tank has a capacity between 75 m$^3$ and 151 m$^3$;
- a fixed roof and an internal floating roof, with the floating roof resting on the waste at all times except during filling or when the tank is emptied.  The floating roof must be foam- or liquid-sealed, with 2 seals mounted 1 above the other, or have a mechanical shoe seal (40 CFR 265.1091(a)(1));
- an external floating roof with 2 seals of either the mechanical-shoe type or the liquid-mounted type, mounted 1 above the other (40 CFR 265.1091(a)(2)); or
- a pressure tank designed to operate as a closed system.

Σ  Ensure that tanks meet the following requirements:

- Covers with vents must be designed, installed, maintained, and operated so that they give off no detectable emissions.
- All openings must be in a closed, sealed position except when inspecting or adding waste.
- Vents must meet standards (40 CFR 265.1088).
- Piping systems must exist for transfer to and from the units, and piping must be maintained.
- If tanks are equipped with venting safety devices, they are not to be used for routine venting, remaining closed except during unplanned events.

Σ

Verify that tanks with a fixed roof and an internal floating roof are inspected and monitored (40 CFR 265.1091(b)(1)) and the information is recorded. Inspections must be carried out on primary and secondary seals at least every 12 months. Repairs must be made within 45 days or the tank removed from service. An exception exists for double-seal systems, which require inspection every 5 years (40 CFR 265.1091(b)(1)). Inspection records must include the design of equipment and certification of specifications, dates inspected, observed conditions, the nature of any defects, and the dates repairs were made or tanks were removed from service (40 CFR 265.1091(c)(1)).

Σ Verify that tanks with an external floating roof are inspected and monitored (40 CFR 265.1091(b)(2)) and the information recorded. Also determine the following:

- gap distances between the primary seal and the wall of tank, and the secondary seal and the wall of the tank. Measurements must be conducted within 60 days of bringing the tank into service. Subsequently, measurements must be conducted every 5 years for the primary seal and every year for the secondary seal (40 CFR 265.1091(b)(2));
- that records are maintained on the gap distances, and that tanks are repaired or taken out of service within 45 days should the gaps become too extensive (40 CFR 265.1091(b)(2));
- that inspection records include the design of equipment and certification of specifications, gap measurements, dates inspected, observed conditions, the nature of any defects, and the dates repairs were made or tanks were removed from service (40 CFR 265.1091(c)(2)).

**13.11** Surface impoundments to which volatile organic waste is being added must be controlled. (40 CFR 264.1085 and 265.1086)

**Guide Note**

Σ Ensure that surface impoundments are equipped with a cover that is vented through a closed-vent system to a control device.

Σ Ensure that covers meet the following:

- Covers with vents must be designed, installed, maintained, and operated to operate with no detectable emissions.
- All openings must be in a closed, sealed position except when inspecting or adding waste.
- Vents must meet standards (40 CFR 264.1087 and 265.1088).

Σ Alternatively, a floating membrane may be used if (40 CFR 264.1085 and 265.1086):

- No mixing, stirring, agitation, or circulation is present, the impoundment is not heated, and waste is not treated by stabilization or a process that produces an exothermic reaction.
- The entire surface is covered and the area underneath is not vented to the atmosphere.
- Covers with vents are designed, installed, maintained, and operated to operate with no detectable emissions (40 CFR 265.1086 (e)).
- All openings are in a closed, sealed position except when inspecting or adding waste (40 CFR 265.1086 (e)).
- A synthetic membrane of high-density polyethylene with a thickness greater than 2.5 millimeters (mm) is used (40 CFR 265.1086 (e)).
- Alternatively, a composite material with similar organic permeability and chemical and physical properties to maintain the material integrity is used (40 CFR 265.1086 (e)).

Σ Ensure that an enclosed piping system is installed and maintained (40 CFR 264.1085 and 265.1086).

Σ Ensure that if tanks are equipped with safety devices, they are not used for routine venting and remain closed except during unplanned events.

## *Hazardous Waste Rulebook*

**13.12** All open-vent systems must be installed and operated according to standards. (40 CFR 264.1087 and 265.1088)

**Guide Note**

Σ Verify that the system routes all gases to the control device.

Σ Determine if leak detection monitoring was performed at installation and annually thereafter.

Σ If bypass devices are present, ensure a flow indicator is present and a valve is installed at the inlet to the device.

Σ Verify that the control device is:

- designed to reduce the total organic content by at least 95% by weight;
- an enclosed combustion device;
- a flare;
- a carbon absorption system whose carbon is changed regularly;
- operated at all times when gases are vented;
- tested to demonstrate compliance (40 CFR 264.1087(c)(5) and 265.1088(c)(5)).

**13.13** All closed-vent systems must be installed and operated according to standards. (40 CFR 264.1087 and 265.1088)

**Guide Note**

Σ Verify that all units are visually inspected and monitored for detectable organic emissions. Exemptions exist for tank sections that are buried or underground, and for specified containers.

Σ Verify that inspections are performed before placing the unit into service and every 6 months thereafter.

Σ Determine if inspections are conducted per 40 CFR 60 Appendix A Method 21. If emissions are detected, verify that repairs are attempted within 5 calendar days and repairs are made within 15 days unless the unit is emptied. If repairs could not be made, verify that the unit was prevented from accepting additional hazardous waste.

Σ Determine if a written plan and schedule to perform inspections has been prepared and implemented (40 CFR 264.1088(e) and 265.1089(e)).

**13.14** Records must be maintained for a minimum of 3 years to demonstrate compliance with volatile organic waste regulations. (40 CFR 264.1089 and 265.1090)

**Guide Note**

Σ Review records for the following:

- tank covers (design description and certification of compliance with specifications);
- floating membranes (design description and certification of compliance with specifications);
- enclosures (design description and certification of compliance with specifications);
- closed-vent systems (signed and dated certification stating performance levels, including a design description and certification of compliance with specifications, or performance testing including the test plan and results);
- Method 27 tests;
- visual inspections;
- monitoring for detectable organic emissions;
- the date of each repair attempt, details of repair methods, and the date of successful repair;
- continuous monitoring;
- the management of carbon removed from carbon absorption units;
- inspections of tank covers.

Σ For tanks with fixed roofs only, verify that the following have been recorded:

- the date and time each waste sample was collected for measurement of maximum organic vapor pressure;
- the results of tests;
- tank dimensions and design capacity.

Σ

Persons electing not to use controls must record each waste determination including the time, date, and location of sampling.

Σ Operators of boilers or industrial furnaces must record unit ID numbers.

Σ For equipment that is hard to inspect and monitor, maintain the following:

- ID numbers for covers;
- an explanation of why inspection and monitoring have not been carried out; and
- plans and schedules to resolve the issue.

**13.15** Reports must be filed each time a waste is placed into a unit that has been exempted. (40 CFR 264.1090)

**Guide Note**

Σ Verify that written reports were submitted within 15 days of the time the operator became aware of the occurrence. Report must contain the facility's EPA ID number, its name and address, a description of incidents, dates incidents occurred, and corrective actions.

**13.16** Facilities using closed-vent systems must submit semiannual written reports to the EPA Regional Administrator when instances of noncompliance occur. (40 CFR 264.1090(c))

**Guide Note**

Σ Verify that the report includes:

- all periods of noncompliance that lasted at least 24 hours; and
- all circumstances where flares occurred and visible emissions were present.

Σ Review the report for the facility's EPA ID number, its name and address, an explanation of noncompliance incidents, and corrective actions undertaken.

---

| **14.  TSD Facilities:  Training** |
| --- |

**14.1** All personnel who work in a TSD facility must receive certain training. (40 CFR 264.16 Subpart B or 265.16 Subpart B; 29 CFR 1910.120(p))

**Guide Note**

Σ Review the written description of the training program and interview the training director.

Σ Verify the training program identifies a training director by title and name (40 CFR 264.16(d)(1) and 265.16(d)(1)).

Σ Verify that the responsibilities and duties of the training director are clearly defined and sufficient to implement the program (40 CFR 264.16(d)(2) and 265.16(d)(2)).

Σ Verify there is documentation that shows that the training director has been trained or has gained experience in hazardous waste management procedures (40 CFR 264.16(a)(2) and 265.16(a)(2)).

Σ Interview the TSD facility operators/supervisors to determine if the training program is in place. Look for the following (40 CFR 264.16 and 265.16):

- contingency plan implementation;
- procedures for using, inspecting, repairing, and replacing TSD facility emergency and monitoring equipment;
- operation of communications and alarm systems;
- response to fire or explosion;
- response to groundwater contamination incidents;
- limits for automatic waste systems;
- facility shutdown procedures; and
- proper handling of the hazardous wastes managed at the facility.

Σ Also see 29 CFR 1910.120(p) (HAZWOPER) requirements in the hazardous materials module.

*Hazardous Waste Rulebook*

**14.2** Training records must be maintained for all facility staff at the TSD facility. (40 CFR 264.16 and 265.16)

**Guide Note**

Σ Verify training records are maintained and available.

Σ Examine training records. Verify they include the following:

- job title and description for each employee by name (40 CFR 264.16(d)(1) and 265.16(d)(1));
- a description of introductory and continuing training available (40 CFR 264.16(d)(3) and 265.16(d)(3)); and
- documentation of the type and amount of training received by each employee (40 CFR 264.16(d)(4) and 265.16(d)(4)).

Σ Examine employee records to determine if each employee's training is received in the first 6 months and updated annually, or when a significant job change occurs (40 CFR 264.16(b) and (c) and 265.16(b) and (c)).

Σ Ensure the facility has identified each person who requires training.

Σ Determine if training records are retained for the life of the facility for present employees and for 3 years for former employees.

---

## 15. TSD Facilities: Emergency Preparedness and Prevention

**15.1** Facilities must control entry to the active portion of the facility. (40 CFR 264.14 and 265.14)

**Guide Note**

Σ Inspect each TSD facility. Determine if the following features are in place:

- The facility is surrounded by a fence or natural barrier (40 CFR 264.14(b)(2)(i) and 265.14(b)(2)(i)).
- Entrances are locked or monitored on a 24-hour basis (40 CFR 264.14(b)(2)(ii) and 265.14(b)(2)(ii)).
- Signs with wording "Danger - Unauthorized Personnel Keep Out" are posted at each entrance and other locations as appropriate (40 CFR 264.14(c) and 265.14(c)).
- Signs are legible from at least 25 feet (40 CFR 264.14(c) and 265.14(c)).
- Signs are written English and in any other language predominant in the area (i.e., facilities bordering Mexico must post in Spanish).

**15.2** All TSD facilities must be designed, constructed, maintained, and operated to minimize the possibility of a fire, explosion, or any unplanned release of hazardous waste. (40 CFR 264.30 - 37 and 265.30 - 37)

**Guide Note**

Σ Inspect the site to determine if the following required equipment is easily accessible and in operating condition:

- internal communications or alarm system, such as a horn, klaxon, or public address system (40 CFR 264.32(a) and 265.32(a));
- a telephone or hand-held 2-way radio (40 CFR 264.32(b) and 265.32(b));
- portable fire extinguishers and special extinguishing equipment (foam, inert gas, or dry chemicals) (40 CFR 264.32(c) and 265.32(c));
- spill control equipment (40 CFR 264.32(c) and 265.32(c));
- decontamination equipment (40 CFR 264.32(c) and 265.32(c));
- fire hydrants or other source of water (reservoir, storage tank, etc.) with adequate volume and pressure within 500 feet of the facility, foam-producing equipment, sprinkler, or water spray (40 CFR 264.32(d) and 265.32(d)).

Σ

## Hazardous Waste Rulebook

Make a spot check to assess the condition of equipment (e.g., boots without holes, respirators with unused cartridges, etc.) (40 CFR 264.33 and 265.33).

Σ  Determine if adequate aisle space (3 feet is GMP) is provided between rows of drums to allow checks for leakage, corrosion, proper labeling, etc. (40 CFR 264.35 and 265.35).

Σ  Review letters of correspondence with local authorities or other records, review the emergency response plan, or interview staff of the facility and agencies to determine whether arrangements have been made to familiarize police and fire departments, local hospitals, and emergency response teams with the facility, and whether agreements have been made to coordinate emergency response efforts (40 CFR 264.37 and 265.37).

Σ  Determine if emergency response drills or exercises are held and if a follow-up critique is performed (GMP).

**15.3**  All TSD facilities must take precautions to prevent the accidental ignition or reaction of ignitable or reactive waste. (40 CFR 264.17 and 265.17)

**Guide Note**

Σ  If the facility handles ignitable, reactive, or incompatible wastes, verify the following:

- The waste is separated and confined from sources of ignition or reaction, sparks, spontaneous ignition, and radiant heat (40 CFR 264.17(a) and 265.17(a)).
- Smoking and open flames are confined to specifically designated areas (40 CFR 264.17(a) and 265.17(a)).
- "No Smoking" signs are posted in areas where ignitable or reactive wastes are handled (40 CFR 264.17(a) and 265.17(a)).
- Incompatible wastes are always separated (40 CFR 264.17(b), 264.177, 264.199, and Subparts K-X; see also 40 CFR 265).
- There are written procedures for avoiding commingling of incompatible wastes (40 CFR 264.17(c) and 265.17(c)).
- Flammable/ignitable wastes are grounded (40 CFR 264.17(a) and 265.17(a)).

**15.4**  Facilities with TSD facilities must have a contingency plan. (40 CFR 264.50 - 56 Subpart D and 265.50 - 56 Subpart D)

NOTE: Contingency planning may be addressed in an SPCC Plan or other emergency plan.

**Guide Note**

Σ  Review the contingency plan to determine if it contains:

- descriptions of actions to be taken in response to fires, explosions, or any unplanned release of hazardous waste (40 CFR 264.52(a) and 265.52(a));
- descriptions of arrangements agreed to by local police and fire departments, hospitals, contractors, and emergency response teams to coordinate emergency response services (40 CFR 264.52(c) and 265.52(c));
- an up-to-date list of names, addresses, and telephone numbers (office and home) of primary and alternate emergency coordinators (40 CFR 264.52(d) and 265.52(d));
- a list giving location, description, and an outline of capabilities of emergency and decontamination equipment (40 CFR 264.52(e) and 265.52(e)); and
- an evacuation plan for facility personnel, including evacuation procedures and routes (40 CFR 264.32(f) and 265.32(f).

Σ  Verify that emergency equipment listed in the contingency plan is consistent with what physically is found at the site, including:

- fire extinguishers;
- spill control equipment;
- alarm systems; and
- decontamination equipment (40 CFR 264.54 and 265.54).

*(continued on next page)*

Σ

## Hazardous Waste Rulebook

      Verify that copies of the contingency plan are maintained at the site and also have been submitted to the local police and fire departments, hospitals, and state and local emergency response teams (documentation is required) (40 CFR 264.53 and 265.53).

Σ  Determine if the contingency plan is reviewed and updated, if the permit has been revised, if the plan has ever failed, or a change in the facility or personnel has occurred (40 CFR 264.54 and 265.54).

Σ  Determine if records containing the time, date, and details of any incident that requires implementation of the contingency plan are kept (40 CFR 264.73(b)(4) and 265.73(b)(4)).

**15.5**  Emergency coordinators must follow certain emergency procedures whenever there is an imminent or actual emergency situation.  (40 CFR 264.56 and 265.56)

### Guide Note

Σ  Review the contingency plan for the site.  Verify that the emergency coordinator is required to follow these emergency procedures:

- Identify the character, exact source, amount, and extent of any released materials (40 CFR 264.56(b) and 265.56(b)).
- Assess possible hazards to human health or the environment, including direct and indirect effects (e.g., release of gases, surface runoff from water or chemicals used to control fire or explosions, etc.) (40 CFR 264.56(c) and 265.56(c)).
- Activate site alarms and evacuate personnel, as necessary (40 CFR 264.56(a)(1) and 265.56(a)(1)).

Σ  Take every reasonable measure to avoid recurrence or spreading of hazardous waste including the following actions:

- Stop processes and operations at the site.
- Collect and contain the released waste.
- Remove or isolate containers (40 CFR 264.56(e) and 265.56(e)).
- Monitor for leaks, pressure build-up, gas generation, or ruptures in valves, pipes, or other equipment whenever appropriate (40 CFR 264.56(f) and 265.56(f)).
- Provide for treatment, storage, or disposal of recovered waste, contaminated soil or surface water, or other material (40 CFR 264.56(g) and 265.56(g)).
- Ensure that no waste that may be incompatible with the released material is treated, stored, or disposed of until cleanup is completed (40 CFR 264.56(h)(1) and 265.56(h)(1)).
- Ensure that all emergency response equipment is cleaned or restocked before operations resume (40 CFR 264.56(b)(2) and 265.56(b)(2)).

Σ  Notify appropriate federal, state, and local authorities verbally and in writing) (40 CFR 264.56(a)(2), (d), (i), and (j), and 265.56(a)(2), (d), (i), and (j)).

Σ  Provide a written follow-up within 15 days.

*Hazardous Waste Rulebook*

**15.6** TSD facilities should have specific procedures for responding to spills and accidents. (GMP)

NOTE: The review items listed are not required by regulations, but are considered GMPs. A good contingency plan should have these or similar procedures.

**Guide Note**

Σ  Inspect to determine if appropriate equipment is available in or near spill locations.

Σ  Examine the contingency plan to determine if it contains the following actions, which are typical GMPs:

-   Determine the identity of the chemical involved (including, if possible, formulation, manufacturer, and percentage of active ingredient).
-   Provide immediate first aid and evacuation around the spill area.
-   Secure the spill site by roping off the area and posting warning signs.
-   Contain and control the spilled chemical by:

    1)  putting on protective equipment;
    2)  preventing further leakage by repositioning containers;
    3)  covering the spill (if liquid) with absorbent material or (if solid) with polyethylene bags; and
    4)  trenching or encircling the area with a dike of sand, absorbent material, soil, or rags.

-   Clean up a dry spill by:

    1)  rolling up bags slowly and sweeping; and
    2)  collecting residue in heavy-duty plastic bags.

-   Clean up a liquid spill by:

    1)  working absorbent material into the spill;
    2)  collecting the absorbent material into properly labeled leak-proof containers; and
    3)  removing contaminated soil to a depth 3 inches below the wet surface lining.

-   Carry out decontamination procedures by:

    1)  removing the bulk of the spill;
    2)  applying a decontamination solution;
    3)  allowing 1 to 6 hours reaction time; and
    4)  applying absorbent material.

-   Carry out disposal procedures by:

    1)  removing all contaminated materials ar J placing them in a sealed leak-proof drum; and
    2)  disposing of the drum in appropriate manner.

-   Carry out post-spill procedures by:

    1)  sampling affected areas to ensure effective decontamination;
    2)  investigating the cause of the spill;
    3)  fully documenting the spill for future reference; and
    4)  revising the contingency plan if necessary.

*Hazardous Waste Rulebook*

| 16.   TSD Facilities:   Recordkeeping |
|---|

**16.1** TSD facility operators must keep written operating records at the facility. (40 CFR 264.73, 265.74, 268.7, and 268.8)

**Guide Note**

Σ   Review the operating records for the TSD facility to determine if they include:

- description, quantity, and date of placement for each shipment of waste received (cross-referenced to manifest document numbers that vary according to conditions) (40 CFR 264.73(b) and 265.73(b));
- the location of each waste at the facility (40 CFR 264.73(b)(2) and 265.73(b)(2));
- records and results of waste analyses and trial tests (40 CFR 264.73(b)(3) and 265.73(b)(3));
- reports on incidents (40 CFR 264.73(b)(4), 264.56(j), 265.73(b)(4), and 265.56(j));
- records/results of inspections (40 CFR 264.73(b)(5), 264.15(d), 265.73(b)(5), and 265.15(d));
- monitoring, testing, and analytical data, where required (40 CFR 264.73(b)(6) and 265.73(b)(6));
- copies of land disposal restriction notices and certification (40 CFR 264.73(b)(11) - (16) and 265.73(b)(11) - (16));
- records of quantities of waste placed in land disposal under an extension of the effective date of any land disposal restriction (40 CFR 264.73(b)(10) and 265.73(b)(10));
- closure and, for disposal facilities, post-closure plans and cost estimates (40 CFR 264.73(b)(8) and 265.73(b)(7)); and
- annual waste minimization program certification (40 CFR 264.73(b)(9); GMP for interim status facility).

**16.2** Additional reports are required. (40 CFR 264.75 - 264.77 and 265.75 - 264.77)

**Guide Note**

Σ   Determine if the biennial report to the EPA Regional Administrator, due March 1 of each even-numbered year, has been properly signed and filed (40 CFR 264.75 and 265.75). Note that states may require reporting on a more frequent basis.

Σ   Review recent reports for completeness (40 CFR 264.75 and 265.75).

Σ   Verify that there are provisions for preparing and submitting "unmanifested waste reports" (40 CFR 264.76 and 265.76).

Σ   Verify that reports on the following subjects are prepared, properly signed, and submitted:

- any release from the solid waste management unit (40 CFR 264.77(a), 264.56(j), 265.77(a), and 265.56(j));
- fires and explosions (40 CFR 264.77(a) and 265.77(a));
- the groundwater detection monitoring program (40 CFR 264.77(c) and 265.77(a));
- the compliance monitoring program (40 CFR 264.77(c) and 265.77(a));
- the corrective action program (40 CFR 264.77(c) and 265.77(a));
- surface impoundment, waste pile, land treatment, and land disposal unit monitoring (40 CFR 264.77(c) and 265.77(d)); and
- certification of closure for hazardous waste surface impoundment, waste pile, land treatment, and landfill units (40 CFR 264.77(b) and 265.77(c)).

| 17. TSD Facilities: Groundwater Monitoring |
|---|

**17.1** If a final status facility has solid waste management units, releases from these units must be monitored and corrective action taken as needed. (40 CFR 264.90 - 264.101; for interim status facilities, see paragraphs 17.2, 17.3, and 17.4)

**Guide Note**

Σ Determine if the facility operates waste piles, surface impoundments, land treatment, or land disposal units.

Σ Determine if these units received hazardous waste after July 26, 1982. If so, these "regulated units" must have a groundwater monitoring program in place (40 CFR 264.90(a)(2)).

Σ If groundwater monitoring is required, the following programs may be necessary:

- detection monitoring (40 CFR 264.98);
- compliance monitoring if a hazardous constituent is detected at a compliance point (40 CFR 264.99); and
- corrective action if a groundwater protection standard is exceeded or hazardous constituents are detected between a compliance point and the facility property boundary (40 CFR 264.100).

Σ Verify that sampling for each of the hazardous constituents or monitoring parameters specified in the permit was conducted.

Σ Verify that the facility implemented a groundwater monitoring plan (40 CFR 264.97 and 270.14(c)). Verify that the plan is approved by EPA and state agencies (40 CFR 264.91(b)).

Σ For permitted facilities, verify the groundwater monitoring systems consist of:

- a sufficient number of wells to yield samples representative of background concentrations, and
- a sufficient number of wells to yield samples representative of groundwater quality at the compliance point (40 CFR 264.97(a)).

Σ Verify the facility's files contain:

- reports to EPA of monitoring and other program results (40 CFR 264.97(j) and 264.98(g) and (h));
- the current groundwater protection standard from the facility's permit (40 CFR 264.92);
- a groundwater sampling and analysis plan, including procedures for:

  1) sample collection;
  2) sample preservation and shipment;
  3) analytical procedures;
  4) chain of custody control; and
  5) setting a sampling schedule (40 CFR 264.97(d)).

Σ Review for compliance with the sampling and analysis plan.

## *Hazardous Waste Rulebook*

**17.2** Groundwater monitoring well sampling must be conducted at certain frequencies at interim status TSD facilities. (40 CFR 265.92)

**Guide Note**

Σ Verify that design and construction of the sampling well was overseen by a professional geotechnical engineer or certified hydrogeologist (40 CFR 265.90(d)(1)).

Σ Verify that the groundwater monitoring system for the interim status facility consists of:

- at least 1 upgradient well for sampling of background groundwater unaffected by the facility; and
- at least 3 downgradient wells for immediately detecting any statistically significant amounts of hazardous waste or hazardous waste constituents (40 CFR 265.91(a)).

Σ Verify that quarterly monitoring was conducted for the first year to establish initial background concentrations (40 CFR 265.92(c)).

Σ Verify that annual sampling for groundwater quality was conducted each year after the initial year (40 CFR 265.92(d)(1)).

Σ Verify that semiannual samples were taken and analyzed for groundwater contamination (40 CFR 265.92(d)(2)).

Σ Verify that compliance monitoring samples were taken for facilities with groundwater contamination problems (40 CFR 264.99 and 265.93).

**17.3** An interim status facility (that is required to conduct groundwater monitoring) must prepare an outline of a groundwater quality assessment program. (40 CFR 265.93)

**Guide Note**

Σ Determine if the groundwater monitoring data has been analyzed using statistical techniques (i.e., student's test) (40 CFR 265.93(b) and Appendix IV).

Σ Verify that the assessment program determines the following:

- whether hazardous wastes or constituents have entered groundwater (40 CFR 265.93(a)(1));
- the rate and extent of migration in groundwater of hazardous wastes or constituents; and
- the concentrations of hazardous wastes or constituents in groundwater (40 CFR 265.93(a)(3)).

**17.4** Groundwater monitoring reports of significant changes in the concentration of indicator parameters at interim status facilities must be submitted to the EPA Regional Administrator. (40 CFR 265.93(c)(1) and 265.94(a)(2)(ii))

**Guide Note**

Σ If a comparison of monitoring data from upgradient and downgradient wells shows a significant increase (or decrease for pH), has this information been submitted to the EPA Regional Administrator (40 CFR 265.93(d))?

Σ Verify that if a significant difference in concentrations is found in the downgradient wells. the owner operator has done the following:

- immediately obtained additional groundwater samples, split the samples, and obtained analysis to determine whether the significant difference was a laboratory error (40 CFR 265.93(c)(2));
- if it is not a laboratory error, provided within 7 days of the confirming test written notice to the EPA Regional Administrator that the facility may be affecting groundwater quality (40 CFR 265.93(d);
- submitted to the EPA Regional Administrator a specific groundwater quality assessment plan within 15 days following initial notification (40 CFR 265.93(d)(2)).

Σ Verify the assessment plan included the following (40 CFR 265.93(d)(3)):

- the number, location, and depth of wells, including any additional wells required;
- sampling and analytical methods for hazardous wastes or constituents in the facility:
- evaluation procedures; and
- an implementation schedule.

Σ

Verify that the facility submitted a report to the EPA Regional Administrator containing an assessment of groundwater quality as soon as technically feasible (40 CFR 265.93(d)(5)).

Σ  For interim status facilities, verify groundwater monitoring reports were sent to the EPA Regional Administrator according to the following timetable:

- During the first year, initial background concentration sampling values for each well should be submitted within 15 days of completing each quarterly analysis (40 CFR 265.94(a)(2)(i)).
- Reports of groundwater contamination parameters should be submitted annually (40 CFR 265.94(a)(2)(ii)).
- Groundwater surface evaluation results should be submitted annually (40 CFR 265.94(a)(2)(iii)).
- Facilities that must use a groundwater quality assessment program must submit assessment results to the EPA Regional Administrator annually until final closure (40 CFR 265.94(b)(2)).

## 18.  TSD Facilities:  Land Disposal for Treatment Facilities

**18.1**  If a treatment facility sends a restricted waste to a land disposal facility, a notice must be sent with the waste shipment.  (40 CFR 268.7(b)(4))

**Guide Note**

Σ  Check manifest files to ensure that notices are sent and/or received. Each notice must contain the following:

- EPA hazardous waste number (40 CFR 268.7(a)(1)(i));
- the corresponding treatment standards and all applicable prohibitions (40 CFR 268.7(a)(1)(ii));
- the manifest number associated with the shipment of waste; and
- waste analysis data where available (40 CFR 268.7(a)(1)(iii)).

**18.2**  A treatment facility must submit a certification with each shipment of waste to a land disposal facility.  (40 CFR 268.7(b)(5))

NOTE:  This would also apply to facilities that ship residues that have been treated to Land Disposal Restriction (LDR) standards.

**Guide Note**

Σ  Check manifest files and cross-reference treatment files to determine if certifications have been submitted (40 CFR 268.7(b)(5)).

**18.3**  If the waste or treatment residue will be further managed at a different treatment or storage facility, the TSD facility sending the waste must comply with the notice and certification requirements applicable to generators.  (40 CFR 268.7(b)(6))

**Guide Note**

Σ  Check manifest files to determine if notice and certification requirements have been complied with (40 CFR 268.7(b)(6) and 268.7(a)).

**18.4**  If the wastes are recyclable materials used in a manner constituting disposal, the TSD facility (recycler) must keep records of the name and location of each entity receiving the hazardous waste-derived product.

**Guide Note**

Σ  Check files and interview personnel to determine that these records are being kept (40 CFR 268.7(b)(7)).

*Hazardous Waste Rulebook*

## 19. TSD Facilities: Land Disposal for Land Disposal Facilities

**19.1** A TSD facility disposing of restricted waste must have copies of all notices and certifications pertaining to the land disposal restrictions. (40 CFR 268.7(c)(1))

**Guide Note**

Σ Check records of incoming wastes to determine if these records are kept (40 CFR 268.7(c)(1); see also 268.7(a) and (b), and 268.8).

**19.2** A TSD facility disposing of restricted waste must test the waste or an extract of the waste to ensure that the waste is in compliance with the treatment standards. (40 CFR 268.7(c)(2))

**Guide Note**

Σ Check the waste analysis plan to determine the frequency and methods specified for testing restricted waste (40 CFR 264.13 and 265.13).

Σ Check the operating record to see if the waste analysis plan is being followed.

Σ Check the laboratory analysis to determine if all waste disposed of meets the treatment standards (40 CFR 268.7(c)(2); see also 40 CFR 268 Subpart D).

Σ Interview employees about the procedures used to sample and analyze the waste prior to disposal. Determine if the waste analysis plan being followed.

## 20. TSD Facilities: Closure/Post-Closure

**20.1** Closure and, for disposal facilities, post-closure plans are required. (40 CFR 264.110 - 120 and 265.110 - 120)

**Guide Note**

Σ Verify that there is an up-to-date closure plan maintained at the facility (40 CFR 264.112 and 265.112). The closure plan should include:

- a detailed description of how and when the facility will be partially closed, if applicable, and totally closed (40 CFR 264.112(b) and 265.112(b));
- an estimate of the maximum inventory of wastes in storage and treatment at any time during the life of the facility (40 CFR 264.112(b)(3) and 265.112(b)(3));
- a description of the steps needed to decontaminate facility equipment during closure (40 CFR 264.112(b)(4) and 265.112(b)(4));
- an estimate of the expected year of closure;
- the schedule for final closure (0 CFR 264.112(b)(6) and 265.112(b)(6)); and
- provision for certification of closure by the owner and an independent registered engineer (40 CFR 264.115 and 265.115).

Σ Verify that the closure plan reflects all units currently operating (40 CFR 264.112(c) and 265.112(c)).

Σ Determine that the closure plan has been amended to reflect the following (40 CFR 264.112(c)(2) and 265.112(c)(2)):

- operating changes;
- design changes; or
- changes in the closure plan itself?

Σ Review written requests for permission to amend the closure plan and EPA responses to these requests.

Σ Verify that the post-closure plan, if necessary, is written, and that amendments to it are undertaken in the same manner as for closure plans (40 CFR 264.118(a) and (b) and 265.118(a) and (b)).

## 21.  TSD Facilities:  Financial Assurance and Liability Insurance

**21.1**  The owner or operator must have a detailed cost estimate for closure.  (40 CFR 264.142 and 265.142)

**Guide Note**

Σ  Review written estimates for closure and post-closure.  Are they accurate and realistic (40 CFR 264.142(a) and 265.142(a))?

Σ  Verify that costs are adjusted upward annually within 60 days of the anniversary date of the assurance instrument.

**21.2**  The owner or operator must establish financial assurance for closure of the facility.  (40 CFR 264.143 Subpart H and 265.143)

**Guide Note**

Σ  Verify that 1 of the following mechanisms for financial assurance has been put in place:

- closure trust fund (40 CFR 264.143(a) and 265.143(a));
- financial test and corporate guarantee for closure (40 CFR 264.143(e) and 265.143(e));
- surety bond guaranteeing payment into a closure trust fund (40 CFR 264.143(b) and 265.143(b));
- surety bond guaranteeing performance of closure (for permitted facilities only) (40 CFR 264.143(b));
- closure insurance (40 CFR 264.143(e) and 265.143(e));
- letter of credit (40 CFR 264.143(d) and 265.143(c));
- multiple financial mechanisms (40 CFR 264.143(g) and 265.143(f)); or
- financial mechanisms for multiple facilities (40 CFR 264.143(h) and 265.143(g)).

**21.3**  An owner or operator of a hazardous waste TSD facility or a group of such facilities must demonstrate financial responsibility for bodily injury and property damage to third parties caused by sudden accidental occurrences.  (40 CFR 265.147 and 264.147)

**Guide Note**

Σ  Determine if coverage for sudden accidental occurrences is in place.  Determine if liability is in place in the amount of at least $1 million per occurrence and an annual aggregate of at least $2 million (40 CFR 264.147(a) and 40 CFR 265.147(a)).

Σ  For nonsudden accidental occurrences, is liability in the amount of at least $3 million per occurrence, with an annual aggregate of at least $6 million (40 CFR 264.147(b) and 40 CFR 265.147(b))?

Σ  What mechanism for financial responsibility has been put in place?  1 of the following must be chosen:

- liability trust fund (40 CFR 264.147(j) and 265.147(j));
- financial test and corporate guarantee for liability (40 CFR 264.147(f) and (g), and 265.147(f) and (g));
- surety bond (40 CFR 264.147(i) and 265.147(i));
- liability insurance (40 CFR 264.147(b)(1) and 265.147(b)(1));
- letter of credit (40 CFR 264.147(h) and 265.147(h));
- multiple financial mechanisms (40 CFR 264.147(a)(6) and (b)(6), and 265.147(a)(6) and (b)(6)); or
- financial mechanisms for multiple facilities (40 CFR 264.147(a) and (b), and 265.147(a) and (b)).

*Hazardous Waste Rulebook*

---

## 22. TSD Facilities: Location and Construction

**22.1** TSD facilities must meet location standards. (40 CFR 264.18 and 265.18)

**Guide Note**

Σ  If it is a new facility, determine if any portions of the facility where treatment, storage, or disposal of hazardous waste is conducted are located at least 200 feet from any fault that has had displacement in Holocene times (40 CFR 264.18(a)).

Σ  If it is a final status facility located within a 100-year flood plain, verify that it is designed, constructed, operated, and maintained to prevent washout of any hazardous waste, or that there are procedures in place to remove the waste prior to flooding (40 CFR 264.18(b)).

Σ  Verify that hazardous waste is not placed in a salt dome, salt bed formation, underground mine, or cave (40 CFR 265.18) unless the waste is contained and a final permit has been obtained (40 CFR 264.18(c)).

**22.2** A written construction quality assurance program must be in place for new surface impoundments, waste piles, or landfills (units, lateral expansions, and replacement units constructed after January 29, 1992). (40 CFR 264.19 and 265.19)

**Guide Note**

Σ ˙ Review the construction quality assurance program and determine if construction was directed and certified by a professional engineer (40 CFR 264.19(a)(1) and (d), and 265.19(a)(1) and (d)).

Σ  Determine if the program certification was submitted to the administering agency prior to the receipt of wastes (40 CFR 264.19(d) and 265.19(d)).

Σ  Determine if the plan addresses the applicable physical components of the facility (foundations, dikes, liners, leachate management, cover systems) (40 CFR 264.19(a)(2) and 265.19(a)(2)).

Σ  Determine if the plan identifies waste management units and key personnel (40 CFR 264.19(b) and 265.19(b)).

Σ  Determine if the plan contains procedures for construction, inspection, sampling, testing, corrective action, and recordkeeping (40 CFR 264.19(b) and (c), and 265.19(b) and (c)).

---

## 23. TSD Facilities: Surface Impoundments

**23.1** Facilities that use surface impoundments to treat, store, or dispose of hazardous wastes must comply with certain requirements. (40 CFR 264.220 - 230 and 265.220 - 230)

**Guide Note**

Σ  Determine if the facility treats, stores, or disposes of wastes in surface impoundments. If not, go to paragraph 24.1.

Σ  Has the facility submitted a Part B permit application to EPA (40 CFR 270.17)? Does the application include a certification that the surface impoundment is operated in accordance with all groundwater and financial responsibility requirements?

Σ  If hazardous wastes are chemically treated in the impoundment:

- Are waste analyses and trial treatment tests conducted on these wastes (40 CFR 264.13, 265.13, and 265.225)?
- Does the owner/operator have written, documented information on similar treatment of similar wastes under similar operation conditions (40 CFR 264.13(a)(2) and 265.13(a)(2))?
- Does the impoundment solely neutralize corrosive waste or listed wastes listed solely because of corrosivity (40 CFR 260.10, 265.1(c)(10), and 264.1(g)(6))?

Σ

Inspect each surface impoundment. Ensure the following criteria are met:

- The impoundment has a liner designed, constructed, and installed to prevent any migration of waste out of the impoundment (40 CFR 264.221(a) and 265.221(a)).
- The impoundment has at least 2 feet of freeboard (required for interim status facilities; GMP for final status facilities) and shows no sign of overtopping by overfilling, wave action, or a storm (40 CFR 265.221(f) and (g), and 264.221(g)).
- The impoundment has a containment system, such as an earthen dike, covered with grass, rock, or shale, that shows no signs of erosion (40 CFR 265.223 and 264.221(h)).
- The impoundment does not contain ignitable or reactive waste unless:

    1) the waste has been treated, rendered, or mixed so it is no longer ignitable or reactive (40 CFR 264.229(a) and 265.229(a)); or
    2) the waste is managed in a way that protects it from materials or conditions that may cause it to ignite or react (40 CFR 264.229(b) and 265.229(b)).

- The impoundment is used solely for emergencies (40 CFR 264.229(c) and 265.229(c)).

**23.2** New surface impoundments constructed after January 29, 1992 are subject to new construction requirements under 40 CFR 264.221(a)(4)(c).

NOTE: These requirements are also applicable to interim status facilities by reference in 40 CFR 265.221(a).

**Guide Note**

Σ Does the design of the surface impoundment have a composite bottom liner consisting of at least 2 components? An upper component must be designed and constructed to prevent the migration of constituents into such a component (e.g., a geomembrane) during the active life and post-closure care period. and a lower liner must consist of at least 3 feet of compacted soil with a hydraulic activity of 1 x 10$^{-7}$ cm/second (40 CFR 264.221(c)(1)(B)).

Σ Does the design include a top liner to prevent the migration of constituents into such liner (e.g.. a geomembrane) during the active life and post-closure care period (40 CFR 264.221(c)(1)(A))?

Σ A leachate collection system must be installed between liners (40 CFR 264.221(c)).

**23.3** The facility operator's personnel must conduct inspections while surface impoundments are in operation. (40 CFR 264.226 and 265.226)

**Guide Note**

Σ Determine if the impoundment is inspected at least daily (for interim status facilities) or weekly (for final status facilities) to check the freeboard level (40 CFR 264.226(a)(1) and 265.226(a)(1)).

Σ Determine if inspections are conducted after storms to detect evidence of the following:

- deterioration, malfunctions, or improper operation of overtopping control systems (40 CFR 264.226(b)(1));
- sudden drops in the level of the impoundment contents (40 CFR 264.226(b)(2)); or
- severe erosion or other signs of deterioration in dikes or other containment devices (40 CFR 264.226(b)(3) and 265.226(a)(2)).

Σ Are all inspection records maintained at the facility (40 CFR 264.75(d), 264.73(b)(5). 265.15(d). and 265.73(b)(5))?

Σ Have problems that were noted during inspections been resolved?

*Hazardous Waste Rulebook*

## 24. TSD Facilities: Waste Piles

**24.1** Waste piles are regulated by both the general TSD facility regulations (40 CFR 264 Subparts A - H and 265 Subparts A - H) and by the specific waste pile regulations (40 CFR 264.250 et seq. and 265.250 et seq.).

**Guide Note**

Σ  Has a Part B application (complete with groundwater and financial responsibility certification) been submitted (40 CFR 270.18)?

Σ  Verify that the pile containing hazardous waste is protected from the wind (40 CFR 264.250(c)(3), 264.251(j), and 265.281).

Σ  If water is used as a wind control, does the waste pile have liners and leachate collection systems (GMP for piles without water controls)?

Σ  Verify that incoming shipments of waste are analyzed before the waste is added to the pile to determine the compatibility of the waste, unless the added waste is known to be compatible with the wastes in the pile (40 CFR 264.257 and 252).

Σ  Does the analysis include a visual comparison of color and texture?

**24.2** If leachate and runoff from the pile constitute a hazardous waste, certain types of containment must be provided. (40 CFR 264.251 and 265.253)

**Guide Note**

Σ  Verify that the pile is managed with the following:

- an impermeable base compatible with the waste (40 CFR 264.251 and 265.253(a)(1));
- runon diversion and control systems (40 CFR 264.251(g) and 265.253(a)); and
- leachate and runoff collection (40 CFR 264.251 and 265.253).

Σ  Is the pile protected from precipitation and runon by some other means (40 CFR 265.253(b) for interim status facilities only; see paragraph 24.4 for final status facility requirements)?

**24.3** No liquids or wastes containing free liquids may be placed in interim status waste piles or final status waste piles that are operating without liners or leachate collection systems. (40 CFR 264.250(c) and 265.253(b))

**Guide Note**

Σ  Verify that liquids or wastes containing free liquids are prohibited from being placed in the pile (40 CFR 264.250(c)(1) and 265.253(b)(2)).

**24.4** A final status waste pile is not required to comply with the lining, leachate collection, and groundwater protection requirements if the unit was built prior to May 8, 1985, or if the unit is located indoors or otherwise protected from factors that produce leachate and runon. (40 CFR 264.250 and 264.251(a)(2))

**Guide Note**

Σ  Check that waste piles using this exemption comply with the following:

- Liquids are not placed in the waste pile (40 CFR 264.250(c)(1)).
- The unit is protected from surface water runon (40 CFR 264.250(c)(2)).
- Water dispersal is controlled by a means other than wetting (40 CFR 264.250(c)(3)).
- The pile does not generate leachate through decomposition or reactions (40 CFR 264.250(c)(4)).

Σ

If the waste pile is new as of May 8, 1985, does it meet the minimum technological requirements, including:

- double liners (40 CFR 264.251);
- leak detection (40 CFR 264.251);
- groundwater monitoring (40 CFR 264.90 - 100)?

Σ For waste piles that are required to have leachate collection and removal systems, verify the following (40 CFR 264.251(a)(2)):

- The collection and removal system is located immediately above the liner.
- The leachate collection system is designed to operate without clogging through the entire operating and closure period.
- Leachate levels do not extend 1 foot above the liner.

**24.5** Waste piles containing wastes that are incompatible with any material stored nearby must be separated from the material or protected by a wall or similar device. (40 CFR 264.257(b) and 265.257(b))

**Guide Note**

Σ If wastes in the waste pile are incompatible with any nearby wastes or material stored in containers, open tanks, piles, or surface impoundments, verify the pile is separated from these materials or protected by means of a dike, berm, or wall (40 CFR 264.257(b) and 265.257(b)).

| 25. | TSD Facilities: Landfills |
| --- | --- |

**25.1** Landfills must comply with both the general TSD facility regulations (40 CFR 264 Subparts A - H and 265 Subparts A - H) and with specific landfill regulations (40 CFR 264 Subpart N and 265 Subpart N).

**Guide Note**

Σ Has a Part B application (complete with groundwater and financial responsibility certification) been submitted (40 CFR 270.21)?

Σ Review the permit to determine compliance with any specific conditions.

**25.2** A runoff collection system is required and collection facilities must be emptied or otherwise managed so as to maintain the required holding capacity. Precipitation runoff may or may not be regulated as hazardous waste, depending upon the design and management of the runoff system and the chemical constituents of the runoff material. Because runoff is potentially a hazardous waste, the landfill owner/operator is subject to the generator waste determination requirements. (40 CFR 264.301(g), (h), and (j), 265.301(b) and (c), 261.22, and 261.3)

**Guide Note**

Σ Does the landfill have a runoff diversion and control system capable of managing a 24-hour, 25-year storm (40 CFR 264.301(h) and 265.301(h))?

- Is the collected runoff analyzed to determine if it is hazardous waste (40 CFR 261)?
- If it is a hazardous waste, how is it managed?

### *Hazardous Waste Rulebook*

**25.3**  Landfill location information is required to be keyed to permanent surveyed benchmarks as recorded on an area map.  (40 CFR 264.73(b)(1) and (2), 264.309, 265.73(b)(1) and (2), and 265.309)

**Guide Note**

Σ  Verify the operating record contains the following information (40 CFR 264.73(b)(2) and 265.73(b)(2)):

- on a map, the exact location and dimensions, including depth of each cell with respect to permanent surveyed benchmarks (40 CFR 264.309(a) and 265.309(a)); and
- contents of each cell and the approximate locations of each hazardous waste type within each cell (40 CFR 264.309(b) and 265.309(b)).

Σ  Does the facility employ a grid map showing the location of each waste within the cell (GMP)?

**25.4**  Reactive or ignitable waste may be placed into a landfill under certain conditions.  (40 CFR 264.312(a), 265.312(a), 264.17(b), and 265.17(b))

**Guide Note**

Σ  If reactive or ignitable waste is placed in the landfill:

- Is it treated, rendered, or mixed before or immediately after placement in the landfill so it is no longer reactive or ignitable (40 CFR 264.17(b), 264.312(a), 265.17(b), and 265.312(a))?
- Is ignitable waste landfilled in nonleaking containers that are protected from sources of ignition (i.e., daily soil cover, segregation from heat-generating wastes) (40 CFR 264.312(b) and 265.312(b))?

Σ  Note that there is a land ban restriction on such waste disposal (40 CFR 268).

**25.5**  Incompatible wastes may be placed in the same landfill cell, but only if precautions are taken.  (40 CFR 264.17(b), 264.313, 265.17(b), and 265.313)

**Guide Note**

Σ  If incompatible wastes are placed in the same landfill cell, verify wastes are managed so as to prevent:

- extreme heat, fire, or explosion (40 CFR 264.17(b)(1) and 265.17(b)(1));
- uncontrolled toxic mists, dusts, fumes, or gases (40 CFR 264.17(b)(2) and 265.17(b)(2));
- uncontrolled flammable vapors or gases (40 CFR 264.17(b)(3) and 265.17(b)(3));
- damage to the structural integrity of the landfill (40 CFR 264.17(b)(4) and 265.17(b)(4)); and
- threat to human health and the environment.

**25.6**  Bulk liquids have been banned from disposal in landfills.  A facility waste analysis plan must include testing procedures to ensure that bulk shipments do not contain free liquids.  (40 CFR 264.314(b) and (e), 265.314(b) and (f), 264.13, and 265.13)

**Guide Note**

Σ  Verify that there is a procedure to prevent bulk or noncontainerized liquid waste or waste containing free liquids from being placed in the landfill (40 CFR 265.314(b), 265.13(b)(6), 264.314(b), and 264.13(b)(6)).

Σ  If needed, is the liquid waste treated chemically or physically prior to placement in the landfill so that free liquids are no longer present (40 CFR 264.314 and 265.314)?

Σ  Is there a procedure to prevent nonhazardous liquid (bulk or containerized) from being placed in the landfill (40 CFR 265.314(c), 265.13(b)(6), and 264.314(d))?

**25.7** The placement of containerized liquids into landfills is allowed under certain conditions. (40 CFR 264.314(d) and 265.314(c))

**Guide Note**

Σ If containers holding liquid wastes are placed in the landfill:
- has all free-standing liquid been removed (40 CFR 264.314(d)(1)(i) and 265.314(c)(1)(i)); or
- has waste been mixed with absorbents or solidified so that free-standing liquid is no longer observed (40 CFR 264.314(d)(1)(ii) and 265.314(c)(1)(ii)); or
- is the container very small, such as an ampule (40 CFR 264.314(d)(2) and 265.314(c)(2)); or
- is the container designed to hold free liquids for use other than storage, such as a battery or capacitor (40 CFR 264.314(d)(3) and 265.314(c)(3)); or
- is the container a lab pack (40 CFR 264.314(d)(4) and 265.314(c)(4))?

Σ Are absorbents used to treat free liquids that are to be disposed of in landfills nonbiodegradeable (40 CFR 264.314(e) and 265.314(f)).

**25.8** Empty containers placed in a landfill are required to be reduced in volume. (40 CFR 264.315 and 265.315)

**Guide Note**

Σ Check to see that empty containers are shredded or crushed before landfilling (40 CFR 264.315 and 265.315).

**25.9** Beginning January 29, 1992, each new landfill, replacement landfill, or lateral expansion of an existing landfill that first received waste after November 8, 1984 must meet specific design and construction requirements. Landfills that are not specifically exempt are also required to conduct groundwater monitoring. (40 CFR 265.301(a), 264 Subpart F, and 265 Subpart F)

**Guide Note**

Σ Verify that any landfill cell that is new as of January 29, 1992 meets the minimum technology requirements for:
- double liners;
- leak detection (40 CFR 264.301(c) and 265.301(a)); and
- groundwater monitoring (40 CFR 264.90 - 101 and 265.90 - 265.94).

Σ Verify that there is a written program in place to (40 CFR 270.21):
- Determine that received waste is appropriate for landfilling.
- Ensure the placement of waste in the proper landfill cell.

Σ Review operations for compliance with the program.

## 26. TSD Facilities: Incinerators

**26.1** Incinerators must be operated at steady-state conditions whenever waste is being added to the unit, including all startup and shutdown periods. (40 CFR 264.345(c) and 265.345)

**Guide Note**

Σ Verify that there is a written procedure (GMP) to ensure that the incinerator is operating at steady-state conditions (temperature and air flow) before adding hazardous waste (40 CFR 264.345(c) and 265.345).

Σ Verify that there is a written procedure to ensure that the hazardous waste feed is stopped during steady-state conditions before shutdown (40 CFR 264.345(c) and 265.345).

Σ Verify that there is a procedure and/or operational controls to prohibit charging wastes during upset conditions.

*Hazardous  Waste  Rulebook*

**26.2** The facility is required to have a written waste analysis plan. Permitted facilities are required to conduct additional waste analysis during operating periods. (40 CFR 264.13, 265.13, and 265.341)

**Guide  Note**

Σ  Verify that the waste analysis plan for interim status facilities includes analysis for the following:

- heating value (40 CFR 265.341(a));
- halogen content (40 CFR 265.341(b));
- sulfur content (40 CFR 265.341(b));
- concentration of lead (40 CFR 265.341(c));
- concentration of mercury (40 CFR 265.341(c));
- PCB content (TSCA).

Σ  Is the above information documented in the operating record (40 CFR 264.73(b)(3) and 265.13(b)(3))?

Σ  NOTE:  The analyses for lead and mercury are not required if the facility has written, documented data that show the elements are not present.

Σ  Verify that the waste analysis plan for final status facilities includes analysis for chemical composition and limits specified in the permit (40 CFR 264.341).

**26.3** Interim status incinerator regulations require specific monitoring and inspection. (40 CFR 265.347)

**Guide  Note**

Σ  Verify  the following instruments on the incinerator are monitored at least every 15 minutes when  hazardous waste is burning (40 CFR 265.347(a)):

- waste feed gauge;
- auxiliary fuel feed gauge;
- carbon monoxide gauge;
- air flow gauge;
- incinerator temperature gauge;
- scrubber flow gauge;
- scrubber pH gauge; and
- relevant level controls.

Σ  Verify that the incinerator and associated equipment is monitored at least daily, including the following  (40 CFR 265.347(b)):

- pumps, valves, conveyors, and pipes for leaks, spills, and fugitive emissions;
- emergency shutdown controls; and
- system alarms.

Σ  Verify that records of inspections are recorded in the inspection log (40 CFR 265.15).

**26.4** Final status incinerators require specific monitoring and inspection.

**Guide  Note**

Σ  Verify that the following parameters are continually monitored during incineration (40 CFR 264.347(a)(1)):

- combustion temperature;
- waste feed rate; and
- combustion gas velocity.

Σ

Verify that the following are monitored daily for leaks, spills, or fugitive emissions (40 CFR 264.347(b)):

- pumps;
- valves;
- conveyors; and
- pipes.

Σ  Verify that waste feed cut-off and associated alarms are monitored at least weekly (40 CFR 264.347(c)).
Σ  Verify that monitoring results are recorded in the operating record (40 CFR 264.347(d)).

**26.5**  Solid wastes generated from the treatment, storage, or disposal of hazardous wastes must be disposed of at RCRA-permitted or -approved facilities.  (40 CFR 261.3(c)(2)(i))

**Guide Note**

Σ  Does the facility manage ash residue as hazardous waste?
Σ  Does the facility manage scrubber liquid as hazardous waste?
Σ  Does the facility handle used refractory material as a hazardous waste?

---

## 27.  TSD Facilities:  Thermal Treatment

---

**27.1**  Thermal treatment units must be operated at steady-state conditions whenever waste is being added to the unit, including all startup and shutdown times.  A written plan may be required in the facility's Part B application.  (40 CFR 265.373)

**Guide Note**

Σ  If the procedure is a continuous process, is there a written procedure to ensure that the process is operating at steady-state conditions (including temperature) before adding hazardous waste (40 CFR 265.373)?
Σ  Interview operators to determine compliance with and knowledge of procedures.

**27.2**  The facility is required to have a written waste analysis plan.  (40 CFR 265.73 and 265.375)

**Guide Note**

Υ  Review waste analysis information in the operating record.  Verify that a waste analysis is performed on hazardous waste not previously burned (40 CFR 265.73 and 265.375).
Σ  Verify that there is a written procedure to incorporate the analysis results into operating parameters that establish steady-state conditions (40 CFR 265.73 and 265.375).
Σ  Verify that the waste analysis plan includes analyses for the following:

- heating value (40 CFR 265.375(a));
- halogen content (40 CFR 265.375(b));
- sulfur content (40 CFR 265.375(b));
- concentration of lead (40 CFR 265.375(c));
- concentration of mercury (40 CFR 265.375(c));
- PCB content (TSCA);
- other parameters as specified in the permit (40 CFR 264.13).

Σ  Is the above information documented in the operating record (40 CFR 264.73(b)(3) and 265.73(b)(3))?
Σ  NOTE: The analyses are not required for lead or mercury if the facility has written, documented data that show the elements are not present.

## *Hazardous Waste Rulebook*

**27.3** Thermal treatment units are required to monitor certain parameters. (40 CFR 265.377(a)(1))

**Guide Note**

Σ  Verify that the following instruments relating to combustion and emission control are monitored at least every 15 minutes when treating waste (40 CFR 265.377(a)(1)):

- waste feed gauge;
- auxiliary fuel feed gauge;
- treatment process temperature gauge;
- process flow gauge; and
- other controls (e.g., afterburner and temperature controls, oxygen and carbon monoxide meters, process levels, etc.).

Σ  Verify that the stack plume (emissions) is monitored at least hourly, including (40 CFR 265.377(a)(2)).

- Is the color normal?
- Is the opacity normal?

Σ  Verify that thermal treatment process equipment is monitored at least daily, including (40 CFR 265.377(a)(3)):

- pumps, valves, conveyors, pipes, etc. (for leaks, spills, and fugitive emissions);
- emergency shutdown controls; and
- system alarms.

Σ  Verify inspections are recorded in the inspection log (40 CFR 265.15(d)).

**27.4** A written plan should address procedures to ensure that corrective actions are initiated. (GMP and 40 CFR 265.377)

**Guide Note**

Σ  Are there written procedures in place (GMP) to immediately make any operating corrections that are indicated by the combustion and emission control instruments or the observation of the emission plume (40 CFR 26.377(a)(2))?

Σ  Verify through interviews that procedures are followed.

**27.5** There are minimum distances established for detonation of explosives. (40 CFR 265.382)

**Guide Note**

Σ  If open burning or detonation of waste explosives is conducted, is the detonation performed in accordance with the following table (40 CFR 265.382)?

| Pounds of Waste Explosive | Minimum Distance to Other Properties |
|---|---|
| 0-100 | 670 feet |
| 101-1,000 | 1,250 feet |
| 1,001-10,000 | 1,730 feet |
| 10,001-30,000 | 2,260 feet |

Σ  Is there a written procedure in place (GMP) to prohibit open burning of hazardous waste (except for waste explosives (40 CFR 265.382)?

- Does the site have a current air permit for the thermal treatment process?
- Is the process in compliance with the permit terms?

Σ  NOTE: Thermal treatment of waste explosives may be conducted under an exception allowing for abatement of a fire hazard rather than an air permit.

| 28. TSD Facilities:  Boilers and Industrial Furnaces |
|---|

**28.1**   Owners and operators of boilers and industrial furnaces (BIFs) operating under interim status must comply with requirements of 40 CFR 266.103 and 266.104.

**Guide Note**

Σ  Is the facility operating under interim status as per 40 CFR 266.103? If so, verify the following:

- Combustion temperature is a minimum of 1,800∞F.
- Adequate oxygen is in the combustion gases to combust organic constituents (verify documentation and records).
- If a cement kiln is used, the hazardous waste is fired into the kiln.
- The hydrocarbon controls are in compliance with 40 CFR 266.104.

Σ  Ensure that any unit operating under interim status is not burning dioxin-based wastes (F020, F021, F022, F023, F026, and F027) (40 CFR 266.103 (a)(3)).

**28.2**   Owners and operators of BIFs burning hazardous waste and not operating under interim status must comply with the requirements of 40 CFR 266.100, 270.22, and 270.66.

**Guide Note**

Σ  Is the facility operating under the small quantity burner exemption (40 CFR 266.108)?

Σ  Does the operator have an analysis of the hazardous waste that quantifies the concentrations of constituents identified in 40 CFR 261 Appendix VIII that may reasonably be expected to be in the waste (40 CFR 266.100(c) and 266.102(b)(1))?

Σ  Does the operator conduct sampling throughout the normal operation to ensure that permit limits are being met (40 CFR 266.102(b)(2))?

Σ  Owners and operator must comply with emission standards in 40 CFR 266.104 through 266.107.

Σ  Did the facility submit a trial burn plan (40 CFR 270.66(c))?

Σ  Has the facility conducted a trial burn? If so, did the burn include the following (40 CFR 266.104):

- the designated principal organic hazardous constituents (POHC);
- a destruction and removal efficiency of 99.99% for POHC and 99.9999% for dioxin-based wastes;
- carbon monoxide at a concentration of less than 100 ppm;
- hydrocarbon limits that were previously established;
- continuous monitoring for hydrocarbon emissions.

Σ  Did the facility qualify for a waiver of trial burn under 40 CFR 266.110?

Σ  Verify the following:

- Waste feed rates meet permit conditions.
- Minimum and maximum device production rates are expressed in appropriate units (if applicable).
- The unit has appropriate controls for the hazardous waste firing system.
- Unit variation is within design.
- Combustion gas temperature is measured at an appropriate location.
- Combustion gas velocity is measured to ensure adequate residence time.
- Other permit conditions are met.

Σ  NOTE: The following hazardous wastes are not regulated under BIF regulations:

- used oil that is a hazardous waste solely because it exhibits a characteristic of hazardous waste identified in 40 CFR 261 Subpart C;
- gas recovered from hazardous or solid waste landfills;
- hazardous wastes that are exempt under 40 CFR 261.4 and 261.6(a)(3) (40 CFR 266.100(c)); and
- coke ovens if the only hazardous waste burned has an EPA hazardous waste number of K087.

Σ  Is hazardous waste being properly transferred into the BIF (40 CFR 266.111)?

*Hazardous Waste Rulebook*

| 29. TSD Facilities: Chemical, Physical, and Biological Treatments |
| --- |

**29.1** A written plan should be developed that describes the types of wastes that are not permitted to be added to treatment systems and that specifies all operating and safety procedures. (40 CFR 265.17 and 265.401(b))

**Guide Note**

Σ Verify there is a written procedure in place to prevent the treatment of waste that could cause the process equipment to leak, rupture, corrode, or otherwise fail prior to the end of its intended life (40 CFR 265.401(b)).

Σ Interview operators to review compliance and knowledge of the plan.

**29.2** Continuous feed units must have waste feed cut-off systems. (40 CFR 265.401 (c))

**Guide Note**

Σ If the process is a continuous feed system, is it equipped with a means to stop waste inflow (e.g., waste feed cut-off system or bypass)?

**29.3** Owners/operators of chemical, physical, and biological treatment systems are required to comply with TSD waste analysis requirements. New waste streams are required to be analyzed, and trial treatment tests are required prior to treatment. (40 CFR 265.13, 265.402, 265.17(b), and 265.401(a))

**Guide Note**

Σ If hazardous waste to be treated is substantially different from any hazardous waste previously treated at the facility, or if a substantially different process than any previously used at the facility is to be used to chemically treat hazardous wastes, is 1 of the following obtained (40 CFR 265.13 and 265.402):

- waste analyses and trial treatment tests (e.g., bench scale); or
- written, documented information on similar treatments or similar waste?

**29.4** Owners/operators are required to inspect treatment systems and maintain records. (40 CFR 265.403, 265.15(d), and 265.73(b)(5))

**Guide Note**

Σ Verify the owner/operator inspects the following, where present, at least daily:

- discharge control and safety equipment (e.g., waste feed cut-off, bypass, drainage, or pressure relief systems) (40 CFR 265.403(a)(1)); and
- data gathered from monitoring equipment (e.g., pressure and temperature gauges) (40 CFR 265.403(a)(2)).

Σ Verify that, in order to detect corrosion or obvious signs of leakage, inspections of construction materials used in the treatment process or equipment are conducted at least weekly (40 CFR 265.403(a)(4)).

Σ Are these inspections recorded in the inspection log (40 CFR 265.15 and 265.73)?

**29.5** Incompatible wastes may not be placed in the same treatment process unless precautions are taken to avoid adverse reactions. (40 CFR 265.17(b) and 265.406)

**Guide Note**

Σ If the facility treats incompatible wastes, verify that there a written procedure (GMP) in place to ensure that treatment is done in a manner that will prevent:

- the generation of extreme heat, fire, or explosion (40 CFR 265.17(b)(1));
- uncontrolled toxic mists, dusts, vapors, or gases (40 CFR 265.17(b)(2));
- flammable fumes (40 CFR 265.17(b)(3));
- damage to the structural integrity of the equipment (40 CFR 265.17(b)(4)); and
- threat to human health and the environment (40 CFR 265.17(b)(5)).

Σ If a waste is to be placed in treatment equipment that previously held an incompatible waste, is there a procedure in place (GMP) to ensure that the equipment is washed (40 CFR 265.406(b))?

**29.6** Ignitable or reactive waste must be treated to prevent ignition or reaction. (40 CFR 265.17(b) and 265.405)

**Guide Note**

Σ Verify that ignitable or reactive waste is:

- treated, rendered, or mixed before or immediately after placement in the treatment process so that the waste is no longer ignitable or reactive (40 CFR 265.405(a)(1));
- treated in a manner that does not threaten human health or the environment (40 CFR 265.17(b)); or
- treated so that it is protected from any material or condition that may cause the waste to ignite or react (40 CFR 265.405(a)(2)).

**29.7** Residues from hazardous waste treatment processes are hazardous wastes unless specifically exempt. (40 CFR 261.3(c)(2)(i), 261.3(c) and (d), 260.22, and 265.404)

**Guide Note**

Σ Verify that residue from the treatment process is managed as a hazardous waste or is delisted (40 CFR 261.3(c)(2)(i)).

Σ Review the closure plan and cost estimate to determine if they are current and complete. At closure, verify all hazardous waste and residuals are removed and properly managed as hazardous waste (40 CFR 265.404).

## 30.　TSD Facilities:　Containment Buildings

**30.1** Design and operating standards for containment buildings must ensure containment comparable to RCRA tanks or containers. (40 CFR 265.1100)

**Guide Note**

Σ　Verify that the building:

- is completely enclosed and self-supporting, and can support the waste and daily operating activities (40 CFR 264.1100(a) and 265.1100(a));
- has a primary barrier that is designed to be sufficiently durable to withstand the movement of personnel and equipment (40 CFR 264.1101(a)(4) and 265.1101(a)(4));
- has surfaces that are chemically compatible with wastes that may come into contact with them (40 CFR 264.1101(a)(2) and 265.1101(a)(2));
- has a decontamination area and procedures to prevent tracking waste out of the building (40 CFR 264.1101(c) and 265.1101(c)); and
- has dust control and particulate collection devices (40 CFR 264.1101(c)(1)(iv) and 265.1101(c)(1)(iv)).

Σ　If the building is used to manage liquids:

- The primary barrier is sloped to drain liquids to the collection system, and the liquids are removed from the containment system as necessary (40 CFR 264.1101(b) and 265.1101(b)).
- The leak detection system is constructed with a slope of 1% or more and meets design standards (40 CFR 264.1101(b)(3) and 265.1101(b)(3)).
- The primary barrier must be designed to prevent the migration of hazardous constituents (40 CFR 264.1101(b)(1) and 265.1101(b)(1)).
- The building must have a liquid collection system to minimize the accumulation of liquids on the primary barrier (40 CFR 264.1101(b)(2) and 265.1101(b)(2)).
- A secondary containment system with a leak detection system is constructed of materials suitable to prevent migration of hazardous constituents to the barrier (40 CFR 264.1101(b)(3) and 265.1101(b)(3)).

**30.2** Containment buildings must meet recordkeeping requirements.

**Guide Note**

Σ　Determine if the following records are kept (40 CFR 264.1101 and 265.1101):

- a certification by a qualified professional engineer that the building meets the requirements stated in 40 CFR 264.1101(a) - (c) or 40 CFR 265.1101(a) - (c) (40 CFR 264.1101(c)(2) and 265.1101(c)(2)); or
- a facility operating log that includes the containment building operations and any reported leaks or spills (40 CFR 264.1101(c) and 265.1101(c));
- correspondence with the regulatory agencies regarding the containment building; and
- operating procedures to maintain the integrity of any area without secondary containment (40 CFR 264.1101(d)(3) and 265.1101(d)(3)).

**30.3** Containment buildings must implement inspection programs. (40 CFR 264.1101 and 265.1101(c))

**Guide Note**

Σ  Verify the following:

- The building is inspected once every 7 days and the results of the inspection are recorded in the operating log (40 CFR 264.1101(c)(4) and 265.1101(c)(4)).
- The inspection includes data gathered from the monitoring equipment and leak detection equipment (if applicable), the containment building, and the immediate area surrounding the containment building (40 CFR 264.1101(c)(4) and 265.1101(c)(5)).
- Any release or conditions that could lead to a release resulted in corrective actions and timely reporting (40 CFR 264.1101(c)(3) and 265.1101(c)(3)).

**30.4** Containment buildings must meet requirements for closure and post-closure care. (40 CFR 264.1102 and 265.1102)

**Guide Note**

Σ  Does the facility have a closure plan and closure cost estimate as required under Subparts G and H (40 CFR 264.1102(a) and 265.1102(b))?

Σ  Does the plan include removal or decontamination and disposal of all contaminated materials as a hazardous waste (40 CFR 264.1102(b) and 265.1102(b))?

Σ  If all the contaminated materials cannot be removed or decontaminated, does the plan include closure and post-closure care as described in 40 CFR 265.310? If this is the case, then the building is considered a landfill, and the requirements for landfills specified in Subparts G and H must be met.

---

## 31.  TSD Facilities:  Land Treatment

---

**31.1** Land treatment facilities must be in compliance with their permit conditions. (40 CFR 264.270 - 283) or interim status requirements (40 CFR 265.270 - 282).

**Guide Note**

Σ  Review the hazardous waste facility permit. Determine if wastes are treated in accordance with permit conditions (40 CFR 264.271).

Σ  Determine if the facility is designed and operated in accordance with permit conditions (40 CFR 264.273(a)).

Σ  Determine if the treatment zone is designed, constructed, operated, and maintained to minimize runoff of hazardous constituents (40 CFR 264.273(b) and 265.272(c)).

Σ  Determine if there is a runon control system capable of preventing flow onto the treatment zone during peak discharges from at least a 25-year storm (40 CFR 264.263(c) and 265.272(b)).

Σ  Determine if there is a runoff control system capable of collecting and controlling at least water volume resulting from a 24-hour, 25-year storm (40 CFR 264.273(d) and 265.272(c)).

Σ  Determine if collection and holding facilities associated with runon and runoff control systems are managed to maintain the design capacity of the system (40 CFR 264.273(e) 265.272(d)).

Σ  If the treatment zone contains particulate matter, is wind dispersal being controlled (40 CFR 264.273(f) and 265.272(e))?

Σ  Review inspection records and determine if the unit is inspected weekly and after storms to detect deterioration, malfunctions, or improper operation of the runon and runoff control systems, and improper functioning of wind dispersal control measures (40 CFR 264.273(g)).

Σ  Determine if the facility is conducting unsaturated zone monitoring in accordance with permit conditions or a monitoring plan (40 CFR 264.278 and 265.278).

*(continued on next page*

Σ

## *Hazardous Waste Rulebook*

Review records and determine if the administering agency has been notified within 7 days, and an application for permit modification has been submitted within 90 day of any significant increase in hazardous constituents below the treatment zone (40 CFR 264.278(g)).

Σ  Determine if records are kept of hazardous waste application dates and rate (40 CFR 264.279 and 265.279).

Σ  If ignitable or reactive wastes are treated, are they immediately incorporated into the soil so that they no longer meet the definition of ignitability or reactivity, or are the wastes otherwise managed to prevent ignition or reaction (40 CFR 264.281 and 265.281)?

Σ  If incompatible wastes are treated, are they treated in separate treatment zones, or are other adequate precautions taken to prevent reactions from occurring (40 CFR 264.282 and 265.282)?

Σ  If waste codes F020 - F023 or F026 - F027 are treated, determine if the facility has a management plan approved by the administering agency for the treatment of these wastes (40 CFR 264.283).

**31.2**  Land treatment facilities must be properly closed. (40 CFR 264.280 and 265.280)

**Guide Note**

Σ  Determine if all operations necessary to maximize degradation, transformation, or immobilization of the waste, and to minimize runon, runoff, and wind dispersal have been continued through the closure and post-closure care periods (40 CFR 264.280).

Σ  Determine if an adequate vegetative cover has been established and maintained (40 CFR 264.280(a)(8) and (c)(2), and 265.280(c)).

Σ  Determine if unsaturated zone monitoring has been continued (40 CFR 264.280(c)(7) and 265.280(d)(1)).

Σ  Determine if soil pore monitoring has been continued for 90 days after the last waste application (40 CFR 264.280(c)(7) and 265.280(d)(1)).

Σ  Determine if the closure has been certified by an independent qualified soil scientist or an independent registered professional engineer (40 CFR 264.280(b) and 265.280(e)).

---

## 32.  TSD Facilities:  Management of Remediation Wastes

---

**32.1**  CAMUs must be in compliance with the permit or order issued by the administering agency. (40 CFR 264.552)

**Guide Note**

Σ  Review the permit or order.  Determine if the CAMU is in conformance with the specified areal configuration (40 CFR 264.552).

Σ  Determine if the CAMU is in conformance with specified design and operating requirements (40 CFR 264.552).

Σ  Determine if the CAMU is in conformance with specified groundwater monitoring requirements (40 CFR 264.552).

Σ  Determine if procedures are in place to ensure closure and post-closure care in conformance with permit or order requirements.

**32.2**  Temporary units (tanks and container storage areas) used for treatment and storage of remediation wastes during remedial activities required under RCRA may operate under alternative requirements that are protective of human health and the environment.  (40 CFR 264.553)

**Guide Note**

Σ  Determine if the unit is located within facility boundaries (40 CFR 264.553(b)(1)).

Σ  Review records and determine if the unit is used only for treatment or storage of remediation wastes (40 CFR 264.553(b)(1)).

Σ  Determine if the unit is in compliance with the conditions of a permit or order issued by the administering agency (40 CFR 264.553(c)).

## 33. TSD Facilities: Drip Pads

**33.1** Drip pads that convey treated wood drippage, precipitation, and/or surface water runon to an associated collection system must meet certain design and operating requirements. (40 CFR 264.573 and 574, and 265.440 - 445)

**Guide Note**

Σ Determine if the unit is constructed solely of nonearthen materials (i.e., no wood or unsupported asphalt) (40 CFR 264.573(a)(1) and 265.443(a)(1)).

Σ Determine if there is an intact curb or berm around the perimeter of the pad, and the pad is sloped to drain liquids into a collection system (40 CFR 264.573(a)(2) and (3), and 265.443(a)(2) and (3)).

Σ Determine if the pad is either covered or capable of preventing runon and runoff from a 24-hour, 25-year storm (40 CFR 264.573(e) and (f)), and if collecting and holding units are emptied as soon as possible after storms (40 CFR 264.573(h) and (l)).

Σ Determine if the pad has hydraulic conductivity of $1 \times 10^{-7}$ cm/sec or less and is free of cracks and gaps, or there is (40 CFR 264.573(a)(4), (b)(1), and (b)(2), and 265.443(a)(4), (b)(1), and (b)(2)):

- a synthetic liner below the drip pad;
- a leak detection system above the liner; and
- a leak collection system immediately above the liner.

Σ Determine if there is a written independent professional engineering assessment and annual certification (40 CFR 264.573(a)(4)(ii) and (g), 264.574(a), 265.441, 265.443(a)(4)(ii), and 265.443(g)).

Σ Determine if past operating and waste handling practices are documented in the facility records (40 CFR 264.573(o) and 265.443(n)).

Σ Review the closure plans, and determine if there are plans to remove or decontaminate all waste residues, contaminated equipment, subsoils, and structures, or close the facility as a landfill (40 CFR 264.575 and 265.445).

Σ Review procedures and documentation to verify that all wastes are removed from the drip pad and collection system at least every 90 days (40 CFR 262.34(a)(1)).

**33.2** Drip pads and their components must be inspected. (40 CFR 264.574 and 265.444)

**Guide Note**

Σ Examine construction records and determine if liners and cover systems were inspected during and after installation (40 CFR 264.574(a) and 265.444(a)).

Σ Determine if drip pads are inspected weekly while in operation and after storms to detect deterioration, malfunction, or leakage of runon and runoff control systems, leak detection systems, and the drip pad surface (40 CFR 264.574(b) and 265.444(b)).

Σ Determine if drip pads are sufficiently cleaned to allow weekly inspections, and if the date and time of cleaning are documented in the facility records (40 CFR 264.573(l) and 265.443(i)).

Σ If leakage is detected, is the administering agency notified within 24 hours, a written report submitted within 10 days, and an independent engineering certification submitted upon completion of repairs and cleanup (40 CFR 264.573(m) and 265.443(m))?

*Hazardous Waste Rulebook*

## 34. TSD Facilities: Miscellaneous Units

**34.1** Miscellaneous hazardous waste management units must be located, constructed, operated, maintained, and closed to ensure protection of human health and the environment. (40 CFR 264.600 - 603)

**Guide Note**

Σ Review facility design, operation, and permit conditions. Determine if the facility is in compliance with permit conditions regarding prevention of migration of waste constituents in the groundwater or subsurface environment (40 CFR 264.601(a)).

Σ Determine if the facility is in compliance with permit conditions regarding prevention of migration of waste constituents in surface water, wetlands, or soil surface (40 CFR 264.601(b)).

Σ Determine if the facility is in compliance with permit conditions regarding prevention of migration of waste constituents in air (40 CFR 264.601(c)).

Σ Determine if the facility is in compliance with permit conditions and procedures regarding monitoring and analysis, inspection, emergency response, spill reporting, and corrective action (40 CFR 264.602).

Σ Determine if there is a plan in place for closure, and for disposal facilities or if contamination cannot be completely removed, post-closure care (40 CFR 264.603).

## 35. TSD Facilities: Air Emission Standards

**35.1** Facilities that treat, store, or dispose of hazardous waste must comply with the Air Emissions Standards for Process Vents. These standards apply to vents associated with distillation, fractionation, thin-film evaporation, solvent extraction, or air or steam stripping that manage hazardous waste with at least 10 ppm organics and have a permit that was issued after December 21, 1990. (40 CFR 264.1030 - 1036 and 265.1030 - 1035)

**Guide Note**

Σ Review process throughput and emissions records and determine if total organic emissions from all affected process vents are reduced below 3 pounds per hour and 3.1 tons per year (tpy), or controlled to at least 95% as measured by percentage of weight throughput (40 CFR 264.1032(a) and 265.1032(a)).

Σ If emissions from process vents are controlled, determine whether all devices meet the applicable requirements of 40 CFR 264.1033 for closed-vent systems and control devices for thermal vapor incinerators, flares, boilers, process heaters, condensers, or carbon adsorption systems, or if documentation has been developed describing the control system and process parameters (40 CFR 264.1031(b) and 265.1032(b)).

Σ For closed-vent systems, verify that repairs were made within 15 days of any detectable emissions (40 CFR 264.1033(k)(3) and 265.1033(j)(3)).

Σ Determine if performance test have been performed to determine total organic compound concentrations and mass flow rates entering and exiting the system (40 CFR 264.1034(c)(1) and 265.1034(c)(1)).

Σ If the facility has not installed a closed-vent system and control device, verify that the operating record includes an implementation schedule (40 CFR 264.1035(b) and 265.1035(b)).

Σ Examine records and determine whether the records include identification of vents, their annual throughput and operating hours, their locations and estimated emissions rates, the estimated emissions rates for the facility, and data supporting the emissions determinations (40 CFR 264 1035(b)(2)(i) and (ii), and 265.1035(b)(2)).

Σ Determine if records include a performance test plan and data, and engineering specifications and calculations (40 CFR 264.1035(b)(3) and (4), and 265.1035(b)(3) and (4)).

Σ Determine if records include design documentation and monitoring, and operating and inspection information (40 CFR 264.1035(c) and 265.1035(c)).

Σ For final status facilities, review the semiannual report and determine if it contains the data specified by the administering agency (40 CFR 264.1036).

**35.2** Facilities that treat, store, or dispose of hazardous waste must comply with the Air Emission Standards for Equipment Leaks as set forth in 40 CFR 264.1050 - 1056, or 265.1050 - 1064. This paragraph applies to equipment that contains or contacts hazardous waste that has an organic concentration of at least 10% and is managed in accordance with the permitting requirements of 40 CFR 270.

**Guide Note**

Σ Ensure that each pump in light liquid service is checked at least weekly for implications of drips from seals (40 CFR 264.1052(a)(2) and 265.1052(a)(2)).

Σ Are all pumps in light liquid service monitored monthly to detect leaks as stated in 40 CFR 265.1063 (40 CFR 264.1052(a)(1) and 265.1052(a)(1))?

Σ Are testing instruments calibrated before use (40 CFR 264.1063(b))?

Σ For pumps, an instrument reading above 10,000 ppm indicates that a leak is present (40 CFR 264.1052(b)(1), 264.1058(b), 265.1052(b), and 265.1058(b)).

Σ Repairing of leaks will be within 15 calendar days of the detection of the leak, and the first attempt at repair will be made within 5 calendar days (40 CFR 264.1052(c), 264.1058(c), 265.1052(c), and 265.1058(c)).

Σ Ensure that pressure relief devices conform with the standards listed in 40 CFR 264.1054 or 265.1054.

Σ Does the facility employ any open-end valves or lines.? If yes, does it comply with 40 CFR 264.1056 or 265.1056?

Σ Does the facility employ any sampling connecting systems? If yes, does it comply with 40 CFR 264.1055 or 265.1055?

Σ Verify that each piece of equipment is distinguished by marking (40 CFR 264.1050(d) and 265.1050(c)).

Σ Are all units permitted according to the requirements of 40 CFR 270 (40 CFR 264.1050(b) and 265.1050(b))?

Σ Note that vacuum units are generally excluded from the requirements of this paragraph if they are identified as being required in 40 CFR 264.1064(g)(5) or 265.1064 (g)(5) (40 CFR 264.1050(e) and 265.1050(d)).

## 36. Universal Wastes

*This section applies to hazardous wastes that are subject to the universal waste requirements, including batteries, pesticides, and thermostats.*

**36.1** Facilities must determine whether they generate universal wastes. (40 CFR 273.6)

**Guide Note**

Σ Verify that the site has identified all universal wastes in the following categories:

- batteries;
- pesticides;
- thermostats.

**36.2** All sites must determine if they are conditionally exempt, SQGs, or LQGs of universal waste. (40 CFR 273.6)

**Guide Note**

Σ Determine quantities of universal waste generated at the site.

Σ Classify the facility as a CESQG (same for universal waste generators as for hazardous waste generators), SQG (accumulates 5,000 kilograms or less total at any time), or an LQG (accumulates more than 5,000 kilograms total at any time).

Σ Determine if the site was an LQG at any time during the calendar year.

**36.3** SQGs must ensure that all universal wastes are handled properly while being stored onsite. ( 40 CFR 273 Subpart B)

**Guide Note**

Σ Verify that no universal wastes are being disposed of or treated onsite (40 CFR 273.11).

Σ Verify that all universal wastes are being handled to prevent releases to the environment and that all containers are closed, structurally sound, and compatible (40 CFR 273.13 and 14).

Σ Verify that all universal waste batteries are (40 CFR 273.13 and 14):

- contained, if leaking or damaged;

## Hazardous Waste Rulebook

- labeled "Universal Waste - Battery(ies)", "Waste Battery(ies)", or "Used Battery(ies) ";
- labeled individually (along with containers).

Σ Verify that all waste pesticides are (40 CFR 273.13 and 14):

- if not in an approved container, in an overpack;
- in a tank that meets hazardous waste requirements;
- in a transport vehicle that is closed, structurally sound, compatible, and shows no signs of damage or leakage;
- labeled with the product label (the DOT label may be used if the product label is not available) <u>and</u> "Universal Waste - Pesticides "or ""Waste Pesticides".

Σ Verify that all waste thermostats are (40 CFR 273.13 and 14):

- contained, if leaking or damaged;
- labeled individually or in containers with "Universal Waste - Mercury Thermostat(s) ", "Waste Mercury Thermostat(s)", or "Used Mercury Thermostats ".

Σ Verify that wastes are not stored for more than 1 year from the date they are generated, unless being accumulated to reach critical mass for recovery, treatment, or disposal. Verify that 1 of the following methods is being used to show compliance (40 CFR 273.15):

- Containers are marked with the earliest date that any waste in them became a waste.
- Containers are marked with the date the individual item became a waste.
- An inventory system is used.
- There is a specific accumulation area with the date of first accumulation noted in the area.

Σ Ensure employees have been informed of the proper waste handling methods and emergency procedures (40 CFR 273.16).
Σ Ensure all releases are immediately handled (40 CFR 273.17).
Σ Verify that all universal wastes are sent to an authorized location (40 CFR 273.18).
Σ Verify that if an SQHUW self-transports universal wastes, the SQHUW is in compliance with transportation regulations (40 CFR 273.18).
Σ Ensure that all shipments of universal wastes that are hazardous are shipped per DOT regulations (40 CFR 273.18).
Σ Verify that rejected shipments are either returned or an alternate destination identified (40 CFR 273.18).
Σ Verify that records concerning the shipments are maintained (GMP).

**36.4** LQHUW must ensure that wastes are handled properly while being stored onsite. (40 CFR Subpart C)

### Guide Note

Σ Verify that no universal wastes are being disposed of or treated onsite (40 CFR 273.31).
Σ Verify that all waste handlers either have an EPA ID number or have notified EPA of their universal waste activity (40 CFR 273.32).
Σ Verify that all universal wastes are being handled to prevent releases to the environment and that all containers are closed, structurally sound, and compatible (40 CFR 273.33 and 34).
Σ Verify that all universal waste batteries are (40 CFR 273.33 and 34):

- contained, if leaking or damaged;
- labeled "Universal Waste - Battery(ies) ", "Waste Battery(ies) ", or "Used Battery(ies) ";
- labeled individually, along with containers.

Σ Verify that all waste pesticides are (40 CFR 273.33and 34):

- if not in an approved container, in an overpack;
- on a tank that meets hazardous waste requirements;
- in a transport vehicle that is closed, structurally sound, compatible, and shows no signs of damage or leakage;
- labeled with the product label (the DOT label may be used if the product label is not available) <u>and</u> "Universal Waste - Pesticides" or "Waste Pesticides"

Σ Verify that all waste thermostats are (40 CFR 273.33 and 34):

- contained, if leaking or damaged;
- individually labeled or their containers labeled "Universal Waste - Mercury Thermostat(s)", "Waste Mercury Thermostat(s)", or "Used Mercury Thermostats".

Σ  Verify that wastes are not stored for more than 1 year from the date they are generated, unless being accumulated to reach critical mass for recovery, treatment, or disposal. Verify that 1 of the following methods is being used to show compliance (40 CFR 273.35)

- Containers are marked with the earliest date that any waste in them became a waste.
- Containers are marked with the date the individual item became a waste.
- An inventory system is used.
- There is a specific accumulation area with the date of first accumulation noted in the area.

Σ  Ensure employees are thoroughly familiar with proper waste handling methods and emergency procedures (40 CFR 273.36).

Σ  Ensure all releases are immediately handled (40 CFR 273.37).

Σ  Verify that all universal wastes are sent to an authorized location (40 CFR 273.38).

Σ  Verify that if an LQHUW self-transports universal wastes, the LQHUW is in compliance with transportation regulations (40 CFR 273.38).

Σ  Ensure that all shipments of universal wastes that are hazardous are shipped per DOT regulations (40 CFR 273.38).

Σ  Verify that rejected shipments are either returned or an alternate destination identified (40 CFR 273.38).

Σ  Verify that records concerning shipments received or shipped are maintained for at least 3 years (40 CFR 273.39).

**36.5**  Transporters of universal wastes must ensure that wastes shipped are handled and transported properly.  (40 CFR 273 Subpart D)

**Guide Note**

Σ  Ensure that all transporters comply with DOT requirements (40 CFR 273.52).

Σ  Verify that transporters are not treating or disposing of universal wastes (40 CFR 273.51).

Σ  Verify that wastes are not being stored for more than 10days at a transfer facility (40 CFR 273.52).

Σ  Verify that all releases have been contained and reported (40 CFR 273.53).

Σ  Ensure that all wastes are delivered to a universal waste handler or an approved destination facility (40 CFR 273.55).

**36.6**  Destination facilities for universal wastes must treat, store, or dispose of wastes properly.  (40 CFR Subpart E)

**Guide Note**

Σ  Verify that the facility complies with all applicable hazardous waste requirements (40 CFR 273.60).

Σ  If a shipment contains a hazardous waste that is not a universal waste but is stated as such, the owner or operator must immediately notify the regional EPA office (40 CFR 273.61).

Σ  Verify that a record of each shipment of universal waste is maintained for at least 3 years (40 CFR 273.62).

**36.7**  SQGs and LQGs who remove mercury-containing ampules from thermostats must take precautions.  (40 CFR 273.13 and 33)

**Guide Note**

Σ  Verify that ampules are removed over or in containment devices

Σ  Ensure that a mercury cleanup system is readily available

Σ  Verify that all spillage is collected in an approved hazardous waste container

Σ  Verify that the area is well vented and employees are thoroughly familiar with mercury handling and emergency response procedures

Σ  Verify that all ampules removed are packed with packaging materials to prevent breakage.

# Appendix D

# SAMPLE TRAINING TOOLS

# CLASS EXERCISES

**Audit Verification**
**Interviewing Skills**
**Management Systems Assessments**
**Report Writing**
**Audit Conferences**

# Audit Verification

ENVIRONMENTAL, HEALTH & SAFETY AUDITS
PROBLEM SOLVING EXERCISE

CONFINED SPACES

## BACKGROUND

You have been assigned to review the confined spaces program during the audit of a facility. The site provided you with the following information about on-site confined spaces:

- Thirty-five (35) confined spaces
- Twenty (25) spaces are permit-required
- Thirty (30) employees are trained in/qualified for confined spaces
- Twenty (20) permits issued all of last year
- Ten (15) permits issued in the first quarter of this year.

The OSHA confined space standard (29 CFR 1910.146) defines a confined space as a space that is large enough for an employee to enter, has restricted means of entry or exit, and is not designed for continuous employee occupancy. Examples include ship compartments, missile fuel tanks, vats, silos, sewers, tunnels and vaults. Confined space hazards include physical hazards, oxygen deficiency, combustibility and toxic air contaminants.

## ASSIGNMENT

1. List the following activities in the order in which you would do them:

    ___ Review written permit entry program
    ___ Inspect training records
    ___ Review emergency procedures
    ___ Inspect the confined spaces
    ___ Review canceled permits
    ___ Review results of annual review
    ___ Review lockout/tagout program
    ___ Interview employees

2. Using the attached sampling methodology, how many permit-required confined spaces and non-permit confined spaces would you review. What kind of sampling strategy (e.g., random, block, stratified, interval) would you use?

ENVIRONMENTAL, HEALTH & SAFETY AUDITS
PROBLEM SOLVING EXERCISE

CONFINED SPACES

3a. Upon inspecting a confined space entry that is underway you discover that no entry permit has been issued. A certificate is posted at the entry portal that explains that all hazards have been eliminated. What conclusion can you make with regard to the status of this confined space?

3b. What if welders enter the space to make a repair?

4. At a minimum, which type of employees would you need to interview to understand the program and verify compliance? How do you distinguish a good program from a bad one?

5. Based on a review of the confined space activities at the site over the past couple of years, what management issue might you want to investigate? What types of questions would you ask?

## ENVIRONMENTAL AUDITS
## PROBLEM SOLVING EXERCISE

### SELECTING A SAMPLE SIZE
### ON ENVIRONMENTAL AUDITS

| Population Size | Suggested Minimum Sample Size | | | |
|---|---|---|---|---|
| | A | B | C | D |
| 10 | 98% | 88% | 72% | 50% |
| 25 | 94% | 74% | 49% | 28% |
| 50 | 89% | 58% | 32% | 18% |
| 100 | 80% | 41% | 19% | 9% |
| 250 | 61% | 21% | 8% | 4% |
| 500 | 43% | 12% | 4% | 2% |
| 1000 | 28% | 6% | 2% | 1% |
| 2000 | 16% | 3% | 1% | 0.5% |

A -  Suggested minimum sample size for a population(s) being reviewed which is considered to be extremely important in terms of verifying compliance with applicable requirements, and/or is of critical concern to the corporation in terms of potential or actual impacts associated with non-compliance. A confidence interval of 95% is assumed.

B -  Suggested minimum sample size for a population(s) being reviewed that will provide additional information to substantiate compliance or non-compliance and/or is of considerable importance to the corporation in terms of potential or actual impacts associated with non-compliance. A confidence interval of 90% is assumed.

C -  Suggested minimum sample size for a population(s) being reviewed that will provide ancillary information in terms of verifying overall compliance with a requirement. A confidence interval of 85% is assumed.

D -  Suggested minimum sample size for a population(s) being reviewed that will provide ancillary information in terms of verifying overall compliance with a requirement. A confidence interval of 80% is assumed.

## ENVIRONMENTAL AUDITS
## PROBLEM SOLVING EXERCISE

### AUDIT SAMPLING

Sampling theory can provide a means to identifying the number of items to be reviewed to be fairly confident the results are representative. That is, when a confidence interval of a given width is desired for an unknown proportion, p (in this case the proportion of sources that are in compliance), the sample size, n, required can be obtained using a formula given by Inman and Conover[1]. However, the formula applies to an infinite population. A correction for finite population is given by Cochran[2]. The finite population correction is close to unity when the sampling fraction n/N remains low (approximately 5% or less). Combining the two formulae gives:

$$n = (4z[\alpha/2]^2 pqN)/(w^2N - w^2 + 4z[\alpha/2]^2 pq)$$

Where,

n is the required number of sources in the sample

$z[\alpha/2]$ is the $(1-\alpha/2)$ quantile from the standard normal probability function; that is:

| Confidence Level | $\alpha$ | $1-\alpha/2$ | $z[\alpha/2]$ |
| --- | --- | --- | --- |
| 95% | 0.05 | 0.975 | 1.96 |
| 90% | 0.10 | 0.950 | 1.65 |
| 85% | 0.15 | 0.925 | 1.44 |

p is the proportion of interest (proportion of sources in compliance); when p is unknown it is set equal to 0.5 to give the most conservative possible value of n

q is (1-p)

N is the total population being sampled
w is $2\alpha$

---

[1] Inman, R.L. and Conover, W.J., *A Modern Approach to Statistics*, John Wiley and Sons, New York, NY.

[2] Cochran, W.G., *Sampling Techniques*, Second Edition, John Wiley and Sons, New York, NY.

# ENVIRONMENTAL AUDITS
## PROBLEM SOLVING EXERCISE

### HAZARD COMMUNICATION

# BACKGROUND

You have been assigned the review of the hazard communication program during the audit of a facility. The written program has procedures for requesting a MSDS if one is not received automatically from a supplier of a hazardous chemical. It also includes the OSHA guide for Reviewing MSDS Completeness, which is attached.

# ASSIGNMENT

1.  Using questions 4 and 5 of the OSHA Guide, perform a review of the MSDS for BIO BETTY™ to determine whether the facility is implementing this hazard communication procedure. What, if any, obvious inadequacies were found with the MSDS for BIO BETTY™?

2.  Review the label on the container of BIO BETTY™ for the identity of the material, appropriate hazard warnings, and the name and address of the producer. What, if any, obvious inadequacies were found with the label for BIO BETTY™?

3.  Is the label prepared below adequate for a container into which you might subsequently transfer the BIO BETTY™? What, if any, are its obvious inadequacies?

BIO BETTY™

Contains Water Base Willie
and dry chlorine

OSHA Instruction CPL 2-2.38C
Office of Health Compliance Assistance

**Appendix D**

Guide for Reviewing MSDS Completeness

NOTE: This guide has been developed for use as an optional aid during inspections.

During CSHO review for Material Safety Data Sheet completeness, the following questions may be helpful:

1. Do chemical manufacturers and importers have an MSDS for each hazardous chemical produced or imported into the United States?

2. Do employers have an MSDS for each hazardous chemical used?

3. Is each MSDS in at least English?

4. Does each MSDS contain at least the:

 (a) Identity used on the label?

 (b) Chemical and common name(s) for single substance hazardous chemicals?

 (c) For mixtures tested as a whole:

  (1) Chemical and common name(s) of the ingredients which contribute to the known hazards?

  (2) Common name(s) of the mixture itself?

 (d) For mixtures not tested as a whole:

  (1) Chemical and common name(s) of all ingredients which are health hazards (1 percent concentration or greater), including carcinogens (0.1 percent concentration or greater)?

  (2) Chemical and common name(s) of all ingredients which are health hazards and present a risk to employees, even though they are

present in the mixture in concentrations of less than 1 percent of 0.1 percent for carcinogens?

(e) Chemical and common name(s) of all ingredients which have been determined to present a physical harm when present in the mixture?

(f) Physical and chemical characteristics of the hazardous chemical (vapor pressure, flash point, etc.)?

(g) Physical hazards of the hazardous chemical including the potential for fire, explosion, and reactivity?

(h) Health hazards of the hazardous chemical (including signs and symptoms and medical conditions aggravated)?

(i) Primary routes of entry?

(j) OSHA permissible exposure limit (PEL)? The American Conference of Governmental Industrial Hygienists (ACGIH) Threshold Limit Value (TLV)? Other exposure limit(S) (including ceiling and other short term limits)?

(k) Information on carcinogen listings (reference OSHA regulated carcinogens, those indicated in the National Toxicology Program (NTP) Annual Report on Carcinogens and/or those listed by the International Agency for Research on Carcinogens (IARC))?

NOTE: Negative conclusions regarding carcinogenicity, or the fact that there is no information, do not have to be reported unless there is a specific space or blank for carcinogenicity on the form.

(l) Generally applicable procedures and precautions for safe handling and use of the chemical (hygienic practices, maintenance and spill procedures)?

(m) Generally applicable control measures (engineering controls, work practices and personal protective equipment)?

(n) Pertinent emergency and first aid procedures?

(o) Date that the MSDS was prepared or the date of the last change?

(p) Name, address and telephone number of the responsible party?

5. Are all sections of the MSDS completed?

# BIO BETTY™

American Health & Safety, INC.
6250 Nesbitt Rd.
Madison, WI 53719
800-522-7554

## SECTION 1 - CHEMICAL PRODUCT AND COMPANY INFORMATION

Date Prepared:          November 8, 1989
Product Name:           BIO BETTY™
Chemical Name(s):       Sodium Polyacrylate and Calcium Hypochlorite
Emergency Telephone:    608-273-4000

## SECTION 2 - COMPOSITIONAL INFORMATION

| Name: | CAS# | Approximate Weight (% wt.): |
|-------|------|------------------------------|
| Sodium Polyacrylate | 9003-04-07 | 99.9 |
| Calcium Hypochlorite | 7778-54-3 | .1 |

## SECTION 3 - POTENTIAL HEALTH EFFECTS

**Effects of Overexposure:**
Excessive exposure to dust may irritate eyes, nose and respiratory

**Acute effects:**      not available

**Chronic effects:**

Medical Conditions Generally Aggravated by Exposure:      None known

Chemical Listed as Carcinogen or Potential Carcinogen:      Not Available

NTP:      No
IARC:     No
OSHA:     No

## SECTION 4 - FIRST AID MEASURES

**Eye Contact:**
Flush with gently running water for 20 minutes. Consult a physician.

**Skin Contact:**
Wash with soap and water. If irritation persists consult a physician.

**Inhalation:**
Remove to fresh air. Consult a physician if necessary.

**Ingestion:**
Drink large quantities of water. Do not induce vomiting, consult a physician.

**Emergency and First Aid Procedures:**
Obtain medical advice except for minor exposure.

## SECTION 5 - FIRE FIGHTING MEASURES

| | |
|---|---|
| Flash Point: | Not Applicable |
| Lower Explosive Limit: | Not Applicable |
| Upper Explosive Limit: | Not Applicable |
| Autoignition Temperature: | Not Applicable |

**Extinguishing Media:**
Carbon Dioxide, Dry Chemical or Foam.

**Unusual Fire Hazards: and explosion hazards:**
Not applicable

**Fire Fighting Procedures:**
Wear self contained breathing apparatus when fighting fire.

## SECTION 6 - ACCIDENTAL RELEASE MEASURES

**Releases:**
> In case of a release or spill, vacuum, sweep or shovel up material and dispose of in a waste container in accordance with local, state, federal and provincial authorities.

> Water may make the area slippery.

## SECTION 7 - HANDLING AND STORAGE

Store in a dry place under 100°F. Avoid exposure to direct sunlight and High Humidity.

> **Other Precautions:**

>> Keep dust levels low, wear personal protective equipment if excessive dust is present.

## SECTION 8 - EXPOSURE CONTROLS/PERSONAL PROTECTION

**ACGIH Threshold Limit Value**
> N/A

**OSHA Permissible Exposure Limit**
> N/A

**Other Exposure Limits**
> N/A

**Engineering Controls:**

> **Ventilation Requirements:**
> Local exhaust suggested.

**Personal Protective Equipment:**

> **Respiratory Protection:**
> Dust or filter type respirator when working with excessive amounts of product.

> **Eye Protection:**
> Chemical goggles recommended.

**Skin Protection:**
Plastic or rubber chemical resistant.

**Hygienic Practices:**
Wash hands with soap and water after using product.

## SECTION 9 - PHYSICAL AND CHEMICAL PROPERTIES

| | |
|---|---|
| Appearance: | White Powder |
| Color: | White |
| Odor: | Chlorine Fragrance |
| Vapor Pressure (mm Hg): | Not Applicable |
| Vapor Density (Air=1): | Not Applicable |
| Boiling Point: | Not Available |
| Melting Point: | Decomposition at 350°F |
| Specific Gravity (Water=1): | Not Available |
| Solubility in Water: | Insoluble (Swells in Water) |
| Reactivity in Water | Swells in Water |

## SECTION 10 - STABILITY AND REACTIVITY

**Stability:**
This material is stable.

**Conditions to Avoid:**
None Known

**Hazardous Decomposition Products:**
Thermo decomposition may produce Carbon Monoxide/Carbon Dioxide.

**Incompatibility (Materials to Avoid):**
Strong Oxidizing Agents

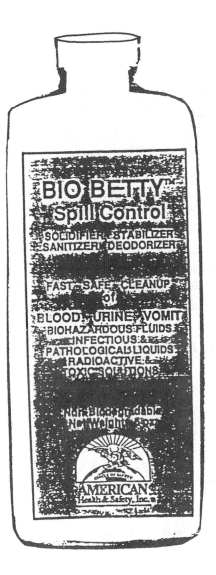

Bio Betty™ is a unique, fast acting encapsulator with stabilized chlorine available at 10,000 ppm. Hospital grade disinfectant. The application of a chlorine compound is consistently recommended for use on spilled body fluids. Aggressively attacks: AIDS, Hepatitis B and any other blood borne virus.

## Read DIRECTIONS

1. Use appropriate protective clothing, respirators, gloves and goggles before attempting spill clean-up.
2. Generously sprinkle Bio Betty™ over spilled area until completely covered.
3. When liquid is congealed, remove with a disposable scoop, dust pan or cardboard.
4. After disposing of waste properly, wash hands thoroughly with soap and water or alcohol antiseptic.

**Precautions:**

1. Put on disposable plastic or rubber gloves for clean-up.
2. Clean and disinfect the contaminated area with a hospital grade germicidal detergent.

**Use Ratio:**

Use 5 oz. of Bio Betty™ to handle up to 15 gallons of spill. Amount may vary depending on sodium content.

## WARNING: Please Read Carefully!

First Aid Requirements: If inhaled, remove person to fresh air. If ingested, drink large quantities of water. Do not induce vomiting, consult a physician. If in contact with skin, wash with soap and water. If in contact with eyes, flush with water for 15 minutes. Consult a physician if necessary.

Dispose of per Federal and State regulations.
Contents: Water Base Willie™ and dry chlorine.

## KEEP OUT OF REACH OF CHILDREN

Manufactured exclusively for:

AMERICAN Health & Safety, Inc.®
6250 Nesbitt Road
Madison, WI 53719
800-522-7554

## ENVIRONMENTAL AUDITS
## PROBLEM SOLVING EXERCISE

### Air Emissions

# BACKGROUND

In the course of preparing to audit air emissions compliance you discover that the facility has several sources which are regulated for opacity (fuel burners, incinerators). The facility also has about 20 stacks and vents on the roofs of three buildings, some of which may be exempted from permitting. The actual audit is scheduled for the next day and you are working by yourself.

# ASSIGNMENT

1.  How would you schedule activities for the next day to perform the audit efficiently and accurately.

2.  Would you physically go up on the roofs and record ALL the stacks and attempt to match them with equipment/operations inside the buildings?

3.  How would you best determine what stacks or vents were exempted from any licenses or permits?

4.  How would you determine if all sources of volatile organic compounds (VOC) emissions are identified at the facility? What would you do to verify compliance with VOC emission standards?

5.  Later in the day, you observe asbestos is being removed from a boiler and piping system by an outside contractor. You had not anticipated this because your escort forgot to tell you it was happening. What would you do? What would some of your key questions be? Who would you address them to?

## ENVIRONMENTAL AUDITS
## PROBLEM SOLVING EXERCISE

### Wastewater Discharges

## BACKGROUND

The facility you are about to audit has an industrial wastewater that discharges *directly* into the local river. The facility's wastewater discharge permit, which has been issued by the regulatory authority, requires monthly sampling. The permit expires six months from now. Permitted effluent parameters include flow, BOD, TSS, TOC, Ph, and several priority pollutants. The treatment plant includes primary clarifiers, activated sludge tanks, and settling tanks.

The facility also has recently received an *indirect* discharge permit from the Onondaga County. Two pages from this permit are included with this exercise. You notice during your tour that the facility is testing the fire water systems and the discharge is going into the County Sewer System. Your facility contact tells you they test the fire water systems twice per month.

The facility also has an unpermitted stormwater outfall. Your job is to audit wastewater and stormwater compliance at the site.

## ASSIGNMENT

1. What is the logical sequence of tasks you would undertake to evaluate wastewater discharge permit compliance at the facility?

    _____ Observe technician taking a sample for proper technique

    _____ Determine if lab is certified

    _____ Inspect effluent discharge

    _____ Inspect treatment plant/interview operator

    _____ Inspect wastewater discharge monitoring reports

    _____ Review permit from the regulatory authority

2. You notice from the monitoring reports that the flow and pH exceeded the limits specified in the permit twelve times during the last 12 months. How would you react to this? What would you ask the facility contact?

ENVIRONMENTAL AUDITS
PROBLEM SOLVING EXERCISE

Wastewater Discharges

3. What key scheduling question might you want to ask about the wastewater discharge Permit?

4. Is the fire water system discharge to the Municipal sewer system permitted or prohibited?

5. How would you verify stormwater discharge compliance?

6. What general wastewater *management issue* might you want to discuss with the facility contact based on your overall review of the wastewater discharges?

## ONONDAGA MUNICIPAL SEWER ORDINANCE

I.  *AUTHORITY*

A.  This permit is hereby promulgated by the Commissioner of the Onondaga Municipal Department of Drainage and Sanitation (OMDDS) to regulate the discharge of wastewater, polluted or unpolluted, to the sanitary sewer system, under the authority of The Onondaga Municipal Rules and Regulations Relating to the Use of the Public Sewer System (the Rules and Regulations) and the Onondaga Municipal Administrative Code.

B.  Article VII of the Rules and Regulations provides that any violation of this permit may subject the permittee to a fine of one thousand dollars per day per violation. In addition Articles VI and VII of the Rules and Regulations specify other penalties and procedures the Department may employ for any violation of this permit or the Rules and Regulations.

II.  *PERMITTED WASTEWATER DISCHARGE*

A.  The permittee is authorized to discharge the following to the Municipal sewer system:

1.  Sanitary Wastewater;

2.  Boiler blowdown, wastewater including that from vehicle washing, EP 10 Outdoor Sump and EP 9 Basement Sump discharged via the Building 10 oil/water separator.

3.  Filtered wastewater with the addition of peroxide originating from production and operation of ceramic sonar transducer element cylinders and lids (Transducer Products Operations (TPO) - Building #5);

4.  Wastewater generated from the fabrication of semiconductors, cleaning of facilities and equipment and the cleaning of parts in the Electronics Laboratory (Building 3), and the cleaning of parts in the Electronics Lab Model Shop in Building 7.

5.  Contact cooling water blowdown from the cooling towers located outside of Buildings 7 and 10; and non-contact cooling water associated with the operation of air conditioners and process equipment and basement sump in EP-6 until such time as

## ONONDAGA MUNICIPAL SEWER ORDINANCE

permission from the state is obtained to reroute through the EP storm drains.

6.  Photographic laboratory wastewater.

7.  Wastewater from Projection Display Products Operations (PDPO) group production, parts cleaning and silk screening in Building 7.

8.  Wastewater discharges from the Environmental Test Chambers, the Conformal Coat Process, and the cleaning of air filters in the calibration laboratory in EP-5.

9.  Sump water contaminated with low levels of trichloroethene (less than 25 parts per billion) from building EP-6. This discharge shall be prohibited after May 1, 1993.

B.  All wastewater discharged to the sanitary sewer system must comply with the effluent limitations set forth in Section IV of this permit and Article III of the Onondaga Municipal Rules and Regulations Relating to the Use of the Public Sewer System, unless otherwise indicated in this permit expressly or by implication.

III.    PROHIBITED DISCHARGES

A.  In accordance with Article III of the Rules and Regulations, the following wastes shall not be introduced into the Onondaga Municipal Sanitary Sewer System.

1.  Wastewater constituents which by their introduction to the sewer system, cause pass-through (in accordance with Sections 3.01(d), 3.01(f), and 3.01(g)).

2.  Wastewater constituents which by their introduction to the sewer system, cause interference (in accordance with Sections 3.01(b), 3.01(d), 3.01(i), and 3.01(j)).

3.  Wastewater which has the potential to create a fire or explosion hazard in the publicly-owned treatment works (POTW), including wastewater having a closed-cup flashpoint less than 140°F, (pursuant to Section 3.01(a)).

4.  Wastewater having a pH lower than 5.5 or higher than 9.5 Standard Units (in accordance with Section 3.01(c)).

5.  Wastewater constituents which result in the presence of toxic gases, vapors or fumes within the POTW in a quantity that may

## ONONDAGA MUNICIPAL SEWER ORDINANCE

cause acute worker health and safety problems (pursuant to Sections 3.01(a), 3.01(e)).

6. The introduction of wastewater having a temperature greater than 150°F into the sanitary sewer system <u>or</u> at a quantity such that the temperature at the headworks of the POTW exceeds 104°F (Section 3.01(i)).

7. Batch discharges without prior written approval from the Commissioner. Any request to discharge such wastewater must be submitted in writing to this office and is subject to the approval of the Commissioner on a case-by-case basis.

8. Any wastewater which will subject the receiving POTW to reporting and permitting requirements of the federal hazardous waste rules.

9. Non-contact cooling water and other unpolluted wastewater (in accordance with Section 3.02).

10. All other wastes which are prohibited by Article III of the Rules and Regulations.

B. In addition to the above prohibitions, dilution shall not be used as a substitute for pretreatment.

ENVIRONMENTAL AUDITS
PROBLEM SOLVING EXERCISE

Hazardous Waste Management

## BACKGROUND

You are on a three-person team about to conduct an audit of a major manufacturing facility. Your responsibility is hazardous waste. From the pre-audit questionnaire you have determined that the facility is a large quantity generator (LQG) with a wide variety of characteristic and listed hazardous wastes coming from laboratory, maintenance and manufacturing operations. There are five satellite accumulation points, two 90-day accumulation points and one permitted storage area on the site.

## ASSIGNMENT

1.  How would you begin your hazardous waste audit of the site? That is, what specifically would you do in the first couple of hours to assure an efficient review?

2.  What are the key items you need to evaluate at the 90-day accumulation point? What are some of the key differences among the requirements for satellite, 90-day and storage areas?

3.  Review the attached manifest and identify any deficiencies. Are any of these significant? What other things should you be looking for in the manifest file?

ENVIRONMENTAL AUDITS
PROBLEM SOLVING EXERCISE

Hazardous Waste Management

## ASSIGNMENT (Continued)

4. You have given yourself only two hours to review the manifest file, which contains 1000 manifests. Using the attached Sample Size table, how many manifests would you look at to obtain a representative sample? If, as you conduct your review you discover some major discrepancies in the first few manifests, would you (1) increase your sample size, (2) stop where you are, or (3) retain the original sampling scheme? Pick only one and explain your rationale.

5. Review the attached two-page inspection log for the storage area and identify any deficiencies. Speculate on the likely underlying causes for the deficiencies.

Bureau of Waste Management
P. O. Box 8550
Harrisburg, PA 17105-8550

Form approved.
OMB No. 2050-0039
Expires 9-30-91

ER-WM-51 REV. 11/88

| UNIFORM HAZARDOUS WASTE MANIFEST | 1. Generator's US EPA ID No. PAD 000000789 | | 2. Page 1 of 1 | Information in the shaded areas is not required by Federal law but is required by State law. |
|---|---|---|---|---|

**3. Generator's Name and Mailing Address**
XYZ Company
123 Anystreet
Ourtown, PA 11131

A. State Manifest Document Number
**PAC 3025131**

B. State Gen. ID
DAD 000000789

**4. Generator's Phone ( )**

**5. Transporter 1 Company Name**
WE DRIVE TRUCKS

**6. US EPA ID Number**
PAD 555 12347

C. State Trans. ID
PA- 1AH1 123451

**7. Transporter 2 Company Name**
Specialty Transporters

**8. US EPA ID Number**
NJD 555 123489

D. Transporter's Phone ( )

E. State Trans. ID
PA- 1AH1 34512

**9. Designated Facility Name and Site Address**
ABC Disposal
Industrial Park
Somewhere, PA 11132

**10. US EPA ID Number**
PAT 111 454 769

F. Transporter's Phone (609) 222-7777

G. State Facility's ID

H. Facility's Phone (412) 000-3131

| 11. US DOT Description (Including Proper Shipping Name, Hazard Class, and ID Number) | 12. Containers No. | Type | 13. Total Quantity | 14. Unit Wt/Vol | I. Waste No. |
|---|---|---|---|---|---|
| a. Paint, Flammable Liquid UN 1263 | 10 | DM | 550 | G | D001 |
| b. Waste Flammable Liquid, N.O.S. (Xylene) Flammable Liquid UN 1993 | 10 | DM | | G | D001 |
| c. Waste Corrosive Liquid, N.O.S. (Sulfuric Acid, Chromic Acid) Corrosive Material | 15 | DF | 1000 | P | D003 |
| d. | | | | | |

**J. Additional Descriptions for Materials Listed Above**

| Lab Pack | Physical State | | Lab Pack | Physical State |
|---|---|---|---|---|
| a. ☐ | L | c. ☐ | ☐ | |
| b. ☐ | L | d. ☐ | ☐ | |

**K. Handling Codes for Wastes Listed Above**

a. S01  c. S01

b. S01  d.

**15. Special Handling Instructions and Additional Information**
15- empty Drums (previously contained Flammable solvents)
1- Drum of unknown Material

**16. GENERATOR'S CERTIFICATION:** I hereby declare that the contents of this consignment are fully and accurately described above by proper shipping name and are classified, packed, marked, and labeled and are in all respects in proper condition for transport by highway according to applicable international and national government regulations.

If I am a large quantity generator, I certify that I have a program in place to reduce the volume and toxicity of waste generated to the degree I have determined to be economically practicable and that I have selected the practicable method of treatment, storage, or disposal currently available to me which minimizes the present and future threat to human health and the environment; OR, if I am a small quantity generator, I have made a good faith effort to minimize my waste generation and select the best waste management method that is available to me and that I can afford.

| Printed/Typed Name John C. Generator | Signature JC Gen | MONTH 5 | DAY 15 | YEAR 93 |
|---|---|---|---|---|

**17. Transporter 1 Acknowledgement of Receipt of Materials**

| Printed/Typed Name Bill M. Driver | Signature Bill M. Driver | MONTH 5 | DAY 15 | YEAR 93 |
|---|---|---|---|---|

**18. Transporter 2 Acknowledgement of Receipt of Materials**

| Printed/Typed Name Mark Milvords | Signature Mark Milvords | MONTH 5 | DAY 25 | YEAR 93 |
|---|---|---|---|---|

**19. Discrepancy Indication Space**

**20. Facility Owner or Operator: Certification of receipt of hazardous materials covered by this manifest except as noted in Item 19.**

| Printed/Typed Name Mike T. Owner | Signature Mike T. Owner | MONTH 5 | DAY 11 | YEAR 93 |
|---|---|---|---|---|

EPA Form 8700-22 (Rev. 9/88) Previous editions are obsolete

21

ENVIRONMENTAL AUDITS
PROBLEM SOLVING EXERCISE

SELECTING A SAMPLE SIZE
ON ENVIRONMENTAL AUDITS

| Population Size | Suggested Minimum Sample Size | | | |
|---|---|---|---|---|
| | A | B | C | D |
| 10 | 98% | 88% | 72% | 50% |
| 25 | 94% | 74% | 49% | 28% |
| 50 | 89% | 58% | 32% | 18% |
| 100 | 80% | 41% | 19% | 9% |
| 250 | 61% | 21% | 8% | 4% |
| 500 | 43% | 12% | 4% | 2% |
| 1000 | 28% | 6% | 2% | 1% |
| 2000 | 16% | 3% | 1% | 0.5% |

A - Suggested minimum sample size for a population(s) being reviewed which is considered to be extremely important in terms of verifying compliance with applicable requirements, and/or is of critical concern to the corporation in terms of potential or actual impacts associated with non-compliance. A confidence interval of 95% is assumed.

B - Suggested minimum sample size for a population(s) being reviewed that will provide additional information to substantiate compliance or non-compliance and/or is of considerable importance to the corporation in terms of potential or actual impacts associated with non-compliance. A confidence interval of 90% is assumed.

C - Suggested minimum sample size for a population(s) being reviewed that will provide ancillary information in terms of verifying overall compliance with a requirement. A confidence interval of 85% is assumed.

D - Suggested minimum sample size for a population(s) being reviewed that will provide ancillary information in terms of verifying overall compliance with a requirement. A confidence interval of 80% is assumed.

ENVIRONMENTAL AUDITS
PROBLEM SOLVING EXERCISE

AUDIT SAMPLING

Sampling theory can provide a means to identifying the number of items to be reviewed to be fairly confident the results are representative. That is, when a confidence interval of a given width is desired for an unknown proportion, p (in this case the proportion of sources that are in compliance), the sample size, n, required can be obtained using a formula given by Inman and Conover[1]. However, the formula applies to an infinite population. A correction for finite population is given by Cochran[2]. The finite population correction is close to unity when the sampling fraction n/N remains low (approximately 5% or less). Combining the two formulae gives:

$$n = (4z[\alpha/2]^2pqN)/(w^2N - w^2 + 4z[\alpha/2]^2pq)$$

Where,

n is the required number of sources in the sample

$z[\alpha/2]$ is the $(1-\alpha/2)$ quantile from the standard normal probability function; that is:

| Confidence Level | $\alpha$ | $1-\alpha/2$ | $z[\alpha/2]$ |
|---|---|---|---|
| 95% | 0.05 | 0.975 | 1.96 |
| 90% | 0.10 | 0.950 | 1.65 |
| 85% | 0.15 | 0.925 | 1.44 |

p is the proportion of interest (proportion of sources in compliance); when p is unknown it is set equal to 0.5 to give the most conservative possible value of n

q is (1-p)

N is the total population being sampled
w is $2\alpha$

---

[1] Inman, R.L. and Conover, W.J., *A Modern Approach to Statistics*, John Wiley and Sons, New York, NY.

[2] Cochran, W.G., *Sampling Techniques*, Second Edition, John Wiley and Sons, New York, NY.

**ABC Chemical Company**

Hazardous Waste Storage Facility
Weekly Inspection Log

| Date | Time | Inspector | No. of Drums | Status OK | Status Needs Att'n | Comments |
|---|---|---|---|---|---|---|
| 4/12/93 | 8 AM | C.Moore | 20 | ✓ | | |
| 4/19/93 | 8 AM | C.Moore | 20 | ✓ | | |
| 4/26/93 | 8 AM | C.Moore | 20 | ✓ | | |
| 5/3/93 | 9:00 | L.Jones | 21 | | X | Acid waste drum leaking pumped into new drum on 5/3 |
| 5/17/93 | 8:00 AM | J.Smith | 21 | ✓ | | |
| 5/24/93 | 8:00 AM | J.Smith | 29 | | ✓ | Missing bung on drum of spent paint gun cleaner Bung replaced on 5/28 |
| 6/4 | 8:00 | L.Jones | 29 | X/✓ | | |
| 6/11 | 8:00 | D.Jones | 29 | ✓ | | |
| 6/18 | 9:00 | D.J | 29 | ✓ | | |
| 6/25 | 8:00 | Jones | 29 | ✓ | | |
| 7/2 | 8:00 AM | J.S. | 11 | ✓ | | |
| 7/9 | 8:15 | Smith | 11 | ✓ | | |
| 7/16 | 8 AM | C.M. | 10 | ✓ | | |

**ABC Chemical Company**

Hazardous Waste Storage Facility
Weekly Inspection Log

| Date | Time | Inspector | No. of Drums | Status OK | Status Needs Att'n | Comments |
|------|------|-----------|--------------|-----------|--------------------|----------|
| 1/4/93 | 8:00 AM | J. Smith | 10 | ✓ | | |
| 1/11/93 | 8:00 AM | J. Smith | 12 | ✓ | | |
| 1/19/93 | 8:00 | G. Jones | 12 | X | | |
| 1/25/93 | 8:00 | G. Jones | 12 | X | | |
| 2/1/93 | 8 AM | C. Meyer | 11 | ✓ | | |
| 2/5/93 | 8:15 | G. Jones | 12 | X | | |
| 2/12/93 | 8:15 | G. Jones | 12 | X | | |
| 2/19/93 | 8:20 | G. Jones | 12 | X | | |
| 2/29/93 | 8:00 AM | J. Smith | 13 | ✓ | | |
| 3/5/93 | 8:00 AM | J. Smith | 21 | | ✓ | Acid waste drum leaking / pumped into new drum on 3/8 |
| 3/15/93 | 8:00 AM | J. Smith | 21 | ✓ | | |
| 3/22/93 | 8 AM | E. Meyer | 20 | ✓ | | |
| 3/30/93 | 8:00 | G. Jones | 21 | ✓ | | |
| 4/5/93 | 8:00 AM | J. Smith | 21 | | ✓ | wall between acid waste and cyanide waste area damaged by forktruck |

# ENVIRONMENTAL AUDITS
# PROBLEM SOLVING EXERCISE

## Spill Control

### BACKGROUND

As part of a multimedia comprehensive environmental audit scheduled to last five days, you will be examining the facility's Spill Prevention Control and Countermeasures (SPCC) Plan. After reviewing the entire facility layout and SPCC Plan, you note that, according to the Plan, there are 20 oil storage tanks of various capacities that have a system of diking installed. You have decided that you will physically inspect only five (5) tanks, which you consider to be a representative sample.

You look at the first tank and discover that, contrary to the Plan, there is no diking. You confirm that you are at the right tank and that you are reading the Plan accurately.

You continue on to the second oil storage tank. Here you find a somewhat different situation. While there is diking installed, your rough calculation indicates that it is about half the capacity indicated in the SPCC Plan which calls for secondary containment of 150% of tank volume. Additionally, you note that the dike walls are made of concrete and the base (inside the diked area) is natural ground. Your escort tells you that he has repeatedly put in work requests to management to get the base of the dike paved but nothing seems to happen.

### ASSIGNMENT

1.  Would you at this point modify your original audit plan? How? Have you seen enough? Would you decide to look at more than the 5 original tanks?

2.  What would be your reactions relative to the first tank?

3.  How would you react to the problems at the second tank? What specifically would you look for, do or ask at this tank?

4.  How and when would you discuss this with the Plant Manager?

## ENVIRONMENTAL AUDITS
## PROBLEM SOLVING EXERCISE

### Calling Regulatory Authorities

## BACKGROUND

You have just completed an environmental review of a major process unit. On this unit you have discovered a small incinerator burning hydrogen sulfide off-gas. You later determine that hydrogen sulfide is regulated as a hazardous waste. This incinerator is presently covered under the facility's air permit but does not have a hazardous waste permit.

To determine the need for a hazardous waste permit you make a "blind" phone call to the regulatory authority. They tell you in no uncertain terms that the incinerator must have a permit and it is your legal and ethical duty to report the non-compliance instance immediately to them for possible enforcement. You are now beginning to understand the true meaning of the phrase "What did he know and when did he know it?"

## ASSIGNMENT

1.     How would you end your phone call with the regulatory authority representative?

2.     What should your next steps be?

3.     How would you determine your personal liability? How will this affect your actions?

## ENVIRONMENTAL AUDITS
## PROBLEM SOLVING EXERCISE

### Working Papers

## BACKGROUND

You are the team leader an audit team of four(4) people. Part of your responsibility is to review the working papers of the team members. It's 10:00pm on the second day of a four day audit. You're in your hotel room and begin to review John's notes. John is responsible for assessing site compliance with hazardous waste regulations. The first page is as follows:

> Hazardous Waste Audit
> May 6, 1991
>
> Noted some empty drums stacked on their side. Oily substance on the ground; probably a major source of groundwater contamination. Locked tractor trailer; should come back later and open it; could have chemicals inside. Unlabeled transformers: not a hazardous waste issue. Probably some violations here; review regulations. Site should clean this place up immediately; don't want a Regulatory Authority to see this.

Your job is to give counsel to John the next morning at breakfast about improving his working papers. John has been with the Company for twenty (20) years and is currently an Environmental Coordinator at another Plant. He is a headstrong and quite opinionated individual.

## ASSIGNMENT

1.  How would you begin your conversation with John?

2.  When would you have the discussion? Before, during or right after breakfast?

3.  The audit has not been going that smoothly so far. Would you consider not having the discussion with John until after the audit is over to avoid a potentially messy scene and possible non-performing auditor?

4.  When you do get around to having your conversation, what are some of the specific statements in his notes that you might comment on?

*Interviewing Skills*

## ENVIRONMENTAL AUDITS
## PROBLEM SOLVING EXERCISE

### Interviewing

## BACKGROUND

Interviewing is a central part of any environmental audit. During the course of an audit, the auditors may be talking with plant managers, environmental managers, purchasing agents, laboratory staff with doctoral degrees, and unit operators with 35 years seniority with the company. Information gained from these discussions is vital to making valid and persuasive findings and recommendations. However, obtaining valuable information through the interview is the most difficult way to gather evidence. In some ways, audits will always seem like performance evaluations to those being audited; no matter how much they are assured that "we're from corporate and we're here to help you." Let's explore some key interviewing issues by answering the four following questions:

## ASSIGNMENT

1.  What percent of the interviewer's time is typically spent:

    | | |
    |---|---|
    | Listening? | _____ % |
    | Taking Notes? | _____ % |
    | Talking? | _____ % |
    | Observing? | _____ % |
    | Thinking? | _____ % |
    | | |
    | TOTAL | _____ % |

    Is the interviewer doing anything else? What should the interviewer do once the interview is over?

2.  Under what circumstances do you writeup a finding solely based on the results of an interview?

3.  What do you do if an interviewee is non-responsive?

## ENVIRONMENTAL AUDITS
## PROBLEM SOLVING EXERCISE

4.    What is right or wrong about the following interview questions?

(1)    Do you inspect weekly as required by hazardous waste regulations?
(2)    Tell me about your hazardous waste Preparedness and Prevention Program?
(3)    Do you have any concerns about maintaining compliance at this site?
(4)    What concerns do you have about maintaining compliance at this site?
(5)    Have you ever been cited by a regulatory agency?

(6)    Do you have any PCB's at this site?
(7)    Why is housekeeping so poor in this area?
(8)    Could you find that exemption letter that you mentioned you received from the regulatory authority eight years ago?
(9)    Does the plant look any different during other times of the year? Is there more activity at other times? Does the plant ever have planned maintenance shutdowns?
(10)   Who is responsible for the labels missing from these drums?

(11)   What haven't you told me that might help me be complete in my review?
(12)   What do you think is the best way to begin our review of this part of the facility?
(13)   Can we go over this again? I need help in understanding how the regulations apply to this operation.
(14)   Would you mind if I recorded our conversation? It would be a great help?
(15)   I've heard you have a nice operation here. Can we talk a bit about what happens on a day-to-day basis?

## ENVIRONMENTAL AUDITS
## INTERVIEWING EXERCISE (INT'L)

AUDITOR: *Good morning Mr. _____, my name is _____ and I am here to conduct an environmental audit of the hazardous waste storage facility here at your plant. The purpose of this audit is to uncover poor practices and negligent operations and report them to officials responsible so that corrective actions can be taken. Those actions might be to fund needed maintenance and repair or arrange for personnel action to remove or replace incompetent staff. Do you understand this, or have any questions?*

ENGINEER: Well, no ... EXCEPT that the hazardous waste regulations are overwhelming and I am doing the best that I can. Mr. _____, he's my supervisor, said you were coming over, but I figured this was more of a regulatory authority type of inspection. We had one of those last month and we did very well, except for three items which I'm working on right now. They said we should have more specific training and I've arranged for a company to come in and help with ...

AUDITOR: *(Interrupting) That is all very interesting Mr. _____, but I have my own inspection protocol which I have to follow and I'm sure some of it will cover all of the same areas that the regulatory authority touched on.*

ENGINEER: Okay. It was kind of interesting, what we did but what ever you want to do is okay with me.

AUDITOR: *Well, can you tell me how your storage facility complies with the hazardous waste regulations. Specifically, how do you comply with the security requirements?*

ENGINEER: Well, the place is secure, you know. It's inside the plant and it has a fence and we keep people out; so it's secure.

AUDITOR: *How about signs?*

ENGINEER: We have signs and we keep people out of the storage area.

AUDITOR: *What do the signs say?*

## ENVIRONMENTAL AUDITS
## INTERVIEWING EXERCISE (INT'L)

ENGINEER: They say to keep out ... I don't remember the exact wording. It's whatever the hazardous waste regulations require. I remember having to get the exact wording from the regulations to give to the sign shop. They made up the signs with the wording from the regulations.

AUDITOR: *How big are the signs and the lettering?*

ENGINEER: Oh, they're about like this (uses hands). I don't know!

AUDITOR: *How far away can the signs be read?*

ENGINEER: I guess that depends on how good your eyes are ... (acting annoyed)

AUDITOR: *Mr. _____, I'm trying to conduct a serious effort here and your present attitude is not helping at all.*

ENGINEER: Well, Mr. _____, your questions are not exactly to the point. You seem to be playing some sort of game here. If you could be more specific in your questions, I could probably give you better answers.

AUDITOR: *All right, Mr. _____, how about this: Are the signs you erected legible from a distance of 10 meters?*

ENGINEER: Yes, of course. What if we go outside and look at the signs so you can see for yourself?

AUDITOR: *I have to finish this check list first and then I want to review all your permits, plans and shipping manifests. Another auditor will look at the physical layout.*

ENGINEER: Well, I would judge that the average person could read the signs at 10 meters, but I really can't say for sure.

AUDITOR: *Let's move on. With regard to inspections: Do you inspect the facility at least weekly?*

ENGINEER: Yes.

AUDITOR: *Do you have a written inspection plan?*

ENGINEER: Yes.

## ENVIRONMENTAL AUDITS
## INTERVIEWING EXERCISE (INT'L)

AUDITOR: *Can I see the plan?*

ENGINEER: (Reaches for PROP) Well, let me see. Here it is.

AUDITOR: *This shows that you inspect the facility daily; why do you do that? You just said that you inspected weekly.*

ENGINEER: Well you said weekly and I figured that was what I was supposed to do.

AUDITOR: *Okay, Mr. _____, let's go over this again. You actually inspect the storage area daily?*

ENGINEER: That's right. Is something wrong with that?

AUDITOR: *Well, you only have to do it weekly. Where do you record the results of the inspections?*

ENGINEER: We write-up the problems on this form and send it over to plant maintenance for them to repair. We keep copies here in this folder.

AUDITOR: *So, you only note exceptions on the inspection report?*

ENGINEER: Yes, I guess you could say that.

AUDITOR: *Mr. _____, you should be aware that the hazardous waste regulations require you to keep a log of inspections which includes, at a <u>minimum</u>, a separate entry for each inspection including the date and time of each inspection, the name of the inspector - not simply the initials as you have done here - a notation of observations made - both positive and negative - and the date and nature of any repairs. Mr. _____, the inspection recording procedures you have here are totally inadequate and are in violation of the regulations.*

ENGINEER: Hey! Wait a minute! I'm doing my best. We know what the problems are and we fix anything we find. If you would go out and see the facility, you would see that we put any leaking drums in overpacks, we have our spill clean-up equipment in place, the facility is dry, and everything is acceptable. So, I don't have the proper forms. What are you concerned with anyway, the paperwork or how we manage the waste?

## ENVIRONMENTAL AUDITS
## INTERVIEWING EXERCISE (INT'L)

AUDITOR: *Mr. _____, the hazardous waste program is a paper intensive effort and if you can't deal with these requirements, I don't see how this facility can ever be brought into compliance.*

ENGINEER: What else do you want to see?

AUDITOR: *Well, we're way behind schedule here but I would like to discuss personnel training. Has everyone who works here completed a program of training to teach them to perform their duties?*

ENGINEER: Yes.

AUDITOR: *The person who directs the training must also be trained in hazardous waste management. Who directs the training?*

ENGINEER: I do.

AUDITOR: *Are you yourself trained Mr. _____?*

ENGINEER: I am.

AUDITOR: *Where did you get trained?*

ENGINEER: I took a course with _____.

AUDITOR: *Do you feel that such a course qualified you to train people?*

ENGINEER: I do. Why, do you have some kind of problem with that?

AUDITOR: *No, No, Mr. _____. One last question and then I really must leave. Do any newly hired people - people who have not completed training - do any of these people work alone or unsupervised?*

ENGINEER: No. We closely supervise all of our new employees.

AUDITOR: *Good. What about that young fellow over there. Is he a new employee?*

ENGINEER: Yes.

AUDITOR: *Is anyone supervising him?*

## ENVIRONMENTAL AUDITS
## INTERVIEWING EXERCISE (INT'L)

**ENGINEER:** Well, he's only eating his lunch. I think he can do that by himself.

**AUDITOR:** *Well, Mr. _____, thank you for your time. My associate Mr. _____ will be coming along later to discuss container storage standards. This has been most interesting.*

**ENGINEER:** Okay. Watch your step on the way out. I wouldn't want anything unpleasant to happen to you.

## ENVIRONMENTAL AUDITS
## INTERVIEWING EXERCISE (INT'L)

*AUDITOR:*    *Well, hello Mr. _____, _____ said he had finished his portion of the interview and that you were ready to talk about container standards.*

ENGINEER:   Okay, I guess I can spare some time. How long will this take?

*AUDITOR:*    *I know you're a busy man and I really won't take any more of your time than necessary. Probably no more than one hour.*

ENGINEER:   Okay.

*AUDITOR:*    *Let's see ... You know there is so much involved in these regulations, I can hardly keep up and I deal with this everyday. How do you keep track of all these regulations?*

ENGINEER:   Well, the Corporate Environmental Staff sends us notices on changes and \_\_\_\_ \_\_\_\_\_, the environmental coordinator, and I talk about changes every few weeks. I have a copy of the Hazardous Waste Regulations and I refer to that.

*AUDITOR:*    *Do you keep copies of that around here?*

ENGINEER:   Yes. I have to in order to conduct my inspections and everything. It's a big job to keep up to date.

*AUDITOR:*    *Yes, it really is. Well, I want to see how you store containers, so I guess the best thing to do is to go out and see the drum storage area.*

ENGINEER:   Okay. You're the boss, Mr. _____.

*AUDITOR:*    *Well, is there anything else you think we should do first? This is your facility and you know it best, so I guess I'm asking you if that would be a good place to start.*

ENGINEER:   You're right. We should probably see the facility first.

*AUDITOR:*    *What do you store here?*

ENGINEER:   Mostly waste solvents, paint stripper, and paint sludge.

*AUDITOR:*    *Are those flammable?*

## ENVIRONMENTAL AUDITS
### INTERVIEWING EXERCISE (INT'L)

ENGINEER: Yes. We had some the wastes tested and they are flammable.

*AUDITOR:* *Is there anything special about any of the waste?*

ENGINEER: No.

*AUDITOR:* *What do you think is in this drum labeled "solvent and sludge?"*

ENGINEER: Oh, those guys over at the metal shop have a small plating line and they often times drop the sludge into a solvent drum and mix the whole mess together. Now looks like they've done it again.

*AUDITOR:* *Will you have to do anything special with this waste?*

ENGINEER: Yes. I will probably have to pay the waste hauler more money, since he can't reclaim the solvent with all the metal hydroxides in it now. Those guys have no idea what problems they cause.

*AUDITOR:* *It sure makes your job difficult. Those drums over there that are all dented; are they full of hazardous waste?*

ENGINEER: Yes, that's one of our designated accumulation points.

*AUDITOR:* *Can we look at them?*

ENGINEER: Yes, I inspected it just the other day.

*AUDITOR:* *I notice you have a funnel on the top of the drums. That's good to prevent spillage. Do you keep the funnels in the drums all the time?*

ENGINEER: Yes, you never know when these guys are going to put something in them and half the time they are always rushing and would probably spill the waste if it wasn't for the funnel.

*AUDITOR:* *Yes, I know what you mean. That's a good idea. I think it would be wise to use the funnel that has the spring loaded top instead of the open top one you are using here. That would keep the drum sealed and prevent emissions of any volatiles.*

## ENVIRONMENTAL AUDITS
## INTERVIEWING EXERCISE (INT'L)

ENGINEER: Yes, you're right. I never realized that. We've been doing it like this for years. But I'll change the funnel.

AUDITOR: *I see you've put the accumulation start date on all the drums.*

ENGINEER: Yes, we're careful to always do that.

AUDITOR: *Why do you think two drums over there have been stored for greater than 90 days?*

ENGINEER: Where? Oh, those drums. Those are just old labels from last time we used them. The new labels are on this side of the drum. See, they show April 18, that is less than 90 days. (Note: Actor should substitute for April 18, a date that is less than 90 days).

AUDITOR: *Okay. I think you understand the requirement, but that could be confusing. I think you should make sure only the correct start dates are on the drum.*

ENGINEER: Yes, I'll do that.

AUDITOR: *Also, these other drums are damaged. A few are dented and some corrosion is starting to work its way through. It is probably a good idea to change these as soon as you can.*

ENGINEER: Yes, looks like I'm failing this test.

AUDITOR: *Don't look at this audit as a test. We're trying to focus your attention on some things that could cause problems if not managed properly. These little problems I'm finding are typical at other plants, too. They may seem trivial but they are important. These are potential risks to the company. This region is highly dependent on groundwater for drinking water and with these sandy soils, contamination from leaking drums could put us all in a bad position. A lot of companies are paying out a lot of money for major problems that all started out as small items like these.*

ENGINEER: Yes, I think we all need a little refresher training on some of these procedures.

AUDITOR: *Yes, I think so, too. Okay, that completes this part of the audit. Thanks for your time. Is there anything you want to tell me before I go?*

## ENVIRONMENTAL AUDITS
### INTERVIEWING EXERCISE (INT'L)

ENGINEER: No, I don't think so.

AUDITOR: *Ok, I'll be summarizing my findings at the closing meeting on Friday for the plant manager. If you want to talk to me about anything else before then, let me know. Also, I think you might find it useful to come to the meeting.*

ENGINEER: There is one thing, though.

AUDITOR: *What's that?*

ENGINEER: The man who audited me earlier this morning, does he work for you?

AUDITOR: *Yes, he does.*

ENGINEER: Well, no offense intended but, I think he could use a little refresher training, too.

AUDITOR: *Really, in what area?*

ENGINEER: In the proper way to conduct these audits.

# *Management Systems Assessments*

### ENVIRONMENTAL MANAGEMENT SYSTEMS AUDITS
### PROBLEM SOLVING EXERCISE

#### Conducting Process Flow Analyses

## BACKGROUND

One of the most valuable techniques used in identifying root causes and systemic failures on environmental, health and safety audits, is to develop a flow diagram for the process being reviewed. An example process flow diagram for the generation of a solvent waste by the maintenance department is provided with this exercise. This charting effort can help in identifying and defining inefficient systems, critical path constraints and process gaps. Evaluating processes in this way is what re-engineering is all about. EHS auditors must develop this charting skill if they are to be responsive to the new requirements to evaluate management systems.

## ASSIGNMENT

1.  Examine the attached flow diagram and identify any parts of the process that have not been included, which would be critical in determining if the *management system* is deficient in any way. That is, what else would you need to know to complete a management systems audit?

2.  Review the photograph provided with this exercise. Pose a series of questions to your instructor about the management of the drums in the picture so that you can accurately draw a process flow diagram for the delivery, storage and disposal of the drums. Be prepared to share the diagram with the class and discuss your view of the system deficiencies.

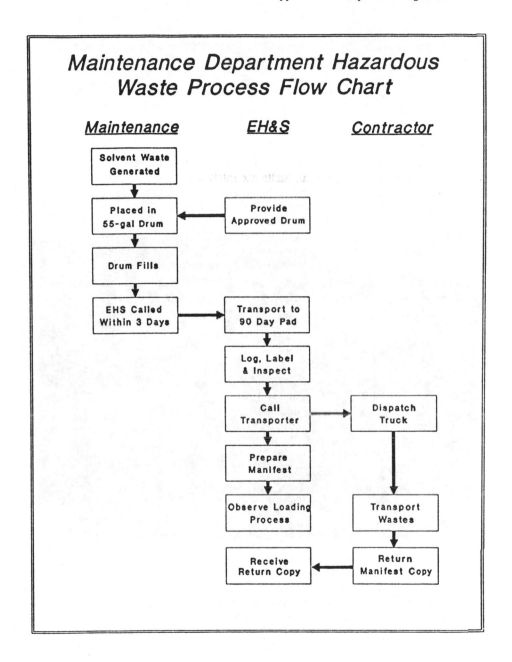

Inspection of Facilities

## ENVIRONMENTAL MANAGEMENT SYSTEMS AUDITS
## PROBLEM SOLVING EXERCISE

### Root Cause Analysis

### BACKGROUND

Root cause analysis is an important part of what auditors do on environmental management systems audits. Unfortunately for auditors, root cause analysis is both difficult to do and relatively time consuming. Yet, significant value is added to the process when auditors can identify the underlying cause of a problem (e.g., lack of training) as opposed to the symptom (e.g., improperly labeled drums). Corrective actions can then be designed to resolve not just the observed deficiency but what caused it to begin with (e.g., strengthening the training program as well as fixing the labels on the drums).

Undertaking root cause analysis, while both difficult and time consuming, is nonetheless conceptually quite simple. It involves little more than continuously asking the question "why?" as opposed to "what?". The attached hypothetical investigation models the process that an auditor would go through in doing a root cause analysis. The symptom or deficiency is traced back through its derivative causes until a root cause is identified. The hypothetical investigation ends in a startling conclusion that is deliberately far-fetched (or is it?).

### ASSIGNMENT

Your assignment is to do a similar analysis for the symptoms listed on the attached page. The idea is to analyze the symptoms and speculate on what the causes might be and work your way back through the chain. Use a "fishbone" diagram if that is helpful. Attempt to identify enough underlying causes to get you through at least three or four levels of analysis. Once you have identified the "root" cause, propose a corrective action that you believe will resolve the problem.

As opposed to the hypothetical case, don't worry about being too "cute" in identifying the ultimate causes; treat the symptoms as though they were real audit findings. Be prepared to formally discuss your analysis with the class.

## ENVIRONMENTAL MANAGEMENT SYSTEMS AUDITS
## PROBLEM SOLVING EXERCISE

### Root Cause Analysis
### Sample Symptoms

**Symptom No. 1:**  A newly built process unit, that has been up and running for two months, does not have the required air emissions permits for construction and operation.

**Symptom No. 2:**  The wastewater treatment plant is routinely exceeding its permit limits for pH in four months out of the year.  There are no exception reports in the files.

**Symptom No. 3:**  The site has a confined space entry permit program. However, the audit team observed an employee entering a tank for cleanout with no other employee around and no evidence a permit was obtained.

**Symptom No. 4:**  The site's 5000 gallon above ground diesel oil tank does not have the secondary containment that is described in the Spill Prevention Control and Countermeasure (SPCC) Plan.  It has no apparent containment at all.

**Symptom No. 5:**  The site's Environmental Coordinator was not aware that he will be required to have permits for several sources under the Title V Permit Program of the 1990 Clean Air Act Amendments.  He has done nothing to prepare for this eventuality and it is likely that the site will not meet the regulatory deadline for permit applications.

**Symptom No. 6:**  It was recommended in the previous audit report, which is now two years old, that the site should have a written respiratory protection program.  This is required by company policy and OSHA regulations.  The quarterly corrective-action status reports indicated that this was done, and the audit was formally closed-out.  You observed on the current audit that it was, indeed, never done.

## ENVIRONMENTAL MANAGEMENT SYSTEMS AUDITS
## PROBLEM SOLVING EXERCISE

### Root Cause Analysis
### Hypothetical Investigation

| LEVEL | DEFICIENCY/SYMPTOM |
|---|---|
| I (Baseline) | WHAT??? A label is missing from a hazardous waste drum at the site's accumulation point. WHY??? |
| II | The label weathered, fell to the ground and blew away. WHY??? |
| III | The label's adhesive was of poor quality. WHY??? |
| IV | The site's Purchasing Department recently bought less expensive labels without notifying the site environmental coordinator. WHY??? |
| V | The Department was told by the Plant Manager to quickly reduce the annual costs of goods purchased by 20%. WHY??? |
| VI | The Plant Manager is under pressure to improve profitability and has responded by cutting costs where he can. WHY??? |
| VII | The CEO has asked all Plant Managers in this Division to improve profitability by 15%. WHY??? |
| VIII | The Company's overall profits are declining and large institutional investors (i.e., pension funds) are putting pressure on the Board of Directors and CEO to improve performance. WHY??? |
| IX | The individual investors in the pension funds are concerned about its performance and are putting pressure on fund managers to get a better return. WHY??? |
| X | The investors are concerned that Social Security (SS) will not be there when they retire, so they must rely solely on their pensions. WHY??? |
| XI | They know the SS System is in trouble and the U.S. Government "can't manage its way out of a paper bag!" |
| **Root Cause:** | **It's the Government's fault!!!** |
| **Corrective Action:** | **Vote out all the Incumbents!!!** |

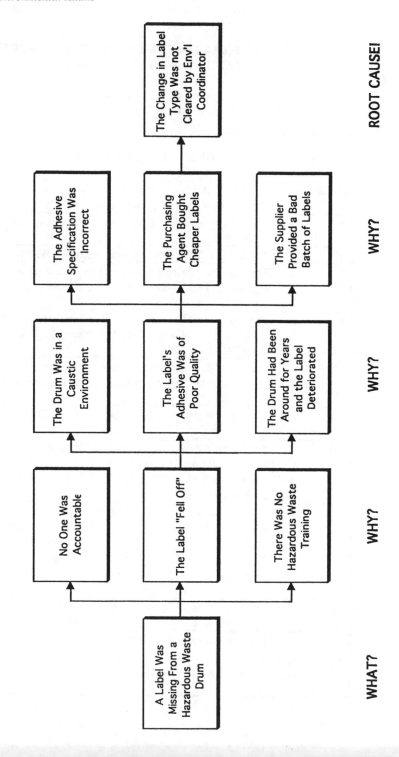

FISHBONE DIAGRAM ANALYSIS
HAZARDOUS WASTE LABEL

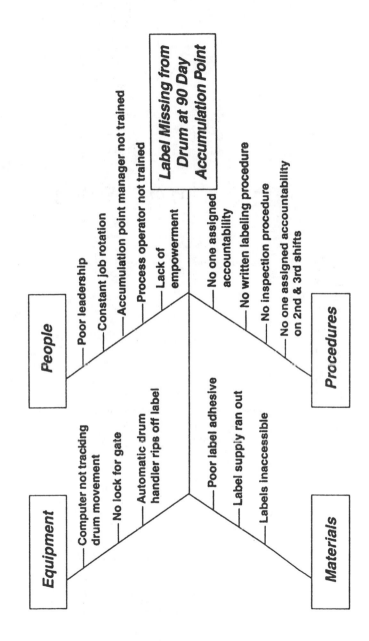

Example Fishbone Diagram for a "Missing Hazardous Waste Label"

**Equipment**
- Computer not tracking drum movement
- No lock for gate
- Automatic drum handler rips off label

**People**
- Poor leadership
- Constant job rotation
- Accumulation point manager not trained
- Process operator not trained
- Lack of empowerment

**Label Missing from Drum at 90 Day Accumulation Point**

**Procedures**
- No one assigned accountability
- No written labeling procedure
- No inspection procedure
- No one assigned accountability on 2nd & 3rd shifts

**Materials**
- Poor label adhesive
- Label supply ran out
- Labels inaccessible

14803 5.10 95 DL

*Report Writing*

**EH&S AUDITS**
**PROBLEM SOLVING EXERCISE**
**Audit Reports**
**"Tips for Success"**

1. Separate the Process from the Product

2. Use pyramid/logical thinking/topic sentences

3. Present activities/findings/conclusions:

   • What you saw

   • What you found

   • What it means

4. Look for underlying causes

5. Keep findings in perspective

6. Organize daily

7. Write findings as you go

8. Aim for direct, factual, informative style

9. Hunt for bad words

10. Avoid conjecture ("conclusions deduced by guesswork")

11. Use evidence (inquiry, observation, test)

12. Use proper grammar

13. Bottom line your interviews

14. Copy important documents

15. Complete at least an annotated outline before you leave

16. Start early

17. Challenge each other

18. Agree on report format

19. Don't indict the staff

20. KISS

## THE EVOLUTION OF AN EH&S AUDIT FINDING

1.  Hazardous waste management at the site needs improvement.

2.  Hazardous waste management at the site was deficient.

3.  Hazardous waste drum management at the site's accumulation point was not adequate.

4.  The drum storage area contained a drum that had a funnel in it.

5.  The site's accumulation point contained a drum that had a funnel in it.

6.  One of the four hazardous waste drums at the site's 90-day accumulation point was not closed and no operators were nearby.  RCRA requires drums to be closed except when adding or removing waste.  (40 CFR 265.14)

*Combining the management system finding with the compliance finding is probably the best approach, but.....*

## EH&S AUDITS
## PROBLEM SOLVING EXERCISE

### Audit Reports
### "What's Wrong With This Picture"

1.  The respiratory protection program was deficient and could be improved. This is a serious concern.

2.  The audit team inspected 20 transformers and 10 capacitors, of which 3 showed past signs of leaks.

3.  Several of the drums at the hazardous waste storage area had no labels, as required by RCRA.

4.  Three fire extinguishers at the site did not have the required inspection tags.

5.  The hazardous waste 90 day accumulation area had no secondary containment.

6.  It is possible that some previous spillage occurred at the bulk loading transfer station.

7.  A unit operator reported that the wastewater treatment plant is bypassed during severe storms events. This violates the site's wastewater discharge permit.

8.  Because of the possibility of solvent releases into the sewers, the wastewater treatment ponds might have to be permitted as a hazardous waste surface impoundment.

9.  The blood-borne pathogen program should be improved to reflect the most recent OSHA regulations.

10. The permitted hazardous waste storage facility at Building 51 was recently inspected by the New Jersey DEP and found to be in compliance with environmental standards. Used oil at the site is accumulated at Building 52 and disposed of monthly by Used Oil Reclamation Company. Used oil is exempt from hazardous waste regulations.

11. The fire training area has reportedly been contaminated with waste solvents, used as recently as 5 years ago as supplementary fuel for the fires. The audit team recommends that the facility file for a hazardous waste permit as a treatment/disposal facility and immediately excavate the contaminated soils.

12. As part of the hazardous waste audit we went to the point where the hazardous wastes are accumulated at the site and we reviewed whether the drums had labels and whether there was enough aisle space for a fork truck and whether the hazardous waste accumulation point was inspected weekly and whether all the other hazardous waste regulations were complied with. Their were a plethora of violations.

### ENVIRONMENTAL AUDIT PROGRAM
### Audit Findings Form

**Description of Finding:**

| Statement of Observation: |
| --- |
|  |

| Statement of Requirement: |
| --- |
|  |

**Citation/Source:**      **Compliance Area (e.g., Wastewater, HazCom):**

|  |  |
| --- | --- |

**Type:**    [ ] Regulatory      [ ] Policy        [ ] Management Practice
**Level:**    [ ] Level I       [ ] Level II         [ ] Level III

**Recommendation (Immediate/Long-term):**

|  |
| --- |

**Code:**    **Auditor:**           **Date:**

|  |  |  |
| --- | --- | --- |

# *Audit Conferences*

## ENVIRONMENTAL AUDITS
### Opening Conference Questions

1.  I've been called out of town for the rest of the week. I won't be available for the closing conference. Is that okay?

2.  I've been having problems with my accumulation point manager. Could you give me a third-party evaluation of his performance?

3.  I'm sure we've removed all of our PCB equipment and Underground Storage Tanks. You could eliminate your review of these areas and save time. Okay? What do you think?

4.  We just received notice this morning that we're to have an EPA inspection this week, starting today. Can you cancel or postpone the audit?

5.  I've got some bad news. Due to the torrential rains we've had this week, we've had to shut down the plant because the lagoons are overflowing. Does that mean you can't do the audit?

## ENVIRONMENTAL AUDITS
### Closing Conference Questions

### Plant Manager

1.  You've been on many audits this year. Tell me, how does my plant stack up?

2.  On a scale of 1 to 10 how would you rate our performance on the audit?

3.  Which of the findings are high priority requiring my immediate attention?

4.  Can you make sure that you indicate all the positive things we are doing at the plant in the report?

5.  It seems some of the findings are repeated from the last audit. We just haven't had the resources. How will you handle those in the report?

## ENVIRONMENTAL AUDITS
### Closing Conference Questions

**Environmental Manager**

1. I don't think I can get the corrective action plan completed in the required time frame. Can I have another three to four weeks to get it done?

2. Will I be given a chance to review the audit report and make comments? Will there be anything new in the report that has not been addressed at the closing conference?

3. If I fix some of the problems today, can you go back out and look at them? If they're okay, do the findings have to go in the report?

4. Do you have any feelings about our people? Performance is crucial as we restructure. Any staff who are doing an outstanding job?

5. We're having an agency inspection next week. Will this audit guarantee a good result if we fix everything?

# Appendix E

# SAMPLE REGULATORY INSPECTION PROTOCOL

EXAMPLE OF A REGULATORY INSPECTION PROCEDURE


The common elements of a compliance inspection can be grouped into the following five procedural categories:

1. Pre-inspection Preparation
2. Entry of Inspector
3. Opening Conference
4. Physical Inspection
5. Closing Conference


These elements are common to all inspections, but the emphasis given to the separate elements will vary with the needs of the individual inspection. Each of the procedural categories is discussed in detail below.

Each facility will designate a Facility Inspection Coordinator (FIC). The FIC is responsible for coordinating all procedures for dealing with inspections, organizing and maintaining the necessary documentation, and advising appropriate personnel of essential actions required before and during an inspection.


## 1. Pre-inspection Preparation

Notification of inspection may or may not be provided to the plant depending on the circumstances and specific regulatory agency policy. Pre-notification is generally granted except where significant violations are suspected. In such instances, prior planning becomes even more important. In either case, the primary consideration in anticipation of a compliance inspection is preparation of the necessary documentation and notification of key personnel at the facility. At a minimum, the pre-inspection items the FIC is responsible for are listed below.


o Determine the objective and scope of the inspection and the authority under which the inspection is being made. The motivation for the inspection visit should also be identified.

o Determine the composition of the facility inspection team.

o Establish the protocol and planned schedule of the inspection.

o Determine the legal significance and potential need for legal counsel to participate in the inspection.

o Compile all appropriate records/reports/files necessary for the inspection.

If time is available, a meeting of key plant inspection participants should be held to review all pertinent items prior to the inspection.

Any personnel at the facility receiving notification from a regulatory agency that a compliance inspection will be held should immediately inform the Facility Inspection Coordinator.

2. Entry of Inspector

Authorized agency inspectors are required to present proper credentials and a written notice of inspection at the time of the entry. The facility has the option of denying entry at any time before or during inspection, however, this practice is to be avoided unless special circumstances warrant denying access to inspectors. The FIC should meet the inspector upon arrival at the plant and escort him to his office for an opening conference.

3. Opening Conference

The opening conference provides an opportunity for cooperation in meeting inspection objectives. At this conference, the inspector should be appraised of the facility's environmental audit program and in particular, the results of any recent audits related to the scope of the inspection. The opening conference should be attended by the Plant Manager, if possible, in addition to the Inspection Coordinator.

The following items should be discussed during the opening conference.

o Facilities to be inspected
o Records/documentation to be reviewed
o Environmental samples to be taken
o Results of plant internal audit inspections
o Key personnel to be interviewed
o Length/duration/schedule for inspection
o Schedule for closing conference

4. Physical Inspections

Each inspector from different regulatory agencies will likely have his own procedure or protocol for conducting the inspection. The inspection will normally include three main elements:

o Records review - An inspector is entitled to review and have copies made of all appropriate records. The Inspection Coordinator should ensure that all requested records are made available.

o Sampling - If an inspector desires to take environmental samples he may do so. The Inspection Coordinator has the right to request and receive duplicates of any samples collected. Copies of analytical results may also be provided if requested. The Inspection Coordinator should ensure that proper handling of documentation and chain of custody records for any samples taken are properly developed.

o Facility Inspection - Compliance inspections will usually involve inspection of facilities equipment or other operational activities. The inspector should be accompanied by the Inspection Coordinator at all times during the inspection. The Inspection Coordinator is responsible to obtain any necessary clearances or permission for the inspector to enter areas that may be sensitive with respect to operational activities.

The Inspection Coordinator should coordinate with the supervisor of the area being inspected to ensure that appropriate staff are available to meet with the inspector.

Also, any special requirements that the inspector
wishes to be included in the inspection, (i.e.,
observing spill clean up procedure) should be
discussed with appropriate staff prior to the
arrival of the inspector.

The Inspection Coordinator should ensure that facility
staff are courteous and professional in responding to the
inspector's inquires.

Any items that cannot be answered immediately should be
documented for subsequent followup.

5. Closing Conference

A final meeting with the inspector prior to his departure
is critical to ensure mutual understanding of findings,
particularly with regard to possible violations.  A written
receipt for all samples and documents taken should be
received, describing each item and its point of origin.
In addition, a number of actions are necessary to properly
close out an inpection after the inspection team has
departed:

   o Advise appropriate Plant Supervisors of the
     preliminary results of the inspection and their
     significance

   o Develop an action plan to correct the deficiencies
     cited (include short and long-term actions)

   o Prepare a memorandum for the record, documenting
     the inspection and planned subsequent actions.
     The memorandum should be reviewed by legal counsel
     prior to distribution.  Send a copy to corporate
     management

   o Hold a "lesson learned" session among key plant
     personnel to disseminate and discuss lessons
     learned from the inspection (if appropriate, include
     lessons learned in the above memorandum for record).

   o Make revisions to audit protocols based on infor-
     mation developed or compliance problems uncovered
     in the inspection.

# Appendix F

# MAJOR REGULATIONS TO BE
# REVIEWED PRIOR TO AN AUDIT

## AIR POLLUTION REGULATIONS

### Federal Regulations

1. National Emission Standards for Hazardous Air Pollutants (40 CFR 61)

### State Regulations

1. Permits and Registration of Air Pollution Sources

2. General Emission Standards, Prohibitions and Restrictions

3. Control of Incinerators

4. Process Industry Emission Standards

5. Control of Fuel Burning Equipment

6. Control of Volatile Organic Compounds (VOC)

7. Sampling, Testing, and Reporting

8. Visible Emissions Standards

9. Control of Fugitive Dust

10. Toxic Air Pollutants Control

11. Removal of Asbestos Materials

12. Vehicle Emissions Inspections and Licensing

## WASTEWATER DISCHARGE REGULATIONS

### Federal Regulations

1. Oil Spill Prevention Control and Countermeasures (SPCC) Requirements (40 CFR 112)

2. Designation of Hazardous Substances (40 CFR 116)

3. Determination of Reportable Quantities for Hazardous Substances (40 CFR 117)

4. NPDES Permit Requirements (40 CFR 122)

5. Toxic Pollutant Effluent Standards (40 CFR 129)

6. General Pretreatment Regulations for Existing and New Sources of Pollution (40 CFR 403)

7. Organic Chemicals Manufacturing Point Source Effluent Guidelines and Standards (40 CFR 414)

8. Inorganic Chemicals Manufacturing Point Source Effluent Guidelines and Standards (40 CFR 415)

9. Plastics and Synthetics Point Source Effluent Guidelines and Standards (40 CFR 416)

### State Regulations

1. Water Quality Standards

2. Effluent Limitations for Direct Discharges

3. Permit Monitoring/Reporting Requirements

4. Classifications and Certifications of Operators and Superintendents of Industrial Wastewater Plants

5. Collection, Handling, Processing of Sewage Sludge of Process Sludge

6. Oil Discharge Containment, Control and Cleanup

7. Standards Applicable to Indirect Discharges (Pretreatment)

## DRINKING WATER

### Federal Regulations

1. Underground Injection Control Regulations, Criteria and Standards (40 CFR 144 and 146)

2. National Primary Drinking Water Standards (40 CFR 141)

3. Community Water Systems, Monitoring and Reporting Requirements (40 CFR 141)

### State Regulations

1. Permit Requirements for Appropriation/Use of Water from Surface or Subsurface Sources

2. Underground Injection Control Requirements

3. Monitoring, Reporting and Recordkeeping Requirements for Community Water Systems

## TOXIC SUBSTANCES

### Federal Regulations

1. Manufacture and Import of Chemicals, Recordkeeping and Reporting Requirements (40 CFR 704)

2. Import and Export of Chemicals (40 CFR 707)

3. Chemical Substances Inventory Reporting Regulations (40 CFR 710)

4. Chemical Information Rules (40 CFR 712)

5. Health and Safety Data Reporting (40 CFR 716)

6. Pre-manufacture Notifications (40 CFR 720)

7. PCB Distribution Use, Storage and Disposal (40 CFR 761)

8. Regulations on Use of Fully Halogenated Chlorofluoroalkanes (40 CFR 762)

9. Storage and Disposal of Waste Material Containing TCDD (40 CFR 775)

### State Regulations

Note:  Only five states have PCB regulations that supplement the federal regulations relative to disposal of PCB's.

## PESTICIDE REGULATIONS

### Federal Regulations

1. Enforcement of FIFRA, Pesticide Use Classification (40 CFR 162)

2. Procedures for Disposal and Storage of Pesticides and Containers (40 CFR 165)

3. Certification of Pesticide Applications (40 CFR 171)

### State Regulations

1. Pesticide Licensing Requirements

2. Labeling of Pesticides

3. Pesticide Sales, Permits, Records, Application and Disposal Requirements

4. Disposal of Pesticide Containers

5. Restricted Use and Prohibited Pesticides

## HAZARDOUS WASTE REGULATIONS

### Federal Regulations

1. Identification and Listing of Hazardous Waste (40 CFR 261)
   - Characteristic waste
   - Listed wastes

2. Standards Applicable to Generators of Hazardous Waste (40 CFR 262)
   - Manifesting
   - Pre-transport requirements
   - Recordkeeping/reporting

3. Standards Applicable to Transporters of Hazardous Wastes (40 CFR 263)
   - Transfer facility requirements
   - Manifest system and record- keeping
   - Hazardous wastes discharges

4. Standards for Owners and Operators of Hazardous Waste TSD Facilities (40 CFR 264)
   - General facility standards
   - Preparedness and prevention
   - Contingency plan and emergency procedures
   - Manifest system, recordkeeping and reporting
   - Groundwater protection
   - Financial requirements
   - Use and management of containers
   - Tanks
   - Waste piles
   - Land treatment
   - Landfills
   - Incinerators

5. Interim Status Standards for Owners and Operators of Hazardous Waste TSD Facilities (40 CFR 265)

6. Interim Standards for Owners and Operators of New Hazardous Waste Land Disposal Facilities (40 CFR 267)

7. Administered Permit Program (Part B) (40 CFR 270)

## HAZARDOUS MATERIALS MANAGEMENT

### Federal Regulations

1.  Control of Pollution by Oil and Hazardous Substances (33 CFR 153).

2.  Designation of Reportable Quantities and Notification of Hazardous Materials Spills (40 CFR 302).

3.  Hazardous Materials Transportation Regulations (49 CFR 172 to 173).

4.  Worker Right to Know Regulations (29 CFR 1910.1200).

5.  Community Right to Know Regulations (Emergency Planning and Community Right to Know Act of 1986).

### State Regulations

Note:    Hazardous Materials Storage and Handling is normally regulated on a local level.

## SOLID WASTE REGULATIONS

### Federal Regulations

1.  Criteria for Classification of Solid Waste Disposal Facilities and Practices (40 CFR 257).

### State Regulations

1.  Permit Requirements for Solid Waste Disposal Facilities.

2.  Installation of Systems of Refuse Disposal.

3.  Solid Waste Storage and Removal Requirements.

4.  Disposal Requirements for Special Wastes (i.e. Asbestos).

# Appendix G

# SUMMARY OF AUDIT PROGRAMS

# SUMMARY OF AN
# ENVIRONMENTAL AUDIT PROGRAM

| | |
|---|---|
| **COMPANY:** | **FORTUNE 500 CHEMICAL COMPANY** |

| | |
|---|---|
| Program Name: | Environmental, Health and Safety Audit Program |
| Start Date: | 1993 in its current form |
| Purpose: | Achieve compliance with regulations and company policies and assess the efficacy of environmental management systems |
| Organization: | One corporate manager who leads management systems audits with consultants on the teams; Compliance audits done by 2 business units, also with consultants as team members; Evolving strategy with no formal plan for integration; Conducted under attorney-client privilege protections |
| Audit Scope: | Global; Environmental, Health and Safety; Multi-media |

| | |
|---|---|
| Staffing: | 2-3 technical staff either consultants or borrowed from line organization; 10-25 years experience; No formal internal training program |
| Duration: | 2-5 days; 3 days on the average |
| Number: | About 15 corporate management systems audits per year; same level of activity for business compliance audits |
| Frequency: | Every 2, 3 or 4 years; No risk evaluation |

| | |
|---|---|
| Methodology: | Standard approach; Management systems oriented checklists at corporate level; Scoring system used |
| Reporting: | Brief exception reports which include recommendations; Legal not typically involved; Goal of leaving the draft report with the site at the closing conference |
| Follow-up: | Corrective action plans and quarterly reporting required; Formal tracking implemented |

# SUMMARY OF AN
# ENVIRONMENTAL AUDIT PROGRAM

| COMPANY: | FORTUNE 200 CONSUMER PRODUCTS COMPANY |
| --- | --- |

| | |
| --- | --- |
| Program Name: | Environmental Audit Program |
| Start Date: | Re-engineered in 1995 |
| Purpose: | Achieve compliance with regulations and company policies and assess the efficacy of environmental management systems; Developed a formal set of environmental standards and guidelines (e.g., secondary containment required for all hazardous materials tanks) as part of re-engineering roll-out |
| Organization: | One corporate manager with four audit program managers responsible for geographical regions; Managers lead every audit; Corporate manager established standards of performance for audits; Annual third-party program evaluation; Conducted under attorney-client privilege protections in the U.S. but not overseas |
| Audit Scope: | Global; Environmental; Multi-media |
| Staffing: | 3-5 technical staff borrowed from other facilities within geographical region; 5-20 years experience; Formal internal training program |
| Duration: | 2-5 days; 3 days on the average |
| Number: | About 30 audits per year; Third party does 10% with corporate program manager in attendance as a quality assurance check |
| Frequency: | Every 3, 4 or 5 years depending on risk evaluation |
| Methodology: | Standard approach; Detailed compliance checklists covering regulations and corporate standards; Detailed management systems checklist as well |
| Reporting: | Brief exception reports which include recommendations; Findings categorized into one of three risk groups; Legal involved in U.S.; Goal of leaving the draft report with the site at the closing conference |
| Follow-up: | Corrective action plans and quarterly reporting required |

# SUMMARY OF AN
# ENVIRONMENTAL AUDIT PROGRAM

| | |
|---|---|
| **COMPANY:** | **FORTUNE 50 CHEMICAL COMPANY** |

| | |
|---|---|
| Program Name: | Environmental Review Program |
| Start Date: | 1988 in its current form |
| Purpose: | Achieve compliance with regulations and company policies and assess the efficacy of environmental management systems |
| Organization: | One corporate manager with program delegated to about 20 business units and 3 geographical regions for implementation; Corporate manager established standards of performance for business programs and undertakes formal quality assurance review of 1/3 of business programs each year; Also, annual third-party evaluation; Not conducted under attorney-client privilege protections |
| Audit Scope: | Global; Environmental; Multi-media |
| Staffing: | 3-5 technical staff borrowed from corporate, other facilities and business units; 10-25 years experience; Formal internal training program |
| Duration: | 2-5 days; 3 days on the average |
| Number: | About 50 audits per year |
| Frequency: | Annual or every 2, 3 or 4 years depending on risk evaluation |
| Methodology: | Standard approach; Management systems oriented checklists; covers TSCA and CDTA; On occasion allows for community participation |
| Reporting: | Brief exception reports which include recommendations; Findings categorized into one of three risk groups; Legal involved as needed; Goal of leaving the draft report with the site at the closing conference |
| Follow-up: | Corrective action plans and quarterly reporting required; Implementation in the businesses very uneven |

# SUMMARY OF AN
# ENVIRONMENTAL AUDIT PROGRAM

| | |
|---|---|
| **COMPANY:** | **LARGE U.S. CHEMICAL COMPANY SUBSIDIARY OF EUROPEAN PARENT** |

| | |
|---|---|
| Program Name: | Environmental, Health and Safety Audit Program |
| Start Date: | 1988 in its current form |
| Purpose: | Achieve compliance with regulations and company policies and assess the efficacy of environmental management systems |
| Organization: | One corporate Vice President with four full-time audit program managers who lead every audit; Conducted under attorney-client privilege protections; Annual third-party program evaluation |
| Audit Scope: | North America only; Environmental, Health and Safety; Multi-media plus focused audits (e.g., product stewardship) |
| Staffing: | 3-5 technical staff borrowed from corporate, other facilities and business units; 10-25 years experience; Formal internal training program |
| Duration: | 2-5 days; 3 days on the average |
| Number: | About 30 audits per year |
| Frequency: | Every 3 years |
| Methodology: | Standard approach; Management systems oriented checklists, with move towards detailed compliance checklists |
| Reporting: | Brief exception reports which include recommendations; Findings categorized into one of two risk groups; Legal reviews and issues all reports and attends closing conferences; Goal of leaving the draft report with the site at the closing conference |
| Follow-up: | Corrective action plans and quarterly reporting required; Formal computerized tracking system; Formal verification follow-up audits on 10% of audits each year |

# SUMMARY OF AN
# ENVIRONMENTAL AUDIT PROGRAM

| COMPANY: | BILLION DOLLAR MINING AND MINERALS COMPANY |
|---|---|

| | |
|---|---|
| Program Name: | Environmental Audit Program |
| Start Date: | Program re-engineered in 1994 |
| Purpose: | Achieve compliance with regulations and company policies and assess the efficacy of environmental management systems |
| Organization: | One corporate manager with small Audit Core Team made up of rotating Business representatives; Corporate manager established standards of performance; Annual third-party program evaluation; Conducted under attorney-client privilege protections |
| Audit Scope: | Global; Environmental; Multi-media; Will be integrating health and safety in the future |
| Staffing: | 2-4 technical staff borrowed from other facilities and business units; 10-25 years experience; Formal internal training program |
| Duration: | 2-5 days; 3 days on the average |
| Number: | About 10 audits per year |
| Frequency: | Every 2, 3 or 4 years depending on risk evaluation |
| Methodology: | Standard approach; Detailed compliance checklists |
| Reporting: | Brief exception reports which include recommendations; Findings categorized into one of three risk groups; Legal involved; Goal of leaving the draft report with the site at the closing conference |
| Follow-up: | Corrective action plans and quarterly reporting required |

SUMMARY OF AN
ENVIRONMENTAL AUDIT PROGRAM

---

| | |
|---|---|
| Company: | Fortune 100 Chemical Company |

---

| | |
|---|---|
| Program Name: | Environmental Audit Program |
| Start Date: | 1984 |
| Purpose: | To anticipate problems; assure of compliance |
| Organization: | One corporate staff member from Safety, Health and Environmental Department; rotating senior environmental staff from sister divisions |
| Audit Scope: | U.S., environmental, multi-media |

- - - - - - - - - - - - - - - - - - - - - - - - -

| | |
|---|---|
| Staffing: | 3-4 senior technical staff (10-25 years experience) |
| Duration: | 1-3 days; 2 days on the average |
| Number: | 10 facilities/year |
| Frequency: | 18 month cycle |

- - - - - - - - - - - - - - - - - - - - - - - - -

| | |
|---|---|
| Methodology: | standard approach; open-ended detailed checklists; plants do annual vendor audits with corporate direction |
| Reporting: | Good & bad findings; oral reports to legal; draft report to plant management; no recommendations; 3-5 page reports |
| Follow-up: | Informal; no auditing of follow-up no action plan required |

---

SUMMARY OF AN
ENVIRONMENTAL AUDIT PROGRAM

| | |
|---|---|
| Company: | Fortune 500 Mineral Products Company |

| | |
|---|---|
| Program Name: | Environmental Auditing Program |
| Start Date: | 1976 |
| Purpose: | Increase awareness; assure compliance; data collection for litigation |
| Organization: | Six part-time corporate auditors in corporate environmental department; audits done under direction of General Counsel |
| Audit Scope: | U.S., environmental plus right-to-know and natural resources (e.g., mining); multi-media |

| | |
|---|---|
| Staffing: | 2 technical staff (plus an attorney where sites are under litigation) |
| Duration: | One week |
| Number: | 10 audits/year |
| Frequency: | Two year cycle on the average (1-4 year cycle based on risk) |

| | |
|---|---|
| Methodology: | Standard approach; detailed worksheets; some sampling done; some contact with regulatory authorities |
| Reporting: | One to four page reports; very limited distribution; report sent to Division President with plant manager's comments; reports for plants under litigation are sometimes verbal; good & bad findings; no recommendations |
| Follow-up: | Informal; auditors provide technical assistance |

## SUMMARY OF AN
## ENVIRONMENTAL AUDIT PROGRAM

| | |
|---|---|
| Company: | Fortune 200 Chemical Company |

| | |
|---|---|
| Program Name: | Regulatory Audits |
| Start Date: | 1978 |
| Purpose: | Compliance |
| Organization: | Five Full-time Corporate Auditors with at least 10 years experience in Financial Audit Department |
| Audit Scope: | International, comprehensive, multi-media |
| Staffing: | 3 technical staff |
| Duration: | Up to 2 weeks |
| Number: | 60/year |
| Frequency: | Once a year |
| Methodology: | Surprise audits, comprehensive check-lists, standard approach |
| Reporting: | Good & bad findings, brief off-site report with recommendations, no pro-forma legal review |
| Follow-up: | 60 day formal response required |

**SUMMARY OF AN**
**ENVIRONMENTAL AUDIT PROGRAM**

---

| | |
|---|---|
| **Company:** | Fortune 100 Manufacturing Company |

---

| | |
|---|---|
| **Program Name:** | Plant Audits |
| **Start Date:** | 1972 |
| **Purpose:** | Environmental Engineering Evaluation, Awareness, Compliance (GMP's) |
| **Organization:** | 10 Corporate Full-time auditors in Environmental Department |
| **Audit Scope:** | U.S. and Canada, Environmental, Multi-media |
| **Staffing:** | Technical staff, up to 4 auditors |
| **Duration:** | Up to one week |
| **Number:** | Up to 75 facilities/year |
| **Frequency:** | Four year cycle (or less based on risk) |
| **Methodology:** | Detailed checklists and procedures, standard approach |
| **Reporting:** | Bad findings only, brief off-site written report, no recommendations, no pro-forma legal review |
| **Follow-up:** | Formal response required by plant manager |

**SUMMARY OF AN
ENVIRONMENTAL AUDIT PROGRAM**

| | |
|---|---|
| **Company:** | Fortune 100 Diversified Conglomerate |

| | |
|---|---|
| **Program Name:** | Environmental Surveillance |
| **Start Date:** | 1978 |
| **Purpose:** | Compliance (GMP's) |
| **Organization:** | 3 Full-time auditors in Corporate Environmental Affairs Department |
| **Audit Scope:** | International, Comprehensive, Single-medium |
| **Staffing:** | 2-3 Technical staff |
| **Duration:** | 3-4 days |
| **Number:** | 35 plants/year |
| **Frequency:** | 2-3 year cycle |
| **Methodology:** | Detailed checklists and procedures, standard approach |
| **Reporting:** | Bad findings only, brief on-site preliminary written report, no pro-forma legal review |
| **Follow-up:** | 60 day formal response required |

## SUMMARY OF AN
## ENVIRONMENTAL AUDIT PROGRAM

| | |
|---|---|
| **Company:** | Major Electric Utility |

| | |
|---|---|
| **Program Name:** | Assessment Program |
| **Start Date:** | 1976 |
| **Purpose:** | Compliance (GMP's) and Awareness |
| **Organization:** | 5 Full-time Corporate staff reporting to VP-System Power & Engineering |
| **Audit Scope:** | U.S., Environmental, Multi-media |
| **Staffing:** | 2-3 auditors |
| **Duration:** | 3 days |
| **Number:** | 30/year |
| **Frequency:** | 2 year cycle |
| **Methodology:** | Detailed checklists and procedures, some sampling is conducted, some surprise audits |
| **Reporting:** | Good & bad findings, 30 page off-site report with recommendations, no pro-forma legal review |
| **Follow-up:** | 30 day formal response required |

**SUMMARY OF AN
ENVIRONMENTAL AUDIT PROGRAM**

---

| | |
|---|---|
| Company: | Chemical Subsidiary of Fortune 100 Company |

---

| | |
|---|---|
| Program Name: | Environmental Surveillance |
| Start Date: | 1979 |
| Purpose: | Compliance (GMP's) |
| Organization: | Run by Division in concert with separate Corporate program; One full-time auditor with several part-time auditors out of Environmental Affairs Department |
| Audit Scope: | International, Comprehensive, Multi-Media |

- - - - - - - - - - - - - - - - - - - - - - -

| | |
|---|---|
| Staffing: | 2-3 Technical Staff |
| Duration: | 3-4 days |
| Number: | 25/year |
| Frequency: | 2-3 year cycle |

- - - - - - - - - - - - - - - - - - - - - - -

| | |
|---|---|
| Methodology: | Multi-lingual checklists, standard approach |
| Reporting: | Bad findings only, brief off-site written report with recommendations, no pro-forma legal review |
| Follow-up: | 60 day formal response required |

---

**SUMMARY OF AN**
**ENVIRONMENTAL AUDIT PROGRAM**

| | |
|---|---|
| Company: | Fortune 1000 High-technology Company |

| | |
|---|---|
| Program Name: | Regulatory Affairs Audit Program |
| Start Date: | 1981 |
| Purpose: | Compliance (GMP's), Technology Transfer, Prevention |
| Organization: | Half-time Corporate Manager in Regulatory Affairs Department under the Office of General Counsel |
| Audit Scope: | International, Comprehensive, Multi-media |
| Staffing: | 2 Corporate/Division Technical staff with 2-5 plant staff |
| Duration: | 3-5 days |
| Number: | 6 per year |
| Frequency: | Annual audits |
| Methodology: | Detailed checklists and procedures, standard approach |
| Reporting: | Bad findings only; off-site 15 page written report, no recommendations, no pro-forma legal review; space is left for action plan to be filled in |
| Follow-up: | Action plans submitted to corporate within 2-4 weeks; audit manager follows-up on significant items within 30 days |

**SUMMARY OF AN
ENVIRONMENTAL AUDIT PROGRAM**

---

| | |
|---|---|
| **Company:** | Fortune 100 Petrochemical Company |

---

| | |
|---|---|
| **Program Name:** | Environmental Assessment Program |
| **Start Date:** | 1978 |
| **Purpose:** | Compliance (GMP's), Awareness |
| **Organization:** | Run out of Corporate Environmental Department with two staff; auditors come from 2 Divisions and plant being audited |
| **Audit Scope:** | International, Environmental, Multi-media |
| **Staffing:** | 3 Technical Staff |
| **Duration:** | 2-3 days |
| **Number:** | 15/year |
| **Frequency:** | Varied; one year cycle for major facilities |
| **Methodology:** | Detailed checklists and procedures; standard approach |
| **Reporting:** | Good & bad findings with recommendations; pro-forma legal review |
| **Follow-up:** | Computerized action plan tracking system, with monthly reports and scoring of plants responsiveness |

**SUMMARY OF AN
ENVIRONMENTAL AUDIT PROGRAM**

| | |
|---|---|
| **Company:** | Fortune 500 Chemical Company |

| | |
|---|---|
| **Program Name:** | Compliance Audits |
| **Start Date:** | 1981 |
| **Purpose:** | Compliance (GMP's) |
| **Organization:** | Run by Regional Regulatory Manager; with no permanent auditors (participation at all levels) |
| **Audit Scope:** | International, Comprehensive, Multi-media (including product stewardship) |
| **Staffing:** | 3-6 technical/regulatory staff |
| **Duration:** | 3-4 days |
| **Number:** | 15/year |
| **Frequency:** | 2 year cycle |
| **Methodology:** | Detailed checklists and procedures, standard approach |
| **Reporting:** | Good & bad findings, several oral reports before final brief written report with recommendations; legal review |
| **Follow-up:** | Reporting and follow-up to President of Company |

**SUMMARY OF AN
ENVIRONMENTAL AUDIT PROGRAM**

| | |
|---|---|
| Company: | Fortune 500 Specialty Products Company |

| | |
|---|---|
| Program Name: | Environmental Audit Program |
| Start Date: | 1980 |
| Purpose: | Compliance (GMP's) |
| Organization: | Run from Corporate Environmental Department; no permanent auditors, rotating Divisional/plant staff |
| Audit Scope: | U.S., Environmental, Multi-media |
| Staffing: | 4 Technical staff |
| Duration: | 3 days |
| Number: | 10/year |
| Frequency: | 2 year cycle |
| Methodology: | Detailed checklists and procedures |
| Reporting: | Good & bad findings; elaborate drafting and submission of audit report with recommendations; scoring system is used |
| Follow-up: | Informal follow-up procedures |

**SUMMARY OF AN
ENVIRONMENTAL AUDIT PROGRAM**

| | |
|---|---|
| **Company:** | Fortune 500 Oil Company |

| | |
|---|---|
| **Program Name:** | Systems Review Program |
| **Start Date:** | 1980 |
| **Purpose:** | System compliance (GMP's) |
| **Organization:** | Run with large core (50) of part-time auditors from separate Corporate Environmental Council |
| **Audit Scope:** | U.S., Environmental, Multi-media |

| | |
|---|---|
| **Staffing:** | 3-4 Technical staff |
| **Duration:** | Up to 10 days |
| **Number:** | 10 per year |
| **Frequency:** | Four year cycle |

| | |
|---|---|
| **Methodology:** | Checklists based on 10 management systems criteria that cut across media; very detailed instructions to auditors |
| **Reporting:** | Bad findings only; on-site preliminary written report; report has recommendations |
| **Follow-up:** | 45 day formal follow-up required |

SUMMARY OF AN
ENVIRONMENTAL AUDIT PROGRAM

---

| COMPANY | Fortune 100 Diversified Chemical Company |
|---|---|

---

| | |
|---|---|
| Program Name: | Environmental Protection Audit Program |
| Start Date: | 1985 |
| Purpose: | Provide management with verification of compliance and good management practices |
| Organization: | Corporate group with 30 full-time auditors with consultant support |
| Audit Scope: | International, OSHA, Environmental & Product Safety, Multi-Media |

---

| | |
|---|---|
| Staffing: | Varied |
| Duration: | A week or less |
| Number: | Unknown |
| Frequency: | Phased based on risk assessments of facilities |

---

| | |
|---|---|
| Methodology: | Standard approach; detailed checklists, focus on management systems; both compliance and process audits |
| Reporting: | Report good and bad findings; sent to group vice president; no pro-forma legal review |
| Follow-up: | Formal action plans required by plants; loaded onto MIS for computerized tracking |

---

## SUMMARY OF AN
## ENVIRONMENTAL AUDIT PROGRAM

---

Company:        Fortune 100 Food Products Company

---

Program:        Environmental Compliance Survey

Start Date:     1986

Purpose:        Compliance

Organization:   Two part-time corporate auditors supplemented
                by consultants

Audit Scope:    U.S. only, environmental, multi-media

---

Staffing:       1 technical staff

Duration:       2 days

Number:         80/year

Frequency:      Annual

---

Methodology:    Standard approach, comprehensive previsit
                questionnaire  requiring  emissions/effluent
                inventories, summary checklists

Reporting:      Good & bad findings, brief off-site report
                with  recommendations,  no  pro-forma  legal
                review

Follow-up:      Informal follow-up; actions required by both
                corporate and plant personnel

---

SUMMARY OF AN
ENVIRONMENTAL AUDIT PROGRAM

---

**Company:** Defense and Space Equipment Manufacturer

---

**Program:** Environmental Audit Program

**Start Date:** 1983

**Purpose:** Compliance, assist operations, assure management

**Organization:** One full time on-site auditor from corporate staff function

**Audit Scope:** One major facility in U.S., environmental, three compliance areas

---

**Staffing:** 1 technical staff

**Duration:** One-half day

**Number:** 60 auditable units at one site

**Frequency:** Major units quarterly; others semi-annual

---

**Methodology:** No previsit questionnaire; no checklists; reliance on environmental standards and auditor's skill

**Reporting:** Bad findings only; brief two page report; no pro-forma legal review

**Follow-up:** Very formal; unit response plan required in 14 days; if non-responsive "red tag" can be placed on equipment shutting it down.

---

**SUMMARY OF AN
ENVIRONMENTAL AUDIT PROGRAM**

| | |
|---|---|
| **Company:** | Fortune 500 Chemical Company |

| | |
|---|---|
| **Program Name:** | Environmental Audits |
| **Start Date:** | 1979 |
| **Purpose:** | Compliance |
| **Organization:** | Four part-time auditors in Corporate Legal and Enviornmental Departments |
| **Audit Scope:** | U.S., Environmental, Multi-media |

| | |
|---|---|
| **Staffing:** | Technical and legal staff, 2 auditors |
| **Duration:** | 2-3 days |
| **Number:** | 20/year |
| **Frequency:** | Two year cycle |

| | |
|---|---|
| **Methodology:** | Open-ended checklists and procedures, principally review of last audit findings, standard approach |
| **Reporting:** | Bad findings only, off-site written report, some recommendations, legal review |
| **Follow-up:** | Formal response required in 30 days, automated corporate "tickler file" |

### SUMMARY OF AN
### ENVIRONMENTAL AUDIT PROGRAM

---

| | |
|---|---|
| Company: | Large Governmental Agency |

---

| | |
|---|---|
| Program Name: | Environmental Survey |
| Start Date: | 1985-1986 |
| Purpose: | Identification of environmental problems |
| Organization: | Headquarters management with field support by consultants |
| Audit Scope: | U.S, environmental, multi-media (including radioactive) |

---

| | |
|---|---|
| Staffing: | 7-10 technical staff |
| Duration: | 2 weeks |
| Number: | 10/year |
| Frequency: | One time |

---

| | |
|---|---|
| Methodology: | Standard approach, very detailed checklists; pre-survey visit by team leader; sampling included |
| Reporting: | Good and bad findings; comprehensive off-site report; no pro-forma legal review |
| Follow-up: | Major follow-up program anticipated |

---

SUMMARY OF AN
ENVIRONMENTAL AUDIT PROGRAM

---

Company:        Mid-sized chemical company

---

Program Name:   Environmental Audits

Start Date:     1983

Purpose:        Compliance

Organization:   One part-time auditor in the major division

Audit Scope:    International, environmental, multi-media

---

Staffing:       1 technical staff

Duration:       1-2 days

Number:         5/year

Frequency:      Once every 3 years

---

Methodology:    No detailed checklists; reliance on expertise
                of auditor

Reporting:      Good and bad findings, comprehensive off-site
                report, no pro-forma legal legal review

Follow-up:      Very informal; outside consultant used to
                conduct follow-up audits, 1-2 years later

---

**SUMMARY OF AN
ENVIRONMENTAL AUDIT PROGRAM**

---

| | |
|---|---|
| Company: | Fortune 100 Basic Manufacturing Company |

---

| | |
|---|---|
| Program Name: | Environmental Evaluation |
| Start Date: | 1981 |
| Purpose: | "Improve Environmental Performance" |
| Organization: | Two Corporate staff in Environmental Department; Corporate teams with one plant environmental staff |
| Audit Scope: | Primarily U.S., Environmental, Multi-media |

- - - - - - - - - - - - - - - - - - - - - -

| | |
|---|---|
| Staffing: | 2-3 Technical staff |
| Duration: | 1-5 days |
| Number: | 20/year |
| Frequency: | Varied; large plants on one year cycle; small plants on 4 year cycle |

- - - - - - - - - - - - - - - - - - - - - -

| | |
|---|---|
| Methodology: | Detailed checklists and procedures; standard approach |
| Reporting: | (Some) good & bad findings with recommendations; reports vary from 2-50 pages; pro-forma legal review |
| Follow-up: | No formal follow-up approach |

---

SUMMARY OF AN
ENVIRONMENTAL AUDIT PROGRAM

| | |
|---|---|
| **Company:** | Fortune 500 Manufacturer |

| | |
|---|---|
| **Program Name:** | Environmental Compliance Audit Program |
| **Start Date:** | 1987 |
| **Purposes:** | Assure environmental stewardship: reduce liability; increase awareness; transfer information; assess management practices; create good record |
| **Organization:** | Run from corporate environmental/legal departments; no permanent auditors; rotating plant participants |
| **Audit Scope:** | All environmental laws and regulations, both Federal and state, for all media |

| | |
|---|---|
| **Staffing:** | 1 legal, 1 corporate technical, 1 plant technical (rotating) |
| **Duration:** | 2-3 days for environmentally complex plants; 1 day for smaller plants |
| **Number:** | 6/year |
| **Frequency:** | 2 year cycle for complex plants; 4 year cycle for smaller plants |

| | |
|---|---|
| **Methodology:** | Detailed checklist and procedures; scheduled in advance; no sampling; no contact with agencies; detailed instructions to auditors |
| **Reporting:** | Positive and negative findings; no recommendations; oral on-site report to plant management; 3-5 page written report within one month; limited distribution. |
| **Follow-up:** | Informal, with plant action plan; quarterly review between plant management and corporate environmental manager; assurance letters. |

**Internet and the Law:**
**Legal Fundamentals for**
**the Internet User**
This book explains the basic principles
pertaining to laws of copyright,
trademark, trade secret, patent, libel/
defamation and related issues, as well as
the basic principles of licensing.
*Softcover, 264 pages, Dec '95*
*ISBN: 0-86587-506-5* **$75**

**Environmental Law Handbook,**
**13th Edition**
Now in a new 13th Edition, this book
continues to keep pace with important
changes in environmental laws
while detailing how those laws affect
the regulated community.
*Hardcover, Index, 550 pages,*
*Mar '95, ISBN: 0-86587-450-6* **$79**

**Environmental Telephone Directory,**
**1996 Edition**
The newly updated directory not only
provides you with all new *Electronic*
*Mail Addresses and World-Wide Web*
*Sites* for Federal Agencies, but also
contains the most current phone numbers
and addresses of key U.S. Government
and Environmental Contacts.
*Softcover, 256 pages, Nov '95,*
*ISBN: 0-86587-504-9* **$67**

**Health Effects of Toxic Substances**
This comprehensive book provides
you with an excellent understanding
of the toxicology and industrial
hygiene of hazardous materials.
*Softcover, Index, 256 pages, Aug '95,*
*ISBN: 0-86587-471-9* **$39**

**"So You're the Safety Director!"**
*An Introduction to Loss Control*
*and Safety Management*
This book concentrates on your role
in evaluating, managing, and
controlling your company's losses
and handling the OSHA compliance
process.
*Softcover, Index, 186 pages, Oct '95,*
*ISBN: 0-86587-481-6* **$45**

**Directory of Environmental**
**Information Sources, 5th Edition**
Packed with vital information for
your search, this updated directory
gives the current address, phone
number, and a brief description of
over 1,600 hard-to-find resources.
*Softcover, 322 pages, Sept '95,*
*ISBN: 0-86587-475-1* **$79**

**ISO 14000**
**Understanding the**
**Environmental Standards**
This book explains what ISO (International
Standards Organization) is, what the
standards are, and how you can devise plans
for complying with them.
*Softcover, 283 pages, Jan '96,*
*ISBN: 0-86587-510-3* **$69**

**Ergonomic Problems in**
**the Workplace: A Guide to**
**Effective Management**
The valuable insights you'll gain
from this new book will help you
develop and implement your own
successful ergonomics program.
*Softcover, 256 pages, Sept '95,*
*ISBN: 0-86587-474-3* **$71**

**Environmental Guide to the**
**Internet, 2nd Edition**
From environmental engineering to
hazardous waste compliance issues,
you'll have no problem finding it with
the easy-to-use Environmental Guide to
the Internet.
*Softcover, 236 pages, Feb '96,*
*ISBN: 0-86587-517-0* **$49**

**Safety Made Easy: A Checklist**
**Approach to OSHA Compliance**
This new book provides a simpler way of
understanding your requirements under
the complex maze of OSHA's Safety and
Health regulations.
*Softcover, 192 pages, June '95,*
*ISBN: 0-86587-463-8* **$45**

**Energy and Environmental**
**Market Conditions in Mexico**
This book is essential reading for
individuals with business, environmen-
tal, and energy interests in Mexico.
*Softcover, 83 pages, Dec '93,*
*ISBN: 0-86587-373-9* **$58**

**Principles of Environmental,**
**Health and Safety Management**
This book provides you with informa-
tion and advice on how your company
can develop a comprehensive environ-
mental management program.
*Softcover, Aug '95, 320 pages,*
*ISBN: 0-86587-478-5* **$59**

Government Institutes • 4 Research Place • Rockville, MD 20850 • USA • (301) 921-2355

# Environmental and Health/Safety References

**Risk Assessment Methodologies for Toxic Air Pollutants**
Focusing on routine releases from stationary sources, this EPA report covers: Hazard Identification; Dose-Response Assessment; Exposure Assessment; Risk Characterization; Emerging Issues; and Uses of Risk Assessment Methodologies.
*Softcover, 278 pages, Feb '95*
*ISBN: 0-86587-445-X* **$75**

**Control Technologies for Hazardous Air Pollutants**
With extensive tables and charts throughout, this book provides you with a methodology to determine the performance and cost of air pollution control techniques designed to reduce or eliminate emissions of potential hazardous air pollutants (HAPs) from industrial and commercial sources.
*Softcover, 260 pages, June '92*
*ISBN: 0-86587-301-1* **$75**

**Control Techniques for VOC Emissions from Stationary Sources**
This detailed reference for engineers is a valuable guide for information on sources of oxidant precursors and control of these sources, estimates of control costs, and estimates of the emission reductions achievable through control.
*Softcover, 474 pages, Feb '94,*
*ISBN: 0-86587-378-X* **$85**

**State Wildlife Laws Handbook**
An in-depth analysis of wildlife management and protection laws for all fifty states, this book covers laws on hunting and trapping methods, enforcement, habitat protection, endangered or threatened species protection, and other laws affecting wild animals.
*Hardcover, 854 pages, Nov '93,*
*ISBN: 0-86587-357-7* **$94**

**Sampling, Analysis & Monitoring Methods: A Guide to EPA Requirements**
This book provides a guide for determining which chemicals have sampling, analysis, and monitoring requirements under the environmental laws and regulations, and where those testing and sampling methods can be found.
*Softcover, 256 pages, Sept '95,*
*ISBN: 0-86587-477-8* **$65**

**Product Side of Pollution Prevention: Evaluating the Potential for Safe Substitutes**
This report focuses on safe substitutes for products that contain or use toxic chemicals in their manufacturing process.
*Softcover, 240 pages, Sept '95,*
*ISBN: 0-86587-479-4* **$69**

**Complying with the CFC Prohibitions of the Clean Air Act: A Guide for HVAC Compliance**
This practical book provides the legal and technical guidance necessary for building owners and facilities managers to make the right decisions when selecting and buying major pieces of equipment, and in evaluating the operation and maintenance of those systems to ensure compliance.
*Softcover, 196 pages, Nov '94*
*ISBN: 0-86587-423-9* **$75**

**Emergency Planning & Management:**
*Ensuring Your Company's Survival in the Event of a Disaster*
This book will help you assess your exposure to disasters and prepare emergency preparedness, response, and recovery plans for your facilities, both to comply with OSHA and EPA requirements and to reduce the risk of losses to your company.
*Softcover, 192 pages, Nov '95,*
*ISBN: 0-86587-505-7* **$59**

**Chemical Information Manual, 3rd Edition (Book and Disk Format)**
This database contains essential data for over 1,400 chemical substances.
*Softcover, 400 pages, Aug '95,*
*ISBN: 0-86587-469-7* **$99**
Also available on disk!
*3.5" Floppy Disk for Windows,*
*#4070* **$99**

**Exposure Factors Handbook, Review Draft**
EPA uses this document to develop pesticide tolerance levels, assess industrial chemical risks, and to undertake Superfund site assessments and drinking water health assessments.
*Softcover, 866 pages, Nov '95,*
*ISBN: 0-86587-509-X* **$125**

**Wetlands: An Introduction to Ecology, the Law and Permitting**
This book explains to you what wetlands are and how they fit into our complex environmental systems.
*Softcover, Index, 185 pages, Aug '95,*
*ISBN: 0-86587-467-0* **$65**

**OSHA Technical Manual, 4th Edition**
This inspection manual is used nationwide by OSHA Compliance Safety and Health Officers (CSHOs) in checking industry compliance.
*Softcover, 400 pages, Feb '96,*
*ISBN:0-86587-511-1* **$85**

Government Institutes • 4 Research Place • Rockville, MD 20850 • USA • (301) 921-2355

# Environmental and Health/Safety References

**Federal Facility Pollution Prevention: Planning Guide and Tools for Compliance**
This EPA Guide presents pollution prevention tools and provides a step-by-step approach to develop a pollution prevention program plan for federal facilities.
*Softcover, 240 pages, Aug '95 ISBN: 0-86587-476-X* **$75**

**Design for the Environment: Product Life Cycle Design Guidance Manual**
This new life cycle approach examines each step in the life of a product from acquisition of raw materials through processing, manufacture, use, and final disposal of all residuals.
*Softcover, 181 pages, Mar '94, ISBN: 0-86587-384-4* **$65**

**Pollution Prevention Strategies and Technologies**
This book is an indispensible guide to understanding pollution prevention policies and regulatory initiatives designed to reduce wastes.
*Hardcover, Index, 484 pages, Oct '95, ISBN: 0-86587-480-8* **$79**

**Environmental Regulatory Glossary, 6th Edition**
This glossary defines and standardizes more than 4,000 terms, abbreviations and acronyms, compiled directly from the environmental statutes or the U.S. Code of Federal Regulations.
*Hardcover, 544 pages, June '93, ISBN: 0-86587-353-4* **$68**

**Environmental Science and Technology Handbook**
This is the first book to bridge the gap between the latest environmental science and technology available, and compliance with today's complex regulations.
*Hardcover, 389 pages, Jan '94, ISBN: 0-86587-362-3* **$79**

**South Korea Environmental Report: A Resource for Business**
This new, authoritative report provides you with a wealth of information on environmental regulatory programs in South Korea.
*Softcover, Index, 260 pages, Feb '96, ISBN: 0-86587-508-1* **$495**

**Radiation and Mixed Waste Incineration: EPA Research Series**
*By Environmental Protection Agency*
This EPA background document contains two volumes on technology and the risk of radiation exposure.
*Softcover, 352 pages, Feb '94, ISBN: 0-86587-386-0* **$85**

**Bioremediation of Hazardous Waste: EPA Research Series**
*By Environmental Protection Agency*
Compiled from four EPA documents, this book contains over 35 papers discussing approaches to the bioremediation of hazardous waste sites.
*Softcover, 188 pages, Mar '94, ISBN: 0-86587-388-7* **$55**

**Cleaning Up the Nation's Waste Sites: Markets and Technology Trends**
*By Environmental Protection Agency*
With over 60 charts and tables, this EPA report forecasts opportunities for remediation services and new cleanup technologies for all major programs. *Softcover, 164 pages, Feb '94, ISBN: 0-86587-379-8* **$79**

**Total Quality for Safety and Health Professionals**
F. David Pierce, a CSP and a CIH, shows you how to apply concepts - proven successful - to your safety management program to achieve increased productivity, lowered costs, reduced inventories, improved quality, increased profits, and raised employee morale.
*Hardcover, 244 pages, June '95, ISBN: 0-86587-462-X* **$59**

**Bioremediation of Ground Water and Geological Material: A Review of In-Situ Technologies**
*By Environmental Protection Agency*
This EPA manual provides a detailed background of the technologies available, as well as the most recent scientific understanding of the processes involved for the bioremediation of contaminated soil and ground water.
*Softcover, 250 pages, June '94, ISBN: 0-86587-404-2* **$75**

**Environmental Statutes, 1996 Edition**
The complete text of each statute as currently amended is included, with a detailed Table of Contents for quick reference.
*Hardcover, 1,200 pages, Mar '96, ISBN: 0-86587-521-9* **$69**
*Softcover, 1,200 pages, Mar '96, ISBN: 0-86587-522-7* **$59**
Disks include the Folio search Engine.
*3.5" Floppy Disks for Windows (#4058)* **$135**
*Statutes Package, hardcover w/Windows disk (#4059)* **$204**

Government Institutes • 4 Research Place • Rockville, MD 20850 • USA • (301) 921-2355

## Environmental and Health/Safety References

**GI Environmental Database CD ROM**
Our four most popular environmental references — Directory of Environmental Information Sources; Environmental Telephone Directory; Environmental Regulatory Glossary; and Environmental Acronyms — are now included on one CD ROM — a powerful compliance tool for instant information retrieval on your computer. *Single CD ROM, Windows™, Nov '95, Product Code #4073* **$149**

**Multi-Media Investigation Manual**
This EPA manual is used to guide its inspectors in conducting a multi-media compliance audit inspection of facilities that generate effluents, emissions, wastes or materials regulated under several laws such as the Clean Water Act, Clean Air Act, RCRA, and TSCA.
*Softcover, 192 pages, Sept '92, ISBN: 0-86587-300-3* **$79**

**1994 International Symposium on Environmental Contamination in Eastern Europe**
These proceedings from the Budapest '94 symposium contain over 370 technical manuscripts of the papers presented.
*Softcover, 1,020 pages, June '95, ISBN: 0-86587-488-3* **$135**

**Understanding Workers' Compensation: A Guide for Safety and Health Professionals**
This book is designed to help you understand how the workers' comp system works, and provides a basic understanding of injury prevention, types of injuries, and cost containment strategies.
*Softcover, 250 pages, July '95, ISBN: 0-86587-464-6* **$45**

**Environmental, Health & Safety CFRs on CD ROM**
Now you can scan EPA, OSHA and Hazmat regulations in seconds. Search on a regulation number, word, or phrase, and print or save the results!
*Single Issue (most recent quarterly release) #4017* **$395 + $5 Air shipping**

*One-Year Subscription (receive an updated CD quarterly, beginning with the most recent release) #4000* **$980 + $20 shipping & handling**

**Underground Storage Tank Management: A Practical Guide, 4th Edition**
This new edition covers the complete and current regulatory and technical picture – to help you develop or maintain UST management programs that will minimize the risk of a release and reduce the potential for costly repercussions.
*Softcover, 420 pages, Nov '91, ISBN: 0-86587-271-6* **$84**

**Aboveground Storage Tank Management: A Practical Guide**
This manual will provide you with a review of what regulations are on the horizon and help you evaluate how to cost-effectively design, build, manage, operate and maintain an aboveground tank system that meets your storage needs and complies with all federal and state codes and regulations.
*Softcover, 220 pages, Feb '90, ISBN: 0-86587-202-3* **$72**

**Environmental Engineering Dictionary, 2nd Edition**
This dictionary is intended to provide a comprehensive source of environmental engineering definitions along with their origins, and a guidebook for obtaining additional information by using the unique reference provided at the end of the definitions.
*Softcover, 630 pages, Oct '92, ISBN: 0-86587-298-8* **$75**

**Ground Water Handbook, 2nd Edition**
This is EPA's technical guide for assessing and monitoring ground water contamination.
*Softcover, 295 pages, Mar '92, ISBN: 0-86587-279-1* **$75**

**ESAs Made Easy: A Checklist Approach to Phase I Environmental Site Assessments**
Learn how to conduct or review a successful Phase I ESA to identify existing or potential environmental hazards, and "special resources" for a subject property. Through its easy-to-follow checklist format, this new book shows you everything you need to know about ESAs.
*Softcover, 224 pages, Feb '96, ISBN: 0-86587-536-7* **$59**

**Fate and Transport of Organic Chemicals in the Environment: A Practical Guide, 2nd Edition**
This easy-to-use guide provides a unique tool to help you predict the fate and transport of chemicals in air, water, soil, flora and fauna.
*Softcover, Index, 308 pages, Aug '95, ISBN: 0-86587-470-0* **$49**

**TSCA Handbook, 2nd Edition**
Get a comprehensive look at your requirements under the Toxic Substances Control Act (TSCA). Contains charts, figures, tables, and multiple indexes.
*Softcover, 490 pages, Nov '89, ISBN: 0-86587-791-2* **$95**